Carbon Fiber Composites, Volume II

Carbon Fiber Composites, Volume II

Editor

Jiadeng Zhu

Basel • Beijing • Wuhan • Barcelona • Belgrade • Novi Sad • Cluj • Manchester

Editor
Jiadeng Zhu
Brewer Science Inc.
Springfield, MO
USA

Editorial Office
MDPI AG
Grosspeteranlage 5
4052 Basel, Switzerland

This is a reprint of articles from the Special Issue published online in the open access journal *Journal of Composites Science* (ISSN 2504-477X) (available at: https://www.mdpi.com/journal/jcs/special_issues/carbon_fiber_composites_volume_II).

For citation purposes, cite each article independently as indicated on the article page online and as indicated below:

Lastname, A.A.; Lastname, B.B. Article Title. *Journal Name* **Year**, *Volume Number*, Page Range.

ISBN 978-3-7258-2169-3 (Hbk)
ISBN 978-3-7258-2170-9 (PDF)
doi.org/10.3390/books978-3-7258-2170-9

© 2024 by the authors. Articles in this book are Open Access and distributed under the Creative Commons Attribution (CC BY) license. The book as a whole is distributed by MDPI under the terms and conditions of the Creative Commons Attribution-NonCommercial-NoDerivs (CC BY-NC-ND) license.

Contents

About the Editor vii

Jiadeng Zhu
Editorial for the Special Issue on Carbon Fiber Composites, Volume II
Reprinted from: *J. Compos. Sci.* **2024**, *8*, 307, doi:10.3390/jcs8080307 1

Chiemela Victor Amaechi, Cole Chesterton, Harrison Obed Butler, Nathaniel Gillet, Chunguang Wang, Idris Ahmed Ja'e, et al.
Review of Composite Marine Risers for Deep-Water Applications: Design, Development and Mechanics
Reprinted from: *J. Compos. Sci.* **2022**, *6*, 96, doi:10.3390/jcs6030096 4

Jinsil Cheon and Donghwan Cho
Effect of MWCNT Anchoring to Para-Aramid Fiber Surface on the Thermal, Mechanical, and Impact Properties of Para-Aramid Fabric-Reinforced Vinyl Ester Composites
Reprinted from: *J. Compos. Sci.* **2023**, *7*, 416, doi:10.3390/jcs7100416 54

Hongwei Chen, Kaibao Wang, Yao Chen and Huirong Le
Mechanical and Thermal Properties of Multilayer-Coated 3D-Printed Carbon Fiber Reinforced Nylon Composites
Reprinted from: *J. Compos. Sci.* **2023**, *7*, 297, doi:10.3390/jcs7070297 68

Shaoyong Cao, Yan Zhu and Yunpeng Jiang
Notched Behaviors of Carbon Fiber-Reinforced Epoxy Matrix Composite Laminates: Predictions and Experiments
Reprinted from: *J. Compos. Sci.* **2023**, *7*, 223, doi:10.3390/jcs7060223 81

Kazuto Tanaka and Shuhei Kyoyama
The Effect of Pulse Current on Electrolytically Plating Nickel as a Catalyst for Grafting Carbon Nanotubes onto Carbon Fibers via the Chemical Vapor Deposition Method
Reprinted from: *J. Compos. Sci.* **2023**, *7*, 88, doi:10.3390/jcs7020088 95

Hasan Borke Birgin, Antonella D'Alessandro, Andrea Meoni and Filippo Ubertini
Self-Sensing Eco-Earth Composite with Carbon Microfibers for Sustainable Smart Buildings
Reprinted from: *J. Compos. Sci.* **2023**, *7*, 63, doi:10.3390/jcs7020063 108

Matej Gljušćić, Domagoj Lanc, Marina Franulović and Andrej Žerovnik
Microstructural Analysis of the Transverse and Shear Behavior of Additively Manufactured CFRP Composite RVEs Based on the Phase-Field Fracture Theory
Reprinted from: *J. Compos. Sci.* **2023**, *7*, 38, doi:10.3390/jcs7010038 121

Mengfan Li, Yanxiang Wang, Bowen Cui, Chengjuan Wang, Hongxue Tan, Haotian Jiang, et al.
Optimization of Electrical Intensity for Electrochemical Anodic Oxidation to Modify the Surface of Carbon Fibers and Preparation of Carbon Nanotubes/Carbon Fiber Multi-Scale Reinforcements
Reprinted from: *J. Compos. Sci.* **2022**, *6*, 395, doi:10.3390/jcs6120395 141

Veronica Marchante Rodriguez, Marzio Grasso, Yifan Zhao, Haochen Liu, Kailun Deng, Andrew Roberts and Gareth James Appleby-Thomas
Surface Damage in Woven Carbon Composite Panels under Orthogonal and Inclined High-Velocity Impacts
Reprinted from: *J. Compos. Sci.* **2022**, *6*, 282, doi:10.3390/jcs6100282 153

Uday Vaidya, Mark Janney, Keith Graham, Hicham Ghossein and Merlin Theodore
Mechanical Response and Processability of Wet-Laid Recycled Carbon Fiber PE, PA66 and PET Thermoplastic Composites
Reprinted from: *J. Compos. Sci.* **2022**, *6*, 198, doi:10.3390/jcs6070198 . . . . **169**

Mohammad Amir Khozeimeh, Reza Fotouhi and Reza Moazed
Static and Vibration Analyses of a Composite CFRP Robot Manipulator
Reprinted from: *J. Compos. Sci.* **2022**, *6*, 196, doi:10.3390/jcs6070196 . . . . **186**

Ramy Abdallah, Richard Hood and Sein Leung Soo
The Machinability Characteristics of Multidirectional CFRP Composites Using High-Performance Wire EDM Electrodes
Reprinted from: *J. Compos. Sci.* **2022**, *6*, 159, doi:10.3390/jcs6060159 . . . . **215**

Tao Wu, Roland Kruse, Steffen Tinkloh, Thomas Tröster, Wolfgang Zinn, Christian Lauhoff and Thomas Niendorf
Experimental Analysis of Residual Stresses in CFRPs through Hole-Drilling Method: The Role of Stacking Sequence, Thickness, and Defects
Reprinted from: *J. Compos. Sci.* **2022**, *6*, 138, doi:10.3390/jcs6050138 . . . . **231**

Chiemela Victor Amaechi, Nathaniel Gillet, Idris Ahmed Ja'e and Chunguang Wang
Tailoring the Local Design of Deep Water Composite Risers to Minimise Structural Weight
Reprinted from: *J. Compos. Sci.* **2022**, *6*, 103, doi:10.3390/jcs6040103 . . . . **252**

Murat Çelik, Thomas Noble, Frank Jorge, Rongqing Jian, Conchúr M. Ó Brádaigh and Colin Robert
Influence of Line Processing Parameters on Properties of Carbon Fibre Epoxy Towpreg
Reprinted from: *J. Compos. Sci.* **2022**, *6*, 75, doi:10.3390/jcs6030075 . . . . **285**

Sukanta Kumer Shill, Estela O. Garcez, Riyadh Al-Ameri and Mahbube Subhani
Performance of Two-Way Concrete Slabs Reinforced with Basalt and Carbon FRP Rebars
Reprinted from: *J. Compos. Sci.* **2022**, *6*, 74, doi:10.3390/jcs6030074 . . . . **300**

Thiago de Sousa Burgani, Seyedhamidreza Alaie and Mehran Tehrani
Modeling Flexural Failure in Carbon-Fiber-Reinforced Polymer Composites
Reprinted from: *J. Compos. Sci.* **2022**, *6*, 33, doi:10.3390/jcs6020033 . . . . **318**

Hauke Kröger, Stephan Mock, Christoph Greb and Thomas Gries
Damping Properties of Hybrid Composites Made from Carbon, Vectran, Aramid and Cellulose Fibers
Reprinted from: *J. Compos. Sci.* **2022**, *6*, 13, doi:10.3390/jcs6010013 . . . . **328**

About the Editor

Jiadeng Zhu

Jiadeng Zhu received his B.S. degree in Chemical Engineering and Materials Science from Soochow University and his Ph.D. in Fiber and Polymer Science from North Carolina State University. He is now a Principal Investigator/Team Lead in the Discovery and Proof-of-Concept group for emerging polymer/carbon-based sensing materials and technology at Brewer Science Inc. He has published 10 book chapters and over 100 peer-reviewed journal papers with an h-index of 44. He is an enthusiastic, confident, and creative scholar who has a fruitful and consistent history of research on applications of advanced polymers and carbon materials in energy and environmental areas, including, but not limited to, energy storage/conversion, lightweight structural materials, printed/wearable electronics, smart textiles, filtration, sensor fabrication and testing, and sensor integration.

Editorial

Editorial for the Special Issue on Carbon Fiber Composites, Volume II

Jiadeng Zhu

Smart Devices, Brewer Science Inc., Springfield, MO 65810, USA; zhujiadeng@gmail.com

Citation: Zhu, J. Editorial for the Special Issue on Carbon Fiber Composites, Volume II. *J. Compos. Sci.* **2024**, *8*, 307. https://doi.org/10.3390/jcs8080307

Received: 9 July 2024
Accepted: 2 August 2024
Published: 6 August 2024

Copyright: © 2024 by the author. Licensee MDPI, Basel, Switzerland. This article is an open access article distributed under the terms and conditions of the Creative Commons Attribution (CC BY) license (https://creativecommons.org/licenses/by/4.0/).

Fibers with lengths much larger than their widths have been developed over centuries because of their unique properties [1–4]. Therefore, these fibers have been extensively applied in different fields, including wearable electronics, energy storage, sports, and environmental protection [5–8]. Many materials, such as polymers, metals, and metal oxides, can be made into fibers using various approaches [9–12]. Among these materials, carbon fibers (CFs), which contain mostly carbon atoms, have attracted tremendous interest since their discovery in the 1860s because of their mechanical properties, chemical resistance, and electrical/thermal conductivities [13–16]. Additionally, CF-derived composites are lighter in weight and provide superior performance compared with other fibers because of their high strength-to-weight ratio. As such, these CF-derived composites show promise for developing next-generation composites for various practical applications [17–22].

This Special Issue includes papers reporting studies related to the development of approaches to improve the structure and processing design of CF-derived composites [23–25], analyses of these composites from both experimental and computational aspects [26–29], and the various applications of CF-derived composites (i.e., catalysts, sensors, and risers for deep-water) [30–32]. Çelik et al. investigated the effects of line processing parameters on the properties of a carbon fiber epoxy composite, finding that fiber straightness plays a critical role in mechanical performance [33]. The influence of the notch shape and size on carbon-fiber-reinforced epoxy was explored by Cao et al. [34], in which the samples with/without notches were tested for comparison. The features of the failure were identified and compared among these specimens. Khozeimeh et al. found that carbon-fiber-reinforced polymer (CFRP) in a manipulator's structure enhanced the static and vibrational properties of the manipulator, increasing its loading capacity [35]. Some other factors may also affect the final properties of CFRPs, such as stacking sequence and overall thicknesses. Wu et al. studied the effects of the pores generated during the fabrication process on the residual stresses in a composite [36], and the results were experimentally analyzed. In addition to the above-discussed experimental methods, modeling was performed to understand the working mechanisms of CFRPs. For instance, Burgani et al. developed a finite-element model to explain the increase in load-bearing capacity achieved with the use of CFRPs, which was validated with experimental data [37].

Some composite systems other than CFRPs have been studied. Tanaka et al. [38] grafted carbon nanotubes (CNTs) via chemical vapor deposition onto CFs using Ni as the catalyst. Li et al. tuned the intensities of electrochemical anodic oxidation to optimize the surface modification of CFs to increase the loading of catalyst particles and enable CNTs to be uniformly distributed on CFs [39].

Recycled CFs have received considerable interest, and the corresponding supply chain has been well established because of the relatively low cost of and legislation on CFs. For example, Vaidya et al. [40] developed a practical approach to recycle CFs with thermoplastic resins (i.e., polyethylene, polyamide 66, and polyethylene terephthalate). The corresponding mechanical properties of the resultant composites were further explored. The proposed approach could benefit the design of sustainable composites.

In summary, this Special Issue collects studies on different CF composites, including those on the design of their structure, failure analyses, and applications. These studies provide new insights into the fundamental understanding of CF composite systems, which benefit their future design and accelerate their practical application.

Data Availability Statement: No new data were created or analyzed in this study. Data sharing is not applicable to this article.

Conflicts of Interest: The author declares no conflicts of interest.

References

1. Wang, P.; Wang, M.; Zhu, J.; Wang, Y.; Gao, J.; Gao, C.; Gao, Q. Surface engineering via self-assembly on PEDOT:PSS fibers: Biomimetic fluff-like morphology and sensing application. *Chem. Eng. J.* **2021**, *425*, 131551. [CrossRef]
2. Zhu, J. Advanced Separator Selection and Design for High-Performance Lithium-Sulfur Batteries. Ph.D. Thesis, North Carolina State University, Raleigh, NC, USA, 2016.
3. Hou, Z.; Liu, X.; Tian, M.; Zhang, X.; Qu, L.; Fan, T.; Miao, J. Smart fibers and textiles for emerging clothe-based wearable electronics: Materials, fabrications and applications. *J. Mater. Chem. A* **2023**, *11*, 17336–17372. [CrossRef]
4. Zhu, J.; Yildirim, E.; Aly, K.; She, J.; Chen, C.; Lu, Y.; Jiang, M.; Kim, D.; Tonelli, A.E.; Pasquinelli, M.A.; et al. Hierarchical multi-component nanofiber separators for lithium polysulfide capture in lithium-sulfur batteries: An experimental and molecular modeling study. *J. Mater. Chem. A* **2016**, *4*, 13572–13581. [CrossRef]
5. Lang, K.; Liu, T.; Padilla, D.J.; Nelson, M.; Landorf, C.W.; Patel, R.J.; Ballentine, M.L.; Kennedy, A.J.; Shih, W.S.; Scotch, A.; et al. Nanofibers Enabled Advanced Gas Sensors: A Review. *Adv. Sens. Energy Mater.* **2024**, *3*, 100093. [CrossRef]
6. Zuo, X.; Zhang, X.; Qu, L.; Miao, J. Smart fibers and textiles for personal thermal management in emerging wearable applications. *Adv. Mater. Technol.* **2023**, *8*, 2201137. [CrossRef]
7. Gao, Q.; Zhang, Y.; Wang, P.; Zhu, J.; Gao, C. Robust and knittable wet-spun PEDOT: PSS fibers via water. *Compos. Commun.* **2023**, *40*, 101623. [CrossRef]
8. Yin, Z.; Lu, H.; Gan, L.; Zhang, Y. Electronic fibers/textiles for health-monitoring: Fabrication and application. *Adv. Mater. Technol.* **2023**, *8*, 2200654. [CrossRef]
9. Yadav, R.; Zabihi, O.; Fakhrhoseini, S.; Nazarloo, H.A.; Kiziltas, A.; Blanchard, P.; Naebe, M. Lignin derived carbon fiber and nanofiber: Manufacturing and applications. *Compos. Part B Eng.* **2023**, *255*, 110613. [CrossRef]
10. Chen, J.; Zhu, J.; Wei, Z.; Chen, Z.; Zhu, C.; Gao, Q.; Gao, C. Highly stretchable and elastic PEDOT: PSS helix fibers enabled wearable sensors. *J. Mater. Chem. C* **2023**, *11*, 13358–13369. [CrossRef]
11. Beknalkar, S.A.; Teli, A.M.; Shin, J.C. Current innovations and future prospects of metal oxide electrospun materials for supercapacitor technology: A review. *J. Mater. Sci. Technol.* **2023**, *166*, 208–233. [CrossRef]
12. Jang, H.; Ko, K.H.; Kim, S.J.; Basch, R.H.; Fash, J.W. The effect of metal fibers on the friction performance of automotive brake friction materials. *Wear* **2004**, *256*, 406–414. [CrossRef]
13. Zhu, J.; Gao, Z.; Mao, Q.; Li, Y.; Zhang, X.; Gao, Q.; Jiang, M.; Lee, S.; van Duin, A.C.T. Advances in Developing Cost-Effective Carbon Fibers by Coupling Multiscale Modeling and Experiments: A Critical Review. *Prog. Mater. Sci.* **2024**, *146*, 101329. [CrossRef]
14. Zhu, J.; Park, S.W.; Joh, H.I.; Kim, H.C.; Lee, S. Study on the stabilization of isotropic pitch based fibers. *Macromol. Res.* **2015**, *23*, 79–85. [CrossRef]
15. Li, H.; Schamel, E.; Liebscher, M.; Zhang, Y.; Fan, Q.; Schlachter, H.; Köberle, T.; Mechtcherine, V.; Wehnert, G.; Söthje, D. Recycled carbon fibers in cement-based composites: Influence of epoxide matrix depolymerization degree on interfacial interactions. *J. Clean. Prod.* **2023**, *411*, 137235. [CrossRef]
16. Zhu, J.; Park, S.W.; Joh, H.I.; Kim, H.C.; Lee, S. Preparation and characterization of isotropic pitch-based carbon fiber. *Carbon. Lett.* **2013**, *14*, 94–98. [CrossRef]
17. Zhu, J.; Li, G.; Kang, L. Editoral for the Special Issue on Carbon Fiber Composites. *J. Compos. Sci.* **2024**, *8*, 113. [CrossRef]
18. Rani, M.; Choudhary, P.; Krishnan, V.; Zafar, S. A review on recycling and reuse methods for carbon fiber/glass fiber composites waste from wind turbine blades. *Compos. Part B Eng.* **2021**, *215*, 108768. [CrossRef]
19. Zheng, H.; Zhang, W.; Li, B.; Zhu, J.; Wang, C.; Song, G.; Wu, G.; Yang, X.; Huang, Y.; Ma, L. Recent advances of interphases in carbon fiber-reinforced polymer composites: A review. *Compos. Part B Eng.* **2022**, *233*, 109639. [CrossRef]
20. Fu, Y.; Zhang, Y.; Chen, H.; Han, L.; Yin, X.; Fu, Q.; Sun, J. Ultra-high temperature performance of carbon fiber composite reinforced by HfC nanowires: A promising lightweight composites for aerospace engineering. *Compos. Part B Eng.* **2023**, *250*, 110453. [CrossRef]
21. Wu, Q.; Wan, Q.; Yang, X.; Wang, F.; Bai, H.; Zhu, J. Remarkably improved interfacial adhesion of pitch-based carbon fiber composites by constructing a synergistic hybrid network at interphase. *Compos. Sci. Technol.* **2021**, *205*, 108648. [CrossRef]
22. Li, N.; Link, G.; Wang, T.; Ramopoulos, V.; Neumaier, D.; Hofele, J.; Walter, M.; Jelonnek, J. Path-designed 3D printing for topological optimized continuous carbon fibre reinforced composite structures. *Compos. Part B Eng.* **2020**, *182*, 107612. [CrossRef]

23. Cheon, J.; Cho, D. Effect of MWCNT Anchoring to Para-Aramid Fiber Surface on the Thermal, Mechanical, and Impact Properties of Para-Aramid Fabric-Reinforced Vinyl Ester Composites. *J. Compos. Sci.* **2023**, *7*, 416. [CrossRef]
24. Shill, S.K.; Garcez, E.O.; Al-Ameri, R.; Subhani, M. Performance of two-way concrete slabs reinforced with basalt and carbon FRP rebars. *J. Compos. Sci.* **2022**, *6*, 74. [CrossRef]
25. Kröger, H.; Mock, S.; Greb, C.; Gries, T. Damping properties of hybrid composites made from carbon, vectran, aramid and cellulose fibers. *J. Compos. Sci.* **2021**, *6*, 13. [CrossRef]
26. Chen, H.; Wang, K.; Chen, Y.; Le, H. Mechanical and Thermal Properties of Multilayer-Coated 3D-Printed Carbon Fiber Reinforced Nylon Composites. *J. Compos. Sci.* **2023**, *7*, 297. [CrossRef]
27. Gljušćić, M.; Lanc, D.; Franulović, M.; Žerovnik, A. Microstructural analysis of the transverse and shear behavior of additively manufactured CFRP composite RVEs based on the phase-field fracture theory. *J. Compos. Sci.* **2023**, *7*, 38. [CrossRef]
28. Rodriguez, V.M.; Grasso, M.; Zhao, Y.; Liu, H.; Deng, K.; Roberts, A.; Appleby-Thomas, G.J. Surface damage in woven carbon composite panels under orthogonal and inclined high-velocity impacts. *J. Compos. Sci.* **2022**, *6*, 282. [CrossRef]
29. Abdallah, R.; Hood, R.; Soo, S.L. The machinability characteristics of multidirectional CFRP composites using high-performance wire EDM electrodes. *J. Compos. Sci.* **2022**, *6*, 159. [CrossRef]
30. Birgin, H.B.; D'Alessandro, A.; Meoni, A.; Ubertini, F. Self-sensing eco-earth composite with carbon microfibers for sustainable smart buildings. *J. Compos. Sci.* **2023**, *7*, 63. [CrossRef]
31. Amaechi, C.V.; Gillet, N.; Ja'e, I.A.; Wang, C. Tailoring the local design of deep water composite risers to minimise structural weight. *J. Compos. Sci.* **2022**, *6*, 103. [CrossRef]
32. Amaechi, C.V.; Chesterton, C.; Butler, H.O.; Gillet, N.; Wang, C.; Ja'e, I.A.; Reda, A.; Odijie, A.C. Review of composite marine risers for deep-water applications: Design, development and mechanics. *J. Compos. Sci.* **2022**, *6*, 96. [CrossRef]
33. Çelik, M.; Noble, T.; Jorge, F.; Jian, R.; Ó Brádaigh, C.M.; Robert, C. Influence of line processing parameters on properties of carbon fibre epoxy towpreg. *J. Compos. Sci.* **2022**, *6*, 75. [CrossRef]
34. Cao, S.; Zhu, Y.; Jiang, Y. Notched Behaviors of Carbon Fiber-Reinforced Epoxy Matrix Composite Laminates: Predictions and Experiments. *J. Compos. Sci.* **2023**, *7*, 223. [CrossRef]
35. Khozeimeh, M.A.; Fotouhi, R.; Moazed, R. Static and vibration analyses of a composite CFRP robot manipulator. *J. Compos. Sci.* **2022**, *6*, 196. [CrossRef]
36. Wu, T.; Kruse, R.; Tinkloh, S.; Tröster, T.; Zinn, W.; Lauhoff, C.; Niendorf, T. Experimental analysis of residual stresses in CFRPs through hole-drilling method: The role of stacking sequence, thickness, and defects. *J. Compos. Sci.* **2022**, *6*, 138. [CrossRef]
37. Burgani, T.D.; Alaie, S.; Tehrani, M. Modeling Flexural Failure in Carbon-Fiber-Reinforced Polymer Composites. *J. Compos. Sci.* **2022**, *6*, 33. [CrossRef]
38. Tanaka, K.; Kyoyama, S. The Effect of Pulse Current on Electrolytically Plating Nickel as a Catalyst for Grafting Carbon Nanotubes onto Carbon Fibers via the Chemical Vapor Deposition Method. *J. Compos. Sci.* **2023**, *7*, 88. [CrossRef]
39. Li, M.; Wang, Y.; Cui, B.; Wang, C.; Tan, H.; Jiang, H.; Xu, Z.; Wang, C.; Zhuang, G. Optimization of Electrical Intensity for Electrochemical Anodic Oxidation to Modify the Surface of Carbon Fibers and Preparation of Carbon Nanotubes/Carbon Fiber Multi-Scale Reinforcements. *J. Compos. Sci.* **2022**, *6*, 395. [CrossRef]
40. Vaidya, U.; Janney, M.; Graham, K.; Ghossein, H.; Theodore, M. Mechanical response and processability of wet-laid recycled Carbon Fiber PE, PA66 and PET thermoplastic composites. *J. Compos. Sci.* **2022**, *6*, 198. [CrossRef]

Disclaimer/Publisher's Note: The statements, opinions and data contained in all publications are solely those of the individual author(s) and contributor(s) and not of MDPI and/or the editor(s). MDPI and/or the editor(s) disclaim responsibility for any injury to people or property resulting from any ideas, methods, instructions or products referred to in the content.

Review

Review of Composite Marine Risers for Deep-Water Applications: Design, Development and Mechanics

Chiemela Victor Amaechi [1,2,*], Cole Chesterton [3], Harrison Obed Butler [4], Nathaniel Gillet [5], Chunguang Wang [6], Idris Ahmed Ja'e [7,8], Ahmed Reda [9,10] and Agbomerie Charles Odijie [11]

1. Department of Engineering, Lancaster University, Lancaster LA1 4YR, UK
2. Standardisation Directorate, Standards Organisation of Nigeria (SON), 52 Lome Crescent, Wuse Zone 7, Abuja 900287, Nigeria
3. EDF Energy, Power Plant Development, Bridgewater House, Counterslip, Bristol BS1 6BX, UK; cole.chesterton@sky.com
4. DTU Energy, Danmarks Tekniske Universität (DTU), 2800 KGS Lyngby, Denmark; obedbutler@gmail.com
5. Department of Production Engineering, Trident Energy, Wilton Road, London SW1V 1JZ, UK; gillettnathaniel@gmail.com
6. School of Civil and Architectural Engineering, Shandong University of Technology, Zibo 255000, China; cgwang@sdut.edu.cn
7. Department of Civil Engineering, Universiti Teknologi PETRONAS, Seri Iskander 32610, Malaysia; idris_18001528@utp.edu.my
8. Department of Civil Engineering, Ahmadu Bello University, Zaria 810107, Nigeria
9. Engineering Services, Qatar Energy, Doha, Qatar P.O. Box 3212, United Arab Emirates; reda@qatarenergy.qa
10. School of Civil and Mechanical Engineering, Curtin University, Bentley, WA 6102, Australia
11. Department of Engineering, MSCM Limited, Coronation Rd., High Wycombe HP12 3TA, UK; charlesodijie@hotmail.com
* Correspondence: c.amaechi@lancaster.ac.uk

Citation: Amaechi, C.V.; Chesterton, C.; Butler, H.O.; Gillet, N.; Wang, C.; Ja'e, I.A.; Reda, A.; Odijie, A.C. Review of Composite Marine Risers for Deep-Water Applications: Design, Development and Mechanics. *J. Compos. Sci.* **2022**, *6*, 96. https://doi.org/10.3390/jcs6030096

Academic Editor: Jiadeng Zhu

Received: 4 February 2022
Accepted: 22 February 2022
Published: 17 March 2022

Publisher's Note: MDPI stays neutral with regard to jurisdictional claims in published maps and institutional affiliations.

Copyright: © 2022 by the authors. Licensee MDPI, Basel, Switzerland. This article is an open access article distributed under the terms and conditions of the Creative Commons Attribution (CC BY) license (https:// creativecommons.org/licenses/by/ 4.0/).

Abstract: In recent times, the utilisation of marine composites in tubular structures has grown in popularity. These applications include composite risers and related SURF (subsea umbilicals, risers and flowlines) units. The composite industry has evolved in the development of advanced composites, such as thermoplastic composite pipes (TCP) and hybrid composite structures. However, there are gaps in the understanding of its performance in composite risers, hence the need for this review on the design, hydrodynamics and mechanics of composite risers. The review covers both the structure of the composite production riser (CPR) and its end-fittings for offshore marine applications. It also reviews the mechanical behaviour of composite risers, their microstructure and strength/stress profiles. In principle, designers now have a greater grasp of composite materials. It was concluded that composites differ from standard materials such as steel. Basically, composites have weight savings and a comparative stiffness-to-strength ratio, which are advantageous in marine composites. Also, the offshore sector has grown in response to newer innovations in composite structures such as composite risers, thereby providing new cost-effective techniques. This comprehensive review shows the necessity of optimising existing designs of composite risers. Conclusions drawn portray issues facing composite riser research. Recommendations were made to encourage composite riser developments, including elaboration of necessary standards and specifications.

Keywords: composite riser; pipeline; marine riser; marine composite; marine structures; composite structures; advanced composite material; thermoplastic composite pipes (TCP); fibre-reinforced composites (FRP); hybrid composite structures; review

1. Introduction

In recent times, the utilisation of marine composites has grown in popularity [1–5]. This has been considered both as full thermoplastic composite pipes (TCP) or as hybrid composite structures [6–10]. Particular applications of composites are seen in researches on composite risers [11–13]. The reason is that composites differ from standard materials such as steel in a number of ways, and more marine designers now have a better understanding of composite materials from flexible risers [14–16]. Secondly, these composite materials have weight savings and comparative stiffness-to-strength ratio, which could be advantageous in marine engineering [17–19]. Thirdly, the offshore sector has grown in response to newer innovations in composite structures, such as composite risers [20–22]. In principle, more subsea developments enhance successful drilling operations and provide new cost-effective techniques. Thus, marine risers are crucial components of all subsea production systems, termed SURF (subsea umbilicals, risers and flowlines) [23–27]. In composition, composite tubulars like marine hoses and composite risers are composed of various layers, with internal and external polymer sheaths to ensure internal fluid and exterior sea water integrity [28–33]. To ensure the reliability, vortex-induced vibration (VIV) and fatigue studies have been conducted on SURF structures under global loadings and deep-water conditions [34–37]. Figure 1 presents some floating deep-sea offshore platforms and marine riser configurations.

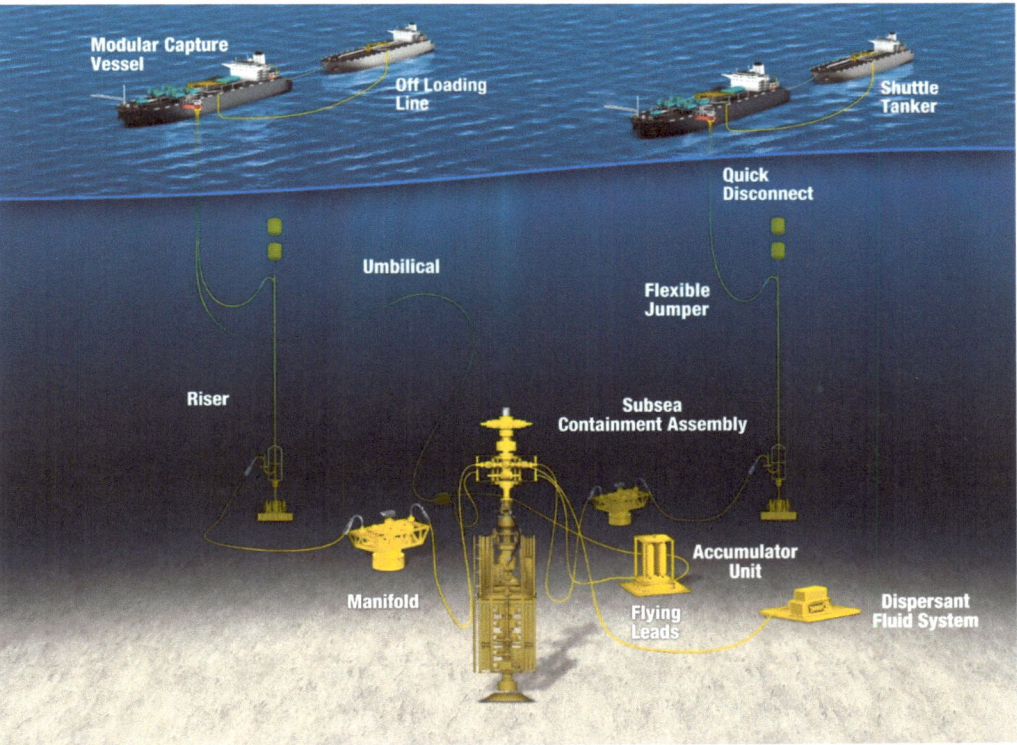

Figure 1. Deep-water facilities showing offshore platforms with the configurations for marine risers. Fluids are directed from the well to the Modular capture vessels (MCV) via flexible pipes and risers in a cap and flow system, which are part of an expanded containment system. (Courtesy: Marine Well Containment Company (MWCC)).

At extreme depths around 3000 m, the risers are induced by external loadings, corrosive fluids, increasing pressures, changing temperatures, etc. [38–43]. As such, it is important to effectively test these composite production risers (CPR) against test limits of steel catenary risers (SCR) and other bonded composite pipes to avoid failure of the marine riser [44–47]. Composite production risers (CPR), hose-lines, pipelines and flowlines, convey hydrocarbons as fluids and production ingredients [48–52]. These fluids include injection fluids, control fluids and gas lift. Risers are primarily used to move fluids or gas from the seabed to a host floating platform or onshore facility or to a transfer vessel [53–55]. However, risers are affected by vibrations and thus require tensioners designed for the control of marine risers [56–58]. Different failure modes have also been recorded on marine risers [59–64]. Additional riser functions, depending on the application, include: conveying fluids between the wells and the floating, production and storage (FPS) units. The fluid types include production, injection, importing, exporting or circulating fluids used for operation between the FPS and remote equipment or pipeline systems; guiding drilling or workover tools and tubulars to the wellbore as well as into the wells; supporting auxiliary lines; and serving as, or incorporating, auxiliary lines [65–67]. However, these ISO standards do not cover composite riser design and analysis. Typical composite riser joint (CRJ) is seen in Figure 2. Developments made are also detailed in Table 1.

Figure 2. The stack of Norske Conoco AS and Kvaerner Oil Field Products (NCAS/KOP) composite drilling risers (CDR) which held the first composite riser joint (CRJ) for Heidrun Platform with an assembly of titanium liner connector. (**a**) CDR Joint Located Over Titanium Drilling Riser Joint, and (**b**) Titanium Liner-Connector Assembly of NCAS/KOP Composite Drilling Riser (Courtesy: OTC's OnePetro Publisher; Norske Conoco AS; and Kvaerner Oil Field Products; Sources: [66,67]).

Table 1. Historical summary of previous joint industry projects on composite risers.

Year	Project Funder/Country	Reference	Riser Type	Materials	Thickness (mm)	ID (m)	Length (m)
1973	Ahlstone Marine Riser		D	Glass fibre/epoxy	–	–	–
1985–1987	Joint industry program (JIP) by Institut Francais du Petrole and Aerospatiale du France/France	[67]	P	Glass and carbon fibres/epoxy	9.57 (carbon) 7.28 (glass) 1.1 (inner layer)	0.2286	4 and 15
1994–2000	National Institute of Standards and Technology (NIST)'s Advanced Technology Program (ATP)/US	[68,69]	D&P	Carbon fibre & E-glass fibers/epoxy	Not specified	0.496/0.255	2.286
1995–1999	JIP—ABB, Vetco Gray and University of Houston.	[70–77]	D	Carbon fibre/epoxy	Not specified	0.5	Not specified
1996–2001	JIP—ABB, Vetco Gray, Aker Kvaerner, Conoco, EU Thermie, Chevron, Hydro, Statoil, Shell and Petrobras/US	[78–81]	D	Carbon fibre/epoxy	Not specified	0.5	Not specified
1999–2000	JIP-NIST/ATP, Shell & BP-Amoco/US	[82–85]	D&P	Carbon fiber/epoxy		0.250	19
1995–2001	CompRiser JIP—Heidrun CDR joint by Norske Conoco AS and Kvaerner Oilfield Products/Norway	[86,87]	D	Glass and carbon fibre/epoxy	Not specified	0.536	14.585
2003	JIP-ConocoPhillips, Kvaerner Oilfield Products & ChevronTexaco/Norway	[88–92]	D&P	carbon fiber/epoxy	Not specified	0.55	14.7
2007	Doris Engineering, Freyssinet, Total and Soficar	[93,94]	P	Carbon fibre/epoxy	–	–	–
2006–2009	Part of NIST Advanced Technology Program by University of Texas/US	[95]	D	Glass and carbon fibre/epoxy	30.5	0.540	4.57
2008–2011	Research Partnership to Secure Energy for America (RPSEA)/US	[96–104]	D&P	Glass and carbon fibre plus epoxy	25.4 (liner) 53.3 (composite)	0.508	10
2009	JIP—Airborne Composite Tubulars, MCS Advanced Subsea Engineering & OTM Consulting	[105–110]	D&P	Glass & carbon fiber/epoxy	Not specified	Not specified	Not specified
2011-date	Magma Global of Technip FMC/UK	[111–115]	D&P	Carbon fibre/epoxy	7–39	0.047–0.6	Up to 27.4
2011-date	Airborne Oil and Gas (now Strohm)/Netherlands	[116–126]	D&P	Glass and carbon fibre/epoxy	Varies	Varies	Varies
2011-date	University of New South Wales/Australia	[127–136]	P	TCP, carbon fiber/epoxy & PEEK	Varies	Varies	Varies
2015-date	Lancaster University/UK	[9–11,137–148]	P	carbon fiber/epoxy & PEEK	Varies	Varies	Varies
2017-date	University of Southampton/UK	[12–15,149–155]	P	carbon fiber/epoxy	Varies	Varies	Varies
2013-date	National University of Singapore/Singapore	[156–162]	P	carbon fiber/epoxy	Varies	Varies	Varies

P—production riser, D—drilling riser, CPR—composite production riser, CDR—composite drilling riser.

As seen in scholarly works and industry research on composite risers as tabulated in Table 1, more novel methods have been employed in recent times. Historically, the progress made in this field was magnified when the first CPR joint was installed on Heidrum Platform in 2002 [88,89], as depicted in Figure 2. Some designs guidelines for marine risers were also given in standards. An earlier analysis of several marine risers, with design methodology, was published in the API 16J Bulletin ([166]), which has been replaced and developed into recent standards like the ISO 13624 [167,168], ISO 13625 [169] and ISO 13628 [170–179], and API [180–184]. Moreira et al. [185] investigated a 0.445 m-ID workover riser system built for a 3000 m water depth and deployed it, excluding any umbilical via an autonomous control system. The findings from earlier studies, application of these standards and advances made led to other cost-effective composite riser joint. In another study, Cederberg [109] found that composite riser joints had common issues that arose were autofrettage and the metal–composite interface on the CPR joint. Similar findings have been examined in different models [78,100–103,114,186–190].

These studies conducted on the stress analysis of composite riser joint and its composite tube showed that there is a connection between the end-fitting and the composite tube. As such, optimisation is recommended to obtain the best orientations, lay-up angles, number of layers, type of materials to use, fibre design, matrix design, resin-coating material, etc. These include investigations on the microstructure of the composite materials conducted to ascertain different failure modes and material strengths of these novel materials. Notable industry players in the field of composite risers include Magma Global of Technip FMC [191,192] and Airborne Oil&Gas (now Strohm) [193,194]. Some qualifications were achieved in recent times on composite riser and TCP tubes. Magma Global had qualified a composite riser tube earlier in 2013, while Airborne Oil & Gas qualified some TCP pipes circa 2017. In the later, the qualified composite riser was deployed as the first subsea TCP flowline to use hydrocarbon fluid services efficiently by Airborne Oil & Gas. Similarly in the former, OCYAN and Magma Global have also effectively configured m-pipes on CPR for pre-salt Brazil, as depicted in Figure 3.

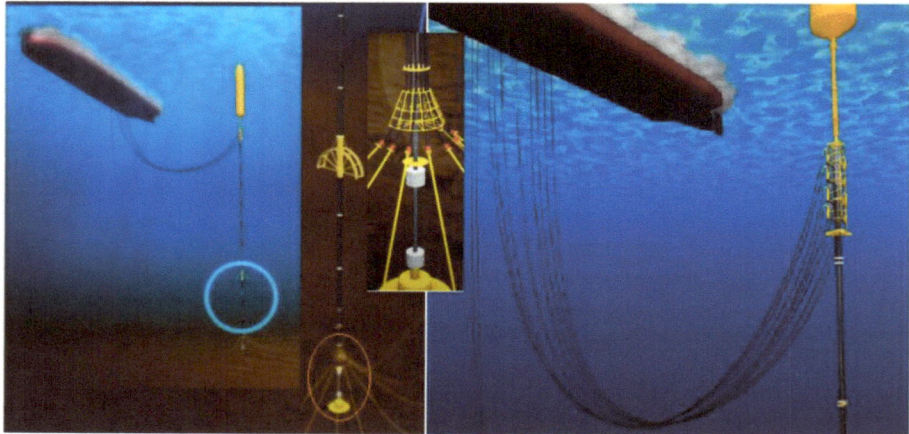

Figure 3. CompRiser as cost-saving composite riser with m-pipe's decoupled solution, which is lightweight, with 7000 tons less of load per FPSO (considering 2 towers) in a collaboration between Magma Global and OCYAN. (Reproduced with permission obtained from Magma Global to reuse image of m-pipe application).

This has also led to advances in the use of TCP pipes on composite risers and related marine composite tubulars for water injection, transport, product transfer and loading and fluid discharge services [31,51,150]. Based on numerical models presented on different CPR models of well-validated composite risers with safety profiles from results of the stress analysis, conducted in ANSYS APDL [137–142] and ANSYS ACP [143–146,149–152], respectively. Additionally, both researchers used different platforms—while the former used a TLP (tension leg platform), the later used a SPAR (single-point anchor reservoir) and a PCSemi (paired-column semisubmersible). However, both models reflected different novelties in modelling approaches and showed that the full application of composites is feasible using multilayered composite structures and effective liner material for layering the structure. Other comparative investigations on steel risers and composite risers have revealed that composite risers offer numerous benefits, particularly a low weight and cost savings [65,98–104,195–199].

Thus, the aim of this review is to conduct a comprehensive overview on the design, developments, and mechanics of composite risers, including their end-fittings, for offshore marine applications. Section 1 presents an overview of the current position on composite risers within the offshore industry, in comparison to related existing applications of composite materials in deep waters. Section 2 presents the design with detailed analysis of the advances in composite riser research for deep waters. Section 3 presents studies that reflect the mechanical behaviour including fatigue. It also includes findings from selected studies on analytical, numerical and experimental studies on composite risers. This study presents the merits and demerits on the application, qualification and utilisation of CPRs.

2. Design and Manufacture

In this section, the advances in mechanics, design, modelling and qualification of CPRs are presented.

2.1. Advances in Composite Risers

Composite materials for offshore applications were first introduced about seven decades ago. They have piqued the interest of the offshore oil and gas industries, owing to their high specific strengths and stiffnesses, which help with weight reduction and cost savings. Earlier investigations on composite risers have been conducted and showed good results [68–76,88–91]. However, composite applications in composite risers have transiting barriers and are thus seen as an enabling technology [65,119]. Composites are currently used in different components such as accumulator vessels, composite tethers, flexible risers, tensioners, buoyancy cans and the topside of platforms as well as on flow-lines, spoolable tubings, spoolable pipes, buoyancy modules and buoyancy floats [5,55,83,200–205]. However, their application in risers has been limited to prototype production and drilling risers to date, despite that they may significantly reduce the weight of deep-water operational systems [93,94,105–110]. Although the material costs of FRP composites are higher than those of steel, many previous studies have shown that their total life-cycle costs will be lower due to the add-on effects of their weight savings for other system components. These components include the reduced stacked volume of BOP, reduced total system weight, top-tension requirements, mooring pretensions, reduced platform sizes and buoyancy weights [55,110]. An increased water depth implies that there will be larger platform payloads and additional load conditions which might be severe, necessitating significant increases in the amount of manufactured steel needed as well as extra mooring pretensions. According to some estimates, a one-pound increase in platform payload costs an additional four to seven dollars [55,199]. The top tension required to apply to a riser grows as its operational depth increases, necessitating more buoyancy in the hull and a larger platform. Due to the increasing hydrostatic pressures experienced, both the needed length of the riser and its thickness increases.

As a result, increasing depth has a two-fold influence on a riser's weight and, as a result, the top tension required. According to research, the size of TLPs increases at a considerably faster pace as their top tension increases [200–202], limiting the number of risers that can be used or the depth to which they can be used [206–210]. Based on the present capabilities, the maximum depth to which a steel riser can be deployed cheaply depends on the available platforms. For production risers, between 1000 and 1800 m is a good range, while drilling risers require depths exceeding 3000 m [23–27,41]. As a result of the weight savings from composite risers, more risers can be built at existing depths attributable to the use of FRP composite materials [211–220]. Also, production and the viable utilisation of petroleum resources will increase to even lower depths by using composites [221–230]. In addition to having a lower density, FRP composites have a higher strength-to-weight ratio [231–240]. Excellent damping, thermal insulation and corrosion and fatigue resistance will result in greater savings by lowering maintenance costs by using composites [241–251]. However, appropriate testing and qualification is required to successfully use composite materials in marine risers. It is also necessary to evaluate the durability of the product in sea water. In the research by Venkatesan et al. [252], carbon-fibre-reinforced polymer (CFRP) composites were observed to have long-term qualities after being exposed to clean water and sea water at the same time, with noteworthy differences under various temperatures. Many researchers also concur with the finding that the CFRP's long-term tensile strength is lowered to between 80% and 95% of its original value in the short term [252–258]. It was also discovered in another study by Bismarck et al. [58] on the evaluation of the carbon/PEEK thermoplastic composite that although its axial tensile strength was unaffected by boiling water, its transverse tensile strength was affected. After being exposed to boiling water, the tensile strength of the material was reduced. It was determined that, in order to avoid failure, thermoplastic composites must have a maximum service temperature that is far below certain polymer matrices' glass transition temperatures. Aside from carbon fibres, there are other fibres, such as glass and aramid, and other synthetic high-performance fibres, such as Toray, Spectra, Dyneema, Zylon, M5 and Victrex PEEK, that are also often used in composites. Nonetheless, the choice of material for subsea conditions (such as water ingress) is sensitively dependent on the performance of composites reinforced with these fibres because a reduction in tensile strength can lead to a severely reduced performance of the composite tubular [259–265].

In addition to studying the mechanical properties of composite materials in sea water, there are effects from fatigue, load distributions, global responses and performances of the full-length composite riser. However, the CPR tube, its stress joint, steel tension joint and other standard joints face global functional and environmental conditions [98,143–146,266]. The effects of functional loadings have also been studied in comparative studies presented on both steel and composite risers, including vortex-induced vibrations (VIVs) and resonances on the CPR [11,55,199–202,246]. Generally, it was discovered that as the water depth increases, the tension force diminishes, and the maximum tension force is reached. The stress joint at the bottom bears the brunt of the bending moment, followed by the joints across the surface of the sea, called the mean water level (MWL), displaying larger bending moments than those in the riser string's midsection [98,101,200–203]. However, when it was compared to an all-steel riser, the axial stress and bending were significantly reduced along a composite riser's full length, comprising tension, standard and stress because of its smaller total weight, it had fewer joints. The fantastic fatigue resistance of FRP composites, particularly carbon-fibre-reinforced composites, contributes to their significance. In principle, the structural composition of the CPR tubular body and the composite riser joint is predicted to have an indefinite fatigue life, depending on the material composition, whereas those of the joint's metal liner, the metal–composite interface (MCI), the steel tension joint and stress joints have been proven to be adequate. It is noteworthy to state that the fatigue characteristics of FRP composites can vary, with respect to the basic materials selected and the manufacturing technique used, whereas the fatigue life of its steel liner welds can be significantly reduced. Based on the VIV and resonance investigation on CPRs, some studies reported a resonant response reflecting that the composite riser's response is due to the high resistance of the system in sea water, as it showed no noticeable resonance in comparison to the steel riser [98,266]. Kim [98] showed that the drag force and vibration amplitudes in its bottom section were very small. This implies that the vibration waves were significantly more damped as they descended from the top of the composite riser than they were as they descended from the top of the composite riser in the steel riser. The fundamental frequency of VIV in composite risers was discovered to be less due to its comparative lower mass, but that its worth was relatively insignificant [98–104]. This behaviour was confirmed in some recent studies [143–146] showing that the fundamental frequency of VIV in composite risers was lower due to the stiffness property of a composite riser because the composite riser was sensitive to structural damping and tension fluctuations. Generally, the VIV could be reduced by increasing tension and damping in the composite riser. According to Omar et al. [200], the maximum VIV strains created in a composite riser were nearly half the weight of a comparable steel riser, indicating that composite risers are more durable. Steel risers have a far shorter fatigue life than aluminium risers. An investigation on VIV and fatigue was demonstrated on composite risers by Huang [203] and opined that in the composite riser, the damage produced by both long-term and severe currents was moderate compared to that inflicted to the riser lacking VIV suppression. However, the addition of strakes could assist in the suppression of VIV with efficient suppression control. As a result, strakes are typically utilised with caution to provide an extra margin of safety in such situations involving VIVs.

The benefits of FRP composites are obvious from the explanation above. Due to their superior qualities, such as high specific stiffness, they have advantages over steel. Thy have high corrosion resistance, fatigue resistance, improved thermal insulation, superior damping and a light weight, resulting in improved global responses and performances. In addition, they have better VIV distributions, better fatigue, better bending profile, better stress distribution and better tension distributions along the length of riser responses. However, these are dependent on the materials chosen for the CPR, the global loading and CPR configuration. These attributes also result in a lower cost, easier installation and a longer service life. However, these factors make composite risers efficient by reducing deck loads, lowing platform size, reducing mooring pretension, reducing top tension, reducing system weight, reducing buoyancy weight and the stacked volume. Since all these factors contribute to a lower total cost, they make composite risers more cost-effective. Furthermore, several design variables, including the liner thickness, the fibre and matrix combinations, stacking sequences, composite lamina thicknesses and fibre orientations, can be modified to customise (or tailor) the design of a composite riser to specific requirements. A customised design that fully optimises these variables can improve the benefits of FRP composites and achieve larger weight reductions [79–81]. Figure 4 depicts a summary of the areas of composite risers reviewed herein.

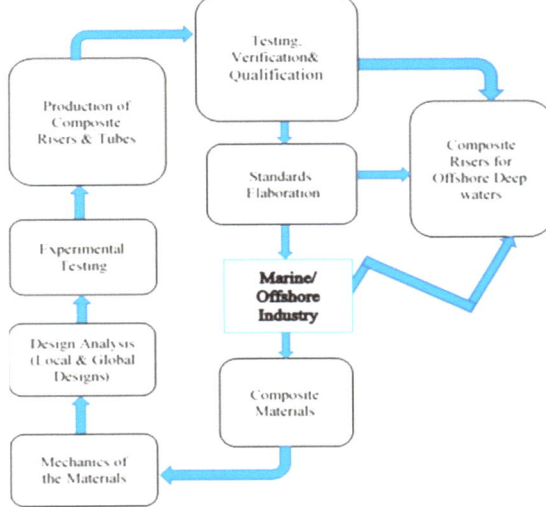

Figure 4. Reflection on sections and aspects covered in this composite riser review.

2.2. Qualification of Composite Risers

Figure 5 shows typical designs of TCP composite pipes, showing model cross-sections from Airborne Oil&Gas and Magma Global. Both pipe designs use TCP materials and have been qualified. Based on different experiments on composite tubes and composite risers, there are attempts to qualify composite tubulars hence elaboration of related standards and regulations by DNV [267–275]. These standards and other presentations on the qualification of TCP composite pipes and composite risers that led to the development of the DNVGL-RP-119 standard [84,85,275]. However, there are still more challenges concerning the qualification. According to an offshore market report on flexible pipes by Lamacchia D. [217], flexible pipes with an inner diameter (ID) of 4–8 inches account for 80% of all flexible pipe applications, as shown in Figure 6. For numerous years, the flexible market limit has been set at PxID = 80,000 psi.inch, with high pressure–high temperature (HPHT) flexible risers qualified up to 20 ksi by TechnipFMC. The bulk of flexible tubing in use has a PID value of less than 50,000 psi-in, which is noteworthy.

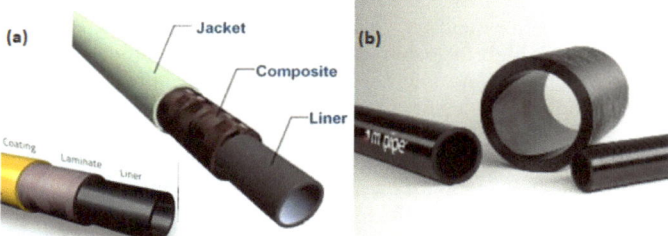

Figure 5. TCP composite pipes by (**a**) Airborne Oil&Gas, and (**b**) MagmaGlobal.

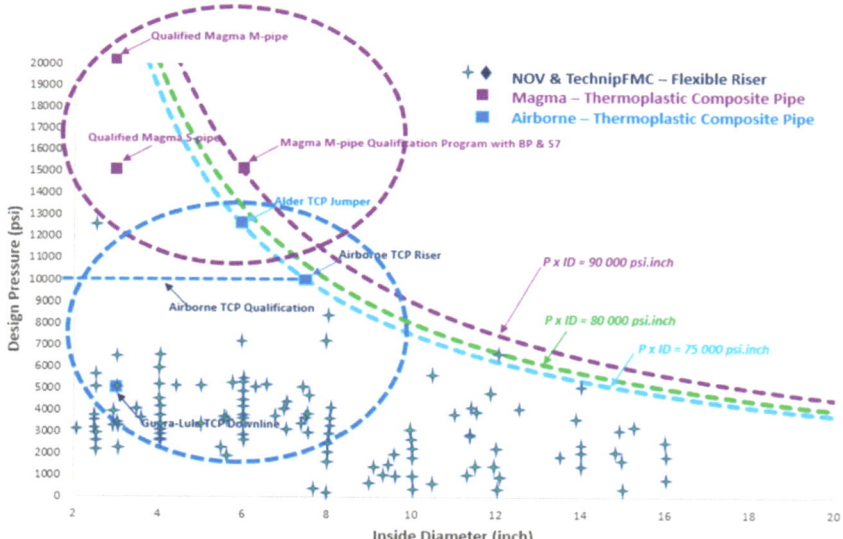

Figure 6. Qualification roadmap of flexible pipes, TCP pipes and composite risers in industry, showing manufacturers including NOV & Technip FMC, Airborne Oil&Gas and Magma Global. (Source: [217,276]; Data extracted from [277]; Courtesy: Leviticus Subsea. Permission to use this image was obtained from Lamacchia D).

TechnipFMC appears to be better positioned for HP projects in the flexible market, whilst NOV appears to be more dedicated to the low-pressure sector. Among these two competitors, GE (formerly Wellstream) flexibles wants to enter the HPHT market with the release of a "hybrid" riser with a composite layer. By using different inner diameter (ID) pipes, the Airborne and Magma Global met offshore market demands for pipes with mid-pressure and mid-temperature applications. These manufactured composite pipes are aiming to reach the same status as the NOV and GE flexible market pipes. Airborne's market limit appears to be PxID = 75,000 psi.inch, based on their expertise and qualifications. The Magma composite pipe is aimed at the HPHT market segment, which is focused on ultra-deep-water applications and is a market that TechnipFMC and GE are pursuing with new research and development as well as the addition of composite layers to lower overall riser weight. Magma's market constraints appear to be at PxID = 90,000 psi.inch, based on their expertise and qualifications [276–280]. With a maximum ID of 7.5 inches, both Magma and Airborne TCP are tailored to small-diameter-ID pipes, which are typical in riser, jumper and downline applications. Larger diameters imply entry into the flowline and pipeline industry, and significant expenditures will be necessary to update real plant capacities, including spooling reels for longer flowlines, in addition to competing not just with flexible pipes but also with inflexible steel pipes. Based on the qualification, the composite riser goes through a validation process once it is designed to guarantee that it can be used on site without endangering the offshore platform. A thorough understanding of the certification methods and risk-based methodologies in DNV-RP-A203 [274] is critical to the success of any high-integrity engineering utilising HPHT.

2.3. Material Characterisation and Metal–Composite Interface (MCI)

An important aspect of the composite riser review is the discussion on the material characterisation and microstructure of composite materials used in developing composite risers, as depicted in Figures 7 and 8. Since the designs of composite risers are based on material designs on their laminas, further investigations on the material strength and effect of the lay-up sequence can be conducted. Each of the laminas (or layers) are made of matrix and fibres, manufactured together, at an orientation angle and with a microstructure, as illustrated in Figure 8a. Different studies have shown that the lay-up sequence has an effect on the strength characteristics of the composite structure [131,257,258]. Another critical aspect of the CPR is the metal–composite interface (MCI). MCI is considered a mechanical attribute of the CPR as well as the thermal conductivity, stiffness and other strength properties of the composite material. The laminate microscopy in Figure 8b shows that the layers of the composite material are structured along different orientations and patterns. In principle, carbon fibres have a higher modulus and strength than glass fibres; however, glass ribbons are more shock-resistant and cost less [237–239]. The resin used is typically chosen based on its ability to absorb oil, water, gas or other fluids. On the other hand, the liner can be made from a thermoplastic extrusion, as in Picard et al. [95], where the liner is the first barrier that the internal fluid encounters, and the thermoplastic material chosen is based on the design requirements. Furthermore, considering the kind of loads faced by a CPR and the variances in material qualities between the orthotropic composite and the isotropic steel, the MCI connection must be cost-effective. Since MCI is a mechanical attribute, it is required that systems that have couplings such as end-fittings undergo HPHT processes for the manufacturing of composite risers, composite tubes, TCP pipes and composite flowlines, depending on the material compositions under high temperatures. Therefore, the MCI is critical, as it provides a secure connection between the composite body and metal couplings at a riser joint's terminations [95,186,187].

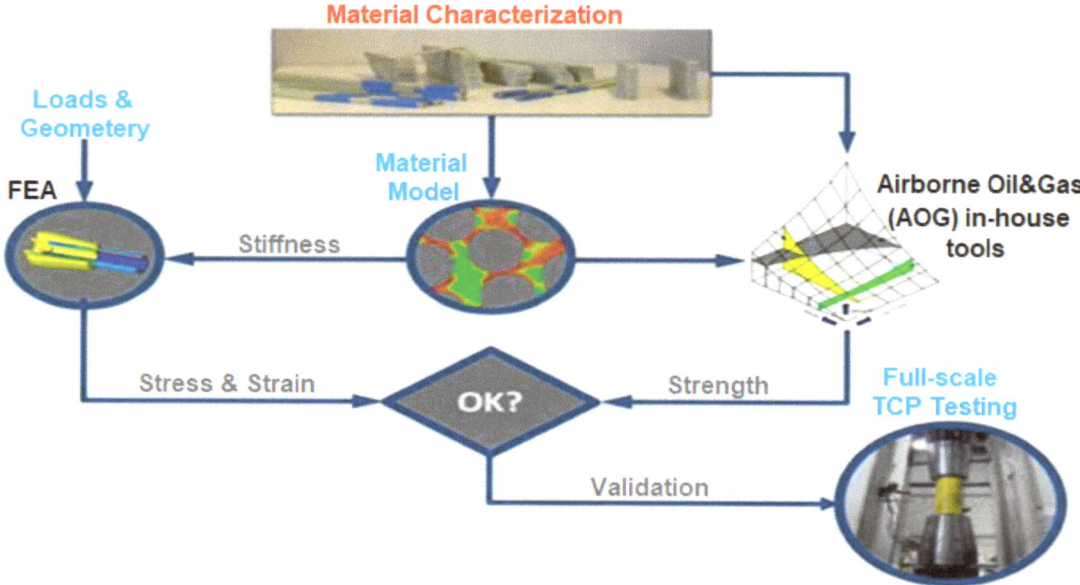

Figure 7. Design procedure for TCP material characterisation. (Courtesy of Airborne Oil&Gas).

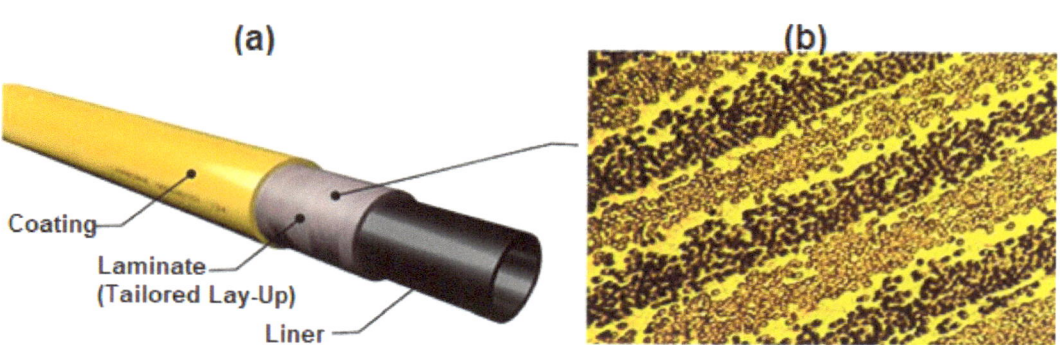

Figure 8. Depiction of (**a**) design concept of TCP pipes and (**b**) laminate microscopy. (Reproduced with permission of Jak Annemarie and Martin van Onna; Courtesy of Airborne Oil&Gas).

2.4. Loading Conditions

The background on marine waves was established by earlier scientists, such as Moskowitz, Pierson, Hasselmann, Kalman, Morison, Bernoulli and Newton [281–285]. They researched the performance of floating structures under various load influencers. These influencers are 'environmental forces', such as ocean waves, external pressures, flows in pipes, flows around cylindrical structures, etc. Hence, the investigation on composite risers using these factors laid the groundwork for hydrodynamics, marine waves and ocean engineering. Figure 9 depicts a typical composite riser with the loadings on it, such as waves and current. Wave forces are regarded as a critical aspect in any riser study since they aid in determining control of top tension, VIV and fatigue prediction [196–199,250,286–297]. Based on industry specifications DNV-RP-F202 [269] and DNV-OS-F201 [270], there are four types of loads on composite risers: accidental loads, environmental loads, functional loads and pressure loads, as tabulated in Table 2.

Furthermore, the pressures exerted by both internal and exterior fluids must be considered while designing composite risers. Under either static or dynamic conditions, the internal fluid pressures, external hydrostatic pressures and the pressures induced by the risers' operating sea depth all contribute to pressure loads. Functional loads, on the other hand, are those that arise as a result of the systems' quiddity, operation and subsistence, with disregard for incidental or environmental impacts. Additionally, it includes the riser's applied top tensions throughout the design and installations, as well as thermal loads. The ocean environment imposes environmental pressures either directly or indirectly, called environmental loads. The loadings that occur by chance and must be analysed against a goal failure probability are called accidental loads. Thermal effects and applied loads cause stress to develop along distinct laminas of the body of the composite tubular structure [278–280].

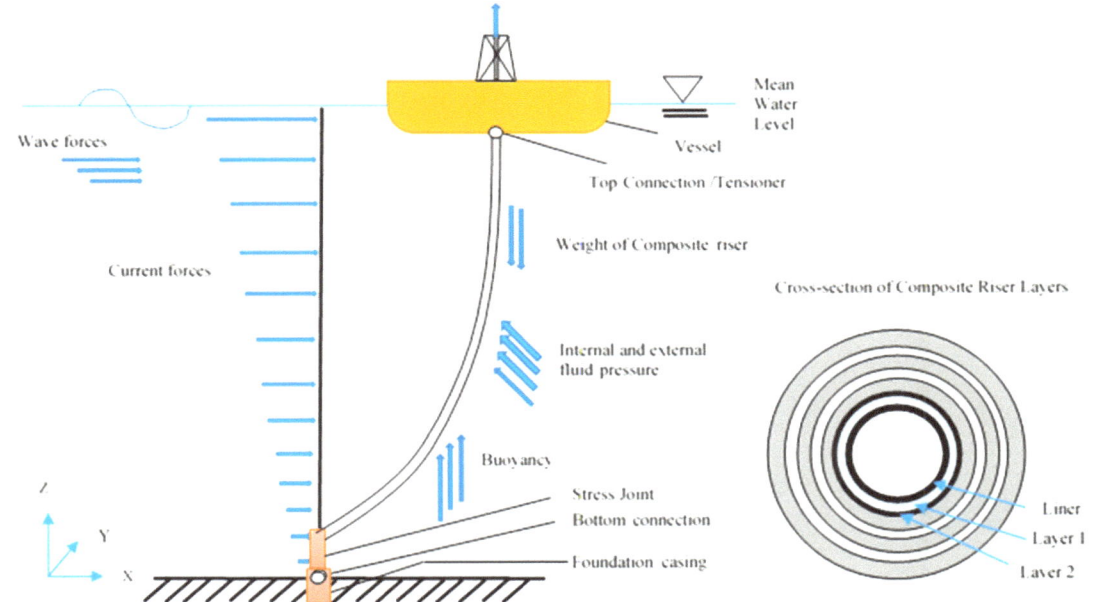

Figure 9. Depiction of composite riser system with loads, layers and cross-sectional cut.

Table 2. Different types of loads (Sources: DNV-OS-F201; 2010 and API RP 2RD).

F-Load or Functional Load	E-Load or Environmental Load	P-Load or Pressure Load	A-Load or Accidental Load
Weight of riser	Floater motions due to currents, waves and wind	Internal fluid pressure (dynamic): global load effects can be generated by both slugs and pressure surges on compliant configurations	Risk analysis related to support systems, such as loss of mooring line and loss of riser.
Weight of the internal fluid	Vessel motions	Internal fluid pressure (static)	A loosened tensioner in the system
Applied tensions on top-tensioned risers (TTR)	Waves	Internal fluid pressure (hydrostatic)	Fire hazards, explosions and riser collisions.

Table 2. Cont.

F-Load or Functional Load	E-Load or Environmental Load	P-Load or Pressure Load	A-Load or Accidental Load
Installation-induced residual loads or prestressing	Current	External hydrostatic pressure	Flow-induced impact between risers
The preloads of connectors	Due to changes in water density, internal waves and other phenomena.	Water levels	Impacts from dropped objects and anchors
Guidance loads and applied displacements, plus support for floater's active positioning system	Dynamic load effects, such as slug flow generated from the fluid pressure (P-Loads)		Naturally occurring environmental issues, such as earthquakes, tsunami, icebergs and hurricanes
Construction loads and loads caused by tools	Icey locations having ice formations or tendency to develop, be slippery or drifts		Failure of lower marine riser package (LMRP)
Soil pressure on buried risers	Seismic effects such as earthquakes (in seismically active regions)		Pressure surge and overpressure of well tubing
Differential settlements	Mean offset including current forces, wind and steady wave drifts		Loss of pressure safety system
Loads from drilling operations	Wave frequency (WF) motion		Seismic effects such as earthquakes (in seismically active regions)
Thermal loads	Low-frequency (LF) motion		Load from anchor, hooks and support systems (hook/snag load)
Inertia			Partial loss of station-keeping capability
Internally run tools			Internal pressure exceeded
Buoyancy of riser (including absorbed water), attachments, fluid contents, anodes, marine growths, buoyancy modules, tubing and coatings.			Risk analysis related to monitoring failure, such as dynamic positioning system (DPS), loss of buoyancy and loss of heave compensating system

2.5. Composite Risers' Layers

Based on the loading conditions discussed in Section 2.4, the stress profiles and safety factor profiles of composite risers can be generated, as given in Figure 10. The industry recommendation in API-RP-2RD [164] stipulates that high-performing composite tubulars have global stiffness as well as stress in various directions, such as fibre, transverse or in-plane shear, as confirmed in Figure 10.

The results profile, using an array of 4 axial layers, 10 off-axis layers and 4 hoop layers with an off-axis orientation of $\pm 53.5°$, shows that the titanium liner used worked effectively under the pure tension, collapse and burst loadings of the composite riser modelled using AS4/Epoxy. The local/global design of multi-layered composite risers has been conducted by different research groups as seen in Table 1. The justification for this is that each composite riser design is based on material attributes; thus, the forces on its laminas are subject to its equivalent properties. Each of the laminas (or layers) are made of matrix and fibres, manufactured together, at an orientation angle. A typical composite tubular structure showing its microstructure is depicted in Figure 8b.

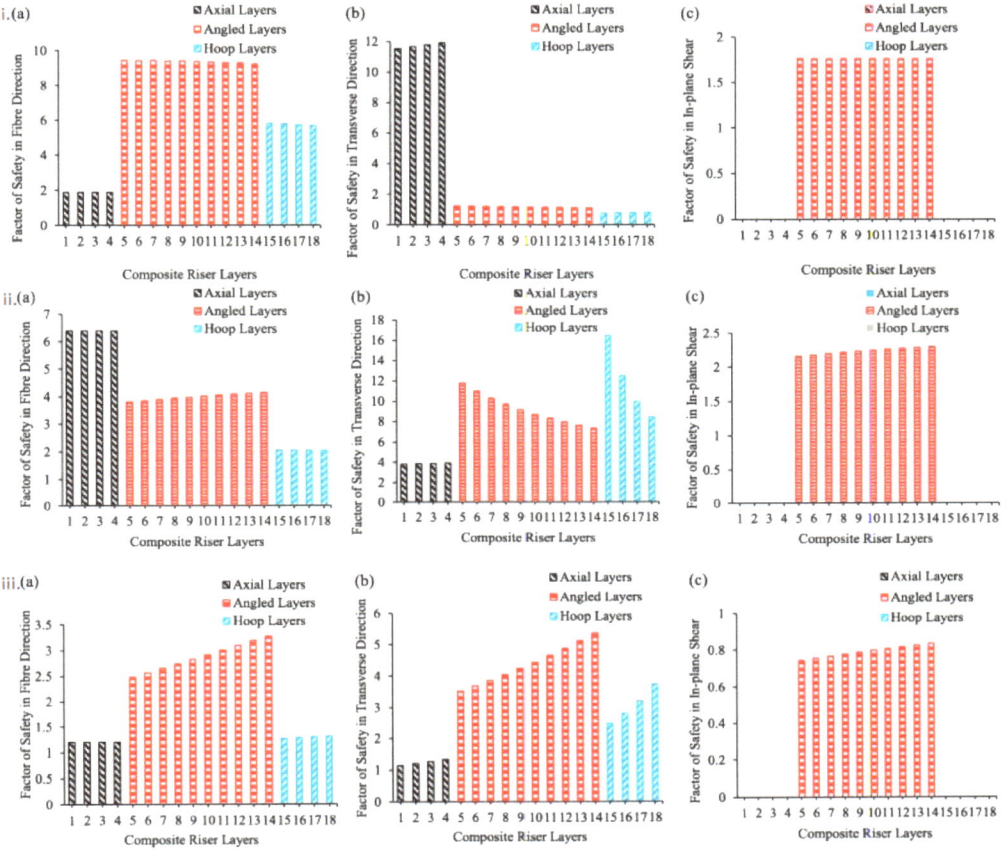

Figure 10. Composite riser with titanium as liners and configured utilising AS4/Epoxy for (**a**) fibre, (**b**) transverse and (**c**) in-plane shear directions, subjected to (**i**) pure tension, (**ii**) collapse loads and (**iii**) burst loads. The results were obtained from ANSYS ACP's FEM. (Permission to reuse was obtained from Elsevier Publishers, Source: [149]; Published: 15 February 2019; Copyright date: 2018).

2.6. Manufacturing Process

There have been different innovative concepts, designs and patents based on the manufacturing process for composite risers. Salama and Spencer [92] patented a method of manufacturing composite risers with liners aligned horizontally in a production line that is filament-wound. Thomas [229] outlined the procedures for making a composite production riser using filament winding. The traplock MCI, which consists of several grooves to trap a series of composite layers and install them on a mandrel, is often the first manufacturing process [92,160,229]. To make the riser's inner liner, an elastomer layer, such as uncured hydrogenated nitrile butadiene rubber (HNBR), is wrapped around the mandrel. Wilkins [122] described the manufacturing method for a thermoplastic composite pipe made from individual composite tapes that are 10 mm broad, 0.2 mm thick and hundreds of metres long. Several glass and carbon fibres were encapsulated into the thermoplastic matrix to form composite tapes, and adjacent tapes were placed in the same direction to form plies oriented at different directions. These fibres were pressed throughout the entire length of the pipe, and a laser heat source was used to apply thermal each incoming tape, which was then cured chemically and subjected to a thermoplastic welding process to enable a high level of cohesion. This technique could be mechanised, and a robot could be utilised, but it is worth noting that the manufacturing process is critical to the finished laminate's structural strength. A continuous process using a thermoplastic liner and composite-reinforced carbon layers is used to manufacture a composite carbon thermoplastic tube, which is fabricated by winding impregnated fibre ribbons around the thermoplastic liner at specific angles, with the thermo-fusion controlled to ensure proper placement on the liner. The CPR manufacturing process, according to Baldwin et al. [75,76], comprises an examination of fibre stress rupture data, which aids in forecasting the composite structure's average stress rupture life, desired service life and reliability. During the manufacturing process, special attention is paid to the end-fittings to guarantee that the connections between the pipe lengths, anchoring and MCI are appropriately blended. It is critical to alternate the hoop and axial reinforcements throughout the composite riser production process, as this influences the mechanical properties of the composite riser. Figure 11 shows an example of composite layers for the composite riser pipe showing lay-up patterns. It is worth noting that the manufacturing method that industry leaders such as Airborne Oil&Gas employ relies on some complexities in the capacity to create long lengths of continuous pipe; meanwhile, the production technology has been certified by DNV for the creation of qualified products. The composite polymers used in Airborne include polyethylene (PE), polypropylene (PP), polyamide (PA), polyvinylidene difluoride (PVDF) and polyether ether ketone (PEEK). According to Osborne [115], when the company creates a PP composite pipe, it starts with a PP liner, then melt-fuses glass/PP tapes onto the PP liner and melt-fuses a PP jacket on the exterior. According to the report, Airborne's technology produces a strong and stiff pipe comprising a single polymer and a single-fibre system. It further reported that it was the world's first manufacturer to succeed in producing a continuous solid thermoplastic composite pipe, which differs from reinforced thermoplastic pipe (RTP) in that RTP is frequently unbonded, requiring loose fibre layers or tapes to be wound around the liner, before Magma Global joined in similar manufacturing technique around 2015. RTP is unable to sustain external pressure and cannot be used in deep waters offshore. The pipe from Airborne is made of a thermoplastic compound and has three parts: a jacket, a composite and a liner, as shown in Figure 5. A thermoplastic is used to make the lining. Strength and stiffness are provided by glass or carbon fibres. The fully bonded flexible pipe is melt-fused during production; the pipe has a single solid wall made of glass fibres, E-glass or/and S-glass. Carbon fibres, but not aramid fibres, are the other material utilised, as the latter do not perform well under compression (So, a pipe made from aramid cannot handle high external pressure.). The tape fibres are impregnated into the plastic liner using the same plastic as the liner. As a result, the tapes are looped around the liner and melt-fused to it, with the tape then being melt-fused to the underlying tape and continued [115].

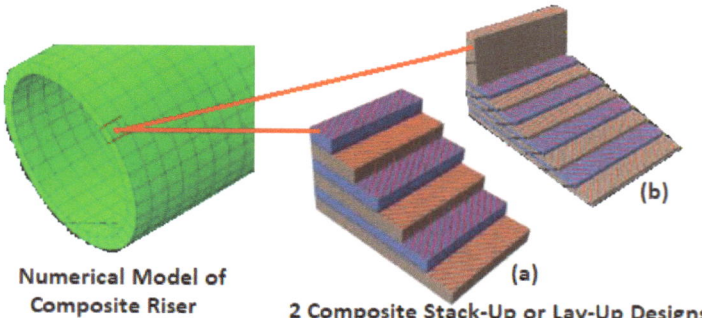

Figure 11. Typical composite riser structure showing (**a**) tubular structure and (**b**) the layers with two different lay-up sequences.

2.7. End-Fitting

In summary, these studies on the end-fittings will enable the designer to understand the behaviour at the joints and connections. Additionally, the pressure exerted at the joints of the composite risers creates some stress at these end-fittings. Lincoln Composites reported numerical and experimental assessments of a composite riser end-fitting joint using a CFRP tube and steel pipe (X80), which included the metal composite interface (MCI) for autofrettage prestressing and a factory acceptance test (FAT) [109,110]. The FEA showed the regions of high stresses on the neck of the end-fitting, mostly on the grooves, as shown in Figure 12. It was conducted by first setting up the end-fitting, then axially stretching the steel pipe. Secondly, a design pressure that exceeded its yield strength was exerted during a burst or internal pressure test. Next, the pressure was relieved and reapplied at a reduced scale. The results were recorded, and finally the pressure was released to relieve the system.

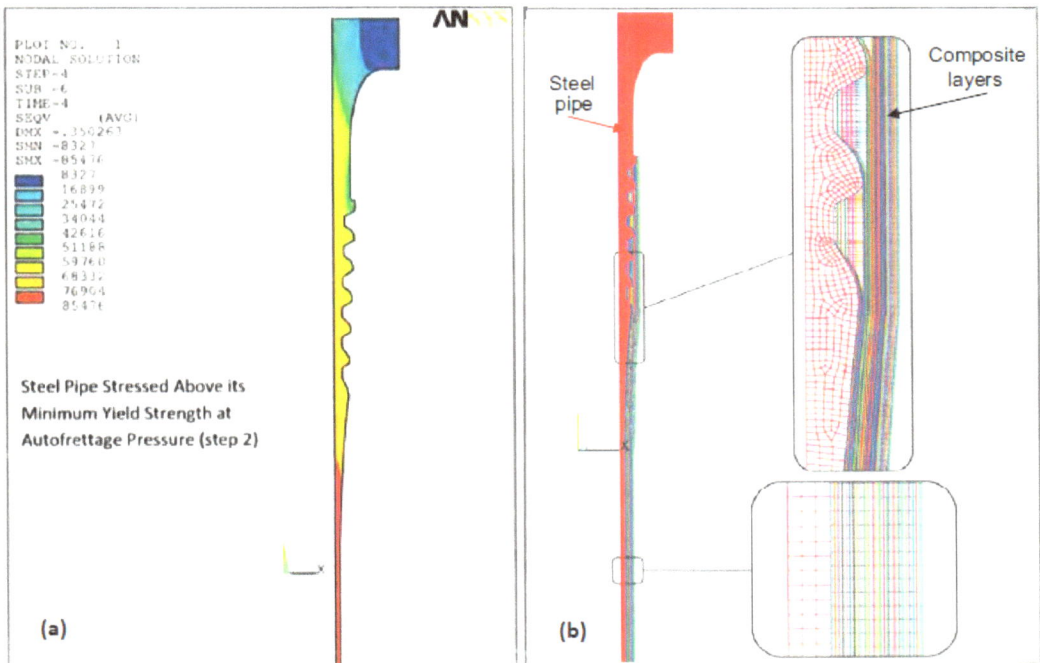

Figure 12. Finite element model of composite riser end-fitting joint in ANSYS, showing (**a**) Compression stress in the steel pipe, and (**b**) the local design for the riser joint model (Permission to reuse was obtained from (i) OTC's OnePetro Publisher and (ii) Elsevier Publishers; Source: [109,110,160]. Published: (i) 06 May 2013; Copyright date: 2013; (ii) 22 December 2015; Copyright date: 2016).

In this review, three end-fittings are considered: the traplock end-fitting, Magma end-fitting and Airborne end-fitting. Baldwin et al. [76] was the first recorded patent with traplock end-fitting device. For thermoset composite risers, a traplock end-fitting was created. A number of grooves are carved into the two ends of the mandrel in this design, allowing load transmission between the composite laminate and the metallic connector ends. The displacement vs. force profile of a traplock end-fitting with double the number of grooves is shown in Figure 13a. Some comparative studies on end fitting designs are carried out in literature [187]. These are based on different inventions and patents on composite risers which are currently been applied in the industry [74,76,92,188].

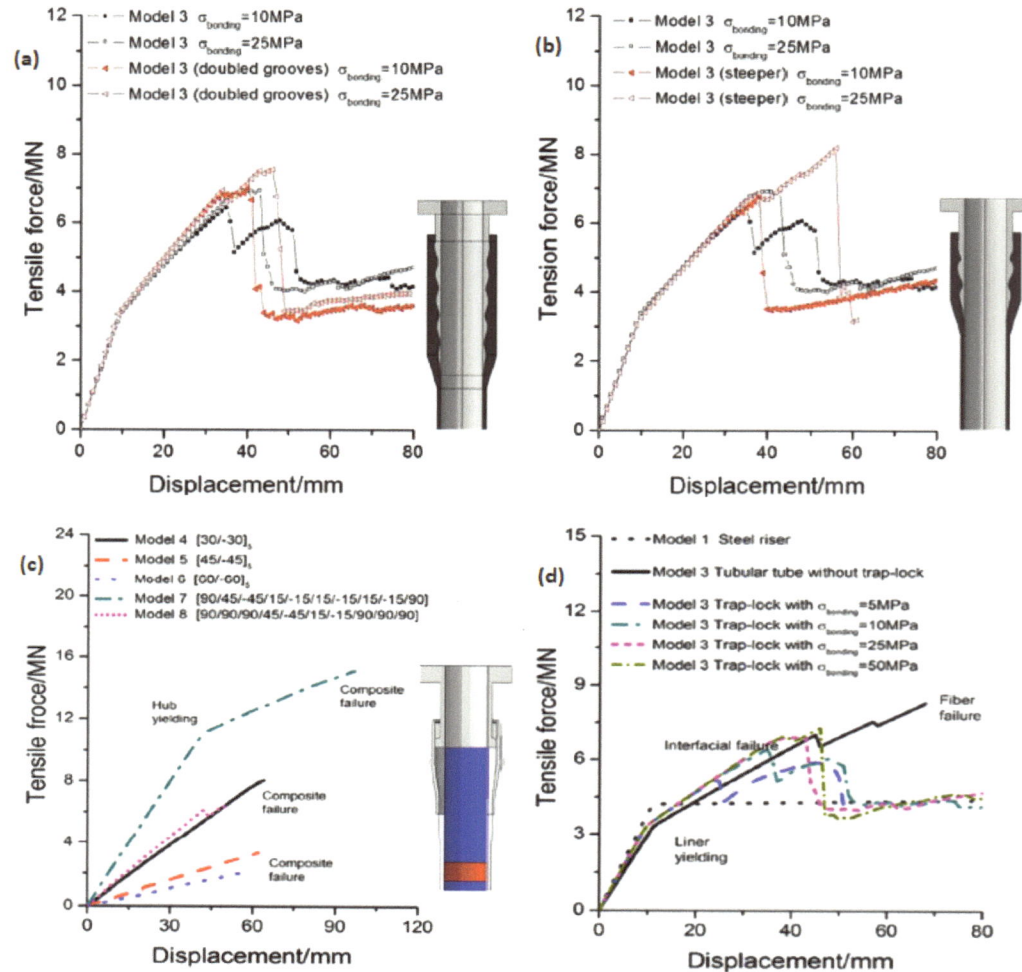

Figure 13. Results of displacement vs. force profiles for different models with load transfer mechanism of traplock and Magma end-fitting models, to present the failure profiles from tension loads on the body of the composite riser pipe. It shows the displacement-force curves of (**a**) trap-lock end fittings with doubled number of grooves, (**b**) trap-lock end fittings with steeper grooves, (**c**) Magma end fitting model with different composite laminates, and (**d**) trap-lock end fitting with steel riser and other composite risers. (Permission to reuse was obtained from Elsevier Publishers, Source: [187]; Published: 4 June 2018; Copyright date: 2018).

As shown in Figure 13a,b, the load transmission was accomplished by a combination of mechanical compression force and adhesive bonding force. Stress concentration and possible debonding between the liner and composite laminate can be caused by the grooves at both ends. The valley of the grooves with thinner walls may experience yielding, necking and subsequent failure as a result of such debonding. To avoid this, the wall thickness at the groove region was purposefully increased from the middle to both ends [74–76,187]. As a result of the thicker wall at the MCI region, the axial stiffness increased slightly, as demonstrated by the force–displacement response. However, it had little effect on the ultimate tensile load (UTL), which is mostly governed by interfacial strength, which is unaffected by extra grooves. The results of another end-fitting are shown in Figure 13b, where the groove interfaces were steeper and saw-toothed, as suggested by Baldwin. This adjustment had no effect on the response's axial stiffness, but it did allow for a higher UTL. The basic lengths of all grooves remained unchanged in all circumstances, suggesting that the end-fitting section length in the cases depicted in Figure 13a is double that of the length shown in Figure 13b, whereas the section length in Figure 13b is equals that shown in Figure 13d, as seen in the cumulative results. For the Magma end-fitting shown in Figure 13c, it can be seen that slope—either gentle or steep—and frictional coefficients had an effect on the strength of the end-fitting.

The qualification of the Magma m-pipe design that incorporates a fibre and matrix of Victrex PEEK to make composite laminates was reported in literature [121,122,126]. It was demonstrated by applying an axial load of 23.9 MPa as an operational pressure load, 42.75 MPa as a test pressure load and an internal pressure load of 38.3 MPa. In an earlier investigation on Magma end-fitting, Hatton [119] avowed carbon fibre/PEEK pipes to be a potential enabler for offshore applications. The study covered the end-fitting design, which enables further manufacturing activity by serving as a stable structural interface with the steel end-fitting. As comparatively shown in Figure 13, each end-fitting has unique features that make them useful for special applications, as seen in the Airborne end-fitting in Figure 14a. The Magma end-fitting is presented in Figure 14b.

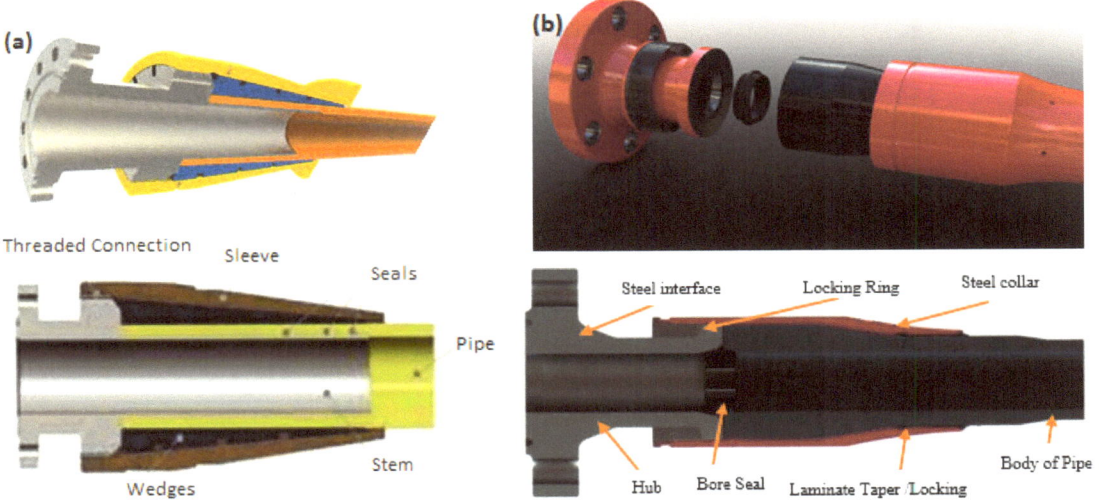

Figure 14. End-fitting designs for (**a**) Airborne end-fitting and (**b**) Magma end-fitting. (Permission to reuse was obtained by the industry designers—Airborne Oil and Gas for Image (**a**) and Magma Global for Image (**b**)).

3. Mechanical Behaviour

In this section, the mechanical behaviour is discussed and reviewed.

3.1. Strength Behaviour

A comprehensive analysis of the strength behaviour and economic consideration of composite risers is carried out here. Economic assessments from literature show that composite risers are comparatively more expensive than steel risers [55,196–199,217,246]. It has also been observed that different CPR models have been developed to ascertain the strength of composite risers, in comparison to steel risers. Table 3 shows various mechanical behaviours and loads that can be applied in a composite riser model when analysing the mechanical behaviour of composite risers. Steel is a universal benchmark material in riser systems because it has tested track records, especially regarding failure behaviour and strength characteristics. This is not the case with composites. Steel also has more design codes than composite materials; thus, it differs from composite materials used for CPR designs, as seen in their physical properties, as shown in Table 4. Additionally, in comparison to composites, steel is heavy, requires high maintenance and fabrication, has a high installation but low material cost and must be coated because it is corrosive [135,235,293–295]. Comparisons based on the mechanical, chemical and physical properties of composites against steel and wood have shown that composites to have excellent behaviour in coastal environments, as opposed to mild steel [295–300].

Table 3. Models for analysis of mechanical behaviour of composite risers.

Type of Model	Problems to Be Solved	Loads to Check for
Composite risers	Global analysis and local analyses	Load distribution on riser ends
Compound infinite anisotropic cylinder	Stress and strength analysis, selection of layer thicknesses and technological parameters	Dead weight, internal pressure, external pressure, residual stresses from force winding and thermal shrinkages
Compound semi-infinite cylinder	Stress concentration and length of boundary effect zone	Load distribution effect on riser ends
Cylindrical sections of different lay-ups on axial coordinate	Selection of reinforcement scheme at different depths of sections	Load variation along axial coordinate
Cylindrical section with lay-up varying in wall thickness	Optimisation of reinforcement scheme of riser sections	Nonuniformity of stress fields on wall thickness
Extensible weighted thread	Estimation of axial strength, effect of extensibility of riser axis on its deflection	Dead weight and flow-past
Flexible rod in linear statement	Calculation of riser deflection and stresses	Flow-past, reactive forces and moments
Flexible rod in nonlinear statement	Stresses, deflection, required top tension at longitudinal-transverse bending and buckling	Dead weight, flow-past, top end tension, reactive forces and moments
Laminated cylindrical tube	Displacements and stresses	Bending loads, torsion, tensile loads, external pressure, and internal pressure.
Multilayered cylindrical shell	Stress, strain and strength analysis	Effect of asymmetric loading (flow-past, concentrated loads)
Repaired cracks in composite pipes	Fracture using stress intensity factor (SIF)	Load distribution
Quasilinear 3D anisotropic elastic cylinder	Refined calculation of stresses	Synergetic effect of different loads

Table 4. Material attributes of composite risers compared to other materials.

Property	Specific Gravity	Young's Modulus (GPa)	Poisson Ratio, v	Density (kg/m^3)
Composite Riser	1.68	(depends)	0.28	1680
Steel	7.8	200	0.30	7850
Titanium	4.43	113.8	0.342	4430
Aluminium	2.78	68.9	0.33	2780
PEEK	1.32	5.15	0.40	1300
P75/PEEK	1.77	33	0.30	1773
P75/Epoxy	1.78	31	0.29	1776
Sea Water	1.0	2.15	0.5	1030
AS4-PEEK	1.56	66	0.28	1561
AS4-Epoxy	1.53	49	0.32	1530

To evaluate the strength of marine risers, effective tension profiles are usually used [11,24–27,98,137–142,149–153]. Details of similar effective tension profiles on marine hose risers are available in earlier studies [48–53]. In the comparative study by Kim and Ochoa [98–104], the findings show that the composite riser had a tension factor of 1.3 with a top tension of 1418 KN, whereas a comparable steel riser requires 3822 KN. Here, the cumulative weight of the submerged weight of the riser and the internal fluids in the riser was used because the effective weight of any riser is very important [98]. An earlier NIST study [99] recommended the replacement of steel joints with composite riser joints; however, the findings of Dikdogmus [43], avow the challenge of replacing steel with composite materials, as the riser design must focus on the variety of loadings and water depths. Saleh [199] opined that composite risers are more expensive than steel risers, but they also have the lowest life cycle cost as composite materials require less maintenance, unlike steel materials. It is noteworthy to add that the composite riser pipe costs about three times the cost of an equivalent steel riser pipe. However, the replacement of the steel pipe with a composite pipe improves the life on site by a factor of more than 100 times. As regards the performance of the riser, it improved significantly due to the fatigue characteristics of the composites and the absence of welds along the riser leg. Another comparison conducted on CPR models by Brown [246] showed that there is high potential for the utilisation of composite risers, as it shows good results from utilisation on flowlines and TCP pipes, as shown in Figure 15. Studies on the use of titanium alloys in CPRs for FPSO, TLP, and Truss SPAR found that titanium alloys have high mechanical attributes suitable for Composite risers [211,212,295–306] but the configurations, fatigue, reliability, large-scale experiments and cost should also be considered [156,157,161,247,303,306–317]. A bibliographical list of the comparative studies on composite risers is in Table 5.

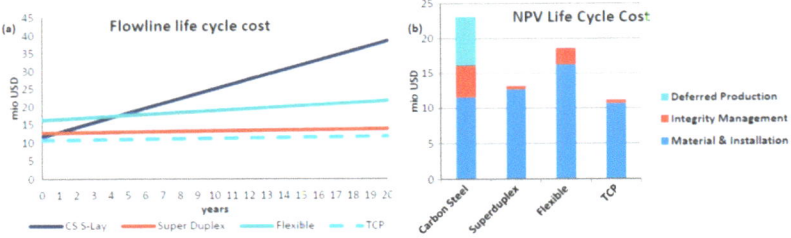

Figure 15. Potential of marine composites in TCP flowlines and composite risers, showing (**a**) flowline life cycle cost, and (**b**) NPV life cycle cost (Courtesy of 2H Offshore, Source: [247]).

Table 5. Bibliographical list of comparative studies on steel risers and composite risers.

Authors	Title	Highlight
Gibson A.G. [306]	Composites for Offshore Applications	Compared steel, protruded FRP against steel and wood
Hopkins P. et al. [196]	Composite Pipe Set to Enable Riser Technology in Deeper Water	Compared both steel and composite riser performances and riser fatigue
Cheldi T. et al. [4]	Use of spoolable reinforced TCP pipes for oil and water transportation	Spoolable reinforced TCP pipes, material design
Toh W. et al. [187]	A comprehensive study on composite risers: Material solution, local end fitting design and global response	End-fitting design for composite risers and global design with VIV responses
Hopkins P. et al. [197]	Composite riser study confirms weight, fatigue benefits compared with steel	Compared the composite risers and steel riser with material attributes
OGJ [211]	Composite riser technology advances to field applications	Compared the composite risers and steel riser with material attributes
Pham D.C. et al. [160]	A review on design, manufacture and mechanics of composite risers	Comparative assessment of the literature
Amaechi C.V. et al. [149]	Composite Risers for Deep Waters Using a Numerical Modelling Approach	Numerically compared composite riser models, compared different liners
Ward E.G. et al. [101]	A Comparative Risk Analysis of Composite and Steel Production Risers	Comparison assessment, local design, global riser analysis and risk analysis for both steel and composite risers.
Pham D.C. et al. [159]	Composite riser design and development—a review	Comparative assessment of the literature
Saleh P. [199]	The benefits if composite materials in deepwater riser applications	Benefits of composites in developing composite risers
Brown T. [246]	The impact of composites on future deepwater riser configurations	Configurations for composite risers, with fatigue of steel and composite risers
Lamacchia D. [217], Lamacchia D. et al. [276],	Thermoplastic Composite Pipe (TCP) Offshore Market 101	Compared MagmaGlobal and Airborne Oil&Gas TCP pipes for composite risers
Saad et al. [55]	Application of composites to deepwater top tensioned riser systems	Economic aspects of composite risers on SPAR and TLP
Mintzas A. et al. [134]	An integrated approach to the design of high performance carbon fibre reinforced risers—from micro to macro scale	Combined bend–burst, burst and compressive tests
Amaechi et al. [153]	Local and Global Design of Composite Risers on Truss SPAR Platform in Deep waters	Comparative assessment of composite risers; local and global design
Wang et al. [141]	Tailored design of top-tensioned composite risers for deep-water applications using three different approaches	Numerically compared composite riser models, compared 3 different approaches
Gibson A.G. [80]	The cost effective use of fiber reinforced composites offshore	Compared the composite risers and steel riser with material attributes
Andersen W.F. et al. [71]	Full-Scale Testing of Prototype Composite Drilling Riser Joints-Interim Report	Full-scale testing on composite riser
Wang et al. [142]	Global design and analysis of deep sea FRP composite risers under combined environmental loads	Numerically compared composite riser models, global and local design
Kim W.K. [98]	Composite production riser assessment	Compared steel and composite risers, local and global design,
Gibson A.G. et al. [82]	Non-metallic pipe systems for use in oil and gas.	Application of composite pipes

3.2. Global Performance

The industry specification DNV-OS-F201 [270] stipulates that reasonable probability must be considered for the marine riser during its design, manufacture, fabrication, operation and maintenance so that it can stay ready until its usage. Taking into account its service life and cost, it should remain suitable for the task for which it was designed.

Furthermore, the riser should be able to withstand all predicted loads and related forces that may be experienced over its service lifetime, as well as have suitable durability in proportion to maintenance costs; thus, it should be built with the right degree of reliability. An illustration of a global analysis model is shown in Figure 16, which illustrates the marine risers and the catenary mooring lines. The global response, structure's geometry, structural integrity, stiffness, trans-sectional properties, material attributes, sea-wave behaviour, supporting methods and prices (or economic savings) are all important factors to consider while developing the composite riser structure.

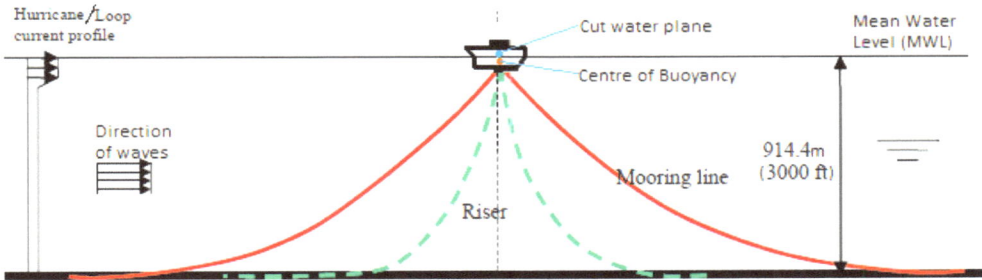

Figure 16. Schematic showing risers and mooring lines on a floating platform.

Hopkins et al. [197] comparatively studied the main criteria for designing composite risers and steel marine risers, operating at a water depth of 2000 m on a conventional single-leg hybrid riser (SLHR) with a steel riser leg. The authors concluded that composite systems have been tested and can be used in offshore risers, as they are further developed using newer riser applications and configurations. In the comparative assessment between composite risers and steel risers by Brown [246], the effect of configuration was also opined as a factor, as there is a shift in the tension from the hang-off when comparing different fluid contents, as shown in Figure 17a. When the riser configuration was ballasted to change the configuration, as depicted in Figure 17b, it added some flexibility and it decoupled the vessel motion and changed the tension distribution, as shown in Figure 17c. This confirms that an increment in the ballasted weight will reduce the bending and angle of the hang-off, while it increases the hang-off loads, based on riser configurations. Roberts & Hatton [126] presented a lightweight design that reduces both drag and weight loadings. This results in a much-enhanced riser response, with all essential factors, such as the foundation loads, buoyancy module size, installation loads, etc., being decreased drastically. This confirms that when designing the model's setup, the host vessel's motion characteristics, specific environmental data, numbers on the field layouts, layout array for the moorings, number of risers required, deck loads on the platform, likelihood of interference, hang-off location, access to a host vessel and the depth of the sea must all be considered, as well as the fact that it is a composite riser. However, debonding would increase the load carried by the matrix; hence, the fibre–matrix interface is critical in achieving a suitable composite structure. Indeed, the debonding phenomenon allows the composite to bear a further strain after the failure of the fibre since the matrix starts to carry the stress around the broken fibres, but after that, the composite is considered to no longer be in working condition. In addition, in comprehensive investigations on the composite material makeup and the stack-up composition of the layers, it is very important to consider the configuration of the composite structure (like TCP flowline or composite riser), as it might be aligned vertically or in S-shape (such as steep-S, lazy-S, lazy wave, etc.), which may vary in wave form, with respect to riser length, the sea bed type and the sea depth [23–25,301–310]. Application of composites have also been extended to marine hoses as marine bonded composite hoses (MBCH) [14,15,154,155,227]. In the global analysis by Amaechi et al. [48–53], the strength of marine bonded hoses was assessed numerically using a coupled model in ANSYS AQWA and Orcaflex, with the hydrodynamic loads from the CALM buoy structure on the riser model (submarine hoses), which was designed in a Chinese lantern configuration. However, there is presently no publicly available literature covering such application on composite risers. Current applications of global design are based on the loadings or deployment as top-tensioned risers [140–142,152,153,156,157]. For fully developed composite risers, this configuration may not be as effective as vertical configurations on a Truss SPAR [152,153], and TLP [140–142,296]. Aside from these, there are other proposed configurations for the hybrid flexible composite riser model and a composite riser using a buoyancy tank, similar to the single-leg hybrid riser (SLHR).

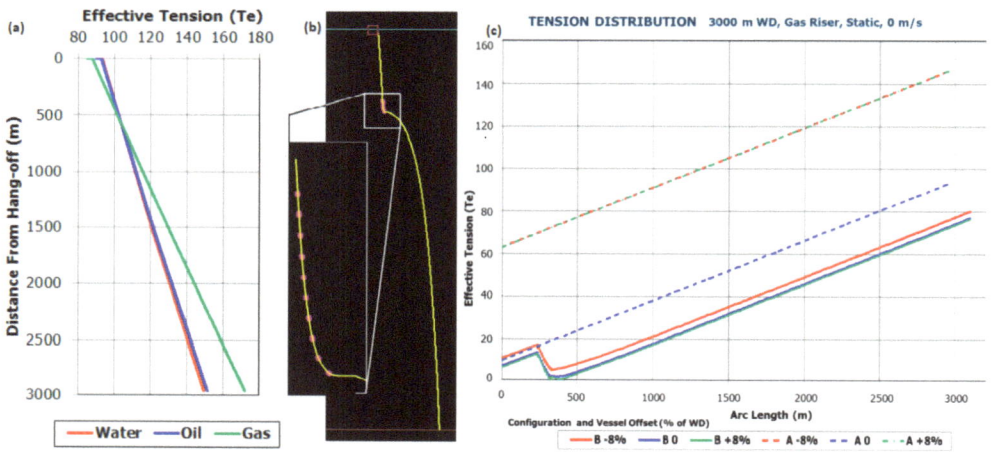

Figure 17. Tension profiles (**a**) during hang-off for different fluids, (**b**) the hang-off configuration when ballasted and (**c**) tension profile when ballasted (Courtesy of 2H Offshore, Source: [247]).

3.3. Vortex-Induced Vibration (VIV)

The importance of vibration control with different methods for offshore structures has been presented in the literature [37]. The flexibility of modern polymer–matrix composites allows for true integrated sections. The thickness, on the other hand, might be adjusted to obtain the required fibre mix with commensurate fluctuation in the stresses. The stack-up of the composite laminate and orientation is critical to the composite tubular's strength, as discussed in Section 2.5. Looking at the mechanical behaviour of composite risers, VIVs on CPRs have been increasingly investigated experimentally with numerical analysis. In an earlier investigation by Omar et al. [200], although the study did not consider a detailed riser analysis, it compared composite TTRs and steel risers, considering dynamic response characteristics in ABAQUS and taking into account the damage rate of each configuration, top tension requirement, VIV amplitudes and VIV-induced local stress. In another study, Wang et al. [143] presented a comparative assessment of the VIV of composite risers using computational fluid dynamics (CFD) in both 2D and 3D models. The vorticity profiles are shown in Figure 18. The natural frequencies and reduced velocities of nine cases were investigated and further examined in [144–146], who presented mode shapes for the FRP composite risers with distinctively different responses to those of steel risers. The study concluded that at 2.13 m/s, the VIV responses of all three risers were almost 30 times those at 0.36 m/s and 3 times those at 1.22 m/s. At 2.13 m/s, the VIV responses of global stresses for all three risers were nearly ten times higher than at 0.36 m/s and three times higher than at 1.22 m/s. "Lock-in" for the steel riser (riser 1) and optimised FRP composite riser (riser 2) only occurred in the 100-year loop current condition at 2.13 m/s. The comparisons reveal that composite risers have better dynamic response characteristics than steel risers.

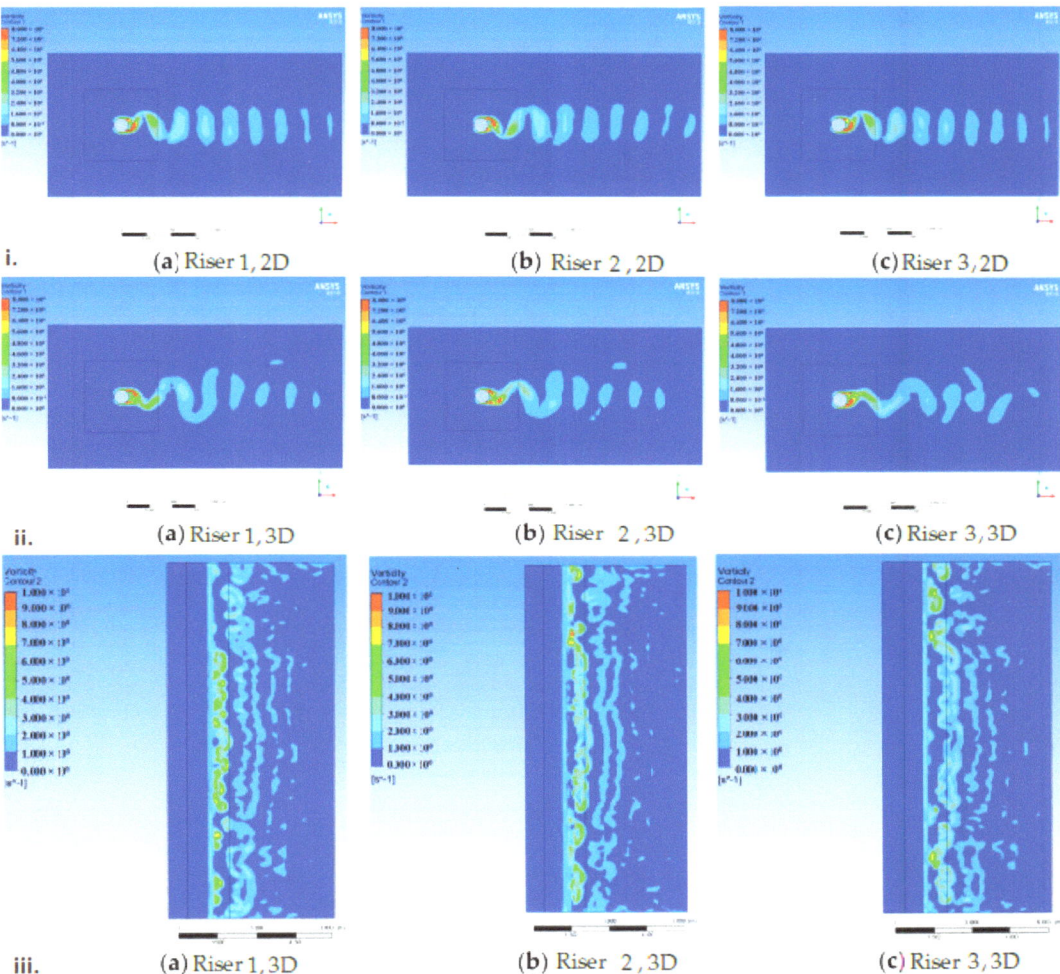

Figure 18. Profile of vorticity in 2D and 3D models for: Riser 1; Riser 2; and Riser 3. Here, (**ia–ic**) presents the vorticities for Risers 1, 2 and 3 using 2D models; (**iia–iic**) illustrates the vorticities for Risers 1, 2 and 3 with 3D models; and (**iiia–iiic**) presents vorticity profile in 3D models for (a) Riser 1; (b) Riser 2; and (c) Riser 3. (Permission to reuse was obtained from MDPI and author 5-C.W., Source: [143]; Published: 27 April 2018; Copyright date: 2018).

Similarly, the earlier study [200] revealed that the maximum VIV stress of a composite riser was substantially lower than that of a steel riser, demonstrating that composite risers have longer fatigue lifetimes. Likewise, Kim [98], Sun et al. [255], Tan et al. [226] and Toh et al. [187] used different techniques like finite element method (FEM) to conduct thorough analyses on both local and global scales on composite and steel risers, finding higher safety factors for composite risers. However, the composite riser models without any VIV suppression exhibited mild fatigue damage. Hence, adding strakes and buoyancy modules to a composite riser could improve VIV responses [203,226,293].

3.4. Dynamic Behaviour

In this section, some studies on dynamic loading on risers are briefly examined. The riser tensioner system for a buoyancy tank was studied by Kang, Z. et al. [311]. Dynamic models were used to assess computing fatigue damage using the fatigue model and compared to benchmarked criteria for maximum permitted stresses coupled with the RAO data [23,162–166]. The dynamic behaviour of risers in seismic designs is different because seismic conditions must be taken into account in the riser design in order to control seismic responses because resonance occurs when the frequency falls within the range of earthquake wave frequencies, causing higher seismic stresses in the riser [312]. Since damping is already included in the response model, the contribution of hydrodynamic damping (which is proportional to the water velocity and compensates for the damping impact of the surrounding water) in VIV within the lock-in region is set to zero (0). According to Sanaati et al. [288], higher applied tensions that result in lower vibration amplitudes can greatly improve the hydrodynamic lift force and are an effective instrument for the active management of riser statics and dynamics. The hydrodynamic interaction is a system-dependent issue in load effect assessments connected to riser interference evaluations (see DNV-RP-F203 standard [313]). Some researchers performed reviews on the modelling and analysis techniques for hydrodynamic assessment of flexible risers to represent the static analysis, frequency domain dynamic analysis and time domain dynamic analysis for a flexible riser, as well as the internal and external pressure effects and information on fluid flows [206–209].

API-RP-2RD [314] examined several riser configurations based on the following criteria: structural integration (integral vs. nonintegral risers), means of support (top-tensioned with tensioners or hard mountings vs. concentrated or distributed buoyancy), structural rigidity (metal vs. flexible risers) and continuity (sectionally jointed vs. continuous tube). Conversely, static analysis was used to determine the riser's configuration; it should be noted that riser configuration design was carried out in accordance with production needs and site-specific environmental circumstances [313–315].

3.5. Experimental Tests

Recent developments have involved JIPs such as Cost Effective Riser Thermoplastic Composite Riser JIP in 2009 and Safe and Cost Effective operation of Flexible Pipes in 2011–2013 [80,306,316,317]. As presented in Table 1, different experiments on composite risers have been conducted to achieve qualification and advances made. In principle, experiments are required for both local and global design, as well as validation of results, because they reveal the test techniques and materials utilised [317–326]. The highlight of related experimental studies on composite risers, composite tubes, cylinders and shell structures are indicated in Table 6. Since marine risers are tubular structures with varied segments, the experimental tests on composite risers were conducted at microscale levels (on composite materials) [257–259] and at macroscale levels (on composite risers with composite tubes) [47,80,307,308], which are benchmarked against those of marine steel risers, such as steel catenary risers [35,286]. It is also crucial to test the failure in risers, as research on carbon-fibre-reinforced composite risers have revealed failure mechanisms [59–63] linked to various loading scenarios [131,134,187,195]. Previous experimental studies on composite risers were based on scale-dependent methods of analysis, as tabulated in Table 6. Presently, the scale tends to shift more towards small-scale methods. The tests conducted include failure tests, fatigue, collapse and burst pressures [148–152]. There are limitations to full-scale testing, as can be seen in the size of composite risers as shown in Figure 19. The first full-scale model examination was an external pressure test to determine the collapse pressure capabilities of the riser main body construction. Andersen et al. [70–72] developed a specially configured test fixture that used a circumferential compressive load rather than just end loads on the specimen during the test since that was designed to meet the standards for a 1500-pound riser's main body structure.

Table 6. Experimental tests on composite risers with the scale of the testing.

Reference	Highlights and Test Modes	Specimen Type	Program/Test Scale
Sparks et al. [68]	Attributes of composite risers on concrete TLP, collapse pressure test, fatigue test	Composite riser	Full-scale and small-scale
Andersen et al. [71]	Burst, and tension tests	Composite drilling riser joint	Full-scale
Gibson [81]	Flexure, tension, fire, durability, blast, impact test, axisymmetric burst, marine composite application, fatigue test	Fibre-reinforced composite pipes and coupons	Full-scale and Small-scale
Picard D. et al. [95]	Tensile test, manufacture of TCP pipe	Composite tube	Large-scale
Ramirez and Engelhardt [96,97]	Collapse pressure test, buckling	Composite tube	Full-scale
Alexander et al. [105]	Burst, bending cycles, impact/drop tests	Composite tube	Full-scale
Cederberg et al. [109,110]	Burst, collapse and impact tests	Composite drilling riser	Full-scale
Mintzas et al. [134]	Tensile test, micro-scale test	Carbon fibre repaired riser	Small-scale
Chen et al. [186]	Burst, and tension tests	Composite riser end fitting	Small-scale
Pham et al. [161]	Bending under transverse loads	Composite pipe and coupons	Full-scale
Sobrinho et al. [224]	Thermal and Mechanical tests		
Ye et al. [257,258]	Tensile test, SEM and CT tests	Glass fibre composite/epoxy	Small-scale
Ellyin et al. [262]	Flexure test, tension test	Composite pipes and coupons	Small-scale
Grant and Bradley [280]	Flexure test, tension test	Composite pipes and coupons	Small-scale
Huang et al. [307,308]	Tensile and fatigue tests	Carbon fibre composite pipe and coupons	Large-scale
Alexander and Ochoa [327]	Burst, tension and 4-point bend tests	Carbon fibre composite repaired steel riser	Full-scale
Rodriguez and Ochoa [328]	4-point flexural test, fatigue test	Carbon and glass fibres/epoxy	Small-scale
Lindsey and Masudi [328,329]	Cyclic test, tension in sea water cases from 25 °C to 75 °C	Graphite epoxy composite	Small-scale
Soden et al. [330]	Flexure test, tension test	Composite pipes and coupons	Small-scale

Another full-scale assessment was an RPSEA projected reported by Alexander et al. [105] on the performance of a composite-reinforced steel drilling riser for HPHT operating conditions using three prototype models with a bore diameter of 495.3 mm. It was subjected to cyclic testing under a service temperature range from 180 °F to 32 °F, a 20-year service life, top tension capacity of 13,333 KN, an internal pressure of 66,667 KN and operation in a water depth of 3048 m. Salama et al. [89] presented an investigation with the test certification for the first installation offshore on the composite drilling riser (CDR) joint, which includes bending fatigue, burst, impact testing and one pressure cycle at 31,712 kPa and two pressure cycles at 42,743 kPa. The Heidrun CDR joint seen in Figure 2, featured titanium connectors and a titanium liner, which was the first application of a composite offshore [88,89]. Titanium was chosen due of its low wear resistance. Secondly, the Heidrun TLP required a lighter marine riser joint so the CDR joint was made of titanium and 3 mm internal hydrogenated nitrile rubber liner seal. The composite drilling joint was visually inspected after the third drilling cycle, and no damage to the interior elastomer liner was found. As a result of this, DNV began developing specifications and guidelines for composite risers and composite fittings. Furthermore, given the lack of specifications and guidelines on composite risers and structural components, such as its end-fittings, extensive research on composite risers is required. As such, it is necessary to consider the pyramid principle for structural analysis and testing, as shown in Figure 20.

Figure 19. Composite riser pressure test samples (Courtesy of Lincoln Composites).

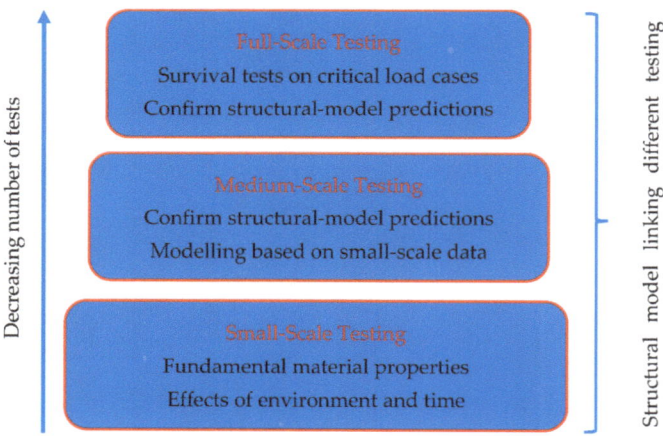

Figure 20. Pyramid principle for structural analysis and testing.

Based on load transfer mechanisms, composite structures are designed to transport loads via the fibre directions; hence, slight changes in the transverse directions have little impact on the load distribution in the structure, as depicted in Figure 13. This was demonstrated in the experimental model [134,257]. Layup sequence of composites are important aspect of the design. Ye et al. [257] carried out microscale and macroscale design of the composite CPR structures, using cylindrical coupons with fibres aligned at +45°/−45° orientations. The study included characterisation of the composite materials and tested them with Imetrum test equipment and SEM and CT digitisers [257].

It is noteworthy to state that both thermal and mechanical analyses are critical for characterising polymeric systems, as demonstrated by Sobrinho et al. [224]. In that study, composite risers were developed from composite tubes that were 1.8 m in length, had an internal diameter of 0.1016 m and were 5.6 mm thick, with findings that the influence of matrix toughening on mechanical behaviour has a significant impact on composite fracture processes. The composite tubes were filament-wound, with the reinforcement fibres neatly tucked into the polymeric structures. They noticed that the tensile strength of the hoop layers increased to 709.05 MPa from 572.20 MPa, having a jump of around 20%, indicating that the addition of rubber to that polymeric matrix encouraged a simultaneous increment in elongation under both fracture and stress conditions. In another study on characterisation, Elhajjar et al. [322] created a profiled 3D model of the superficial resin pockets using the hyperspectral spectroscopy–infrared technique. The large ply approach was used to establish the random wrinkle distribution in the specimen, with the capacity to quantify resin thickness from around 125 to 2500 µm. His findings reveal varied failure modes when the specimen transitioned from a restricted fibre kink band configuration to a random wrinkle distribution. In a recent study by Chen et al. [186], experiments on scale-down prototypes with similar tensile-bending capacity to the tubular model were carried out to evaluate composite riser joints, and it indicated progress in tensile-bending capacity, similar to the findings by Toh et al. [187] and Meniconi et al. [78]. The implication is that the traplock end-fitting and the central tubular segment are both equally strong.

3.6. Numerical Analysis

Numerical analysis on composite risers is critical for ensuring the validity and accuracy of the chosen discretisation by examining its properties [140–144,148–151,331]. There have been a number of investigations on composite risers, including nonlinear failure analysis, mechanics and fatigue due to compression, tension, bending, torsions and their combined loading, internal fluid pressure and sea water hydrostatic pressure, as tabulated in Table 7. Bai et al. [248] investigated the internal pressure effect of a TCP pipe that was reinforced with a steel strip using ABAQUS and introduced a Mises yield failure criterion. Corona and Rodrigues [233] investigated the performance of long, thin-walled cross-ply composite tubes subjected to three phases of pure bending. Bifurcation buckling, material failure and prebuckling reactions were investigated utilising three different material models: AS3501 graphite–epoxy, Kevlar 49–epoxy and E-glass–epoxy. They discovered that the Brazier effect, which causes the cross-section of the tubes to ovalise, results in a nonlinear moment–curvature connection. Brazier [325] observed that the longitudinal stresses and compressions of thin cylindrical shells and other thin sections are orientated towards the toroid's median circumference. This causes flexural distortion by creating a proportional pressure per unit area at all places of the cross-section. Jamal and Karyadi [213] used Novozhilov's nonlinear thin-shell theory to compute the collapse of cylindrical shells under pure bending to investigate material failure. Tatting [228] used the semi-membrane constitutive theory for buckling studies to explore the brazier effect on composite cylinders of finite length under bending nonlinear analysis. He discovered that the tube length parameter has the greatest impact on the load–displacement response and local buckling failure. In another model, Guz et al. [218] solved the problem of the collapse pressure effect on thick-walled composite tubes using the elastic constitutive theory for the interlaminate layers for various loadings and stack-up designs. Elhajjar et al. [322] and Ye et al. [257] looked at composite failure reactions and faults in composite structures since these flaws can cause delamination, matrix debonding, fibre debonding, fibre kinking and matrix cracking. Li et al. [332] investigated the deformation of thin-walled tubes under bending with a small bending radius and a large diameter. To assist trials, Chen et al. [186] presented a numerical analysis for a prototype composite riser joint. The interface between the metallic liner and the composite laminate was modelled with ABAQUS using the traction–separation law to approximate splitting separation on a cohesive surface. Chouchaoui and Ochoa [225] investigated the behaviour of a laminated cylindrical tube with adequate boundary conditions at layer interfaces, i.e., without interfacial friction or sliding, though some presentations on the scaled models were further presented in another study [226]. Rodriguez and Ochoa [328] used both experimental and computational methods to investigate the effects of material systems, lay-up stacking and geometry on the failure behaviour of spoolable composite tubes under bending. They discovered that the tubes' nonlinear bending responses may be dominated by the drop in shear stiffness. External hydrostatic pressure was considered in the collapse investigation of a plastic composite–steel pipe (PSP) with an inner high-density polyethylene layer. A 2D ring model provided by Bai et al. [248,299] was also used to account for wall thickness variation, transverse shear deformation and prebuckling deformation in steel strip reinforced thermoplastic pipe. The numerical study of reinforced composite pipes for RTP made up of an exterior thermoplastic cover (PEEK), an inner thermoplastic liner (PEEK) and carbon-fibre-reinforced PEEK (AS4/PEEK) reinforcement layers was undertaken by Ashraf et al. [9]. They also evaluated the lay-ups ([75/50], [75/25], [50/75], [50/25], [50/25], [25/75], [25/50] and [25/50]). They discovered that pipes with two angle-ply reinforcement layers have a stiffer reaction than pipes with only one angle-ply reinforcement layer when the reinforcement layers are positioned at varied angles. In the finite element analysis (FEA) conducted by Rodriguez and Ochoa [328], shell elements were used in two-dimensional planes (2D-S8R) and an experimental four-point flexural experiment loaded to failure to match the structural features of filament-wound composite tubulars on huge spools to ABAQUS models. Since all of the composite tubes' failure modes were matrices in both tension and compression, and the shear stiffness showed a nonlinearity in flexure, he determined that damage was restricted to the bottom part and progressed slowly. Zhang et al. [66] looked at an analytical mechanical model of a composite riser pipe in both the global and local assessments and found the corresponding material characteristics. They concluded that the hoop layers generally bear internal pressures, while the helical layers mostly bear bend loadings, according to the simplified composite joint local analysis.

Table 7. Numerical analysis on composite risers, showing the test methods and mechanics studied.

Reference	Numerical Methods	Highlights
Bai et al. [248,299]	Numerical model, von Mises failure criteria	TCP Pipe, internal pressure ABAQUS
Andersen [8]	Minimum potential energy approach; failure criteria; progressive damage	Analysis of transverse cracks in composites
Rodriguez and Ochoa [328]	Numerical and experimental, spoolable tube bending; material failure mode; 2D shell element	Flexural response of spoolable composite tubular
Toh et al. [187]	Tensile strength assessment, mode shape from global response	Analysis of 2 composite riser end-fittings—taplock and Magma
Chen et al. [186]	Tensile strength, prototype design and analysis; composite riser joints	Numerical and test analysis of composite riser end-fitting; mechanical tests, tension and combined tension-bending loading tests of composite riser joints
Amaechi et al. [149–151]	Novel numerical approach in ANSYS ACP to model composite riser; netting theory; for 18 layers of composite riser	Buckling, burst, collapse, tension; under 6 load conditions, presented stress profiles for F.S of different layers in 3 stress directions, presented buckling modes
Jamal and Karyadi [213]	Collapse test; under pure bending; LR-739 composite cylindrical tube	Material failure using Novozhilov's nonlinear thin-shell theory
Corona et al. [233]	Nonlinear analysis using material failure criteria and constitutive modelling	Bending response of long and thin-walled cross-ply composite cylinders
Wang C. et al. [137–142]	Design of composite risers for minimum weight; Numerical method using ANSYS APDL to model composite riser.	Local design and global design of composite riser; design on min. weight, factor of safety results in 3 stress directions; Under combined loadings and global responses
Elhajjar R. et al. [322]	A hybrid numerical and imaging approach for characterizing defects in composite structures	Structural and elastic failure responses of composites; hybrid approach coupling with a progressive FEA

Table 7. Cont.

Reference	Numerical Methods	Highlights
Tatting, B. F. et al. [228]	The Brazier effect for finite-length composite cylinders under bending	Numerical nonlinear analysis using semi-membrane constitutive theory for the analyses.
Brazier L.G. [325]	Analysed the flexural behaviour of thin cylindrical shells and other thin sections	The flexural behaviour of thin cylindrical shells and nonlinear bending analysis

3.7. Fatigue Behaviour

Reliability of composite risers has been a discussion in recent studies due to uncertainty about the fatigue and other mechanical behaviours of the riser [156–158]. Fatigue is a type of failure that occurs in constructions subjected to dynamic and changing forces [333]. Risers are dynamic structures that respond to a wide range of loads and pressures. Dropped objects, anchor strikes, anchor dragging, trawling and boat collisions can all cause fatigue damage to risers during operation [23–25,334]. Fatigue has been a key design difficulty for ultra-deep-water risers due to irregular waves produced by varying amplitudes in the sea, as well as the influence of friction as the tubes slide against their conduits. Various research has been carried out to investigate the fatigue behaviour of composite risers using basic approaches that were originally developed for composite tubes and pipes made of other materials based on durability in water [335–341]. A summary of these fatigue studies are given in Table 8. Fatigue damage analysis using Miner's rule summing produces a permissible fatigue damage ratio of 0.1 for production and export risers and 0.3 for drilling risers, with S-N curves being those most typically utilised [23,24]. Connaire et al. [334] used the Newton–Raphson method and quasi-rotations to investigate the path-dependent nature of rotations in three dimensions in subsea risers, as well as the sensitive load cases of nonlinear loading regimes. In some other studies, composite riser repair systems were investigated [6,7]. In the study by Chan [6], the flexural strength of composite-repaired pipe risers was assessed with the laminate orientation of carbon–epoxy-fibre-reinforced polymer. Since the simulation and testing results differed, they realised that the bonding between the CFRP and the steel pipe surface needed to be examined.

Table 8. Fatigue tests on composite risers with the scale of the testing.

Reference	Highlights and Test Modes	Method Used	Specimen Type	Program/Test Scale
Thomas (2004)	Fatigue test	S-N approach	Composite riser	Full-scale
Huybrechts [302]	Fatigue test, fatigue life estimation	S-N approach	Composite tube	Full-scale
Salama et al. [89–91]	Fatigue test	S-N approach	Composite riser	Full-scale
Chouchaoui and Ochoa [225,226]	Fatigue test	S-N approach	Composite coupons	Small-scale
Sobrinho et al. [223]	Application-based test	S-N approach	Composite coupons	Small-scale
Mertiny et al. (2004)	Fatigue test	S-N approach	Composite coupons	Small-scale
Cederberg [109]	Fatigue test, fatigue life estimation	Strain-life model	Composite riser and steel-reinforced drilling riser	Large-scale
Kim [98]	Fatigue life estimation	Semi-log S-N approach, Power law S-N approach	Composite riser tube	Large-scale
Echtermeyer et al. (2002)	Fatigue life estimation	S-N approach	Composite tube	Small-scale

Table 8. *Cont.*

Reference	Highlights and Test Modes	Method Used	Specimen Type	Program/Test Scale
Liu K. et al. [291]	Fatigue life estimation	S-N approach	Composite tube	—
Yu K. et al. [148]	Fatigue life estimation	S-N approach	Composite tube	—
Sun S.X. et al. [163]	Fatigue life estimation	S-N approach	Composite tube	—

Two general fatigue techniques mentioned in the literature are cumulative fatigue damage (CFD) and fatigue crack propagation (FCP). Riser fatigue is a DNV-recommended practise. The design and evaluation of riser fatigue, as well as its global analysis and usage of S-N curves, are all covered by DNV-RP-204:2010. Crack initiation is when a small crack appears at a high-stress-concentration location; fracture propagation is when the crack increases progressively with each stress cycle; and final failure is when the progressing crack reaches a critical size and fails quickly [333]. Hybrid composites improve composite functionality by allowing the designer to add selectivity, including stiffer, more-expensive fibres where stresses are more critical and less-expensive fibres where stresses are less critical [3,102,117,255]. As water depths have increased and service conditions have become more demanding, manufacturers of nonbonded flexible pipes are attempting to build pipe cross-sections incorporating carbon fibre parts to solve these concerns [342–350]. Carbon fibre's most notable application has been in the substitution of heavy and expensive steel parts in nonbonded flexible riser pipes [5,22,119,215]. In an early design of a composite riser design with an ID of 220 mm, the analysis of lamina stresses compared to lamina strengths at each phase was progressive, as failure was discovered when the elastic constants of the lamina involved were modified [78]. Another study of nonlinear dynamic analysis and fatigue damage assessment using random loads on a deep-water test string found that fatigue damage varied with water depth, and the von Mises stress was higher in the string portions near the top of the test string and the flex joints [291]. Strength and the modulus are two key structural properties in load-bearing applications. By definition, the modulus of a material is a measure of its stiffness or resistance to elastic deformation, whereas the strength is the maximum force it can withstand before breaking.

3.8. Comparative Case Study

An earlier comparative study is the Heidrun composite drilling riser joint shown in Figures 2 and 19. One joint was tested for burst and two joints for bending fatigue as part of the qualification programme conducted by DNV. In addition, a number of joints were constructed to demonstrate the manufacturing process and to test impact resistance. All of the test joints had a diameter of 56 cm. Figure 21 depicts a model diagram of the MCI end fitting as well as one of the test specimens manufactured as an aspect of the qualification programme. The full-length (15 m) composite riser joint was installed on Heidrun TLP in July 2001, and pressure tested on well A-41. After that the joint was utilised to drill more than ten wells while being installed at varying positions located along the riser string. As shown in the results presented in Table 9, it can be observed that some findings were made under different tests including fatigue.

In a comparative assessment between composite risers and steel risers by 2H Offshore [199], there were benefits obtained from the application of composites on deep-water risers. The highlights of the fatigue investigation are given in Figure 22; comparing the steel riser and the composite riser shows that there was an improvement in the use of composite risers. Due to the highly rated fatigue performance of the composites and the elimination of welds along the structure, the riser experienced significant improvement in its fatigue performance. The steel stub positioned below the buoyancy tank suffered a 33.0% drop in fatigue life due to the reduced buoyancy. In addition, regarding the upper section, it was observed that the weld nearest to the upper-riser assembly (URA) interface was discovered to be the steel riser's fatigue hot point. On the other hand, the fatigue hot zone for the composite riser lies below the buoyancy tank steel stub, where the fatigue life was 200% higher than the steel riser minimum fatigue life. Details on the results of the comparative study are presented in Tables 10 and 11.

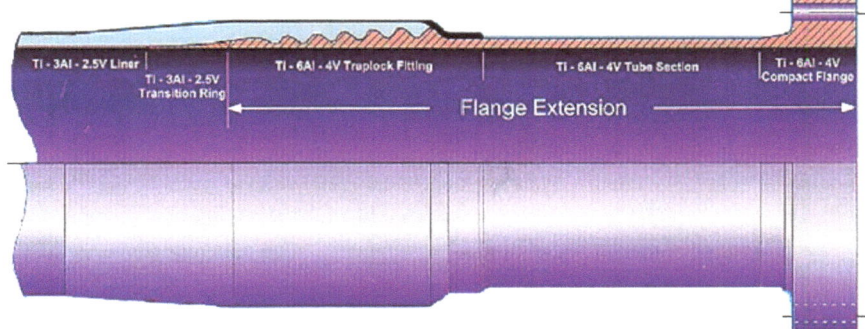

Figure 21. The Traplock end-fitting of the first composite riser joint showing the MCI sketch of the composite drilling riser by NCAS/KOP on the Heidrun Platform. [Permission was obtained from Elsevier Publishers, Source: [65,66,199]; Published: 12 July 2005; Copyright date: 2005].

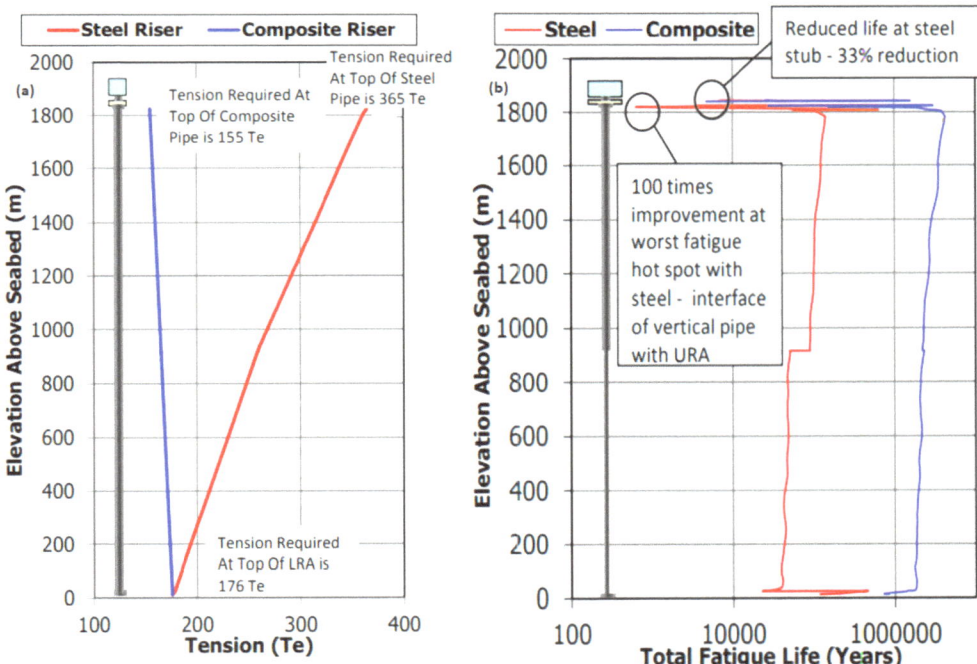

Figure 22. Comparative study on composite riser and steel riser, showing (**a**) tension and (**b**) fatigue life (Courtesy of 2H Offshore; Source: [199]).

Table 9. Results of composite riser joints with 22″ ID by ConocoPhillips/AkerKvaerner (Source: [65]).

Type of Test	Prediction Result	Measured Result	Failure Location
Burst pressure test with closed end loads	14,800 psi	15,850 psi	Body failure
Impact with 5000 kg m (36,170 ft-lb) dropped casing	No structural damage	No structural damage	No failure
Cyclic bending stress range of 850 kN m (627,000 lb-ft), cycles	140,000 psi	160,000 psi -180,000 psi	Circ weld in the Titanium (Ti) liner

Table 10. Comparison of the main design aspects of composite riser and steel riser. (Source: [199]).

Particulars	Steel	Composite	Observations
Max Hang-Off Load (te)	94	93	In an SLHR, the flexible jumper to the vessel acts as interface between the vessel and the vertical riser leg, thus keeping the two isolated. Therefore, negligible change in hang-off loads was seen
Max Hang-Off Bending Moment (kNm)	261	282	
Max Stress Utilisation	0.63	—	While stress is the driving criterion for steel, strain is the driving criterion for composites
Max Safety Factor	—	2.76	MBR is larger than minimum acceptable value
Max Tension Utilisation	—	0.14	Tension is low in comparison to the allowable tension
Max Buoyancy Tank Displacement (m)	247	211	Smaller drag area causes smaller buoyancy tank displacement
Max Buoyancy Tank Tension (Te)	451	258	43% less tension required
Max Bending Moment at Base of URA (kNm)	116	62	Approximately 50% lower bending moment from URA and LRA.
Max Bending Moment at Top of LRA (kNm)	581	270	

Table 11. Main results obtained for composite risers designed for 4000 m depth. (Source: [199]).

Particulars	Values	Observations
Pipe ID	8 in	This is the maximum recommended, and is driven by the collapse criteria
Pipe Max OD	11.9 in	The wall thickness can vary, and thus a smaller pipe OD can be used at shallower depths
Tension at Top	257 Te	Similar level to composite at 2000 m water depth. Note pipe size is different
MBR Safety Factor	2.84	Acceptable MBR
Max Tension Utilisation	0.17	Very low utilisation
Bending Moment At top of LRA	237.70	Similar to composite pipe at 2000 m
Bending Moment At base of URA	32.20	Similar to composite pipe at 2000 m
Maximum Flexible Joint Rotation	8.1 Degrees	Slight increase in comparison to 2000 m

LRA—lower-riser assembly, URA—upper-riser assembly, ID—inner diameter, OD—outer diameter, MBR—minimum bend radius.

4. Conclusions

This review has comprehensively looked at the design, mechanics and development of composite risers and their end-fittings for offshore marine applications. It has also shown current practices in the design of CPRs both from experimental and numerical perspectives. It is imperative that this review looked at different designs of CPRs to evaluate the new and challenging aspects. This review has shown that composites are a viable option for this application, as well as being technically feasible and showing certain improvements. Although there may not be a compelling case for using composites only on the basis of cost, newer configurations are advised. The end-fitting is a key factor in composite riser design, which has been successfully achieved. Overall, it is concluded that the cost of adopting a composite pipe in the existing SLHR system is comparable to that of steel. In addition, the mechanical performance was enhanced. According to several studies, the low use of composites in the offshore sector is due to the comparatively low technological maturity of polymeric composites. The real kicker is the dearth of appropriate test data and well-established design codes. Finally, unlike CPRs, continuous incremental advancements in steel and titanium alloys, as well as nonbonded riser pipes, have kept up with mammalian evolution. Due to their superior fatigue, high corrosion resistance, low specific weight, high stiffness along the reinforcing fibre orientations and high-strength features, these innovative composite materials could be beneficial for use in a variety of offshore structures. However, there is a pressing need for fully adaptive standards to be developed that are uniquely suited to risers. Currently, ABS, API, DNV and ISO standards have successfully published some related recommended specifications for composite risers [167–171,269–272,313–315]. They were developed in response to concerns about composite riser design and the use of composite materials, but they do not all agree, as they are based on the findings from technical committee (TC) meetings and prior composite research. However, a unifying standard for composite risers is expected in future.

Author Contributions: Conceptualization, C.V.A.; methodology, C.V.A., C.C., H.O.B., N.G., C.W., I.A.J., A.R. and A.C.O.; software, C.V.A., C.C., H.O.B., N.G.; validation, C.V.A., C.C., H.O.B., N.G., C.W., I.A.J., A.R. and A.C.O.; formal analysis, C.V.A., C.C., H.O.B., N.G. and I.A.J.; investigation, C.V.A., C.C., H.O.B., N.G., C.W., I.A.J., A.R. and A.C.O.; resources, C.V.A. and N.G.; data curation, C.V.A. and N.G.; writing—original draft preparation, C.V.A.; writing—review and editing, C.V.A., C.C., H.O.B., N.G., C.W., I.A.J., A.R. and A.C.O.; visualization, C.V.A., N.G., C.W. and I.A.J., A.R. and A.C.O.; supervision, C.V.A.; project administration, C.V.A. and N.G.; funding acquisition, C.V.A. All authors have read and agreed to the published version of the manuscript.

Funding: Financial support from the Department of Engineering, Lancaster University, UK, and Engineering and Physical Sciences Research Council (EPSRC)'s Doctoral Training Centre (DTC), UK, is highly appreciated. In addition, the funding from the Overseas Postgraduate Scholarship by the Niger Delta Development Commission (NDDC), Nigeria, is also appreciated. The support of the Standards Organisation of Nigeria (SON), F.C.T Abuja, Nigeria is also recognised. Lastly, the funding support for the APC charges by Author 1—C.V.A., and MDPI's *JCS* journal are appreciated.

Institutional Review Board Statement: Not applicable.

Informed Consent Statement: Not applicable.

Data Availability Statement: The data for this study cannot be shared because they are still part of a present study.

Acknowledgments: The authors acknowledge the technical support from the Lancaster University Library. We also acknowledge the Lancaster University Engineering Department for their assistance with funding for this research. The authors also acknowledge the feedback of Jianqiao Ye on earlier versions of this paper. The permission granted by various authors and publishers for image use is appreciated. Also, the authors acknowledge industry inputs from Magma Global personnel - Stephen Hatton and Andrew Kerry-Bedell; and Airborne Oil and Gas personnel Martin van Onna, Annemarie Jak and Jan van der Graaf, for providing necessary resources for the review. The authors also appreciate the anonymous reviewers for reviewing this manuscript, with comments that helped improve the quality of the submission.

Conflicts of Interest: The authors declare no conflict of interest. The funders had no role in the design of the study; in the collection, analyses, or interpretation of data; in the writing of the manuscript; or in the decision to publish the results.

Abbreviations

2D	two-dimensional
3D	three-dimensional
6DoF	six degrees of freedom
ABS	American Bureau of Shipping
API	American Petroleum Institute
ATP	advanced technology program
BOEM	Bureau of Ocean Energy Management
BOP	blow-out preventer
CALM	catenary anchor leg mooring
CDR	composite drilling riser
CFD	computational fluid dynamics
CFRP	carbon-fibre-reinforced polymer
CPR	composite production riser
CT	computed tomography
D	drilling riser
D&P	drilling and production
DNV	Det Norske Veritas
FAT	factory acceptance test
FCP	fatigue crack propagation
FEA	finite element analysis
FEM	finite element model
FOS	floating offshore structure
FPSO	floating production storage and offloading
FPS	floating production storage
FRP	fibre-reinforced polymer
HNBR	hydrogenated nitrile butadiene rubber
HPHT	high pressure
ID	inner diameter
ISO	International Organization for Standardization
JIP	joint industry program
LRA	lower-riser assembly
MBR	minimum bend radius
MCI	metal–composite interface
MWL	mean water level
NASA	National Aeronautics and Space Administration
NIST	National Institute of Standards and Technology
OD	outer diameter
OTC	Offshore Technology Conference
P	production riser
PCSemi	paired-column semisubmersible
PA	polyamide
PE	polyethylene
PEEK	polyether ether ketone
PP	polypropylene
PSA	Petroleum Safety Authority
PSP	plastic composite–steel pipe
PVDF	polyvinylidene difluoride
RAO	response amplitude operator
RPSEA	Research Partnership to Secure Energy for America
SCR	steel catenary riser

SEM	scanning electron microscope
SLHR	single-leg hybrid riser
SON	Standards Organisation of Nigeria
SPAR	single-point anchor reservoir
SURF	subsea umbilicals, risers and flowlines
SURP	subsea umbilicals, risers and pipelines
TC	technical committee
TCP	thermoplastic composite pipes
TLP	tension leg platform
UK	United Kingdom
USA	United States of America
URA	upper-riser assembly
UTL	ultimate tensile load
VIV	vortex-induced vibration

References

1. Hassan, A.; Khan, R.; Khan, N.; Aamir, M.; Pimenov, D.Y.; Giasin, K. Effect of Seawater Ageing on Fracture Toughness of Stitched Glass Fiber/Epoxy Laminates for Marine Applications. *J. Mar. Sci. Eng.* **2021**, *9*, 196. [CrossRef]
2. Kinawy, M.; Rubino, F.; Canale, G.; Citarella, R.; Butler, R. Face Damage Growth of Sandwich Composites under Compressive Loading: Experiments, Analytical and Finite Element Modeling. *Materials* **2021**, *14*, 5553. [CrossRef] [PubMed]
3. Albino, J.C.R.; Almeida, C.A.; Menezes, I.F.M.; Paulino, G.H. Dynamic response of deep-water catenary risers made of functionally graded materials. *Mech. Res. Commun.* **2021**, *111*, 103660. [CrossRef]
4. Cheldi, T.; Cavassi, P.; Serricchio, M.; Spenelli, C.M.; Vietina, G.; Ballabio, S. Use of spoolable reinforced thermoplastic pipes for oil and water transportation. In Proceedings of the 14th Offshore Mediterranean Conference (OMC) and Exhibition, Revenna, Italy, 27–29 March 2019.
5. Zhang, H.; Tong, L.; Addo, M.A. Mechanical Analysis of Flexible Riser with Carbon Fiber Composite Tension Armor. *J. Compos. Sci.* **2021**, *5*, 3. [CrossRef]
6. Chan, P.H. Design Study of Composite Repair System for Offshore Riser Applications. Ph.D. Thesis, Department of Mechanical, Materials and Manufacturing Engineering, The University of Nottingham, Malaysian Campus, Semenyih, Malaysia, 2015. Available online: http://eprints.nottingham.ac.uk/33455/1/CHAN%20PARK%20HINN%20-%20Design%20Study%20of%20Composite%20Repair%20System%20for%20Offshore%20Riser%20Applications.pdf (accessed on 15 February 2022).
7. Alexander, C.R. Development of Composite Repair System for Reinforcing Offshore Risers. Ph.D. Thesis, Department of Mechanical Engineering, Texas A&M University, College Station, TX, USA, 2007. Available online: http://oaktrust.library.tamu.edu/bitstream/handle/1969.1/ETD-TAMU-2534/ALEXANDER-DISSERTATION.pdf?sequence=1 (accessed on 15 February 2022).
8. Andersen, R. Analysis of Transverse Cracking in Composite Structures. Ph.D. Thesis, Norwegian Institute of Technology of Science, Trondheim, Norway, 1996. Available online: http://www.diva-portal.se/smash/get/diva2:998909/FULLTEXT01.pdf (accessed on 15 February 2022).
9. Ashraf, M.A.; Morozov, E.V.; Shankar, K. Flexure analysis of spoolable reinforced thermoplastic pipes for offshore oil and gas applications. *J. Reinf. Plast. Compos.* **2014**, *33*, 533–542. [CrossRef]
10. Yu, K. Nonlinear Modelling and Analysis of Reinforced Thermoplastic Pipes for Offshore Applications. Ph.D. Thesis, School of Engineering and Information Technology, The University of New South Wales, Australian Defence Force Academy, Canberra, Australia, 2015. Available online: http://unsworks.unsw.edu.au/fapi/datastream/unsworks:36255/SOURCE02?view=true (accessed on 15 February 2022).
11. Wang, C. Tailored Design of Composite Risers for Deep Water Applications. Ph.D. Thesis, School of Engineering and Information Technology, The University of New South Wales, Canberra, Australia, 2013. Available online: http://unsworks.unsw.edu.au/fapi/datastream/unsworks:11345/SOURCE01?view=true (accessed on 15 February 2022).
12. Amaechi, C.V.; Ye, J. A numerical modeling approach to composite risers for deep waters. In *Structural and Computational Mechanics Book Series, Proceedings of the 20th International Conference on Composite Structures (ICCS20), Paris, France, 4–7 September 2017*; Ferreira, A.J.M., Larbi, W., Deu, J.-F., Tornabene, F., Fantuzzi, N., Eds.; Societa Editrice Esculapio: Bologna, Italy, 2017; pp. 262–263. Available online: https://www.google.co.uk/books/edition/ICCS20_20th_International_Conference_on/MPItDwAAQBAJ?hl=en&gbpv=1&dq=A+numerical+modeling+approach+to+composite+risers+for+deep+waters&pg=PR19&printsec=frontcover (accessed on 15 February 2022).
13. Amaechi, C.V. A review of state-of-the-art and meta-science analysis on composite risers for deep seas. *Ocean. Eng.* **2022**. under review.
14. Amaechi, C.V.; Chesterton, C.; Butler, H.O.; Gu, Z.; Odijie, A.C.; Wang, F.; Hou, X.; Ye, J. Finite Element Modelling on the Mechanical Behaviour of Marine Bonded Composite Hose (MBCH) under Burst and Collapse. *J. Mar. Sci. Eng.* **2022**, *10*, 151. [CrossRef]
15. Amaechi, C.V.; Chesterton, C.; Butler, H.O.; Gu, Z.; Odijie, A.C.; Hou, X. Numerical Modelling on the Local Design of a Marine Bonded Composite Hose (MBCH) and Its Helix Reinforcement. *J. Compos. Sci.* **2022**, *6*, 79. [CrossRef]

16. Hanonge, D.; Luppi, A. Challenges of flexible riser systems in shallow waters. Paper No: OTC 20578. In Proceedings of the Offshore Technology Conference, Houston, TX, USA, 3–6 May 2010; pp. 1–10.
17. Amaechi, C.V.; Odijie, C.; Etim, O.; Ye, J. Economic Aspects of Fiber Reinforced Polymer Composite Recycling. In *Encyclopedia of Renewable and Sustainable Materials*; Elsevier: Amsterdam, The Netherlands, 2019. [CrossRef]
18. Amaechi, C.V.; Odijie, C.; Sotayo, A.; Wang, F.; Hou, X.; Ye, J. Recycling of Renewable Composite Materials in the Offshore Industry. In *Encyclopedia of Renewable and Sustainable Materials*; Elsevier: Amsterdam, The Netherlands, 2019. [CrossRef]
19. Saiful Islam, A.B.M. Dynamic characteristics and fatigue damage prediction of FRP strengthened marine riser. *Ocean Syst. Eng.* **2018**, *8*, 21–32. [CrossRef]
20. Vedernikov, A.; Safonov, A.; Tucci, F.; Carlone, P.; Akhatov, I. Pultruded materials and structures: A review. *J. Compos. Mater.* **2000**, *54*, 4081–4117. [CrossRef]
21. Rubino, F.; Nisticò, A.; Tucci, F.; Carlone, P. Marine Application of Fiber Reinforced Composites: A Review. *J. Mar. Sci. Eng.* **2020**, *8*, 26. [CrossRef]
22. Costache, A. Anchoring FRP Composite Armor in Flexible Offshore Riser Systems. Ph.D. Thesis, Technical University of Denmark (DTU), Department of Mechanical Engineering, Lyngby, Denmark, 2015. Available online: https://backend.orbit.dtu.dk/ws/portalfiles/portal/123357751/Anchoring_FRP_Composite_Armor.pdf (accessed on 15 February 2022).
23. Bai, Y.; Bai, Q. *Subsea Pipelines and Risers*, 1st ed.; 2013 Reprint; Elsevier Ltd.: Oxford, UK, 2005.
24. Bai, Y.; Bai, Q. *Subsea Engineering Handbook*; Elsevier: Oxford, UK, 2010.
25. Chakrabarti, S.K. *Handbook of Offshore Engineering*, 1st ed.; Elsevier: Plainfield, IL, USA, 2005.
26. Dareing, D.W. *Mechanics of Drillstrings and Marine Risers*; ASME Press: New York, NY, USA, 2012; pp. 1–396. Available online: https://doi.org/10.1115/1.859995 (accessed on 15 February 2022).
27. Sparks, C. *Fundamentals of Marine Riser Mechanics: Basic Principles and Simplified Analyses*, 2nd ed.; PennWell Books: Tulsa, OK, USA, 2018.
28. Shayan, N. Nonlinear Behaviour of Offshore Flexible Risers. Master's Thesis, Department of Mechanical, Aerospace and Civil Engineering, Brunel University, London, UK, 2014. Available online: https://bura.brunel.ac.uk/bitstream/2438/9488/1/FulltextThesis.pdf (accessed on 15 February 2022).
29. Bahtui, A. Development of a Constitutive Model to Simulate Unbonded Flexible Riser Pipe Elements. Ph.D. Thesis, Department of Mechanical Engineering, Brunel University, London, UK, 2008. Available online: https://citeseerx.ist.psu.edu/viewdoc/download?doi=10.1.1.426.6164&rep=rep1&type=pdf (accessed on 15 February 2022).
30. Amaechi, C.V.; Chesterton, C.; Butler, H.O.; Wang, F.; Ye, J. An overview on bonded marine hoses for sustainable fluid transfer and (un)loading operations via floating offshore structures (FOS). *J. Mar. Sci. Eng.* **2021**, *9*, 1236. [CrossRef]
31. Amaechi, C.V. Novel Design, Hydrodynamics and Mechanics of Marine Hoses in Oil/Gas Applications. Ph.D. Thesis, Lancaster University, Lancaster, UK, 2021.
32. Amaechi, C.V.; Wang, F.; Ye, J. Mathematical modelling of bonded marine hoses for single point mooring (SPM) systems, with Catenary Anchor Leg Mooring (CALM) buoy application: A review. *J. Mar. Sci. Eng.* **2021**, *9*, 1179. [CrossRef]
33. Amaechi, C.V.; Wang, F.; Ja'e, I.A.; Aboshio, A.; Odijie, A.C. A literature review on the technologies of bonded hoses for marine application. *Ships Offshore Struct.* **2022**, 1–32. [CrossRef]
34. Akpan, V.; Ossia, C.V.; Fayemi, F. On the Study of Wellhead Fatigue due to Vortex Induced Vibration in the Gulf of Guinea. *J. Mech. Eng. Autom.* **2017**, *7*, 8–15. [CrossRef]
35. Chibueze, N.O.; Ossia, C.V.; Okoli, J.U. On the Fatigue of Steel Catenary Risers. *J. Mech. Eng.* **2016**, *62*, 751–756. [CrossRef]
36. Udeze, K.U.; Ossia, C.V. Vortex Induced Vibration of Subsea Umbilicals: A Case Study of Deep Offshore Nigeria. *Univers. J. Mech. Eng.* **2017**, *5*, 35–46. [CrossRef]
37. Kandasamy, R.; Cui, F.; Townsend, N.; Foo, C.C.; Guo, J. A review of vibration control methods for marine offshore structures. *Ocean. Eng.* **2016**, *127*, 279–297. [CrossRef]
38. Ja'E, I.A.; Ali, M.O.A.; Yenduri, A.; Nizamani, Z.; Nakayama, A. Optimisation of mooring line parameters for offshore floating structures: A review paper. *Ocean Eng.* **2022**, *247*, 110644. [CrossRef]
39. Ali, M.O.A.; Ja'e, I.A.; Yenduri, A.; Hwa, M.G.Z. Effects of water depth, mooring line diameter and hydrodynamic coefficients on the behaviour of deepwater FPSOs. *Ain Shams Eng. J.* **2020**, *11*, 727–739. [CrossRef]
40. Amaechi, C.V.; Wang, F.; Odijie, A.C.; Ye, J. Numerical investigation on mooring line configurations of a Paired Column Semisubmersible for its global performance in deep water condition. *Ocean. Eng.* **2022**, *250*, 110572. [CrossRef]
41. Odijie, A.C.; Wang, F.; Ye, J. A review of floating semisubmersible hull systems: Column stabilized unit. *Ocean Eng.* **2017**, *144*, 191–202. [CrossRef]
42. Sadeghi, K. An Overview of Design, Analysis, Construction and Installation of Offshore Petroleum Platforms Suitable for Cyprus Oil/Gas Fields. *GAU J. Soc. Appl. Sci.* **2007**, *2*, 1–16. Available online: https://cemtelecoms.iqpc.co.uk/media/6514/786.pdf (accessed on 6 January 2022).
43. Dikdogmus, H. Riser Concepts for Deep Waters. Master's Thesis, Norwegian University of Science and Technology NTNU, Trondheim, Norway, 2012.
44. Sævik, S. On Stresses and Fatigue in Flexible Pipes. Ph.D. Thesis, NTH Trondheim, Norwegian Inst Technology, Dept Marine Structures Norway, Trondheim, Norway, 1992. Available online: https://trid.trb.org/view/442338 (accessed on 15 February 2022).

45. Tamarelle, P.J.C.; Sparks, C.P. High-Performance Composite Tubes for Offshore Applications. In Proceedings of the at the Offshore Technology Conference, Houston, TX, USA, 27 April 1987; OnePetro: Houston, TX, USA, 1987. [CrossRef]
46. Wang, S.S. Composites Key to deepwater oil and gas. *High-Perform. Compos.* **2006**, *14*, 7. Available online: https://www.compositesworld.com/columns/composites-key-to-deepwater-oil-and-gas (accessed on 15 February 2022).
47. Wang, S.S.; Fitting, D.W. Composite Materials for Offshore Operations. In Proceedings of the First International Workshop, Houston, TX, USA, 26–28 October 1993; Available online: https://www.govinfo.gov/content/pkg/GOVPUB-C13-49d78a3320ed702b3f13676611e8da41/pdf/GOVPUB-C13-49d78a3320ed702b3f13676611e8da41.pdf (accessed on 15 February 2022).
48. Amaechi, C.V.; Wang, F.; Hou, X.; Ye, J. Strength of submarine hoses in Chinese-lantern configuration from hydrodynamic loads on CALM buoy. *Ocean Eng.* **2019**, *171*, 429–442. [CrossRef]
49. Amaechi, C.V.; Ye, J.; Hou, X.; Wang, F.-C. Sensitivity Studies on Offshore Submarine Hoses on CALM Buoy with Comparisons for Chinese-Lantern and Lazy-S Configuration OMAE2019-96755. In Proceedings of the 38th International Conference on Ocean, Offshore and Arctic Engineering, Glasgow, UK, 9–14 June 2019; Available online: https://eprints.lancs.ac.uk/id/eprint/134404.pdf (accessed on 15 February 2022).
50. Amaechi, C.V.; Wang, F.; Ye, J. Numerical assessment on the dynamic behaviour of submarine hoses attached to CALM buoy configured as lazy-S under water waves. *J. Mar. Sci. Eng.* **2021**, *9*, 1130. [CrossRef]
51. Amaechi, C.V.; Wang, F.; Ye, J. Numerical studies on CALM buoy motion responses, and the effect of buoy geometry cum skirt dimensions with its hydrodynamic waves-current interactions. *Ocean Eng.* **2022**, *244*, 110378. [CrossRef]
52. Amaechi, C.V.; Wang, F.; Ye, J. Investigation on hydrodynamic characteristics, wave-current interaction, and sensitivity analysis of submarine hoses attached to a CALM buoy. *J. Mar. Sci. Eng.* **2022**, *10*, 120. [CrossRef]
53. Amaechi, C.V.; Wang, F.; Ye, J. Understanding the fluid–structure interaction from wave diffraction forces on CALM buoys: Numerical and analytical solutions. *Ships Offshore Struct.* **2022**, 1–29. [CrossRef]
54. Amaechi, C.V.; Wang, F.; Ye, J. Experimental study on motion characterization of CALM buoy hose system under water waves. *J. Mar. Sci. Eng.* **2022**, *10*, 204. [CrossRef]
55. Saad, P.; Salama, M.M.; Jahnsen, O. Application of Composites to Deepwater Top Tensioned Riser Systems. In Proceedings of the ASME 2002 21st International Conference on Offshore Mechanics and Arctic Engineering, 21st International Conference on Offshore Mechanics and Arctic Engineering, Oslo, Norway, 23–28 June 2002; Volume 3, pp. 255–261. [CrossRef]
56. Rustad, A.M.; Larsen, C.M.; Sorensen, A.J. FEM modelling and automatic control for collision prevention of top tensioned risers. *Mar. Struct.* **2008**, *21*, 80–112. [CrossRef]
57. Kang, H.S.; Kim, M.H.; Aramanadka, S.S.B. Tension variations of hydro-pneumatic riser tensioner and implications for dry-tree interface in semisubmersible. *Ocean Syst. Eng.* **2017**, *7*, 21–38. [CrossRef]
58. Morooka, C.K.; Coelho, F.M.; Shiguemoto, D.A. Dynamic behavior of a top tensioned riser in frequency and time domain. In Proceedings of the 16th International Offshore and Polar Engineering Conference, San Francisco, CA, USA, 28 May–2 June 2006.
59. Drumond, G.P.; Pasqualino, I.P.; Pinheiro, B.C.; Estefen, S.F. Pipelines, risers and umbilicals failures: A literature review. *Ocean Eng.* **2018**, *148*, 412–425. [CrossRef]
60. Li, X.; Jiang, X.; Hopman, H. A review on predicting critical collapse pressure of flexible risers for ultra-deep oil and gas production. *Appl. Ocean. Res.* **2018**, *80*, 1–10. [CrossRef]
61. Li, X.; Jiang, X.; Hopman, H. Prediction of the Critical Collapse Pressure of Ultra-Deep Water Flexible Risers—A Literature Review. *FME Trans.* **2018**, *46*, 306–312. [CrossRef]
62. PSA & 4Subsea. Un-Bonded Flexible Risers—Recent Field Experience and Actions for Increased Robustness. 0389-26583-U-0032, Revision 5, For PSA Norway. 2013. Available online: https://www.ptil.no/contentassets/c2a5bd00e8214411ad5c4966009d6ade/un-bonded-flexible-risers--recent-field-experience-and-actions--for-increased-robustness.pdf (accessed on 17 June 2021).
63. PSA & 4Subsea. Bonded Flexibles—State of the Art Bonded Flexible Pipes. 0389-26583-U-0032, Revision 5, For PSA Norway. 2018. Available online: https://www.4subsea.com/wp-content/uploads/2019/01/PSA-Norway-State-of-the-art-Bonded-Flexible-Pipes-2018_4Subsea.pdf (accessed on 17 June 2021).
64. Amaechi, C.V.; Chesterton, C.; Butler, H.O.; Wang, F.; Ye, J. Review on the design and mechanics of bonded marine hoses for Catenary Anchor Leg Mooring (CALM) buoys. *Ocean Eng.* **2021**, *242*, 110062. [CrossRef]
65. Ochoa, O.O.; Salama, M.M. Offshore composites: Transition barriers to an enabling technology. *Compos. Sci. Technol.* **2005**, *65*, 2588–2596. [CrossRef]
66. Zhang, Y.; Gao, W.W.; Xu, S.X.; Duan, M. The Research about the Strength of Composite Riser Pipes Based on Finite Element Method. *Key Eng. Mater. KEM* **2015**, *665*, 177–180. [CrossRef]
67. OGJ. Composite riser technology advances to field applications. *Oil Gas J.* **2001**, *99*, 17220899. Available online: https://www.ogj.com/home/article/17220899/composite-riser-technology-advances-to-field-applications (accessed on 15 February 2022).
68. Sparks, C.P.; Odru, P.; Bono, H.; Metivaud, G. Mechanical Testing of High-Performance Composite Tubes for TLP Production Risers. In Proceedings of the Offshore Technology Conference, Houston, TX, USA, 2–5 May 1988. [CrossRef]
69. Sparks, C.P.; Odru, P.; Metivaud, G.; Christian, L.F.H. Composite Riser Tubes: Defect Tolerance Assessment and Nondestructive Testing. In Proceedings of the Offshore Technology Conference (OTC), Houston, TX, USA, 4–7 May 1992. [CrossRef]
70. Andersen, W.F. *Proposal for Manufacturing Composite Structures for the Offshore Oil Industry*; Westinghouse Marine Division: Sunnyvale, CA, USA, 1994.

71. Andersen, W.F.; Anderson, J.J.; Landriault, L.S. Full-Scale Testing of Prototype Composite Drilling Riser Joints-Interim Report. In Proceedings of the Offshore Technology Conference, Houston, TX, USA, 4–7 May 1998. [CrossRef]
72. Andersen, W.F.; Anderson, J.J.; Mickelson, C.S.; Sweeney, T.F. The Application of Advanced Composite Technology to Marine Drilling Riser Systems: Design, Manufacturing and Test. In Proceedings of the Offshore Technology Conference, Houston, TX, USA, 5–8 May 1997. [CrossRef]
73. Andersen, W.F.; Burgdorf, J.O.; Sweeney, T.F. Comparative Analysis of 12,500 ft. Water Depth Steel and Advanced Composite Drilling Risers. In Proceedings of the Offshore Technology Conference, Houston, TX, USA, 4–7 May 1998. [CrossRef]
74. Salama, M.M.; Spencer, B.E. Multiple Seal Design for Composite Risers and Tubing for Offshore Applications. U.S. Patent 6,719,058, 13 April 2004. Available online: https://patentimages.storage.googleapis.com/c9/c9/d6/2818d2e7dd5155/US6719058.pdf (accessed on 15 February 2022).
75. Baldwin, D.D.; Newhouse, N.L.; Lo, K.H.; Burden, R.C. Composite Production Riser Design. In Proceedings of the Offshore Technology Conference, Houston, TX, USA, 5–8 May 1997.
76. Baldwin, D.D.; Reigle, J.A.; Drey, M.D. Interface System between Composite Tubing and End Fittings. U.S. Patent 6,042,152, 28 March 2000. Available online: https://patents.google.com/patent/US6042152A/en (accessed on 15 February 2022).
77. Drey, M.D.; Salama, M.M.; Long, J.R.; Abdallah, M.G.; Wang, S.S. Composite Production Riser—Testing and Qualification. In Proceedings of the Offshore Technology Conference, Houston, TX, USA, 5–8 May 1997. [CrossRef]
78. Meniconi, L.C.M.; Reid, S.R.; Soden, P.D. Preliminary design of composite riser stress joints. *Compos. Part A Appl. Sci. Manuf.* **2001**, *32*, 597–605. [CrossRef]
79. Loreiro, W.C., Jr.; DosSantos, F.C., Jr.; Henriques, C.C.D.; Meniconi, L.C.M. Strategy concerning composite flowlines, risers and pipework in offshore applications. OTC 24049. In Proceedings of the Offshore Technology Conference, Houston, TX, USA, 6–9 May 2013. [CrossRef]
80. Gibson, A.G. *The Cost Effective Use of Fiber Reinforced Composites Offshore*; Research Report for the Health and Safety Executive (HSE); University of Newcastle Upon Tyne: Newcastle, UK, 2003. Available online: https://www.hse.gov.uk/research/rrpdf/rr039.pdf (accessed on 15 February 2022).
81. Gibson, A.G. Engineering Standards for Reinforced Thermoplastic Pipe. Paper No: OTC 14063. In Proceedings of the Offshore Technology Conference, Houston, TX, USA, 5–8 May 2003; pp. 1–10. [CrossRef]
82. Gibson, A.G.; Linden, J.M.; Elder, D.; Leong, K.H. Non-metallic pipe systems for use in oil and gas. *Plast. Rubber Compos.* **2011**, *40*, 465–480. [CrossRef]
83. Murali, J.; Salama, M.M.; Jahnsen, O.; Meland, T. Composite Drilling Riser-Qualification, Testing and Field Demonstration. In *Composite Materials for Offshore Operations-2 (CMOO-2)*; Wang, S.S., Williams, J.G., Lo, K.H., Eds.; Cited 2001; American Bureau of Shipping: New York, NY, USA, 1999; pp. 129–149.
84. Echtermeyer, A.T.; Steuten, B. Thermoplastic Composite Riser Guidance Note, OTC 24095. In Proceedings of the Offshore Technology Conference, Houston, TX, USA, 6–9 May 2013; pp. 1–10. [CrossRef]
85. Echtermeyer, A.T.; Osnes, H.; Ronold, K.O.; Moe, E.T. Recommended Practice for Composite Risers. In Proceedings of the Offshore Technology Conference (OTC), Houston, TX, USA, 6–9 May 2002. [CrossRef]
86. Galle, G. *Proc. Composite Riser Workshop*; Paper 5.4; Statoil Research Centre: Trondheim, Norway, 1999.
87. Slagsvold, L. *Proc. Composite Riser Workshop*; Paper 5.5; Statoil Research Centre: Trondheim, Norway, 1999.
88. Bybee, K. The First Offshore Installation of a Composite Riser Joint. *J. Pet. Technol.* **2003**, *55*, 72–74. [CrossRef]
89. Salama, M.M.; Stjern, G.; Storhaug, T.; Spencer, B.; Echtermeyer, A. The First Offshore Field Installation for a Composite Riser Joint. In Proceedings of the Offshore Technology Conference (OTC), Houston, TX, USA, 6–9 May 2002. [CrossRef]
90. Salama, M.M.; Johnson, D.B.; Long, J.R. Composite Production Riser Testing and Qualification. *SPE Prod. Facil.* **1998**, *13*, 170–177. [CrossRef]
91. Salama, M.M.; Murali, J.; Baldwin, D.D.; Jahnsen, O.; Meland, T. Design Consideration for Composite Drilling Riser. In Proceedings of the Offshore Technology Conference, Houston, TX, USA, 3–6 May 1999. [CrossRef]
92. Salama, M.M.; Spencer, B.E. Method of Manufacturing Composite Riser. U.S. Patent 7,662,251B2, 16 February 2010. Available online: https://patentimages.storage.googleapis.com/34/47/39/fcbdb5b1b1524c/US7662251.pdf (accessed on 15 February 2022).
93. Smith, K.L.; Leveque, M.E. *Ultra-Deepwater Production Systems: Technical Progress Report*; ConocoPhillips Company: Houston, TX, USA, 2003; pp. 3–23. Available online: https://www.osti.gov/servlets/purl/896669 (accessed on 15 February 2022).
94. Smith, K.L.; Leveque, M.E. *Ultra-Deepwater Production Systems: Final Report*; Report No: DEFC26-00NT40964; ConocoPhillips Company: Houston, TX, USA, 2005; pp. 8–81. Available online: https://www.osti.gov/servlets/purl/896668 (accessed on 15 February 2022).
95. Picard, D.; Hudson, W.; Bouquier, L.; Dupupet, G.; Zivanovic, I. Composite Carbon Thermoplastic Tubes for Deepwater Applications. In Proceedings of the Offshore Technology Conference (OTC), Houston, TX, USA, 30 April–3 May 2007. [CrossRef]
96. Ramirez, G.; Engelhardt, M.D. Experimental investigation of a large-scale composite riser tube under external pressure. *ASME. J. Pressure Vessel Technol.* **2009**, *131*, 051205. [CrossRef]
97. Ramirez, G.; Engelhardt, M.D. External Pressure Testing Of A Large-Scale Composite Pipe. In Proceedings of the 12th International Conference on Composite Materials (ICCM12), Paris, France, 5–9 July 1999; Available online: https://www.iccm-central.org/Proceedings/ICCM12proceedings/site/papers/pap631.pdf (accessed on 15 February 2022).

98. Kim, W.K. Composite Production Riser Assessment. Ph.D. Thesis, Texas A&M University, College Station, TX, USA, 2007. Available online: https://core.ac.uk/download/pdf/4272879.pdf (accessed on 15 February 2022).
99. NIST. *NIST GCR 04-863 Composites Manufacturing Technologies: Composite Production Riser Case Study*; USA, NIST-ATP (Advanced Technology Program): Gaithersburg, MD, USA, 2005; Available online: https://citeseerx.ist.psu.edu/viewdoc/download?doi=10.1.1.353.6624&rep=rep1&type=pdf (accessed on 15 February 2022).
100. Ochoa, O.O. *Composite Riser Experience and Design Guidance*; MMS Project Number 490; Offshore Technology Research Center: College Station, TX, USA, 2006. Available online: https://www.bsee.gov/sites/bsee.gov/files/tap-technical-assessment-program//490aa.pdf (accessed on 15 February 2022).
101. Ward, E.G.; Ochoa, O.; Kim, W.; Gilbert, R.M.; Jain, A.; Miller, C.; Denison, E. *A Comparative Risk Analysis of Composite and Steel Production Risers. MMS Project 490, Minerals Management Service (MMS)*; Offshore Technology Research Center: College Station, TX, USA; Texas A&M University: College Station, TX, USA, 2007. Available online: https://www.bsee.gov/sites/bsee.gov/files/tap-technical-assessment-program/490ab.pdf (accessed on 15 February 2022).
102. Ochoa, O.O. *Structural Characterization and Design Optimization of Hybrid Composite Tubes for TLP Riser Applications*; Offshore Technology Research Center: College Station, TX, USA, 1995.
103. Johnson, D.B.; Salama, M.M.; Long, J.R.; Wang, S.S. Composite Production Riser—Manufacturing Development and Qualification Testing. In Proceedings of the Offshore Technology Conference, Houston, TX, USA, 4–7 May 1998. [CrossRef]
104. Baldwin, D.D.; Douglas, B.J. Rigid Composite Risers: Design for Purpose Using Performance-Based Requirements. In Proceedings of the Offshore Technology Conference, Houston, TX, USA, 6–9 May 2002. [CrossRef]
105. Alexander, C.; Vyvial, B.; Cederberg, C.; Baldwin, D. Evaluating the performance of a composite-reinforced steel drilling riser via full-scale testing for HPHT service. In Proceedings of the 6th International Offshore Pipeline Forum (IOPF 2011), Houston, TX, USA, 19–20 October 2011; Available online: https://www.chrisalexander.com/wp-content/uploads/2020/05/4-1.pdf (accessed on 15 February 2022).
106. Carpenter, C. Composite Flowlines, Risers, and Pipework in Offshore Applications. *J. Pet. Technol.* **2014**, *66*, 101–103. [CrossRef]
107. Carpenter, C. Qualification of Composite Pipe. *J. Pet. Technol. JPT* **2016**, *68*, 56–58. [CrossRef]
108. Bybee, K. Design Considerations for a Composite Drilling Riser. *J. Pet. Technol.* **2000**, *52*, 42–44. [CrossRef]
109. Cederberg, C. *Design and Verification Testing Composite-Reinforced Steel Drilling Riser*; Final Report, RPSEA 07121-1401; Lincoln Composites, Inc.: Huntington Beach, CA, USA, 2011.
110. Cederberg, C.A.; Baldwin, D.D.; Bhalla, K.; Tognarelli, M.A. Composite-Reinforced Steel Drilling Riser for Ultra-deepwater High Pressure Wells. In Proceedings of the Offshore Technology Conference, Houston, TX, USA, 6–9 May 2013. [CrossRef]
111. OffshoreEngineer. Airborne Begins TCP Qualifications. Offshore Engineer, October Issue, Published 3 October 2016. Available online: https://www.oedigital.com/news/448257-airborne-begins-tcp-qualifications (accessed on 15 February 2022).
112. OceanEnergy. Airborne Oil & Gas to Qualify TCP for Total's Deepwater Jumper Spools. Ocean Energy Resources. 2016. Available online: https://ocean-energyresources.com/2016/10/04/airborne-oil-gas-to-qualify-tcp-for-totals-deepwater-jumper-spools/ (accessed on 15 February 2022).
113. EnergyOilGas. *Airborne Oil & Gas: Profile Gallery. Energy, Oil & Gas*; Schofield Publishing: Nowich, UK, 2009; Available online: https://energy-oil-gas.com/profiles/airborne-oil-gas/ (accessed on 15 February 2022).
114. Mason, K. Thermoplastic composite pipe on the rise in the deep sea. *Composite World*, 3 August 2019. Available online: https://www.compositesworld.com/articles/thermoplastic-composite-pipe-on-the-rise-in-the-deep-sea (accessed on 15 February 2022).
115. Osborne, J. Thermoplastic Pipes—Lighter, More Flexible Solutions for Oil and Gas Extraction. Materials Today, Published on 26 February 2013. & Reinforced Plastics Magazine, January/February 2013 Issue. Available online: https://www.materialstoday.com/surface-science/features/thermoplastic-pipes-lighter-more-flexible/ (accessed on 15 February 2022).
116. MagmaGlobal. *HWCG Selects M-Pipe for Next Generation Emergency Well Containment Riser*; MagmaGlobal: Portsmouth, UK, 2019; Available online: https://www.magmaglobal.com/hwcg-selects-m-pipe-for-next-generation-emergency-well-containment-riser/ (accessed on 15 February 2022).
117. MagmaGlobal. *Qualification of M-Pipe®and Hybrid Flexible Pipe For Deployment In Brazil's Pre-Salt Region: Composite Material Selection*; MagmaGlobal: Portsmouth, UK, 2019; Available online: https://www.magmaglobal.com/qualification-of-m-pipe-and-hybrid-flexible-pipe-for-deployment-in-brazils-pre-salt-region-composite-material-selection/ (accessed on 15 February 2022).
118. Calash. *Commercial Review of 8 Riser SLOR System: Magma M-Pipe versus Steel Pipe*; Magma Global Report 776; Magma Global: Portsmouth, UK, 2015; pp. 1–16.
119. Hatton, S. Carbon fibre—A riser system enabler. *Offshore Eng.* **2012**, *37*, 42–43. Available online: https://www.oedigital.com/news/459619-carbon-fibre-a-riser-system-enabler (accessed on 15 February 2022).
120. Hatton, S. Lightweight Riser Design. 2015. Available online: https://www.magmaglobal.com/lightweight-riser-design-approach/ (accessed on 15 February 2022).
121. Cottrill, A. *Where m-pipe Is Claiming the Edge on Cost*; Upstream Technology: New Brighton, MN, USA, 2015; pp. 16–19. Available online: http://www.upstreamonline.com/upstreamtechnology/?hashedzmagsid=0e43a229&magsid=805641 (accessed on 7 May 2016).
122. Wilkins, J. Qualification of Composite Pipe. In Proceedings of the Offshore Technology Conference, Houston, TX, USA, 2–5 May 2016. [CrossRef]

123. MagmaGlobal. *The M-Pipe Lightweight Riser Solution*; Magma Global Fact Sheet: Portsmouth, UK, 2016; Available online: https://www.magmaglobal.com/lightweight-riser-design/ (accessed on 15 February 2022).
124. MagmaGlobal. Composite Riser OCYAN-Magma Global. 2016. Available online: https://www.youtube.com/watch?v=FIrOP6PbUIQ (accessed on 22 May 2021).
125. MagmaGlobal. Ocyan—Magma CompRisers. 2016. Available online: https://www.magmaglobal.com/risers/ocyan-compriser/ (accessed on 23 May 2021).
126. Roberts, D.; Hatton, S.A. Development and Qualification of End Fittings for Composite Riser Pipe. In Proceedings of the Offshore Technology Conference (OTC), Houston, TX, USA, 6–9 May 2013. [CrossRef]
127. van Onna, M.; Giaccobi, S.; de Boer, H. Evaluation of the first deployment of a composite downline in deepwater Brazil. In *Rio Oil & Gas Expo and Conference 2014*; Brazilian Petroleum, Gas and Biofuels Institute: Rio de Janeiro, Brazil, 2014; pp. 1–9.
128. van Onna, M. Installation of the World's First Thermoplastic Flowline for Hydrocarbon Service. In Proceedings of the MCEDD Deepwater Development, Milan, Italy, 9–11 April 2018; Available online: https://mcedd.com/wp-content/uploads/2018/04/MCEDD21-2.pdf (accessed on 15 February 2022).
129. van Onna, M. Thermoplastic Composite Pipe: Enabler for Enhanced Oil Recovery. MCEDD Conference. 2017. Available online: https://www.subseauk.com/documents/presentations/martin%20von%20onna.pdf (accessed on 15 February 2022).
130. Jak, A. Thermoplastic Composite Pipe Proven for Hydrocarbon Service. *Airborne Oil Gas*, 2 August 2018. Available online: https://airborneoilandgas.com/home/thermoplastic-composite-pipe-hydrocarbon (accessed on 22 October 2018).
131. Namdeo, S.; de Boer, H.; de Kanter, J. A micromechanics approach towards delamination of thermoplastic composite pipe for offshore applications. In Proceedings of the 21st International Conference on Composite Materials (ICCM-21), Xi'an, China, 20–25 August 2017; pp. 20–25. Available online: http://www.iccm-central.org/Proceedings/ICCM21proceedings/papers/3324.pdf (accessed on 15 February 2022).
132. Francis, S. Airborne Oil & Gas begins TCP Riser qualification program in South America. Published 7 May 2018. Available online: https://www.compositesworld.com/news/airborne-oil-gas-begins-tcp-riser-qualification-program-in-south-america (accessed on 15 February 2022).
133. Latto, J. Ultra-deep water Thermoplastic Composite Pipe—From Installation to Operation. Virtual MCE Deepwater Development Conference. 22 April 2021. Available online: https://strohm.eu/en/exhibitions/ultra-deep-water-thermoplastic-composite-pipe-tcp-from-installation-to-operation (accessed on 15 February 2022).
134. Mintzas, A.; Hatton, S.; Simandjuntak, S.; Little, A.; Zhang, Z. An integrated approach to the design of high performance carbon fibre reinforced risers—From micro to macro—Scale. In Proceedings of the Deep Offshore Technology (DOT) International Conference, Houston, TX, USA, 22–24 September 2013; Available online: https://researchportal.port.ac.uk/en/publications/an-integrated-approach-to-the-design-of-high-performance-carbon-f (accessed on 15 February 2022).
135. Steuten, B.; van Onna, M. Reduce Project and Life Cycle Cost with TCP Flowline. In Proceedings of the Offshore Technology Conference Asia, Kuala Lumpur, Malaysia, 22–25 March 2016. [CrossRef]
136. Spruijt, W. Installation of the World's First Subsea Thermoplastic Composite Flowline for Hydrocarbon Service. In Proceedings of the Offshore Technology Conference Asia, Kuala Lumpur, Malaysia, 20–23 March 2018. [CrossRef]
137. Wang, C.; Shankar, K.; Morozov, E.V. Local Design of composite riser under burst, tension, and collapse cases. In Proceedings of the 18th International Conference on Composite Materials (ICCM), Jeju Island, Korea, 21–26 August 2011; Available online: www.iccm-central/Proceedings/ICCM18Proceedings/ (accessed on 15 February 2022).
138. Wang, C.; Shankar, K.; Morozov, E.V. Design of composite risers for minimum weight. Publisher: World Academy of Science, Engineering and Technology. *Int. J. Mech. Aerosp. Ind. Mechatron. Manuf. Eng.* **2012**, *6*, 2627–2636. Available online: https://publications.waset.org/4236/pdf (accessed on 15 February 2022).
139. Wang, C.; Shankar, K.; Ashraf, M.A.; Morozov, E.V.; Ray, T. Surrogate-assisted optimization design of composite riser. *J. Mater. Des. Appl.* **2016**, *230*, 18–34. [CrossRef]
140. Wang, C.; Shankar, K.; Morozov, E.V. Tailored local design of deep sea FRP composite risers. *Adv. Compos. Mater.* **2015**, *24*, 375–397. [CrossRef]
141. Wang, C.; Shankar, K.; Morozov, E.V. Tailored design of top-tensioned composite risers for deep-water applications using three different approaches. *Adv. Mech. Eng.* **2017**, *9*, 1687814016684271. [CrossRef]
142. Wang, C.; Shankar, K.; Morozov, E.V. Global design and analysis of deep sea FRP composite risers under combined environmental loads. *Adv. Compos. Mater.* **2017**, *26*, 79–98. [CrossRef]
143. Wang, C.; Sun, M.; Shankar, K.; Xing, S.; Zhang, L. CFD Simulation of Vortex Induced Vibration for FRP Composite Riser with Different Modeling Methods. *Appl. Sci.* **2018**, *8*, 684. [CrossRef]
144. Wang, C.; Ge, S.; Sun, M.; Jia, Z.; Han, B. Comparative Study of Vortex-Induced Vibration of FRP Composite Risers with Large Length to Diameter Ratio Under Different Environmental Situations. *Appl. Sci.* **2019**, *9*, 517. [CrossRef]
145. Wang, C.; Cui, Y.; Ge, S.; Sun, M.; Jia, Z. Experimental Study on Vortex-Induced Vibration of Risers Considering the Effects of Different Design Parameters. *Appl. Sci.* **2018**, *8*, 2411. [CrossRef]
146. Wang, C.; Ge, S.; Jaworski, J.W.; Liu, L.; Jia, Z. Effects of Different Design Parameters on the Vortex Induced Vibration of FRP Composite Risers Using Grey Relational Analysis. *J. Mar. Sci. Eng.* **2019**, *7*, 231. [CrossRef]
147. Yu, K.; Morozov, E.V.; Ashraf, M.A.; Shankar, K. A review of the design and analysis of reinforced thermoplastic pipes for offshore applications. *J. Reinf. Plast. Compos.* **2017**, *36*, 1514–1530. [CrossRef]

148. Yu, K.; Morozov, E.V.; Ashraf, M.A.; Shankar, K. Numerical analysis of the mechanical behaviour of reinforced thermoplastic pipes under combined external pressure and bending. *Compos. Struct.* **2015**, *131*, 453–461. [CrossRef]
149. Amaechi, C.V.; Gillet, N.; Odijie, A.C.; Hou, X.; Ye, J. Composite Risers for Deep Waters Using a Numerical Modelling Approach. *Compos. Struct.* **2019**, *210*, 486–499. [CrossRef]
150. Amaechi, C.V. Local tailored design of deep water composite risers subjected to burst, collapse and tension loads. *Ocean. Eng.* **2022**, 110196. [CrossRef]
151. Amaechi, C.V.; Gillet, N.; Ja'e, I.A.; Wang, C. Tailoring the local design of deep water composite risers to minimise structural weight. *J. Compos. Sci.* **2022**. under review.
152. Gillett, N. Design and Development of a Novel Deepwater Composite Riser. BEng Thesis, Engineering Department, Lancaster University, Lancaster, UK, 2018.
153. Amaechi, C.V.; Gillett, N.; Odijie, A.C.; Wang, F.; Hou, X.; Ye, J. Local and Global Design of Composite Risers on Truss SPAR Platform in Deep waters. In Proceedings of the 5th International Conference on Mechanics of Composites (MECHCOMP19), Lisbon, Portugal, 1–4 July 2019; pp. 1–3. Available online: https://eprints.lancs.ac.uk/id/eprint/136431 (accessed on 15 February 2022).
154. Chesterton, C. A Global and Local Analysis of Offshore Composite Material Reeling Pipeline Hose, with FPSO Mounted Reel Drum. Bachelor's Thesis, Lancaster University, Engineering Department, Lancaster, UK, 2020.
155. Butler, H.O. An Analysis of the Failure of Composite Flexible Risers. Bachelor's Thesis, Lancaster University, Engineering Department, Lancaster, UK, 2021.
156. Ragbey, H.; Sobey, A. Effects of extensible modelling on composite riser mechanical responses. *Ocean. Eng.* **2021**, *220*, 108426. [CrossRef]
157. Ragbey, H.; Goodridge, M.; Pham, D.C.; Sobey, A. Extreme response based reliability analysis of composite risers for applications in deepwater. *Mar. Struct.* **2021**, *78*, 103015. [CrossRef]
158. Ragbey, H.A.; Grudniewski, P.A.; Sobey, A.J.; Weymouth, G.D. Composite risers design and optimisation using Multi-Level Selection Genetic Algorithm. In *Structural and Computational Mechanics Book Series, Proceedings of the 20th International Conference on Composite Structures (ICCS20), Paris, France, 4–7 September 2017*; Ferreira, A.J.M., Larbi, W., Deu, J.-F., Tornabene, F., Fantuzzi, N., Eds.; Societa Editrice Esculapio: Bologna, Italy, 2017; pp. 249–250. Available online: https://www.google.co.uk/books/edition/ICCS20_20th_International_Conference_on/MPItDwAAQBAJ?hl=en&gbpv=1&dq=composite+risers+design+and+optimisation+using+Multi-Level+Selection+Genetic+Algorithm+-+Hossam+A.+Ragheb&pg=PA249&printsec=frontcover (accessed on 15 February 2022).
159. Pham, D.C.; Narayanaswamy, S.; Qian, X.; Sobey, A.; Achintha. M.; Shenoi, A. Composite Riser Design and Development—A Review. In *Analysis and Design of Marine Structures V*; Soares, C.G., Shenoi, R.A., Eds.; CRC Press: Boca Raton, FL, USA, 2015; Chapter 72. [CrossRef]
160. Pham, D.C.; Sridhar, N.; Qian, X.; Sobey, A.J.; Achintha, M.; Shenoi, A. A review on design, manufacture and mechanics of composite risers. *Ocean. Eng.* **2016**, *112*, 82–96. [CrossRef]
161. Pham, D.C.; Su, Z.; Narayanaswamy, S.; Qian, X.; Huang, Z.; Sobey, A.; Shenoi, A. Experimental and numerical studies of large-scaled filament wound T700/X4201 composite risers under bending. In Proceedings of the ECCM17—17th European Conference on Composite Materials, Munich, Germany, 26–30 June 2016; Available online: https://www.researchgate.net/publication/307631336_Experimental_and_numerical_studies_of_large-scaled_filament_wound_T700X4201_composite_risers_under_bending (accessed on 15 February 2022).
162. Sobey, A.J.; Ragheb, H.; Shenoi, R.A.; Pham, D.C. Composite Riser Reliability Under Harsh Environmental Conditions. In Proceedings of the 2nd International Conference on Safety and Reliability of Ships, Offshore and Subsea Structures, Glasgow, UK, 25–29 September 2016; Available online: https://www.researchgate.net/publication/309425538_COMPOSITE_RISER_RELIABILITY_UNDER_HARSH_ENVIRONMENTAL_CONDITIONS (accessed on 15 February 2022).
163. Sun, X.S.; Tan VB, C.; Tan, L.B.; Chen, Y.; Jaiman, R.K.; Tay, T.E. Fatigue Life Prediction of Composite Risers Due To Vortex-Induced Vibration (VIV). In *International Journal of Fracture Fatigue and Wear, Proceedings of the 3rd International Conference on Fracture Fatigue and Wear, Kitakyushu, Japan, 1–3 September 2014*; Springer: Dordrecht, The Netherlands, 2014; Volume 2, pp. 207–213. Available online: https://www.academia.edu/11559018/FATIGUE_LIFE_PREDICTION_OF_COMPOSITE_RISERS_DUE_TO_VORTEX_INDUCED_VIBRATION_VIV_ (accessed on 15 February 2022).
164. Tan, L.B.; Chen, Y.; Jaiman, R.K.; Sun, X.; Tan, V.B.C.; Tay, T.E. Coupled fluid–structure simulations for evaluating a performance of full-scale deepwater composite riser. *Ocean Eng.* **2015**, *94*, 19–35. [CrossRef]
165. Sun, X.S.; Tan, V.B.C.; Chen, Y.; Jaiman, R.K.; Tay, T.E. An Efficient Analytical Failure Analysis Approach for Multilayered Composite Offshore Production Risers. In Proceedings of the 1st International Conference on Advanced Composites for Marine Engineering (ICACME 2013), Beijing, China, 10–12 September 2013; Available online: https://www.academia.edu/11558989/An_Efficient_Analytical_Failure_Analysis_Approach_for_Multilayered_Composite_Offshore_Production_Risers (accessed on 15 February 2022).
166. API. *Bulletin on Comparison of Marine Drilling Riser Analyses*; API 16J Bulletin; American Petroleum Institute: Washington, DC, USA, 1992.
167. ISO. *ISO 13624-1:2009*; Petroleum and Natural Gas Industries—Drilling and Production Equipment—Part 1: Design and Operation of Marine Drilling Riser Equipment. International Organization for Standardization (ISO): Geneva, Switzerland, 2009.

168. ISO. *ISO/TR 13624-2:2009*; Petroleum and Natural Gas Industries—Drilling and Production Equipment—Part 2: Deepwater Drilling Riser Methodologies, Operations, and Integrity Technical Report. International Organization for Standardization (ISO): Geneva, Switzerland, 2009.
169. ISO. *ISO 13625:2002*; Petroleum and Natural Gas Industries—Drilling and Production Equipment—Marine Drilling Riser Couplings. International Organization for Standardization (ISO): Geneva, Switzerland, 2002.
170. ISO. *ISO 13628-1:2005*; Petroleum and Natural Gas Industries—Design and Operation of Subsea Production Systems—Part 1: General Requirements and Recommendations. International Organization for Standardization (ISO): Geneva, Switzerland, 2005.
171. ISO. *ISO 13628-2:2006*; Petroleum and Natural Gas Industries—Design and Operation of Subsea Production Systems—Part 2: Unbonded Flexible Pipe Systems for Subsea and Marine Applications. International Organization for Standardization (ISO): Geneva, Switzerland, 2006.
172. ISO. *ISO 13628-3:2000*; Petroleum and Natural Gas Industries—Design and Operation of Subsea Production Systems—Part 3: Through Flowline (TFL) Systems. International Organization for Standardization (ISO): Geneva, Switzerland, 2000.
173. ISO. *ISO 13628-4:2010*; Petroleum and Natural Gas Industries—Design and Operation of Subsea Production Systems—Part 4: Subsea Wellhead and Tree Equipment. International Organization for Standardization (ISO): Geneva, Switzerland, 2010.
174. ISO. *ISO 13628-5:2009*; Petroleum and Natural Gas Industries—Design and Operation of Subsea Production Systems—Part 5: Subsea Umbilicals. International Organization for Standardization (ISO): Geneva, Switzerland, 2009.
175. ISO. *ISO 13628-6:2006*; Petroleum and Natural Gas Industries—Design and Operation of Subsea Production Systems—Part 6: Subsea Production Control Systems. International Organization for Standardization (ISO): Geneva, Switzerland, 2006.
176. ISO. *ISO 13628-7:2005*; Petroleum and natural gas industries—Design and Operation of Subsea Production Systems—Part 7: Completion/Workover Riser Systems. International Organization for Standardization (ISO): Geneva, Switzerland, 2005.
177. ISO. *ISO 13628-8:2002*; Petroleum and Natural Gas Industries—Design and Operation of Subsea Production Systems—Part 8: Remotely Operated Vehicle (ROV) Interfaces on Subsea Production Systems. International Organization for Standardization (ISO): Geneva, Switzerland, 2002.
178. ISO. *ISO 13628-9:2000*; Petroleum and Natural Gas Industries—Design and Operation of Subsea Production Systems—Part 9: Remotely Operated Tool (ROT) Intervention Systems. International Organization for Standardization (ISO): Geneva, Switzerland, 2000.
179. ISO. *ISO 13628-10:2005*; Petroleum and Natural Gas Industries—Design and Operation of Subsea Production Systems—Part 10: Specification for Bonded Flexible Pipe. International Organization for Standardization (ISO): Geneva, Switzerland, 2005.
180. API. *Recommended Practice for Design and Operation of Marine Drilling Riser Systems*, 2nd ed.; API RP 2Q; American Petroleum Institute: Washington, DC, USA, 1984.
181. API. *Recommended Practice for Fitness-for-Service*; API 579; American Petroleum Institute: Washington, DC, USA, 2000.
182. API. *Design, Selection, Operation and Maintenance of Marine Drilling Riser Systems*; API RP 16Q; American Petroleum Institute: Washington, DC, USA, 2010.
183. API. *Qualification of Spoolable Reinforced Plastic Line Pipe*; API 15S; American Petroleum Institute: Washington, DC, USA, 2013.
184. API. *Specification for Unbonded Pipe*; API 17J; American Petroleum Institute: Washington, DC, USA, 2013.
185. Moreira, J.R.F.; Oliveira, M.F.D.; Paulo, P.C.S.; Branca, M.; Mateus, F.J. An innovative workover riser system for 3000 m water depth. In Proceedings of the Offshore Technology Conference, Houston, TX, USA, 5–8 May 2003.
186. Chen, Y.; Seemann, R.; Krause, D.; Tay, T.-E.; Tan, V.B. Prototyping and testing of composite riser joints for deepwater application. *J. Reinf. Plast. Compos.* **2016**, *35*, 95–110. [CrossRef]
187. Toh, W.; Taan, L.B.; Jaiman, R.K.; Tay, T.E.; Tan, V.B.C. A comprehensive study on composite risers: Material solution, local end fitting design and global response. *Mar. Struct.* **2018**, *61*, 155–169. [CrossRef]
188. Williams, J.G.; Sas-Jaworsky, A. Spoolable Composite Tubular Member with Energy Conductors. U.S. Patent 5,913,337, 22 June 1999. Available online: https://patentimages.storage.googleapis.com/fe/6a/ed/c870014090b475/US5913337.pdf (accessed on 15 February 2022).
189. Lassen, T.; Eide, A.L.; Meling, T.S. Ultimate Strength and Fatigue Durability of Steel Reinforced Rubber Loading Hoses. In Proceedings of the ASME 2010 29th International Conference on Ocean, Offshore and Arctic Engineering, 29th International Conference on Ocean, Offshore and Arctic Engineering: Volume 5, Parts A and B, Shanghai, China, 6–11 June 2010; pp. 277–286. [CrossRef]
190. Lassen, T.; Lem, A.I.; Imingen, G. Load Response and Finite Element Modelling of Bonded Loading Hoses. In Proceedings of the ASME 2014 33rd International Conference on Ocean, Offshore and Arctic Engineering, Volume 6A: Pipeline and Riser Technology, San Francisco, CA, USA, 8–13 June 2014. [CrossRef]
191. MagmaGlobal. *The M-Pipe: Overview, Applications and Manufacturing*; Magma Global Insight: Portsmouth, UK, 2015; Available online: https://www.magmaglobal.com/m-pipe/ (accessed on 15 February 2022).
192. MagmaGlobal. Carbon fiber pipe for risers. In Proceedings of the SUT Conference, London, UK, 12–14 September 2012.
193. Strohm. TCP Risers. Strohm, Netherlands. 2022. Available online: https://strohm.eu/tcp-risers (accessed on 15 February 2022).
194. Ajdin, A. Airborne Oil & Gas becomes Strohm. Offshore Energy, Published on 8 October 2020. Available online: https://www.offshore-energy.biz/airborne-oil-gas-becomes-strohm/ (accessed on 15 February 2022).

195. De Kanter, J.; Steuten, B.; Kremers, M.; de Boer, H. Thermoplastic Composite Pipe; Operational Experience in Deepwater and Technology Qualification. In Proceedings of the 20th International Conference on Composite Materials (ICCM-20), Copenhagen, Denmark, 19–24 July 2015; ICCM: Copenhagen, Denmark, 2015; pp. 1–11. Available online: http://www.iccm-central.org/Proceedings/ICCM20proceedings/papers/paper-1120-4.pdf (accessed on 15 February 2022).
196. Hopkins, P.; Saleh, H.; Jewell, G. Composite Pipe Set to Enable Riser Technology in Deeper Water. In Proceedings of the MCE Deepwater Development Conference, London, UK, 24–26 March 2015.
197. Hopkins, P.; Saleh, H.; Jewell, G. Composite Riser Study Confirms Weight, Fatigue Benefits Compared with Steel. Offshore Magazine, Article 16758323. 2015. Available online: https://www.offshore-mag.com/pipelines/article/16758323/composite-riser-study-confirms-weight-fatigue-benefits-compared-with-steel (accessed on 15 February 2022).
198. Hassan, S. Benefits of Composite Materials in Deepwater Risers. (2H Offshore Presentation). In Proceedings of the MCE Deepwater Development Conference, London, UK, 26 March 2015; Available online: https://2hoffshore.com/technical-papers/the-benefits-of-composite-materials-in-deepwater-riser-applications/ (accessed on 15 February 2022).
199. Saleh, H. The Benefits of Composite Materials in Deepwater Riser Applications. (2H Offshore Presentation). In Proceedings of the MCE Deepwater Development Conference, London, UK, 26 March 2015; Available online: https://2hoffshore.com/wp-content/uploads/2016/01/2015-MCE-The-Benefits-Of-Composite-Materials-In-Deepwater-Riser-Applications.pdf (accessed on 15 February 2022).
200. Omar, A.F.; Karayaka, M.; Murray, J.J. A Comparative Study of the Performance of Top-Tensioned Composite and Steel Risers under Vortex-induced Loading. In Proceedings of the Offshore Technology Conference, Houston, TX, USA, 3–6 May 1999. [CrossRef]
201. Karayaka, M.; Wu, S.; Wang, S.; Lu, X.; Ganguly, P. Composite Production Riser Dynamics and Its Effects on Tensioners, Stress Joints, and Size of Deep Water Tension Leg Platform. In Proceedings of the Offshore Technology Conference, Houston, TX, 4–7 May 1998. [CrossRef]
202. Karayaka, M.; Steen, A.; Shilling, R.; Edwards, R. Characterization of the Dynamic Loads Between Spar Top-Tensioned Riser Buoyancy Cans and Hull: Horn Mountain Field Data Measurements and Predictions. In Proceedings of the ASME 2004 23rd International Conference on Offshore Mechanics and Arctic Engineering, 23rd International Conference on Offshore Mechanics and Arctic Engineering, Volume 1, Parts A and B, Vancouver, BC, Canada, 20–25 June 2004; pp. 411–416. [CrossRef]
203. Huang, K.Z. Composite TTR design for an ultradeepwater TLP. In Proceedings of the Offshore Technology Conference, Houston, TX, USA, 2–5 May 2005. [CrossRef]
204. Neha, C. Combining Passive and Active Methods for Damage Mode Diagnosis in Tubular Composites. Ph.D. Thesis, Department of Materials, The University of Manchester, Manchester, UK, 2019. Available online: https://www.research.manchester.ac.uk/portal/files/184632302/FULL_TEXT.PDF (accessed on 15 February 2022).
205. Odijie, A.C. Design of Paired Column Semisubmersible Hull. Ph.D. Thesis, Lancaster University, Lancaster, UK, 2016. Available online: https://eprints.lancs.ac.uk/id/eprint/86961/1/2016AgbomeriePhD.pdf (accessed on 14 June 2021).
206. Patel, M.H.; Seyed, F.B. Review of flexible riser modelling and analysis techniques. *Eng. Struct.* **1995**, *17*, 293–304. [CrossRef]
207. Ertas, A.; Kozik, T.J. A review of current approaches to riser modelling. *J. Energy Resour. Technol.* **1987**, *109*, 155–160. [CrossRef]
208. Bernitsas, M.M. Problems in marine riser design. *Mar. Technol.* **1982**, *19*, 73–82. [CrossRef]
209. Chakrabarti, S.K.; Frampton, R.E. Review of riser analysis techniques. *Appl. Ocean. Res.* **1982**, *4*, 73–90. [CrossRef]
210. Ahlstone, A.G. Well Casing Running, Cementing and Flushing Apparatus. U.S. Patent 3885625A, 27 May 1975. Available online: https://patentimages.storage.googleapis.com/78/73/73/c87a604324c2af/US3885625.pdf (accessed on 15 February 2022).
211. Ahlstone, A.G. Light Weight Marine Riser Pipe. U.S. Patent 3768842A, 30 October 1973. Available online: https://patentimages.storage.googleapis.com/a4/08/e6/0a054a58e51d97/US3768842.pdf (accessed on 15 February 2022).
212. OGJ. Composite materials provide alternatives for deepwater projects. *Oil Gas J.* **2008**, *103*, 17236136. Available online: https://www.ogj.com/general-interest/companies/article/17236136/composite-materials-provide-alternatives-for-deepwater-projects (accessed on 15 February 2022).
213. Jamal, A.; Karyadi, E. Collapse of composite cylindrical under pure bending; Report LR-739. In Proceedings of the 5th Conference of the Indonesian Students in Europe, Jerusalem, Israel, 14–19 February 1993.
214. Liu, D.; Yun, F.; Jiao, K.; Wang, L.; Yan, Z.; Jia, P.; Wang, X.; Liu, W.; Hao, X.; Xu, X. Structural Analysis and Experimental Study on the Spherical Seal of a Subsea Connector Based on a Non-Standard O-Ring Seal. *J. Mar. Sci. Eng.* **2022**, *10*, 404. [CrossRef]
215. Kalman, M.; Blair, T.; Hill, M.; Lewicki, P.; Mungall, C.; Russell, B. Composite Armored Flexible Riser System for Oil Export Service. In Proceedings of the Offshore Technology Conference, Houston, TX, USA, 3–6 May 1999.
216. Hisherik, A. Carbon composite riser and integrated deployment system to reduce the cost and risk of hydraulic light well intervention. In Proceedings of the 4th Subsea Expo 2016: The World's Largest Annual Subsea Exhibition and Conference (AECC), Aberdeen, UK, 3–5 February 2016; Available online: https://www.subseauk.com/documents/presentations/asaf%20hisherik%20-%20magma%202016.pdf (accessed on 15 February 2022).
217. Lamacchia, D. Thermoplastic Composite Pipe (TCP) Offshore Market 101. LinkedIn Pulse. Published on 30 January 2018. 2018. Available online: https://www.linkedin.com/pulse/thermoplastic-composite-pipe-tcp-offshore-market-101-diego/ (accessed on 15 February 2022).
218. Guz, I.A.; Menshykova, M.; Paik, J.K. Thick-walled composite tubes for offshore applications: An example of stress and failure analysis for filament-wound multi-layered pipes. *Ships Offshore Struct.* **2015**, *12*, 304–322. [CrossRef]

219. Ha, H. An Overview of Advances in Flexible Riser and Flowline Technology. In *4th Offshore Convention Myanmar*; 2H Offshore: Yangon, Myanmar, 2016; Available online: https://2hoffshore.com/technical-papers/advances-in-flexible-riser-flowline-technology/ (accessed on 15 February 2022).
220. Pauchard, V.; Boulharts-Campion, H.; Grosjean, F.; Odru, P.; Chateauminois, A. Development Durability Model Applied to Unidirectional Composites Beams Reinforced with Glass Fibers. *Oil Gas Sci. Technol.-Rev. IFP* **2001**, *56*, 581–595. [CrossRef]
221. Penati, L.; Ducceschi, M.; Favi, A.; Rossin, D. Installation Challenges for Ultra-Deep Waters. In Proceedings of the Offshore Mediterranean Conference and Exhibition, Ravenna, Italy, 25–27 March 2015; Available online: https://onepetro.org/OMCONF/proceedings-abstract/OMC15/All-OMC15/OMC-2015-441/1767 (accessed on 15 February 2022).
222. Skaugset, K.; Gronlund, P.K.; Melve, B.K.; Nedrelid, K. Composite Choke and Kill Lines: Qualification and Pilot Installation. In Proceedings of the Offshore Technology Conference, Houston, TX, USA, 6–9 May 2013. [CrossRef]
223. Sobrinho, L.L.; Calado, M.A.; Bastian, F.L. Development of composite pipes for riser application in deepwater. *Proc. Am. Soc. Mech. Eng. Press. Vessel. Pip. Div.* **2010**, *6*, 293–301. [CrossRef]
224. Sobrinho, L.L.; Calado, V.M.D.A.; Bastian, F.L. Development and characterization of composite materials for production of composite risers by filament winding. *Mater. Res.* **2011**, *14*, 287–298. [CrossRef]
225. Chouchaoui, C.S.; Ochoa, O.O. Similitude study for a laminated cylindrical tube under tensile, torsion, bending, internal and external pressure. Part I: Governing equations. *Compos. Struct.* **1999**, *44*, 221–229. [CrossRef]
226. Chouchaoui, C.S.; Parks, P.; Ochoa, O.O. Similitude study for a laminated cylindrical tube under tension, torsion, bending, internal and external pressure. Part II: Scale models. *Compos. Struct.* **1999**, *44*, 231–236. [CrossRef]
227. Gao, Q.; Zhang, P.; Duan, M.; Yang, X.; Shi, W.; An, C.; Li, Z. Investigation on structural behavior of ring-stiffened composite offshore rubber hose under internal pressure. *Appl. Ocean. Res.* **2018**, *79*, 7–19. [CrossRef]
228. Tatting, B.F.; Gürdal, Z.; Vasiliev, V.V. The Brazier effect for finite length composite cylinders under bending. *Int. J. Solid Struct.* **1997**, *34*, 1419–1440. [CrossRef]
229. Thomas, P. *Composites Manufacturing Technologies: Applications in Auto-Motive, Petroleum, and Civil Infrastructure Industries: Economic Study of a Cluster of ATP-Funded Projects*; NIST Report; Delta Research Co.: Chicago, IL, USA, 2004.
230. ThunderSaidEnergy. Thermo-Plastic Composite: The Future of Risers? 2019. Available online: https://thundersaidenergy.com/downloads/thermo-plastic-composite-pipe-costs/ (accessed on 12 July 2021).
231. Valenzuela, E.D.; Andersen, W.F.; Burgdorf, O.; Mickelson, C.S. Comparative Performance of a Composite Drilling Riser in Deep Water. In Proceedings of the Offshore Technology Conference, Houston, TX, USA, 3–6 May 1993. [CrossRef]
232. Valenzuela, E.D.; Moore, N.B. Dynamic Response of Deepwater Drilling Risers Using Composite Materials. In Proceedings of the Offshore Technology Conference, Houston, TX, USA, 27–30 April 1987. [CrossRef]
233. Corona, E.; Rodrigues, A. Bending of long crossply composite circular cylinders. *Compos. Eng.* **1995**, *5*, 163–182. [CrossRef]
234. Adam, S.; Ghosh, S. Application of Flexible Composite Pipe as a Cost Effective Alternative to Carbon Steel—Design Experience. In Proceedings of the Offshore Technology Conference Asia, Kuala Lumpur, Malaysia, 22–25 March 2016. [CrossRef]
235. Anderson, T.A.; Fang, B.; Attia, M.; Jha, V.; Dodds, N.; Finch, D.; Latto, J. Progress in the Development of Test Methods and Flexible Composite Risers for 3000 m Water Depths. In Proceedings of the Offshore Technology Conference, Houston, TX, USA, 3–6 May 2016. [CrossRef]
236. Odru, P.; Poirette, Y.; Stassen, Y.; Offshore, B.; Saint-Marcoux, J.F.; Litwin, P.; Abergel, L. Technical and Economical Evaluation of Composite Riser Systems. In Proceedings of the Offshore Technology Conference, Houston, TX, USA, 6–9 May 2002. [CrossRef]
237. Mirdehghan, S.A. Chapter 1-Fibrous polymeric composites. In *Engineered Polymeric Fibrous Materials*; Elsevier Publishers: Amsterdam, The Netherlands; Woodhead Publishing: New York, NY, USA, 2021. [CrossRef]
238. Price, J.C. The "State of the Art" in Composite Material Development and Applications for the Oil And Gas Industry. In Proceedings of the Twelfth International Offshore and Polar Engineering Conference, Kitakyushu, Japan, 26–31 May 2002.
239. Quigley, P.; Stringfellow, W.D.; Fowler, S.H.; Nolet, S.C. JIP Status Report: Advanced Spoolable Composites for Offshore Applications. In Proceedings of the Offshore Technology Conference, Houston, TX, USA, 4–7 May 1998.
240. Barbaso, T. Thermoplastic Composite Pipes. In Proceedings of the Advancing Sustainable Energy—5th GRE Open Days: Technical, Scientific and Business Energy Forum, Singapore, 29–30 October 2018.
241. Bertoni, F. End Fitting for Unbonded Flexible Pipes. Simeros Technologies. 2017. Available online: http://simeros.com/end-fitting-for-unbonded-flexible-pipes/?lang=en (accessed on 7 July 2021).
242. Beyle, A.I.; Gustafson, C.G.; Kulakov, V.L.; Tarnopol'skii, Y.M. Composite risers for deep-water offshore technology: Problems and prospects. 1. Metal-composite riser. *Mech. Compos. Mater.* **1997**, *33*, 403–414. [CrossRef]
243. Blanc, L.L. Composites cut riser weight by 30–40%, mass by 20–30%. *Offshore Mag.* **1998**, *58*. Available online: https://www.offshore-mag.com/deepwater/article/16756540/composites-cut-riser-weight-by-3040-mass-by-2030. (accessed on 15 February 2022).
244. Tarnopol'skii, Y.M.; Beyle, A.I.; Kulakov, V.L. Composite Risers For Offshore Technology. In Proceedings of the 12th International Conference on Composite materials (ICCM 12), Paris, France, 5–9 July 1999; Available online: https://www.iccm-central.org/Proceedings/ICCM12proceedings/site/papers/pap927.pdf (accessed on 15 February 2022).
245. Chan, P.; Tshai, K.; Johnson, M.; Li, S. The flexural properties of composite repaired pipeline: Numerical simulation and experimental validation. *Compos. Struct.* **2015**, *133*, 312–321. [CrossRef]

246. Bøtker, S.; Storhaug, T.; Salama, M.M. Composite Tethers and Risers in Deepwater Field Development: Step Change Technology. In Proceedings of the Offshore Technology Conference, Houston, TX, USA, 30 April–3 May 2001. [CrossRef]
247. Brown, T. The Impact of Composites on Future Deepwater Riser Configurations. (2H Offshore Presentation). In Proceedings of the SUT Evening Meeting, Sepang, Malaysia, 28 September 2017; Available online: https://www.sut.org/wp-content/uploads/2017/09/SUT_170928_presentation2-2H.pdf (accessed on 15 February 2022).
248. Bai, Y.; Chen, W.; Xiong, H.; Qiao, H.; Yan, H. Analysis of steel strip reinforced thermoplastic pipe under internal pressure. *Ships Offshore Struct.* **2016**, *11*, 766–773. [CrossRef]
249. Burke, B.G. An Analysis of Marine Risers for Deep Water. *J. Pet. Technol.* **1974**, *26*, 455–465. [CrossRef]
250. Brouwers, J.J.H. Analytical methods for predicting the response of marine risers. communicated by W.T. Koiter. *Proc. K. Ned. Akad. Van Wetenschappen. Ser. BNPhys. Sci.* **1982**, *85*, 381–400. Available online: https://pure.tue.nl/ws/files/2805365/344354714903364.pdf (accessed on 15 February 2022).
251. Burdeaux, D. API 15S Spoolable Composite Pipeline Systems. In *Pennsylvania Public Utilities Commission Pipeline Safety Seminar*; Pennsylvania State College: Pennsylvania, PA, USA, 2014; Available online: http://www.puc.state.pa.us/transport/gassafe/pdf/Gas_Safety_Seminar_2014-PPT-Flexsteel.pdf (accessed on 15 February 2022).
252. Venkatesan, R.; Dwarakadasa, E.S.; Ravindran, M. Study on behavior of carbon fiber-reinforced composite for deep sea applications. In Proceedings of the Offshore Technology Conference, Houston, TX, USA, 6–9 May 2002. [CrossRef]
253. Balazs, G.L.; Borosnyoi, A. Long-Term Behavior of FRP. In Proceedings of the International Workshop on Composites in Construction, Capri, Italy, 20–21 July 2001. [CrossRef]
254. Ross, G.R.; Ochoa, O.O. Environmental Effects on Unsymmetric Composite Laminates. *J. Thermoplast. Compos. Mater.* **1991**, *4*, 266–284. [CrossRef]
255. Ross, G.R.; Ochoa, O.O. Micromechanical Analysis of Hybrid Composites. *J. Reinf. Plast. Compos.* **1996**, *15*, 828–836. [CrossRef]
256. Ye, J. *Laminated Composite Plates and Shells: 3D Modelling*; Springer: London, UK, 2003.
257. Ye, J.; Cai, H.; Liu, L.; Zhai, Z.; Amaechi, C.V.; Wang, Y.; Wan, L.; Yang, D.; Chen, X.; Ye, J. Microscale intrinsic properties of hybrid unidirectional/woven composite laminates: Part I: Experimental tests. *Compos. Struct.* **2021**, *262*, 113369. [CrossRef]
258. Ye, J.; Wang, Y.; Wan, L.; Li, Z.; Saafi, M.; Jia, F.; Huang, B.; Ye, J. Failure analysis of fiber-reinforced composites subjected to coupled thermo-mechanical loading. *Compos. Struct.* **2020**, *235*, 111756. [CrossRef]
259. Bismarck, A.; Hofmeier, M.; Dörner, G. Effect of hot water immersion on the performance of carbon reinforced unidirectional poly(ether ether ketone) (PEEK) composites: Stress rupture under end-loaded bending. *Compos. Part A Appl. Sci. Manuf.* **2007**, *38*, 407–426. [CrossRef]
260. d'Almeida, J.R.M. Fibre-matrix interface and natural fibre composites. *J. Mater. Sci. Lett.* **1991**, *10*, 578–580. [CrossRef]
261. d'Almeida, A.L.F.S.; Barreto, D.W.; Calado, V.; d'Almeida, J.R. Thermal analysis of less common lignocellulose fibers. *J. Therm. Anal. Calorim.* **2008**, *91*, 405–408. [CrossRef]
262. Ellyin, F.; Maser, R.V. Environmental effects on the mechanical properties of glass-fiber epoxy composite tubular specimens. *Compos. Sci. Technol.* **2004**, *64*, 1863–1874. [CrossRef]
263. Huang, G.; Sun, H.Q. Effect of water absorption on the mechanical properties of glass/polyester composites. *Mater. Des.* **2007**, *28*, 1647–1650. [CrossRef]
264. Aktas, A.; Uzun, I. Sea water effect on pinned-joint glass fibre composite materials. *Compos. Struct.* **2008**, *85*, 59–63. [CrossRef]
265. Afshari, M.; Sikkema, D.J.; Lee, K.; Bogle, M. High Performance Fibers Based on Rigid and Flexible Polymers. *Polymer Reviews* **2008**, *48*, 230–274. [CrossRef]
266. Rakshit, T.; Atluri, S.; Dalton, C. VIV of a Composite Riser at Moderate Reynolds Number Using CFD. *ASME J. Offshore Mech. Arct. Eng.* **2008**, *130*, 011009. [CrossRef]
267. DNV. *Design of Titanium Risers: Recommended Practice*; DNV-RP-F201; Det Norske Veritas: Oslo, Norway, 2002.
268. DNV. *Environmental Conditions and Environmental Loads: Recommended Practice*; DNV-RP-C205; Det Norske Veritas (DNV): Oslo, Norway, 2007.
269. DNV. *Composite Risers: Recommended Practice*; DNV-RP-F202; Det Norske Veritas: Oslo, Norway, 2010.
270. DNV. *Dynamic Risers: Recommended Practice*; DNV-OS-F201; Det Norske Veritas: Oslo, Norway, 2010.
271. DNV. *Composite Components: Recommended Practice*; DNV-OS-C501; Det Norske Veritas (DNV): Oslo, Norway, 2013.
272. DNVGL. *Recommended Practice: Thermoplastic Composite Pipes*; DNVGL-RP-F119; Det Norske Veritas & Germanischer Lloyd (DNVGL): Oslo, Norway, 2015; Available online: https://www.dnvgl.com/oilgas/download/dnvgl-st-f119-thermoplastic-composite-pipes.html (accessed on 15 February 2022).
273. DNV. *Riser Fatigue: Recommended Practice*; DNV-RP-F204; Det Norske Veritas (DNV): Oslo, Norway, 2010.
274. DNV. *Offshore Classification Projects—Testing and Commissioning: Class Guideline*; DNVGL-CG-0170; Det Norske Veritas (DNV): Oslo, Norway, 2015.
275. DNVGL. *Recommended Practice: Technology Qualification*; DNVGL-RP-A203; Det Norske Veritas (DNVGL): Oslo, Norway, 2019.
276. Lamacchia, D.; Choudhary, S.; Mockel, M.; Ulechia, F. *Thermoplastic Composite Pipe (TCP) Market Study*; LVTQS Doc. No: [LVTQS-BD-RPT-0001-0]; Leviticus Subsea: Houston, TX, USA, 2017.
277. OffshoreMagazine. 2015 Deepwater Production Riser Systems & Components. Offshore Magazine, Poster No. 118 Issue April 2015. Available online: https://cdn.offshore-mag.com/files/base/ebm/os/document/2019/06/0415_RiserPoster_032315_Final.5cf68e0c62dee.pdf (accessed on 15 February 2022).

278. Elanchezhian, C.; Ramnath, B.V.; Hemalatha, J. Mechanical behaviour of glass and carbon fibre reinforced composites at varying strain rates and temperatures. In Proceedings of the 3rd International Conference on Materials Processing and Characterisation (ICMPC 2014), Hyderabad, India, 8–9 March 2014; pp. 1405–1418. [CrossRef]
279. Christine, D.M. Comparison of Carbon Fiber, Kevlar® (Aramid) and E Glass Used in Composites for Boatbuilding. ChristineDeMerchant. 2021. Available online: https://www.christinedemerchant.com/carbon-kevlar-glass-comparison.html (accessed on 15 February 2022).
280. Grant, T.S.; Bradley, W.L. In-Situ observations in SEM of degradation of graphite/epoxy composite materials due to sea water immersion. *J. Compos. Mater.* **1995**, *29*, 852–867. [CrossRef]
281. Hasselmann, K.; Barnett, T.P.; Bouws, E.; Carlson, H.; Cartwright, D.E.; Enke, K.; Ewing, J.A.; Gienapp, H.; Hasselmann, D.E.; Kruseman, P.; et al. Measurements of wind-wave growth and swell decay during the Joint North Sea Wave Project (JONSWAP). *Ergnzungsheft Zur Dtsch. Hydrogr. Z. Reihe* **1973**, *8*, 1–95.
282. Torres, L.; Verde, C.; Vázquez-Hernández, O. Parameter identification of marine risers using Kalman-like observers. *Ocean Eng.* **2015**, *93*, 84–97. [CrossRef]
283. Sarpkaya, T. A critical review of the intrinsic nature of vortex-induced vibrations. *J. Fluids Struct.* **2004**, *19*, 389–447. [CrossRef]
284. Morison, J.R.; Johnson, J.W.; Schaaf, S.A. The Force Exerted by Surface Waves on Piles. *J. Pet. Technol.* **1950**, *2*, 149–154. [CrossRef]
285. Pierson, W.J.; Moskowitz, L. A proposed spectral form for fully developed wind seas based on the similarity theory of S. A. Kitaigorodskii. *J. Geophys. Res.* **1964**, *69*, 5181–5190. [CrossRef]
286. Rivero-Angeles, F.J.; Vázquez-Hernández, A.O.; Sagrilo, L.V.S. Spectral analysis of simulated acceleration records of deepwater SCR for identification of modal parameters. *Ocean. Eng.* **2013**, *58*, 78–87. [CrossRef]
287. Rustard, A.M. Modeling and Control of Top Tensioned Risers. Ph.D. Thesis, Department of Marien Technology, Norwegian University of Science and Technology (NTNU), Trondheim, Norway, 2007. Available online: http://hdl.handle.net/11250/237625 (accessed on 15 February 2022).
288. Sanaati, B.; Kato, N. Vortex-induced vibration (VIV) dynamics of a tensioned flexible cylinder subjected to uniform cross-flow. *J. Mar. Sci. Technol.* **2013**, *18*, 247–261. [CrossRef]
289. Wiercigroch, M.; Keber, M. Dynamics of a vertical riser with weak structural nonlinearity excited by wakes. *J. Sound Vib.* **2008**, *315*, 685–699. [CrossRef]
290. Young, R.D.; Fowler, J.R.; Fisher, E.A.; Luke, R.R. Dynamic Analysis as an Aid to the Design. *J. Press. Vessel. Technol.* **1978**, *100*, 200–205. [CrossRef]
291. Liu, K.; Chen, G.M.; Chang, Y.J.; Zhu, B.R.; Liu, X.Q.; Han, B.B. Nonlinear dynamic analysis and fatigue damage assessment for a deepwater test string subjected to random loads. *J. Pet. Sci.* **2016**, *13*, 126–134. [CrossRef]
292. Brouwers, J.J.H.; Verbeek, P.H.J. Expected fatigue damage and expected extreme response for Morison-type wave loading. *Appl. Ocean. Res.* **1982**, *5*, 129–133. [CrossRef]
293. Huang, C. Structural Health Monitoring System for Deepwater Risers with Vortex-Induced Vibration: Nonlinear Modeling, Blind Identification Fatigue/Damage Estimation and Local Monitoring Using Magnetic Flux Leakage. Ph.D. Thesis, Final Report of RPSEA Project, 07121-DW1603D. Rice University, 2012. Available online: http://citeseerx.ist.psu.edu/viewdoc/download?doi=10.1.1.259.7046&rep=rep1&type=pdf (accessed on 15 February 2022).
294. Deka, D.; Hays, P.R.; Raghavan, K.; Campbell, M.; ASME. Straked riser design with VIVA. In Proceedings of the ASME 29th International Conference on Ocean, Offshore and Arctic Engineering, Shanghai, China, 6–11 June 2010; Volume 6, pp. 695–705.
295. Baxter, C.; Pillai, S.; Hutt, G. Advances in Titanium Risers for FPSO's. In Proceedings of the Offshore Technology Conference, Houston, TX, USA, 5–8 May 1997. [CrossRef]
296. Sauer, C.W.; Sexton, J.B.; Sokoll, R.E.; Thornton, J.M. Heidrun TLP Titanium Drilling Riser System. In Proceedings of the Offshore Technology Conference, Houston, TX, USA, 6–9 May 1996. [CrossRef]
297. Schutz, R.W. Guidelines for Successful Integration of Titanium Alloy Components into Subsea Production Systems. In Proceedings of the CORROSION 2001, Paper Number: NACE-01003, Houston, TX, USA, 11–16 March 2001; Available online: https://onepetro.org/NACECORR/proceedings-abstract/CORR01/All-CORR01/NACE-01003/112239 (accessed on 15 February 2022).
298. Sevillano, L.C.; Morooka, C.K.; Mendes, J.R.P.; Miura, K.; ASME. Drilling riser analysis during installation of a wellhead equipment. In Proceedings of the ASME 32nd International Conference on Ocean, Offshore and Arctic Engineering, Nantes, France, 9–14 June 2013; Volume 4A: Pipeline and Riser Technology, V04AT04A040. ASME: New York, NY, USA, 2013. [CrossRef]
299. Bai, Y.; Tang, G.; Wang, P.; Xiong, H. Mechanical behavior of pipe reinforced by steel wires under external pressure. *J. Reinf. Plast. Compos.* **2016**, *35*, 398–407. [CrossRef]
300. Chen, X.H.; Yu, T.P.; Wang, S.S. *Advanced Analytical Models and Design Methodology Developments for Ultra-Deepwater Composite Risers*; CEAC-TR-04-0106; Research Partnership to Secure Energy for America (RPSEA): Houston, TX, USA, 2004.
301. Chen, Y.; Tan, L.B.; Jaiman, R.K.; Sun, X.; Tay, T.E.; Tan, V.B.C. Global–Local analysis of a full-scale composite riser during vortex-induced vibration. In Proceedings of the ASME 2013 32nd International Conference on Ocean, Offshore and Arctic Engineering, Nantes, France, 9–14 June 2013; Volume 7: CFD and VIV, V007T08A084. ASME: New York, NY, USA, 2013. [CrossRef]
302. Huybrechts, D.G. Composite riser lifetime prediction. In Proceedings of the Offshore Technology Conference, Houston, TX, USA, 6–9 May 2002. [CrossRef]

303. Melot, D. Present and Future Composites Requirements for the Offshore Oil and Gas Industry. In *Durability of Composites in a Marine Environment 2. Solid Mechanics and Its Applications*; Davies, P., Rajapakse, Y., Eds.; Springer: Cham, Switzerland, 2018; Volume 245. [CrossRef]
304. Lindefjeld, O.; Murali, J.; Martinussen, E.; Wiken, H.; Paulshus, B.; Kristiansen, R. Composite Research: Composite Tethers and Risers in Deepwater Field Development (First Joint Successfully Installed). Offshore Magazine, Issue: 1 September 2001. Available online: https://www.offshore-mag.com/deepwater/article/16758680/composite-research-composite-tethers-and-risers-in-deepwater-field-development (accessed on 15 February 2022).
305. Melve, B.; Fjellheim, P.; Raudeberg, S.; Tanem, S.A. First Offshore Composite Riser Joint Proven on Heidrun. Offshore Magazine, Issue: 1 March 2008. 2001. Available online: https://www.offshore-mag.com/business-briefs/equipment-engineering/article/16761859/first-offshore-composite-riser-joint-proven-on-heidrun (accessed on 15 February 2022).
306. Gibson, A.G. Chapter 11—Composites in Offshore Structures. In *Composite Materials in Maritime Structures*, 1st ed.; Shenoi, R.A., Wellicome, J.F., Eds.; Volume 2: Practical Considerations, Cambridge Ocean Technology Series; Cambridge University Press: Cambridge, UK, 1993. [CrossRef]
307. Huang, Z.; Zhang, W.; Qian, X.; Su, Z.; Pham, D.-C.; Sridhar, N. Fatigue behaviour and life prediction of filament wound CFRP pipes based on coupon tests. *Mar. Struct.* **2020**, *72*, 102756. [CrossRef]
308. Huang, Z.; Qian, X.; Su, Z.; Pham, D.-C.; Sridhar, N. Experimental investigation and damage simulation of large-scaled filament wound composite pipes. *Compos. Part B Eng.* **2020**, *184*, 107639. [CrossRef]
309. Aboshio, A.; Uche, A.O.; Akagwu, P.; Ye, J. Reliability-based design assessment of offshore inflatable barrier structures made of fibre-reinforced composites. *Ocean Eng.* **2021**, *233*, 109016. [CrossRef]
310. Wreden, C.; Macfarlan, K.; Giroux, R.; LoGiudice, M. Reliability Is Key to Developing Deepwater RCPC. EPMagazine. 2014. Available online: http://www.epmag.com/reliability-key-developing-deepwater-rcpc-712796#p=full (accessed on 15 February 2022).
311. Kang, Z.; Jia, L.; Sun, L.; Liang, W. Design and analysis of typical buoyancy tank riser tensioner systems. *J. Marine. Sci. Appl.* **2012**, *11*, 351–360. [CrossRef]
312. Duan, M.; Wang, Y.; Yue, Z.; Estefen, S.; Yang, X. Dynamics of Risers for Design Check against Earthquakes. *Pet. Sci.* **2010**, *7*, 272–282. [CrossRef]
313. DNV. *Riser Interference: Recommended Practice; DNV-RP-F203*; Det Norske Veritas (DNV): Oslo, Norway, 2010; Available online: https://rules.dnv.com/docs/pdf/dnvpm/codes/docs/2009-04/RP-F203.pdf (accessed on 15 February 2022).
314. API. *Design of Risers for Floating Production Systems (FPSs) and Tension-Leg Platforms (TLPs)*; Errata of First Edition; API Recommended Practice API-RP-2RD; American Petroleum Institute: Washington, DC, USA, 2009.
315. ABS. *Subsea Riser Systems: Guide for Building and Classing*; American Bureau of Shipping (ABS): Houston, TX, USA, 2017; Available online: https://ww2.eagle.org/content/dam/eagle/rules-and-guides/current/offshore/123_guide_building_and_classing_subsea_riser_systems_2017/Riser_Guide_e-Mar18.pdf (accessed on 15 February 2022).
316. Fergestad, D.; Lotveit, S.A. *Handbook on Design and Operation of Flexible Pipes*; Document Reference: OC2017 A-001; NTNU, 4Subsea and SINTEF Ocean: Trondheim, Norway, 2017; ISBN 978-82-7174-285-0. Available online: https://www.4subsea.com/wp-content/uploads/2017/07/Handbook-2017_Flexible-pipes_4Subsea-SINTEF-NTNU_lo-res.pdf (accessed on 15 February 2022).
317. Berge, S.; Olufsen, A. (Eds.) *Handbook on Design and Operation of Flexible Pipes*; SINTEF Report STF70 A92006; SINTEF: Trondheim, Norway, 1992.
318. Frieze, P.A.; Barnes, F.J. Composite Materials for Offshore Application—New Data and Practice. In Proceedings of the Offshore Technology Conference, Houston, TX, USA, 6–9 May 1996. [CrossRef]
319. Melve, B.; Nedrelid, K.; Tanem, S.A.; Kroknes, L.; Myrmel, S.H. Composite Drilling Riser on Heidrun: A Decade in Operational Experience. In Proceedings of the ASME 2012 31st International Conference on Ocean, Offshore and Arctic Engineering, Rio de Janeiro, Brazil, 1–6 July 2012; Volume 6: Materials Technology, Polar and Arctic Sciences and Technology, Petroleum Technology Symposium. ASME: New York, NY, USA, 2012; pp. 229–234. [CrossRef]
320. Chetwynd, G.; Hatton, S. *New Composite Contenders Eye Flexpipe Ranks*; Upstream Technology: New Brighton, MN, USA, 2013; pp. 32–35. Available online: https://www.upstreamonline.com/hc-technology/new-composite-contenders-eye-flexpipe-ranks/1-1-991824 (accessed on 15 February 2022).
321. Abdul Majid, M.S.B. Behaviour of Composite Pipes under Multi-Axial Stress. Ph.D. Thesis, School of Mechanical and Systems Engineering. Newcastle University, Newcastle, UK, 2011. Available online: http://theses.ncl.ac.uk/jspui/handle/10443/1351 (accessed on 3 March 2022).
322. Elhajjar, R.; Shams, S.S.; Kemeny, G.J.; Stuessy, G. A Hybrid Numerical and Imaging Approach for Characterizing Defects in Composite Structures. *Compos. Part A Appl. Sci. Manuf.* **2016**, *81*, 98–104. [CrossRef]
323. Ellyin, F.; Carroll, M.; Kujawski, D.; Chiu, A.S. The behavior of multidirectional filament wound fibreglass/epoxy tubulars under biaxial loading. *Compos. Part A Appl. Sci. Manuf.* **1997**, *28*, 781–790. [CrossRef]
324. Ellyin, F.; Martens, M. Biaxial fatigue behaviour of a multidirectional filament-wound glass-fiber/epoxy pipe. *Compos. Sci. Technol.* **2001**, *61*, 491–502. [CrossRef]
325. Brazier, L.G. On the Flexure of Thin Cylindrical Shells and Other "Thin" Sections. *Proc. R. Soc. London. Ser. A Contain. Pap. A Math. Phys. Character* **1927**, *116*, 104–114. [CrossRef]

326. Echtermeyer, A.T.; Sund, O.E.; Ronold, K.O.; Moslemiane, R.; Hassel, P.A. A new Recommended Practice for Thermoplastic Composite Pipes. In Proceedings of the 21st International Conference on Composite Materials, Xi'an, China, 20–25 August 2017; Available online: http://iccm-central.org/Proceedings/ICCM21proceedings/papers/3393.pdf (accessed on 15 February 2022).
327. Alexander, C.; Ochoa, O. Extending onshore pipeline repair to offshore steel risers with carbon–fiber reinforced composites. *Compos. Struct.* **2010**, *92*, 499–507. [CrossRef]
328. Rodriguez, D.E.; Ochoa, O.O. Flexural response of spoolable composite tubulars: An integrated experimental and computational assessment. *Compos. Sci. Technol.* **2004**, *64*, 2075–2088. [CrossRef]
329. Lindsey, C.G.; Masudi, H. Tensile fatigue testing of composite tubes in seawater. In Proceedings of the ASME Energy Sources Technology Conference, Houston, TX, USA, 3–6 May 1999.
330. Soden, P.D.; Kitching, R.; Tse, P.C.; Tsavalas, Y.; Hinton, M.J. Influence of winding angle on the strength and deformation of filament-wound composite tubes subjected to uniaxial and biaxial loads. *Compos. Sci. Technol.* **1993**, *46*, 363–378. [CrossRef]
331. Hirsch, C. *Numerical Computation of Internal and External Flows: Fundamentals of Computational Fluid Dynamics*, 2nd ed.; Butterworth-Heinemann: Oxford, UK, 2007; Volume 1.
332. Li, H.; Yang, H.; Yan, J.; Zhan, M. Numerical study on deformation behaviors of thin-walled tube NC bending with large diameter and small bending radius. *Comput. Mater. Sci.* **2009**, *45*, 921–934. [CrossRef]
333. Callister, W.D.; Rethwisch, D.G. *Material Science and Engineering: An Introduction*, 7th ed.; John Wiley & Sons, Inc.: Hoboken, NJ, USA, 2007.
334. Connaire, A.; O'Brien, P.; Harte, A.; O'Connor, A. Advancements in subsea riser analysis using quasi-rotations and the Newton-Raphson method. *Int. J. Non-Linear Mech.* **2015**, *70*, 47–62. [CrossRef]
335. Lindsey, C.G.; Masudi, H. Stress analysis of composite tubes under tensile fatigue loading in a simulated seawater environment. In Proceedings of the ASME 2002 Engineering Technology Conference on Energy, Parts A and B, Houston, TX, USA, 4–5 February 2002; pp. 1041–1045. [CrossRef]
336. Davies, P.; Choqueuse, D.; Mazeas, F. Composites Underwater. *Prog. Durab. Anal. Compos. Syst.* **1998**, *97*, 19–24.
337. Choqueuse, D.; Davies, P. Durability of Composite Materials for Underwater Applications. In *Durability of Composites in a Marine Environment (Solid Mechanics and Its Applications)*; Davies, P., Rajapakse, Y., Eds.; Springer: Dordrecht, The Netherlands, 2014; Volume 208. [CrossRef]
338. Summerscales, J. Durability of Composites in the Marine Environment. In *Durability of Composites in a Marine Environment (Solid Mechanics and Its Applications)*; Davies, P., Rajapakse, Y., Eds.; Springer: Dordrecht, The Netherlands, 2014; Volume 208. [CrossRef]
339. Reifsnider, K.L.; Dillard, D.A.; Carbon, A.H. Progress in durability analysis of composite systems. In Proceedings of the Third International Conference on Progress in Durability Analysis of Composite Systems, Blacksburg, VA, USA, 14-17 September 1997; Available online: https://apps.dtic.mil/sti/pdfs/ADA359548.pdf (accessed on 15 February 2022).
340. Lo, K.M.; Williams, J.G.; Karayaka, M.; Salama, M. *Progress, Challenges and Opportunities in the Application of Composite Systems*; Report No: CEAC-TR-01-0101; University of Houston: Houston, TX, USA, 2001; pp. 1–17. Available online: https://bsee_prod.opengov.ibmcloud.com/sites/bsee.gov/files/research-reports/230-ap.pdf (accessed on 15 February 2022).
341. Razavi Setvati, M.; Mustaffa, Z.; Shafiq, N.; Syed, Z.I. A Review on Composite Materials for Offshore Structures. In Proceedings of the ASME 2014 33rd International Conference on Ocean, Offshore and Arctic Engineering, San Francisco, CA, USA, 8–13 June 2014; Volume 5: Materials Technology; Petroleum Technology. [CrossRef]
342. Ochoa, O.O.; Alexander, C. Hybrid Composite Repair for Offshore Risers. In Proceedings of the 17th International Conference of Composite Materials (ICCM17), Edinburgh, UK, 27–31 July 2009; pp. 1–7. Available online: https://www.iccm-central.org/Proceedings/ICCM17proceedings/Themes/Industry/OFFSHORE%20APPLICATIONS/A5%206%20Ochoa.pdf (accessed on 15 February 2022).
343. Gautum, M.; Katnam, K.B.; Potluri, P.; Jha, V.; Latto, J.; Dodds, N. Hybrid composite tensile armour wires in flexible risers: A multi-scale model. *Compos. Struct.* **2017**, *162*, 13–27. [CrossRef]
344. Gautum, M. Hybrid Composite Wires for Tensile Armour in Flexible Risers. Ph.D. Thesis, School of Materials, The University of Manchester, Manchester, UK, 2016. Available online: https://www.research.manchester.ac.uk/portal/files/60827532/FULL_TEXT.PDF (accessed on 15 February 2022).
345. Sundstrom, K.A. Stress Analysis of a Hybrid Composite Drilling Riser. Master's Thesis, Texas A&M University, College Station, TX, USA, 1996. Available online: https://oaktrust.library.tamu.edu/handle/1969.1/ETD-TAMU-1996-THESIS-S87. (accessed on 15 February 2022).
346. Loureiro, W.C., Jr.; Sobreira, R.G.; Buckley, A.L. Hybrid Composite Flexible Risers in Free Hanging Catenary Configuration and Flowlines for UDW Projects. In Proceedings of the Offshore Technology Conference Brasil, Rio de Janeiro, Brazil, 29–31 October 2019. [CrossRef]
347. McGeorge, D.; Sødahl, N.; Moslemian, R.; Hørte, T. Hybrid and Composite Risers for Deep Waters and Aggressive Reservoirs. In Proceedings of the Offshore Mediterranean Conference and Exhibition, Ravenna, Italy, 27–29 March 2019; Available online: https://onepetro.org/OMCONF/proceedings-abstract/OMC19/All-OMC19/OMC-2019-1160/1950 (accessed on 15 February 2022).
348. van Onna, M.; Lyon, J. Installation of World's 1st Subsea Thermoplastic Composite Pipe Jumper on Alder. In Proceedings of the Subsea Expo 2017 Conference, Aberdeen, UK, 1–3 February 2017; Available online: https://www.globalunderwaterhub.com/documents/presentations/martin%20van%20onna%20-%20fields%20of%20the%20future%20-%20airborne.pdf (accessed on 15 February 2022).

349. van Onna, M. A new thermoplastic composite riser for deeperwater application. In Proceedings of the Subsea Expo 2011 Conference, Aberdeen, UK, 2011; Available online: https://www.yumpu.com/en/document/read/26877601/a-new-thermoplastic-composite-riser-for-deepwater-subsea-uk (accessed on 15 February 2022).
350. Smits, A.; Neto, T.B.; de Boer, H. Thermoplastic Composite Riser Development for Ultradeep Water. In Proceedings of the Offshore Technology Conference, Houston, TX, USA, 30 April–3 May 2018. [CrossRef]

Article

Effect of MWCNT Anchoring to Para-Aramid Fiber Surface on the Thermal, Mechanical, and Impact Properties of Para-Aramid Fabric-Reinforced Vinyl Ester Composites

Jinsil Cheon [1,2] and Donghwan Cho [1,*]

[1] Department of Polymer Science and Engineering, Kumoh National Institute of Technology, Gumi 39177, Gyeong-buk, Republic of Korea; jscheon@kotmi.re.kr
[2] Composites Convergence Research Center, Korea Textile Machinery Convergence Research Institute, Gyeongsan 38542, Gyeong-buk, Republic of Korea
* Correspondence: dcho@kumoh.ac.kr

Abstract: In the present work, para-aramid fabrics (p-AF) were physically modified via an anchoring process of 0.05 wt% MWCNT to the aramid fiber surfaces by coating the MWCNT/phenolic/methanol mixture on p-AF, and then by thermally curing phenolic resin of 0.01 wt%. Para-aramid fabric-reinforced vinyl ester (p-AF/VE) composites were fabricated using p-AF/VE prepregs by compression molding. The effect of MWCNT anchoring on the thermo-dimensional, thermal deflection resistant, dynamic mechanical, mechanical, and impact properties and the energy absorption behavior of p-AF/VE composites was extensively investigated in terms of coefficient of linear thermal expansion, heat deflection temperature, storage modulus, tan δ, tensile, flexural, and Izod impact properties and a drop-weight impact response. The results well agreed with each other, supporting the improved properties of p-AF/VE composites, which were attributed to the effect of MWCNT anchoring performed on the aramid fiber surfaces.

Keywords: para-aramid fiber; carbon nanotube; anchoring; modification; vinyl ester; composite; properties

Citation: Cheon, J.; Cho, D. Effect of MWCNT Anchoring to Para-Aramid Fiber Surface on the Thermal, Mechanical, and Impact Properties of Para-Aramid Fabric-Reinforced Vinyl Ester Composites. *J. Compos. Sci.* **2023**, *7*, 416. https://doi.org/10.3390/jcs7100416

Academic Editor: Jiadeng Zhu

Received: 8 September 2023
Revised: 25 September 2023
Accepted: 3 October 2023
Published: 6 October 2023

Copyright: © 2023 by the authors. Licensee MDPI, Basel, Switzerland. This article is an open access article distributed under the terms and conditions of the Creative Commons Attribution (CC BY) license (https://creativecommons.org/licenses/by/4.0/).

1. Introduction

In the field of soft body armors, research challenges have included reducing weight, improving flexibility, and enhancing impact resistance, simultaneously [1–3]. Para-aramid (referred to as p-aramid hereinafter) fiber has been frequently used to protect human beings from physical risks under dangerous military and civilian circumstances, because polymeric chains consisting of the fiber form rigid crystals and are aligned with the fiber direction during fiber spinning [3,4]. Substantially, the surface of p-aramid fiber contains many hydrophilic functional groups, such as amide and hydroxyl groups, which can form a huge number of hydrogen bonds between the polymer chains. Owing to the intrinsic structure of p-aramid fiber, it can resist severe-impact environments without fiber breakages [5].

For this reason, woven fabrics made with p-aramid fibers have been used for protective clothing. According to the energy-dissipating mechanism, the friction between the inter-yarns in the woven fabric with a plain pattern plays a significant role in the ballistic impact response in both direct and indirect manners [6–10]. A direct effect of the inter-yarn friction is that the energy dissipation is increased when the yarns consisting of the fabric begin to displace one another, exhibiting sliding, pulling-out, or re-orientation of the individual fibers. An indirect effect of the inter-yarn friction is that it may influence external forces, which can be transferred and redistributed to the neighboring yarns [11].

P-aramid fabric-reinforced composites have been considered as key materials in many civilian and military applications due to their excellent specific strength, specific stiffness, and lightness in comparison to conventional materials, such as metals and alloys. They exhibit excellent impact resistance and elasticity because of the combination of the viscoelasticity of the polymer matrix and the impact toughness of p-aramid fabrics consisting of the composites [12].

In general, soft armor made with p-aramid fabric requires weak fiber–matrix interfacial adhesion because the friction between the fabric layers critically influences the anti-bulletproofing performance. However, when p-aramid fiber-reinforced polymer composites are used for ballistic protection in the plate form, as in hard armors, the strong interfacial bonding between the fiber and the polymer matrix is important because the composite may experience maximum deformation and can absorb the highest energy upon impact. Therefore, in the case of hard armors, thermosetting polymer matrices have advantages over thermoplastic polymer matrices with low stiffness and high deformation [4].

For the past years, many experimental and theoretical studies have been carried out to understand the material's response and the penetration failure mechanism as well as the energy absorption upon ballistic impact [12–18]. Pandya et al. [17] and Sarasini et al. [18] reported on the ballistic impact behavior of hybrid epoxy composites reinforced with basalt and p-aramid fabrics for hard body armors. Each fabric was laminated with an alternating sequence to investigate the effect of each fabric layer on the composite performance in terms of impact energy absorption capability and enhanced damage tolerance. Davidovitz et al. [19] studied the failure mode and fracture mechanism of Kevlar/epoxy composites under flexural deformation. The failure mode was described in terms of tensile failure and delamination. The tensile failure of the composite was explained by fiber splitting, fiber pull-out, delamination, fiber bending, tearing-off of the fiber skin, and shearing of the individual filaments. Wang et al. [20] studied the crushing behaviors and mechanisms of composite thin-walled structures under quasi-static compression and dynamic impact conditions. They addressed that fiber-reinforced composite structures and materials showed good potential for solving impact problems and energy absorption.

It has been well known that multi-walled carbon nanotubes (MWCNT), which exhibit a high aspect ratio and a large specific volume, are a promising material to improve the mechanical, thermal, electrical, and tribological properties of polymer composites [20–22]. It has been emphasized that a key factor to introduce MWCNT to the polymer matrix is good dispersion. Therefore, many papers have been studied on surface modification of MWCNT to enhance the dispersity and the interfacial bonding between the MWCNT and the polymer matrix, frequently focusing on the chemical functionalization of MWCNT [23–29]. However, chemical functionalization or grafting often requires complicated procedures and large quantities of chemicals.

One of the simplest experimental approaches to incorporate MWCNT into a fiber-reinforced polymer composite material is anchoring [30,31]. Here, the word 'anchoring' refers to a process physically attaching MWCNT nanoparticles on the individual fiber surfaces with the assistance of diluted thermosetting resin at low concentration such that the anchoring process may influence the inter-yarn friction during the pulling-out test of individual yarns. In the case of MWCNT anchoring at high resin concentrations, the fabric drapeability might be lowered to some extent.

Consequently, the objectives of the present work are to physically attach MWCNT to the p-aramid fiber surface by the anchoring process with the assistance of diluted phenolic resin, to fabricate vinyl ester composites reinforced with MWCNT-anchored p-aramid fabrics by a compression molding technique, and finally, to extensively investigate the effect of MWCNT anchoring on the thermo-dimensional, dynamic mechanical, heat deflection temperature, mechanical, and impact properties and the energy absorption behavior of p-aramid fabric-reinforced vinyl ester composites.

2. Materials and Methods
2.1. Materials

P-aramid fabrics (HERACRON®, HT840, Kolon Industries Co., Ltd., Gumi, Republic of Korea) with a plain weave pattern (referred to as p-AF hereinafter) were used as reinforcement in this work. Each fiber yarn has 840 deniers in the warp and weft directions, respectively. The fabric density is 26.7 counts per inch. The areal density is 200 g/cm^2. The commercial p-AF was used 'as-received' without further cleaning and surface treatment. MWCNT (CVD-CM95, Hanhwa Chemical Co., Ltd., Seoul, Republic of Korea) were used 'as-received' without further purification and surface treatment. Resole-type phenolic resin (KRD-HM2, Kolon Industries, Co., Ltd., Gimcheon, Republic of Korea) was used as anchoring agent after being diluted with methanol. The phenolic resin contains the solid contents of about 60%. Methanol (99.95% purity, Daejung Chemicals and Metals, Co., Ltd., Siheung-si, Republic of Korea) was used as diluent of phenolic resin.

2.2. Preparation of MWCNT/Phenolic/Methanol Mixture and MWCNT Anchoring Process

First, prior to preparation of MWCNT-anchored p-AF, MWCNT nanoparticles were well dispersed in methanol. A mixture of MWCNT/phenolic/methanol was prepared by sufficiently mixing with a magnetic stirrer. The MWCNT in the mixture were uniformly dispersed with phenolic resin by ultrasonication. The ultrasonic process was carried out with the frequency of 40 kHz at 50~60 °C for 1 h using an ultrasonic bath (Model Power Sonic 420, 600 W, Hwashin Co., Ltd., Seoul, Republic of Korea). The concentrations of MWCNT and phenolic resin present in the MWCNT/phenolic/methanol mixture were 0.05 wt% and 0.01 wt%, respectively. Phenolic resin of 0.01 wt% was used because it was optimal to have the appropriate fabric drapeability and to cure it with the MWCNT anchored to the fiber surface, as found earlier [29,30]. Accordingly, anchoring process of MWCNT to p-AF was performed with 0.05 wt% MWCNT and 0.01 wt% phenolic resin.

In this work, anchoring of MWCNT to p-AF by using 0.01 wt% phenolic resin means that MWCNT nanoparticles were physically attached on the surface of individual p-aramid fibers by diluted phenolic resin. For MWCNT anchoring, p-AF was immersed in the MWCNT/phenolic/methanol mixture. At this time, the individual fibers in the fabric were surrounded by the MWCNT dispersed in diluted phenolic resin and the MWCNT nanoparticles were physically attached on the fiber surface of p-AF by curing the diluted phenolic resin at 80 °C for 10 min in a convection oven, as shown in Figure 1.

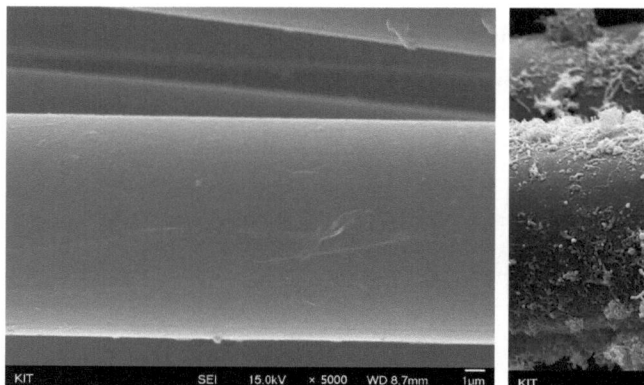

Figure 1. Topography (×5000) showing the pristine p-aramid fiber surface without MWCNT (**left**) and the p-aramid fiber surface with MWCNT anchored by thermally curing dilute phenolic resin (**right**).

2.3. Fabrication of MWCNT-Anchored P-Aramid Fabric/Vinyl Ester Composites

Bisphenol-A modified epoxy acrylate-type vinyl ester resin (Model RF-1001, CCP Composites Korea Co., Ltd., Wanju-gun, Republic of Korea) (referred to as VE hereinafter) was used as a matrix of composites. It contains 45~55 wt% styrene acting as both reactive diluent and curing agent of VE. The resin density is 1.03~1.11 g/cm^3 at 25 °C and the resin viscosity is 250~450 cP. In this work, p-AF-reinforced VE (referred to as p-AF/VE hereinafter) composites were fabricated by a compression molding process.

Two types of peroxides with different molecular sizes were used together as the initiator to cure VE. One was di(4-tert-butylcyclohexyl)peroxydicarbonate (DPDC) and the other was tert-butyl-peroxybenzoate (TBPB). Figure 2 shows the chemical structures of VE, DPDC, and TBPB. The effect of the dual initiators at various concentrations on the VE curing behavior was extensively studied in our earlier report [32]. It was found that 1 pph (parts per hundred) DPDC and 0.75 pph TBPB were optimal to cure VE in the presence of MWCNT. Accordingly, DPDC of 1 pph and TBPB of 0.75 pph were used in the present work.

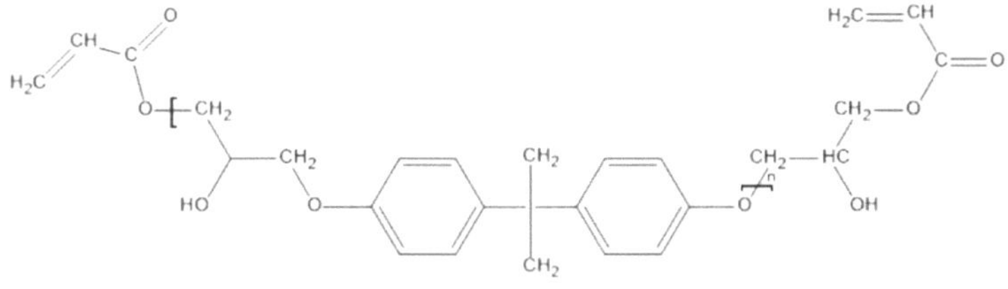

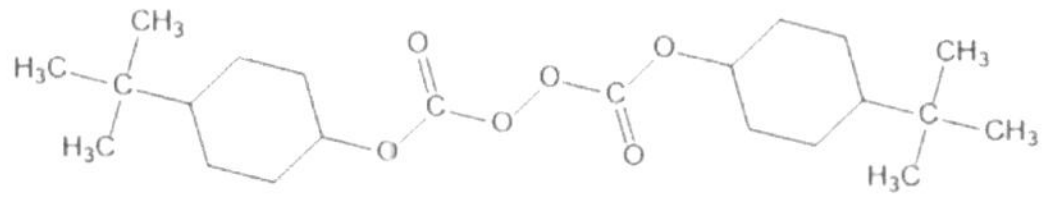

Figure 2. Chemical structures of (**a**) vinyl ester resin, (**b**) *di(4-tert-*butylcyclohexyl)peroxydicarbonate (DPDC), and (**c**) *tert-*butyl-peroxybenzoate (TBPB).

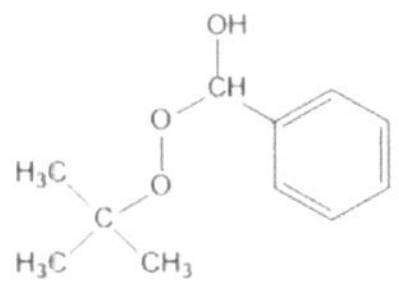

Prior to composite fabrication, p-AF/VE prepregs were prepared. Each prepreg contained excess VE because part of the VE impregnated in p-AF could be squeezed out by the applied pressure upon compression molding. Each prepreg was partially cured at 70 °C for 10 min in a convection oven for B-staging. P-AF/VE prepregs with and without MWCNT anchoring were also prepared for comparison. To prevent possible unintended curing prior to uses, the prepregs were completely sealed and kept in a freezer.

The p-AF/VE prepregs of 14 plies were regularly stacked in a stainless-steel mold and processed using a compression molding machine (GE-122S, Kukje Scien, Daejeon, Republic of Korea). Figure 3 depicts the experimental procedure to prepare p-AF/VE prepregs and to fabricate the composite via prepreg stacking and compression molding. The stacked prepregs in the mold were heated up to 180 °C with the heating rate of 6 °C/min. A pressure of 6.89 MPa was applied from 40 °C. When the mold temperature reached 70 °C, the debulking step was conducted to degas the entrapped air between the prepregs and to evaporate organic volatiles therein. The debulking step was repeated twice until the mold temperature reached 110 °C. The final curing was performed at 180 °C for 10 min. The applied pressure of 6.89 MPa was maintained until the end of compression molding.

Step 1: Pre-impregnation process to prepare of p-AF/VE prepreg

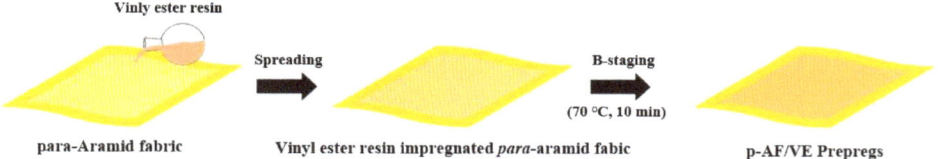

Step 2: Stacking-up of prepregs and charging in the mold

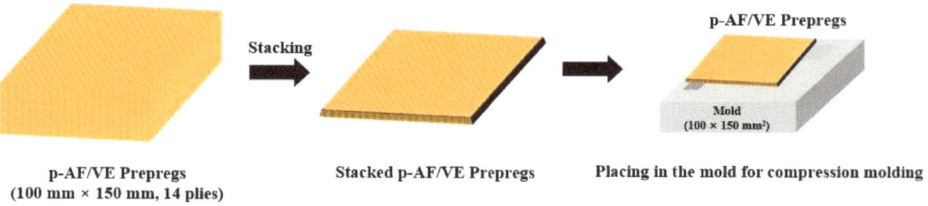

Step 3: Fabrication of p-AF/VE composite by compression molding process

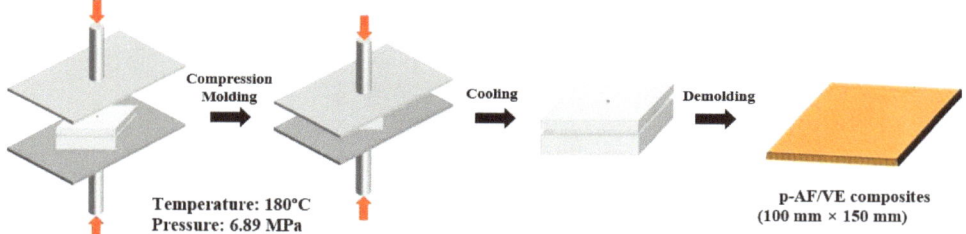

Figure 3. Preparation of p-AF/VE prepregs and fabrication of p-AF/VE composite by compression molding.

The molded composite was cooled down to ambient temperature, and then demolded. Finally, p-AF/VE composites with the dimensions of 150 mm × 100 mm × 3 mm were obtained. P-AF/VE composites with and without MWCNT anchoring were prepared for comparison. The p-AF/VE composite with MWCNT anchoring was designated as MWCNT-p-AF/VE composite. The p-AF/VE composite without MWCNT anchoring was designated as pristine p-AF/VE composite.

2.4. *Microscopic Observation*

A field-emission scanning electron microscope (JSM-6500F, JEOL, Tokyo, Japan) was used to observe the topography of MWCNT anchored on the fiber surfaces. Prior to SEM observations, each sample was uniformly coated with platinum for 3 min by a sputtering method. The acceleration voltage was 15 kV, and the secondary electron image (SEI) mode was used.

2.5. *Thermal Analysis*

Thermomechanical analysis (TMA 2940, TA Instruments, New Castle, DE, USA) was performed to investigate the effect of MWCNT anchoring on the thermo-dimensional stability of p-AF/VE composites. A load of 0.05 N was applied on the specimen using a macro-expansion probe. The thermo-dimensional change was recorded from 30 to 250 °C with the heating rate of 5 °C/min, purging nitrogen gas (50 mL/min).

The effect of MWCNT anchoring on the dynamic mechanical properties of p-AF/VE composites was examined by dynamic mechanical analysis (DMA Q800, TA Instruments, New Castle, DE, USA). The analysis was performed from 30 to 250 °C with the heating rate of 3 °C/min in ambient atmosphere. The dual cantilever mode with a drive clamp and a fixed clamp was used throughout DMA measurement. The oscillation amplitude was 10 µm and the frequency was 1 Hz. The dimensions of composite specimen were 63.5 mm × 12.5 mm × 3 mm.

The heat deflection temperature (HDT) of p-AF/VE composites was measured with a three-point bending mode according to the ASTM D648 standard by using a heat deflection temperature tester (Model 603, Tinius Olsen, Horsham, PA, USA). The dimensions of composite specimen were 127 mm × 12.5 mm × 3 mm. The measurement was performed until the specimen was deflected by 0.254 mm under the bending load of 1.82 MPa. The heating rate of 2 °C/min was used to heat the composite specimen immersed in a silicone oil bath.

2.6. Mechanical Test

A universal testing machine (UTM, AG-50kNX, SHIMADZU, Kyoto, Japan) was used to investigate the effect of MWCNT anchoring on the flexural and tensile properties of p-AF/VE composites. Each specimen was cut to fit the mechanical test requirement by using a low-speed diamond saw. The average values of the flexural and tensile properties were obtained from 10 specimens of each composite.

A three-point flexural test was performed according to the ASTM D790 standard. The dimensions of composite specimen were 127 mm × 12.5 mm × 3 mm. The span-to-depth ratio was 32:1. The load cell of 50 kN and the crosshead speed of 5 mm/min were used. A tensile test was performed according to the ASTM D3039 standard. The dimensions of composite specimen were 140 mm × 12.5 mm × 3 mm. The gage length was 80 mm. The load cell of 50 kN and the crosshead speed of 5 mm/min were used. The flexural and tensile tests were repeated 10 times for each sample. The average values of the flexural modulus, flexural strength, tensile modulus, and tensile strength were obtained from 10 repetitive tests of each composite, respectively.

2.7. Izod Impact Test and Weight-Drop Impact Test

The Izod impact test was carried out according to the ASTM D256 standard using an impact test machine (IT 892, Tinius Olsen, Horsham, PA, USA). The dimensions of composite specimen without notch were 63.5 mm × 12.5 mm × 3 mm. A pendulum energy of 3.17 J was used. The average value of impact strength was obtained from 10 specimens of each composite.

To inspect the effect of MWCNT anchoring on the energy absorption behavior of p-AF/VE composites exposed to high-speed impact energy, drop-weight impact test (CEAST 9350, Instron, Norwood, MA, USA) was performed according to the ASTM D5628 standard. The dimensions of composite specimen were 100 mm × 100 mm × 4 mm. The diameter of the impactor was 20 mm. The height between the specimen and the impactor was 1000 mm. The initial drop-velocity of impactor was 4.41 m/s. The initial impact energy given to each specimen was 205 J.

3. Results and Discussion

Figure 1 exhibits SEM topography of the pristine p-aramid fiber surface without MWCNT and the p-aramid fiber surface with MWCNT anchored by curing dilute phenolic resin. In comparison, at the same magnification, the pristine p-aramid fiber surface without MWCNT was clear with smoothness. On the other hand, the MWCNT nanoparticles were distributed on the fiber surface, increasing the surface roughness. The anchored MWCNT may play a bridging role in connecting between the fiber and the matrix of p-AF/VE composite. It was thought that such a MWCNT anchoring effect may contribute to the resistance of the composite to the applied external force and energy.

In our previous study [29], yarn pull-out tests were performed at high speed (800 mm/min) with p-AF containing MWCNT anchored by phenolic resin at various concentrations. Consequently, the highest pull-out force was obtained when 0.05 wt% MWCNT and 0.01 wt% phenolic resin in the MWCNT/phenolic/methanol mixture prepared for anchoring were used, indicating the synergetic effect of MWCNT anchored on the fiber surface by thermally cured phenolic resin. It was found that the friction between the individual fibers consisting of p-AF influenced the increase in the force required for pulling out the single yarn. The result gave us some indications that MWCNT anchoring to the p-aramid fiber surface would play a positive role in improving the thermal, mechanical, and impact properties of p-AF/VE composites.

3.1. Thermal Expansion Behavior

Figure 4 displays the thermo-dimensional changes of pristine p-AF/VE and MWCNT-p-AF/VE composites measured by means of TMA. Based on the thermo-dimensional changes in each composite, the coefficients of linear thermal expansion (CLTE) were determined from the slope of each TMA curve in the two temperature ranges of 40~100 °C and 150~250 °C, respectively. The temperature ranges were adapted before and after the temperature showing the drastic dimensional changes between 110~140 °C.

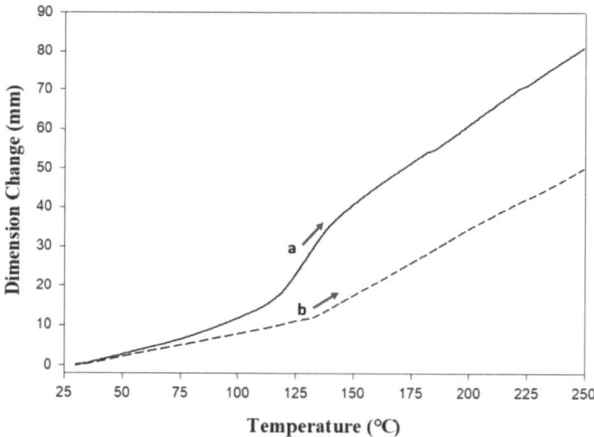

Figure 4. Thermo-dimensional changes as a function of temperature measured with (**a**) pristine p-AF/VE and (**b**) MWCNT-p-AF/VE composites.

In the case of pristine p-AF/VE composite, thermal expansion was obviously found in the temperature range of 110~140 °C. It may be attributed to the glass transition behavior of the VE matrix of p-AF/VE composites. Beyond this temperature range, the CLTE value was highly increased from 57.2 μm/m·°C (between 40 and 100 °C) to 134.0 μm/m·°C (between 150 and 250 °C), as listed in Table 1. As similarly found in pristine p-AF/VE composite, the thermal expansion of MWCNT-p-AF/VE composite was apparently observed between 125~135 °C. It was also ascribed to the glass transition behavior of the VE matrix. The thermal expansion of MWCNT-p-AF/VE composite was smaller than that of pristine p-AF/VE composite, showing the CLTE of 38.2 μm/m·°C (between 40 and 100 °C) and 109.4 μm/m·°C (between 150 and 250 °C). This may be explained considering that the MWCNT anchored to the fiber surface of p-AF contributed to increasing the mechanical interlocking and the interfacial adhesion between the VE matrix and the p-AF due to the increased surface roughness, and consequently contributed to restricting, to some extent, the thermal expansion of the composite.

Table 1. Coefficients of linear thermal expansion (CLTE) of p-AF/VE composites determined from two temperature ranges.

Composite Type	CLTE (µm/m·°C)	
	40~100 °C	150~250 °C
Pristine p-AF/VE	57.2	134.0
MWCNT-p-AF/VE	38.2	109.4

3.2. Dynamic Mechanical Properties

Figure 5 displays the variations of the storage modulus and tan δ of pristine p-AF/VE and MWCNT-p-AF/VE composites as a function of temperature measured by means of DMA. The storage modulus of both pristine p-AF/VE and MWCNT-p-AF/VE composites was distinguishably decreased in the glass transition region between 100~160 °C. In this region, the molecular mobility of the VE matrix was increased because of the weakened molecular interaction with temperature. This resulted in the lowering of the storage modulus. The storage modulus of MWCNT-p-AF/VE composite (curve b) was higher than that of pristine p-AF/VE composite (curve a) in the whole temperature range. The storage modulus of MWCNT-p-AF/VE composite was increased by 11% from 8642 to 9596 MPa at 30 °C. This was attributed to the reinforcing effect of p-AF/VE composite with the increased interfacial adhesion between the p-AF and the VE matrix with the assistance of MWCNT anchoring.

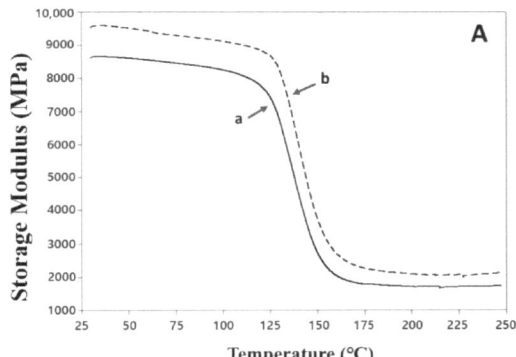
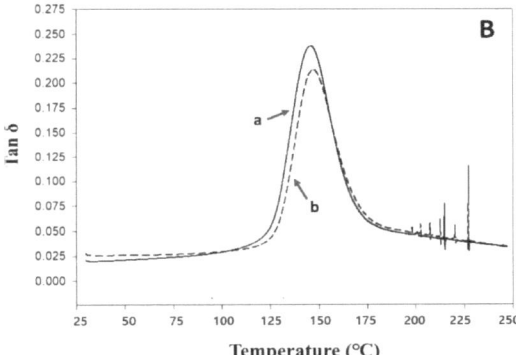

Figure 5. Variations of the (**A**) storage modulus and (**B**) tan δ measured with (**a**) pristine p-AF/VE and (**b**) MWCNT-p-AF/VE composites.

The height of tan δ, which is related to the damping property of a material, was decreased as well. It indicates that MWCNT anchoring to the p-aramid fiber surface played a role in dispersing the external load during DMA measurement. The tan δ peak temperature, which is relevant to the glass transition temperature, was slightly shifted to a high temperature by the anchoring effect.

As a result, it may be said that MWCNT anchoring to p-AF contributed to improving the dynamic mechanical properties as well as the thermo-dimensional stability of p-AF/VE composites, forming the mechanical interlocking at the interface between the p-AF and the VE matrix due to the roughened fiber surface.

3.3. Heat Deflection Temperature

Heat deflection temperature (HDT) can often be measured by applying a three-point flexural load to the specimen until 0.254 mm deflection occurs in the specimen. Accordingly, the HDT of a polymer composite can be affected by its mechanical resistance, strongly depending on the reinforcement. The MWCNT-p-AF/VE composite was further reinforced by MWCNT anchoring. As described above, it was thought that MWCNT anchoring increased the internal friction between the individual yarns consisting of p-AF during the measurement. In addition, the phenolic resin coated and cured on the fiber surface for anchoring somewhat contributed to increasing the stiffness of p-AF.

As listed in Table 2, the HDT of MWCNT-p-AF/VE composite was about 14 °C higher than that of the pristine p-AF/VE counterpart due to the increased mechanical interlocking between the p-AF and the VE matrix by MWCNT anchoring. This indicates that the loads applied to MWCNT-p-AF/VE composite during measurement were effectively distributed to p-AF through the MWCNT bridges connecting between the individual fibers in the matrix. Therefore, it may be described that MWCNT anchoring played a positive role in resisting the deflection of the p-AF/VE composite by heat.

Table 2. A summary of heat deflection temperature, tensile, flexural, and impact properties of pristine p-AF/VE and MWCNT-p-AF/VE composites.

Properties	P-AF/VE Composites	
	Pristine P-AF/VE	MMWCNT-p-AF/VE
Heat Deflection Temperature (°C)	241.8 ± 0.3	255.4 ± 0.3
Tensile Strength (MPa)	266 ± 9	342 ± 10
Tensile Modulus (GPa)	7.5 ± 0.4	10.0 ± 0.5
Flexural Strength (MPa)	112 ± 2	130 ± 4
Flexural Modulus (GPa)	15.1 ± 0.5	16.2 ± 0.6
Izod Impact Strength (J/m)	977 ± 9	1039 ± 8

3.4. Mechanical Properties

The stress–strain curves were measured with pristine p-AF/VE and MWCNT-p-AF/VE composites, respectively, as shown in Figure 6. The initial slope and the highest stress obtained with the MWCNT-p-AF/VE composite were higher than those with the pristine p-AF/VE composite. Substantially, the applied load caused the frictional force between the fiber and the matrix of a fiber-reinforced polymer composite material until the specimen was broken during the tensile test. As seen in Figure 1, the fiber surface of MWCNT-p-AF became roughened with the increased surface area by MWCNT anchoring. Accordingly, it was convinced that the anchored MWCNT nanoparticles existing on the fiber surface may contribute to increasing the frictional force between the fiber and the matrix of the resulting composite.

As a result, the tensile strength and modulus required for deforming the MWCNT-p-AF/VE composite were higher than those required for deforming the pristine p-AF/VE composite until the specimen was broken. The tensile strength was increased by about 29%, and the tensile modulus was increased by about 33% due to the MWCNT anchoring effect, as shown in Table 2. It turns out that the tensile loads applied to the MWCNT-p-AF/VE composite were well transferred from fiber to fiber owing to the increased interfacial bonding between the p-aramid fiber and the VE matrix by MWCNT anchoring.

The area under the stress–strain curve is fundamentally related to the equilibrium toughness absorbing the energy given to a material. It was obvious that the area under the curve obtained with the MWCNT-p-AF/VE composite was greater than that with the p-AF/VE composite. Based on that, it was expected that the dynamic impact toughness of the MWCNT-p-AF/VE composite would be higher than that of the p-AF/VE composite.

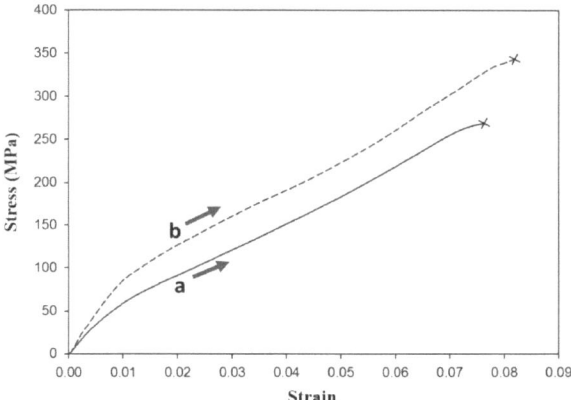

Figure 6. Representative stress–strain curves measured with (**a**) pristine p-AF/VE and (**b**) MWCNT-p-AF/VE composites.

Table 2 compares the flexural strength and modulus of p-AF/VE and MWCNT-p-AF/VE composites. The flexural strength and modulus of the MWCNT-p-AF/VE composite was about 16% and 7% higher than that of the pristine p-AF/VE composite, respectively. Upon flexural deformation, a fiber-reinforced composite material is basically affected by both compressive and tensile stresses at the mid-plane of the specimen. The three-point flexural load normally causes compressive stress through the thickness direction of the specimen, which can be mainly governed by the interfacial bonding between the fiber and the matrix of the composite. Meanwhile, the tensile stress can be generated along with the longitudinal direction of the specimen, being influenced by the frictional force between the fiber and the matrix.

As described above, the anchored MWCNT played a role in increasing the frictional force at the interface between the p-AF and the VE matrix, making the composite stronger and stiffer. In our earlier report [29,30], single-yarn pull-out forces at high speed obtained with p-AF only depended not only on the presence and absence of MWCNT anchored to the fiber surface, but also on the concentration of MWCNT and diluted phenolic resin used for anchoring.

It may be insisted that the MWCNT anchoring to the p-aramid fiber surface performed in this work contributed to increasing the fiber surface roughness, the mechanical interlocking between the fiber and the matrix, and the friction at the fiber–matrix interface without fiber damages. As a result, the tensile and flexural properties of p-AF/VE composites were improved as well.

3.5. Izod Impact Strength

Table 2 also compares the Izod impact strength of the p-AF/VE and MWCNT-p-AF/VE composites. The impact strength of the pristine p-AF/VE composite was increased by about 6% from 977 to 1039 J/m due to the MWCNT anchoring effect. The result indicates that the applied dynamic impact energy can be efficiently absorbed by the MWCNT-p-AF/VE composite, compared to the pristine p-AF/VE composite. It has been found that the impact resistance increases with increasing interfacial bonding between MWCNT and polymer of a polymer composite containing MWCNT [23]. As described above, in the case of the MWCNT-p-AF/VE composite, MWCNT nanoparticles were physically attached to the p-aramid fiber surfaces by thermally curing them with diluted phenolic resin. They increased the interfacial bonding between the aramid fiber and the VE matrix of the composite. Accordingly, the presence of anchored MWCNT was responsible for the increase in the Izod impact strength of the MWCNT-p-AF/VE composite.

3.6. Energy Absorption Behavior by Drop-Weight Impact

Figure 7 displays the results monitored for the variation of the velocity of the impactor as a function of time when the pristine p-AF/VE and MWCNT-p-AF/VE composites were exposed to the drop-weight impact environment, respectively. The initial impactor velocity was 4.41 m/s. The time to reach zero velocity in each composite is relevant to the energy absorption behavior. As shown, the velocity of the impactor was gradually decreased with increasing time, indicating that it was changed from positive to negative after reaching zero velocity. It turns out that the movement of the impactor was changed to the opposite direction relative to the initial direction of the impactor. The time required for reaching zero velocity in the MWCNT-p-AF/VE composite was shorter than that in the pristine p-AF/VE composite.

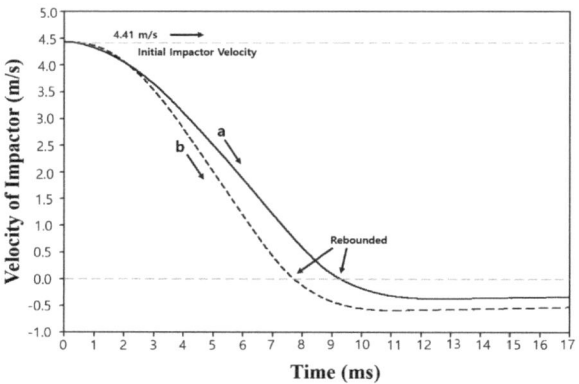

Figure 7. Variation of the velocity of impactor as a function of time occurred during drop-weight impact test: (**a**) pristine p-AF/VE and (**b**) MWCNT-p-AF/VE composites.

This can be explained considering that the anchoring of MWCNT to p-AF contributed to releasing the drop-weight impact energy. In addition, the negative velocity value measured with the MWCNT-p-AF/VE composite was lower than that with the pristine p-AF/VE composite, indicating that the drop-weight impactor can be more rebounding or elastically behaving with the MWCNT-p-AF/VE composite than with the pristine p-AF/VE composite. This indicates that the impact resistance of the p-AF/VE composite was increased by the MWCNT anchoring effect, concordantly with the Izod impact strength results and the stress–strain behavior mentioned above.

Figure 8 exhibits the variation of the impact energy absorption as a function of time measured with pristine p-AF/VE and MWCNT-p-AF/VE composites, respectively. The energy absorbed by the pristine p-AF/VE composite corresponded to the initial energy of the impactor given at 9.3 ms, whereas the energy absorbed by the MWCNT-p-AF/VE composite reached the initial impactor energy after an elapse of 7.7 ms upon drop-weight impact. The result indicates that MWCNT anchoring enhanced the energy absorption capability of the composite, rebounding the impactor in the case of the MWCNT-p-AF/VE composite.

Figure 9 shows the variation of the transferred force as a function of time occurring in the pristine p-AF/VE and MWCNT-p-AF/VE composites during the drop-weight impact test, respectively. The time to reach the peak of the transferred force in the MWCNT-p-AF/VE composite was shorter than that in the pristine p-AF/VE composite. The transferred force (17.0 kN) at the peak of the MWCNT-p-AF/VE composite was about 24% higher than that (13.7 kN) of the pristine p-AF/VE composite. The end time of the force transferred in the MWCNT-p-AF/VE composite was about 2 milli-seconds shorter than that in the pristine p-AF/VE composite, similarly to the impactor velocity and the energy absorption. This revealed that the MWCNT anchoring was good to enhance the energy absorption capability and to toughen the p-AF/VE composite, giving rise to the increase in the fiber surface roughness and the interfacial adhesion between the aramid fiber and the VE matrix of the composite.

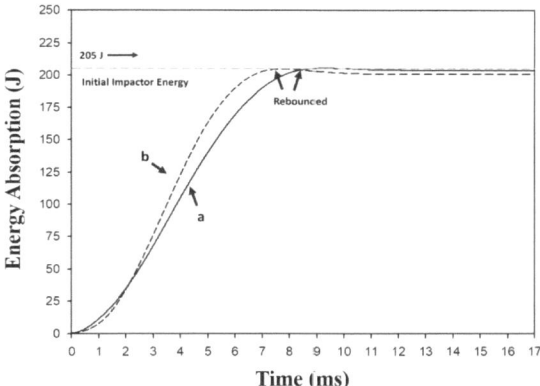

Figure 8. Variation of energy absorption as a function of time occurring during drop-weight impact test: (**a**) pristine p-AF/VE and (**b**) MWCNT-p-AF/VE composites.

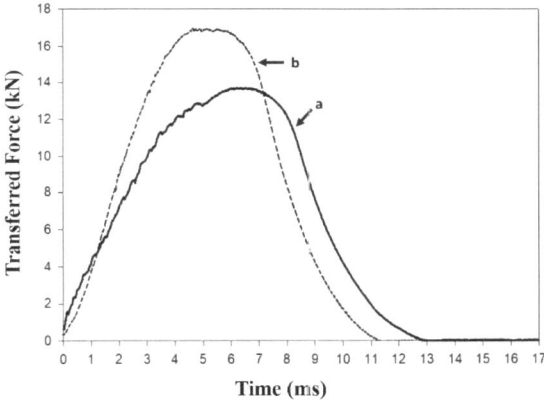

Figure 9. Variation of the transferred force as a function of time occurring during drop-weight impact test; (**a**) pristine p-AF/VE and (**b**) MWCNT-p-AF/VE composites.

4. Conclusions

The thermo-dimensional, dynamic mechanical, tensile, flexural, and impact properties of the p-AF/VE composite were significantly increased with the assistance of the MWCNT anchoring process, which can physically attach MWCN nanoparticles on the fiber surface by applying the MWCNT/phenolic/methanol mixture to p-AF, and then by thermally curing phenolic resin of very low concentration.

In particular, the drop-weight impact test results revealed that the variations in the velocity of the impactor, the energy absorption, and the transferred force as a function of time monitored for the pristine p-AF/VE and MWCNT-p-AF/VE composites during the impact test agreed with each other. The MWCNT-p-AF/VE composite exhibited a toughness higher than the pristine p-AF/VE composite. The result was consistent with the equilibrium toughness based on the stress–strain behavior and the dynamic toughness based on the Izod impact strength. The improvement on the thermal, mechanical, and impact properties of the MWCNT-p-AF/VE composite can be explained considering that the MWCNT anchoring contributed to increasing the interfacial adhesion between the p-aramid fiber and the VE matrix, being attributed to the mechanical interlocking by the roughened surface at the fiber–matrix interface.

The present work addresses that the anchoring of MWCNT to p-AF may be desirable to provide an additional benefit to conventional p-AF/polymer composites, further increasing their thermal, mechanical, and impact properties.

Author Contributions: Conceptualization, D.C.; writing—original draft preparation, D.C.; writing—review and editing, D.C.; supervision, D.C.; Funding acquisition, D.C.; formal analysis, J.C.; methodology, J.C.; investigation, J.C.; data curation, J.C. All authors have read and agreed to the published version of the manuscript.

Funding: This research was supported by the Kumoh National Institute of Technology (2022).

Institutional Review Board Statement: Not applicable.

Informed Consent Statement: Not applicable.

Data Availability Statement: The data presented in this study are available on request from the corresponding author.

Conflicts of Interest: The authors declare no conflict of interest.

References

1. Laha, A.; Majumdar, A. Shear thickening fluids using silica-halloysite nanotubes to improve the impact resistance of *p*-aramid fabrics. *Appl. Clay Sci.* **2016**, *132*, 468–474. [CrossRef]
2. Carrillo, J.G.; Gamboa, R.A.; Jonhnson, E.A.F.; Chi, P.I.G. Ballistic performance of thermoplastic composite laminates made from aramid woven fabric and polypropylene matrix. *Polym. Test.* **2012**, *31*, 512–519. [CrossRef]
3. Majumdar, A.; Butola, B.S.; Srivastava, A. An analysis of deformation and energy absorption modes of shear thickening fluid treated Kevlar fabrics as soft body armour materials. *Mater. Des.* **2013**, *51*, 148–153. [CrossRef]
4. Bhatnagar, A. *Lightweight Ballistic Composites: Military and Law-Enforcement Applications*, 2nd ed.; Woodhead Publishing: Cambridge, UK, 2016; pp. 224–229.
5. Reis, P.N.B.; Ferreira, J.A.M.; Santos, P.; Richardson, M.O.W.; Santos, J.B. Impact response of Kevlar composites with filled epoxy matrix. *Compos. Struct.* **2012**, *94*, 3520–3528. [CrossRef]
6. Nilakantan, G.; Gillespie, J.W. Yarn pull-out behavior of plain woven Kevlar fabrics: Effect of yarn sizing, pullout rate, and fabric pre-tension. *Compos. Struct.* **2013**, *101*, 215–224. [CrossRef]
7. Briscoe, B.J.; Motamedi, F. The ballistic impact characteristics of aramid fabrics: The influence of interface friction. *Wear* **1992**, *158*, 229–247. [CrossRef]
8. Bazhenov, S. Dissipation of energy by bulletproof aramid fabric. *J. Mater. Sci.* **1997**, *32*, 4167–4173. [CrossRef]
9. Duan, Y.; Keefe, M.; Bogetti, T.A.; Cheeseman, B.A. Modeling friction effects on the ballistic impact behavior of a single-ply high-strength fabric. *Int. J. Impact Eng.* **2005**, *31*, 996–1012. [CrossRef]
10. Rao, M.P.; Duan, Y.; Keefe, M.; Powers, B.M.; Bogetti, T.A. Modeling the effects of yarn material properties and friction on the ballistic impact of a plain-weave fabric. *Compos. Struct.* **2009**, *89*, 556–566. [CrossRef]
11. Das, S.; Jagan, S.; Shaw, A.; Pal, A. Determination of inter-yarn friction and its effect on ballistic response of *para*-aramid woven fabric under low velocity impact. *Compos. Struct.* **2015**, *120*, 129–140. [CrossRef]
12. Mayo, J.B.; Wetzel, E.D.; Hosur, M.V.; Jeelani, S. Stab and puncture characterization of thermoplastic-impregnated aramid fabrics. *Int. J. Impact Eng.* **2009**, *36*, 1095–1105. [CrossRef]
13. Lim, C.T.; Tan, V.B.C.; Cheong, C.H. Perforation of high-strength double-ply fabric system by varying shaped projectiles. *Int. J. Impact Eng.* **2002**, *27*, 577–591. [CrossRef]
14. Shim, V.P.W.; Tan, V.B.C.; Tay, T.E. Modelling deformation and damage characteristics of woven fabric under small projectile impact. *Int. J. Impact Eng.* **1995**, *16*, 585–605. [CrossRef]
15. Kirkwood, K.M.; Kirkwood, J.E.; Lee, Y.S.; Egres, R.G.; Wagner, N.J. Yarn pull-out as a mechanism for dissipating ballistic impact energy in Kevlar®KM-2 fabric: Part I: Quasi-static characterization of yarn pull-out. *Text. Res. J.* **2004**, *74*, 920–928. [CrossRef]
16. Kirkwood, J.E.; Kirkwood, K.M.; Lee, Y.S.; Egres, R.G.; Wagner, N.J.; Wetzel, E.D. Yarn pull-out as a mechanism for dissipating ballistic impact energy in Kevlar®KM-2 fabric: Part II: Predicting ballistic performance. *Text. Res. J.* **2004**, *74*, 939–948. [CrossRef]
17. Pandya, K.S.; Pothnis, J.R.; Ravikumar, G.; Naik, N.K. Ballistic impact behavior of hybrid composites. *Mater. Des.* **2013**, *44*, 128–135. [CrossRef]
18. Sarasini, F.; Tirillò, J.; Valente, M.; Ferrante, L.; Cioffi, S.; Iannace, S.; Sorrentino, L. Hybrid composites based on aramid and basalt woven fabrics: Impact damage modes and residual flexural properties. *Mater. Des.* **2013**, *49*, 290–302. [CrossRef]
19. Davidovitz, M.; Mittleman, A.; Roman, I.; Marom, G. Failure modes and fracture mechanisms in flexure of Kevlar-epoxy composites. *J. Mater. Sci.* **1984**, *19*, 377–384. [CrossRef]

20. Wang, J.; Liu, Y.; Wang, K.; Yao, S.; Peng, Y.; Rao, Y.; Ahzi, S. Progressive collapse behaviors and mechanisms of 3D printed thin-walled composite structures under multi-conditional loading. *Thin-Walled Struct.* **2022**, *171*, 108810. [CrossRef]
21. Wang, B.; Fu, Q.; Liu, Y.; Yin, T.; Fu, Y. The synergy effect in tribological performance of paper-based composites by MWCNT and GNPs. *Tribol. Int.* **2018**, *123*, 200–208. [CrossRef]
22. Joshi, S.C.; Dikshit, V. Enhancing interlaminar fracture characteristics of woven CFRP prepreg composites through CNT dispersion. *J. Compos. Mater.* **2012**, *46*, 665–675. [CrossRef]
23. An, Q.; Rider, A.N.; Thostenson, E.T. Electrophoretic deposition of carbon nanotubes onto carbon-fiber fabric for production of carbon/epoxy composites with improved mechanical properties. *Carbon* **2012**, *50*, 4130–4143. [CrossRef]
24. Soliman, E.M.; Sheyka, M.P.; Taha, M.R. Low-velocity impact of thin woven carbon fabric composites incorporating multi-walled carbon nanotubes. *Int. J. Impact Eng.* **2012**, *47*, 39–47. [CrossRef]
25. Meybodi, M.H.; Samandari, S.S.; Sadighi, M. An experimental study on low-velocity impact response of nanocomposite beams reinforced with nanoclay. *Compos. Sci. Technol.* **2016**, *133*, 70–78. [CrossRef]
26. Iqbal, K.; Khan, S.U.; Munir, A.; Kim, J.K. Impact damage resistance of CFRP with nanoclay-filled epoxy matrix. *Compos. Sci. Technol.* **2009**, *69*, 1949–1957. [CrossRef]
27. Davis, D.C.; Wilkerson, J.W.; Zhu, J.; Hadjiev, V.G. A strategy for improving mechanical properties of a fiber reinforced epoxy composite using functionalized carbon nanotubes. *Compos. Sci. Technol.* **2011**, *71*, 1089–1097. [CrossRef]
28. Chi, P.I.G.; Uicab, O.R.; Barrera, C.M.; Calderon, J.U.; Escamilla, G.C.; Pedram, M.Y.; Pat, A.M.; Aviles, F. Influence of aramid fiber treatment and carbon nanotubes on the interfacial strength of polypropylene hierarchical composites. *Compos. Part B* **2017**, *122*, 16–22.
29. Uicab, O.R.; Aviles, F.; Chi, P.I.G.; Escamilla, G.C.; Aranda, S.D.; Pedram, M.Y.; Toro, P.; Gamboa, F.; Mazo, M.A.; Nistal, A.; et al. Deposition of carbon nanotubes onto aramid fibers using as-received and chemically modified fibers. *Appl. Surf. Sci.* **2016**, *385*, 379–390. [CrossRef]
30. Cheon, J.; Yoon, B.I.; Cho, D. The synergetic effect of phenolic anchoring and multi-walled carbon nanotubes on the yarn pull-out force of *para*-aramid fabrics at high speed. *Carbon Lett.* **2018**, *26*, 107–111.
31. Cheon, J.; Cho, D. Enhancement of yarn pull-out force of *para*-aramid fabric at high speed by dispersion and phenolic anchoring of MWCNT on the fiber surfaces in the presence of surfactant and ultrasonic process. *Macromol. Res.* **2020**, *28*, 881–884. [CrossRef]
32. Cheon, J.; Cho, D. Effects of peroxide-based initiators with different molecular sizes on cure behavior and kinetics of vinyl ester resin containing multi-walled carbon nanotubes. *J. Therm. Anal. Calorim.* **2022**, *147*, 11883–11898. [CrossRef]

Disclaimer/Publisher's Note: The statements, opinions and data contained in all publications are solely those of the individual author(s) and contributor(s) and not of MDPI and/or the editor(s). MDPI and/or the editor(s) disclaim responsibility for any injury to people or property resulting from any ideas, methods, instructions or products referred to in the content.

Article

Mechanical and Thermal Properties of Multilayer-Coated 3D-Printed Carbon Fiber Reinforced Nylon Composites

Hongwei Chen, Kaibao Wang, Yao Chen and Huirong Le *

The Future Lab, Tsinghua University, Beijing 100084, China; dananwei@mail.tsinghua.edu.cn (H.C.); kaibaowang@mail.tsinghua.edu.cn (K.W.); yaochen@mail.tsinghua.edu.cn (Y.C.)
* Correspondence: lehr@mail.tsinghua.edu.cn

Abstract: This paper evaluates the mechanical and thermal properties of 3D-printed short carbon fiber reinforced composites (sCFRPs). A numerical analysis was developed to predict the mechanical and thermal properties of the sCFRPs, which were verified via experimental tests. In the experiments, a novel technique was adopted by coating the sCFRPs with carbon fiber fabric and copper mesh to further improve its mechanical and thermal performance. Various copper meshes (60-mesh, 100-mesh and 150-mesh) were integrated with carbon fiber fabric to form a multilayer structure, which was then coated on the surface of Nylon 12-CF composite material (base material) to form a composite plate. The effects of the copper mesh on the mechanical and thermal properties of the composite plate were studied theoretically and experimentally. The results show that the addition of different copper meshes had a significant influence on the mechanical and thermal properties of the composite plate, which contained carbon fiber fabric, copper mesh and the base material. Among them, the mechanical and thermal properties of the composite plate with the 60-mesh copper mesh were significantly improved, while the improvement effect slowly declined with the increase in the thickness of the base material. The composite plate with 100-mesh and 150-mesh copper meshes had improved mechanical properties, whereas the influence on its thermal conductivity was limited. For thermal conductivity calculation, both the thickness and length directions of the heat transfer were considered. The comparative analysis indicated that the calculated values and experimental results are in excellent agreement, meaning that this numerical model is a useful tool for guiding the design of surface lamination for 3D-printed sCFRPs.

Keywords: carbon fiber reinforced nylon composite; 3D printing; multilayer coating; copper mesh; numerical model

Citation: Chen, H.; Wang, K.; Chen, Y.; Le, H. Mechanical and Thermal Properties of Multilayer-Coated 3D-Printed Carbon Fiber Reinforced Nylon Composites. *J. Compos. Sci.* **2023**, *7*, 297. https://doi.org/10.3390/jcs7070297

Academic Editor: Jiadeng Zhu

Received: 29 June 2023
Revised: 11 July 2023
Accepted: 18 July 2023
Published: 20 July 2023

Copyright: © 2023 by the authors. Licensee MDPI, Basel, Switzerland. This article is an open access article distributed under the terms and conditions of the Creative Commons Attribution (CC BY) license (https://creativecommons.org/licenses/by/4.0/).

1. Introduction

Carbon fiber reinforced composites (CFRPs) have the advantages of light weight, high strength, fatigue resistance, corrosion resistance and excellent designability, and are widely used in aerospace, wind power, sports and leisure, the automotive industry and bridge construction [1–7]. With 3D printing technology, the advantages of material properties and the process characteristics of the rapid forming of complex structures can be brought into play simultaneously. This combined application breaks the limitations of traditional winding, laying, lamination and other manufacturing methods on the design of composite materials [8–15]. The 3D printing of short carbon fiber reinforced nylon composites (sCFRPs) has been adopted extensively, and their application in service robots can help lighten the weight of the robots in terms of the aspects of material and structure optimization [16–19]. However, moderate mechanical performance and poor thermal conductivity limit its application in key links with bearing capacity and heat conduction requirements [20]. Therefore, further optimization of the mechanical and thermal performance of the 3D-printed sCFRPs is mandatory to enrich its application in various fields.

Much research has been conducted in the selection of printed raw materials and interface modification, such as the modification of carbon fiber surfaces [21–23], the metallization treatment in the process of composite material molding [24] and the selection of continuous carbon fiber as the reinforcement phase and polyether ether ketone and other high-thermal-performance thermoplastic resins as the matrix [25]. Since composite materials exhibit different characteristics compared with metal materials, the traditional metal surface treatment process may not be suitable for composite materials [26]. Therefore, in order to improve the electrical and thermal conductivity of composite components, surface metallization treatment is adopted [27]. A very thin and dense metal coating is formed on the surface of a composite matrix via chemical plating, magnetron sputtering, arc ion plating and vacuum evaporation, and this coating plays a conductive and protective role. Nevertheless, further investigation is needed for complex structural components. Therefore, in the existing commercial use of 3D-printed short carbon fiber composites, it is still necessary to explore a surface-strengthening treatment method that is straightforward and suitable for complex parts.

Carbon fiber fabric is an excellent reinforcing material which has been widely used in the bridge, construction, hydropower and other industries [28–30]. By applying fiber fabric on the surface of the component for reinforcement, the mechanical properties of the component can be improved without increasing the weight and cross-section size of the structural component [31]. In addition, copper mesh has been used as an antistatic surface layer for CFRPs due to its excellent electrical and thermal conductivity and ductility [32–34]. This paper combines carbon fiber fabric and copper mesh to form a multilayer coating on the surface of a 3D-printed carbon fiber reinforced nylon composite component, and investigates the effects on the mechanical and thermal properties of the component.

2. Materials and Methods

2.1. Materials

The Nylon 12-CF composite filament was purchased from Stratasys with a short carbon fiber content of 35%. Carbon fiber fabric (T300-3K) was a twill fabric purchased from Toray Corporation of Japan (Tokyo, Japan). Copper meshes (60-mesh, 100-mesh and 150-mesh) were purchased from Churui Hardware Products Ltd. (Hengshui, China). The base adhesive (Yini special covering adhesive for carbon fiber) and surface adhesive (Yini epoxy resin adhesive) were purchased from Yini Composite Materials Ltd. (Dongguan, China). The vacuum bag used was PE/PA copolymer purchased from Tang Zheng Machinery Co., Ltd. (Suzhou, China).

2.2. Sample Preparation

2.2.1. Preparation of the Base Material

Stratasys' Fortus 380 mc Carbon Fiber FDM 3D printer was used to print the base material, with dimensions of 120 × 20 × 2.1 mm, 120 × 20 × 3.15 mm, 120 × 20 × 4.15 mm, 10 × 10 × 1 mm and 10 × 10 × 2 mm. An example of the specimens is shown in Figure 1. The 3D printing process parameters of the specimens are shown in Table 1.

Figure 1. An example of the specimens.

Table 1. The 3D printing process parameters.

Process Parameter	Paving Direction (°)	Nozzle Diameter (mm)	Height (mm)	Fill Line Width (mm)	Fill Overlap (mm)	Nozzle Temperature (°C)	Print Speed (mm/s)	Filling Rate (%)
Value	±45	0.5	0.254	0.43	0.01	355	-	100

2.2.2. Coating Sample Preparation

The copper meshes, carbon fiber fabric and plastic film were first trimmed to a specified dimension (slightly larger than the base material). The surface of the base material was then polished with 180-mesh sandpaper and cleaned with alcohol. The two-part epoxy resin base adhesive was prepared according to the mass ratio of 1:1, and after slowly stirring evenly, the adhesive was evenly applied on the surface of the specimen with a nylon brush. The adhesive surface density was roughly 0.1 g/cm^2. Then, the copper mesh and carbon fiber fabric were laid in turn, the whole sample was wrapped with polyethylene (PE) plastic film, and the sample was stored in vacuum bag at room temperature (not less than 25 °C). Once the base adhesive was completely cured, the specimen surface was then polished with 1500-mesh sandpaper, and cleaned with alcohol. The two-part epoxy resin surface adhesive was prepared according to the mass ratio of 1:2, and after stirring evenly, the adhesive was applied on the surface with a nylon brush, and the thickness of the adhesive layer was controlled at about 0.02–0.03 mm via the control of the adhesive amount applied. The sample was then transferred into an oven kept at 60 °C for 1 h, or at room temperature (not less than 25 °C) for 12 h. The adhesive was applied 3~4 times until the desired thickness was reached. The composite plates were cured and polished before the test. Both sides of the base material were coated for the bending test, while only one side of the base material was coated for the thermal conductivity test. The schematic diagram for the preparation of coated sCFRPs and the process illustration are shown in Figures 2 and 3, respectively.

Four kinds of coated sCFRPs were prepared by using the base material with thicknesses of 2.1 mm, 3.15 mm and 4.15 mm: the surface only coated with carbon fiber fabric; the surface coated with 60-mesh copper mesh and carbon fiber fabric (#60); the surface coated with 100-mesh copper mesh and carbon fiber fabric (#100); and the surface coated with 150-mesh copper mesh and carbon fiber fabric (#150).

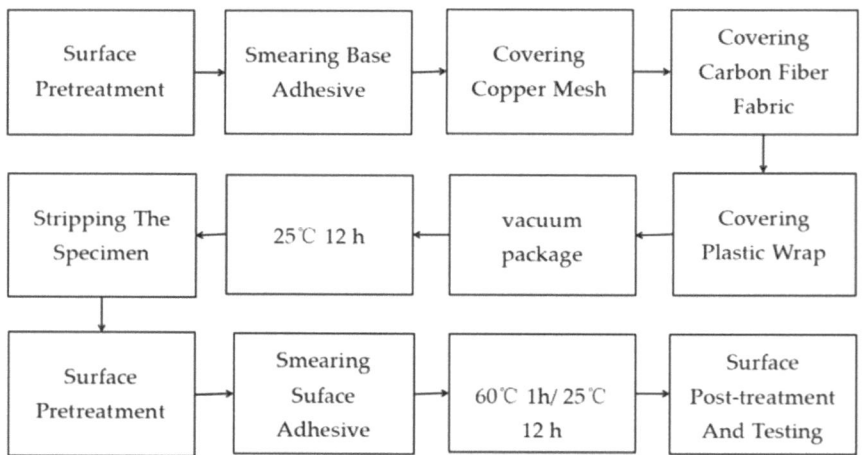

Figure 2. Schematic diagram for the preparation of coated sCFRPs.

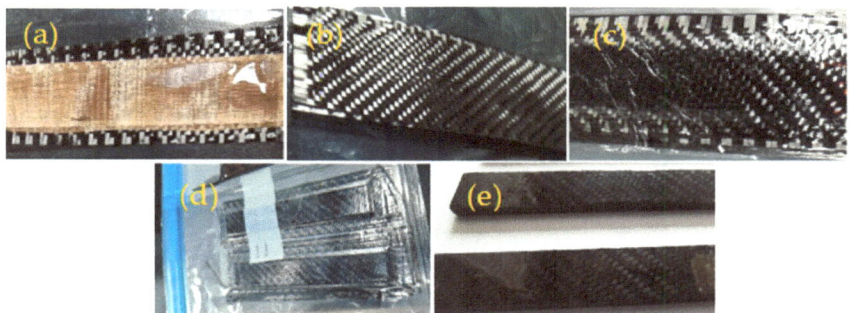

Figure 3. Preparation of coated sCFRPs: (**a**) copper mesh laid flat, (**b**) carbon fiber fabric laid flat, (**c**) wrapped in plastic film, (**d**) stored in a vacuum bag at room temperature, and (**e**) final specimen.

2.3. Sample Characterization

The sides of the #60, #100 and #150 composite plates were polished and cleaned with alcohol before being placed under a microscope (BX53M, Olympus, Toyko, Japan) to observe the microstructure of the sides of the coated sCFRPs.

Three-point bending properties of the base material (without coating) and four kinds of coated sCFRPs were tested using a mechanical testing machine (50ST, Tinius Olsen, Shanghai, China) according to the GB/T1449-2005 standard. The test span was 35 mm, and the loading speed was set at 1 mm/min. The bending modulus E was calculated using Equation (1).

$$E = \frac{L^3 \times \Delta P}{4bh^3 \times \Delta S} \tag{1}$$

where E is the flexural elastic modulus, Pa; ΔP is the load increment in the initial straight section of the deflection curve, N; ΔS is the deflection increment at the midpoint of span corresponding to load increment ΔP, m; L is the span between two supports, m; and b and h are the width and thickness of the sample, respectively, m.

The thermal conductivity of the base material and four kinds of coated sCFRPs were tested using a laser thermal conductivity meter (LFA467, Netzsch, Selb; Germany). The test temperature was 70 °C, and the thermal conductivity was calculated using Equation (2).

$$\lambda = C_p \times D \times \rho \tag{2}$$

where λ is the thermal conductivity, W/(m·K); C_P is the specific heat capacity, J/(kg·K); D is the thermal diffusion coefficient, mm^2/s; and ρ is the material density, kg/m^3.

3. Results and Discussion

3.1. Microstructure Analysis

Figure 4 shows the microstructure of the coated sCFRPs with various copper meshes. As can be seen from Figure 4, the overall thickness is increased by about 0.5 mm, and the carbon fiber fabric forms a dense composite material with the base adhesive and the surface adhesive, which is recorded as the carbon fiber layer with a thickness of about 0.3 mm. The copper mesh and the base material form a dense composite material, which is referred to as the copper mesh layer with a thickness of about 0.2 mm.

At the interface between the carbon fiber layer and the copper mesh layer, it can be found that the carbon fiber fabric and copper mesh are interwoven together to form a composite skin containing carbon fiber fabric and copper mesh as the reinforcement phase and epoxy resin as the matrix. At the reinforcing phase, the volume fraction of the continuous copper wire in the copper mesh can affect the mechanical properties of the composite. The pore size of the copper mesh directly affects the structure of this composite material, which is because the pores of the slightly larger copper mesh can make more carbon fiber fabric interwoven in the larger copper mesh surface and pores, forming a composite material with a complex reinforced phase structure. On the interface between the copper mesh layer and the base material, the copper mesh layer and the base material are directly adhered together through the base adhesive.

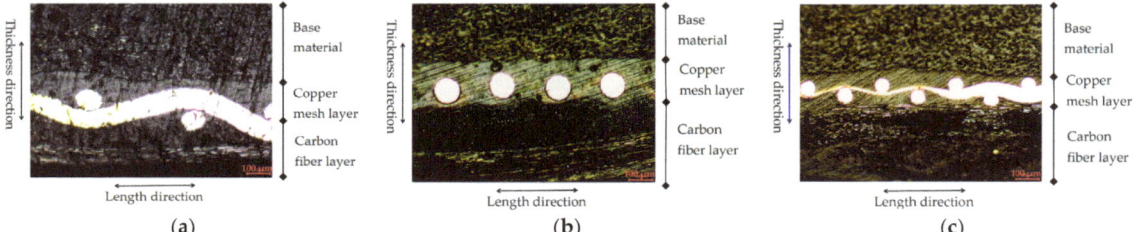

Figure 4. The microstructures of the coated sCFRPs: (**a**) #60 composite plate, (**b**) #100 composite plate, (**c**) #150 composite plate.

3.2. Bending Properties

In the bending process of the sample, the force–displacement curves are nonlinear, as shown in Figure 5. The coating film plays a major role in bearing the load, and the use of different copper meshes affects the bending performance of the sample. When the force is loaded to a certain extent, the curve first exhibits a slight fluctuation, which is a slight fold fracture of the carbon fiber layer on the extruded surface. With the increase in force, the carbon fiber layer and copper mesh layer on the stretched surface also gradually fold and fracture. When the maximum bending force is reached, the amplitude of the curve decreases rapidly, resulting in the obvious fracture of the carbon fiber layer and the copper mesh layer on the stretched surface, and at the same time, the debonding of the coating layer in the stressed area appears.

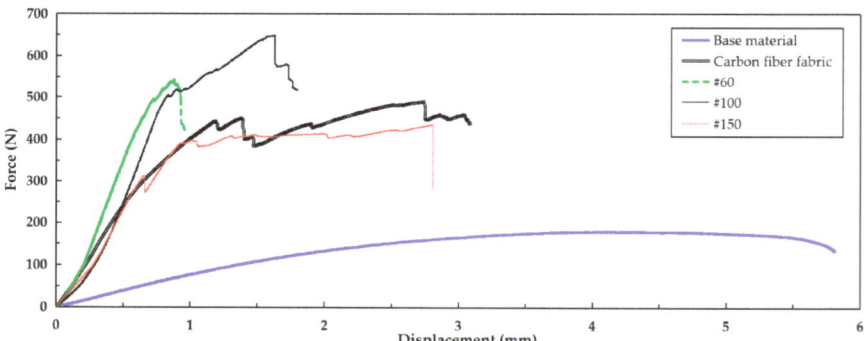

Figure 5. The force–displacement curves of the specimens.

The maximum failure force of the specimen can be obtained through the bending test. The width and thickness of the sample measured before the test are substituted into Equation (1) to obtain the elastic modulus, and the calculated mean value of the three tests of each group of samples is taken. The mean value of the elastic modulus is shown in Figure 6. The results show that the bending property of carbon fiber fabric is greatly improved when the base material is coated with carbon fiber fabric. Moreover, with the addition of copper mesh, its bending property is further increased. When the thickness of the base material is 2.1 mm, the mean bending moduli of the #60, #100 and #150 composite plates are 16.31 GPa, 14.19 GPa and 14.06 GPa, respectively. The bending modulus of the #60 composite plate is the largest, as it is 240% and 79% higher than those of the base material and the sample coated with carbon fiber fabric only, respectively. Generally, the experiment results exhibit good repeatability, except for the #150 composite plate; its error is slightly larger. This may be because the pores of the 150-mesh copper mesh are very small, which is not conducive to the penetration of the base adhesive; the composite of the copper mesh layer and the carbon fiber layer is incomplete; the area of the copper mesh layer adhering to the base material is much smaller; and more of the base adhesive is retained on the surface of the base material, resulting in the thickness of the sample becoming much larger, and these factors cause instability in the performance of the sample. When the thickness of the base material is increased, the bending modulus of the coated sCFRPs decreases slightly.

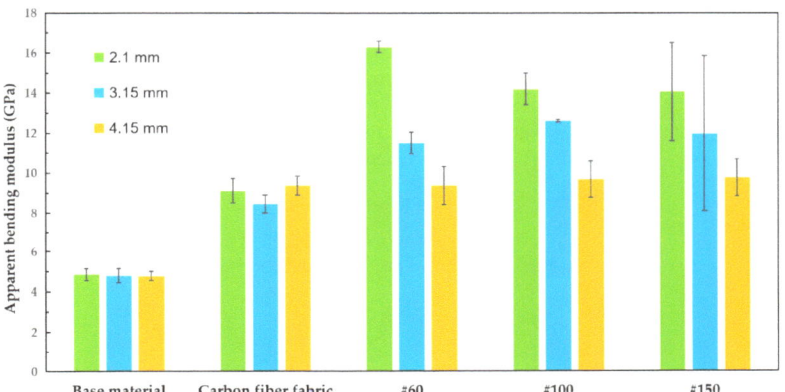

Figure 6. The comparison of bending moduli of the prepared specimens.

3.3. Thermal Conductivity

Figure 7 shows the influence of the addition of different copper meshes on the thermal conductivity of the coated sCFRPs in terms of thickness direction and length direction. The results indicate that only the thermal conductivity of the #60 composite plate is slightly improved. This may be because the pores of the copper mesh in #60 are slightly larger, which can make more carbon fiber fabric interweave with the copper mesh, forming a large area of a continuous heat transfer interface, resulting in higher thermal conductivity.

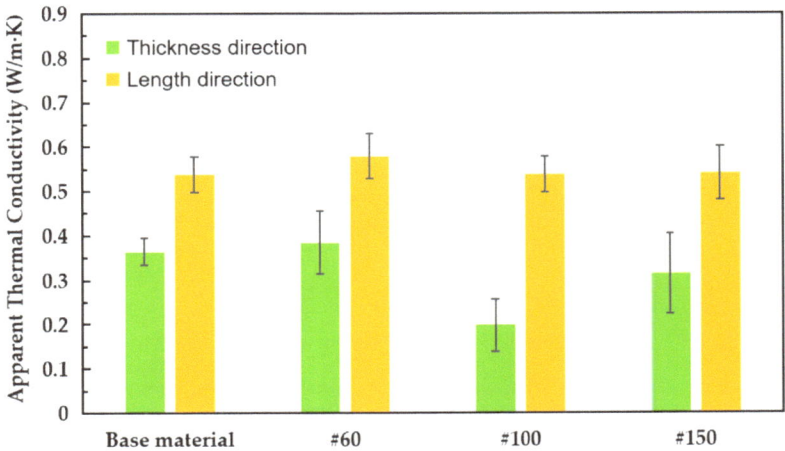

Figure 7. Thermal conductivities of base material and coated sCFRPs.

The thermal conductivity of the specimens in the thickness direction was also examined. The thickness of the base material was 1 mm, and the thermal conductivity of the #60 composite plate was increased by 5.5% compared with that of the base material. To test the thermal conductivity in the direction of length, the thickness of the base material is 9.42 mm, and the thermal conductivity of the #60 composite plate is increased by 7.4% compared with that of the base material.

4. Numerical Model

In order to predict the influence of copper mesh on the mechanical properties and thermal conductivity of composite plates with different thicknesses of base material, the appropriate thicknesses of the base material and copper mesh can be selected for multilayer surface coating to improve the mechanical properties and thermal conductivity of composite plates. For simplicity, several assumptions were made in this model. It was assumed that the interface between the copper mesh layer and the base material does not separate, and the heat loss in the interface is negligible. The composite plate can be simplified into a sandwich structure formed by the base material, the copper mesh layer and the carbon fiber layer; in addition, the interface between the layers is a plane.

4.1. Calculation of Bending Properties

By using Equations (3)–(5), the relationship between the bending modulus of the coated sCFRP and the thickness of the base material can be calculated, as shown in Figure 8

$$I = \frac{bt^3}{12} \tag{3}$$

$$EI = E_1 I_1 + E_2 I_2 + E_3 I_3 \tag{4}$$

$$E = \frac{E_1 t_1^3 + E_2[(t_1 + 2t_2)^3 - t_1^3] + E_3[(t_1 + 2t_2 + 2t_3)^3 - (t_1 + 2t_2)^3]}{t_1 + 2t_2 + 2t_3} \tag{5}$$

where I, I_1, I_2, I_3 are the moments of inertia of the composite plate, base material, copper mesh layer and carbon fiber layer, respectively, m^4; E, E_1, E_2, E_3 are the bending moduli of the composite plate, base material, copper mesh layer and carbon fiber layer, respectively, Pa; t, t_1, t_2, t_3 are the thicknesses of the composite plate, base material, copper mesh layer and carbon fiber layer, respectively, m; L is the distance between the two supports, m; and b is the width of the composite plate, m.

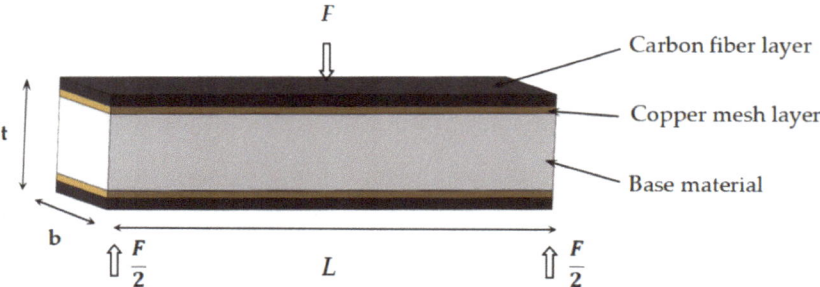

Figure 8. Three-point bending diagram of the composite plate.

The experimental values of the specimens with three thicknesses of base material were substituted into Equation (5) to calculate E_1 and E_2, and their average values were taken into Equation (5) to depict the relationship, as shown in Figure 9 and Table A1.

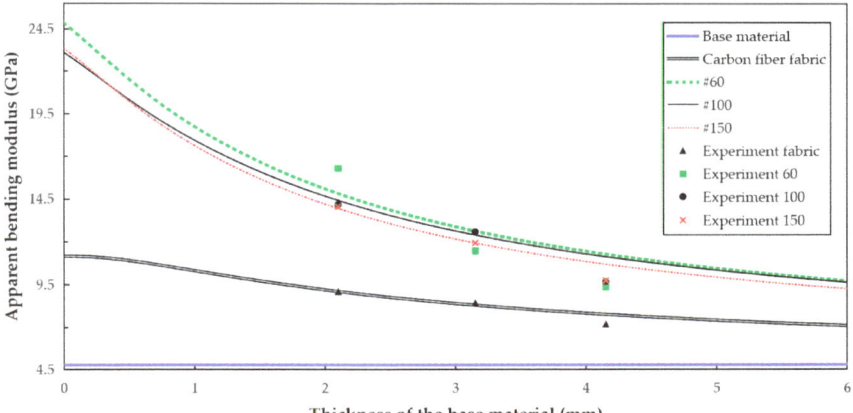

Figure 9. Bending modulus comparison between the calculated and experimental results.

The bending modulus comparison between the calculated and experimental results. The bending modulus tends to decrease with the increase in the thickness of the base material. When the thickness of the base material is 2.1 mm, the #60 composite plate shows the highest modulus. As the thickness of the base material increases, the changes in the bending properties of the coated sCFRPs are negligible.

It is worth mentioning that the largest discrepancy between the calculated data of the #60, #100 and #150 composite plates and the test values is within 10%. When the thickness of the base material is small, the error between the simulated data and the test values is less than 10%, and the data are in good agreement. This can be used as a reference guide for actual production. However, when the thickness of the base material increases, the numerical model overestimates the bending modulus, and this is because the phenomenon of uneven film coating stripping occurs during bending, thus reducing the effect of the coating on the bending performance of the composite plate.

4.2. Calculation of Thermal Conductivity

In this calculation, it is assumed that each layer maintains a continuous form of aggregation to form a continuous block, and the heat flow is conducted in series or parallel mode [35].

4.2.1. Thickness Direction (Series Mode)

According to the series heat transfer, the heat flow through each layer is assumed to be constant, and the temperature difference between the top and bottom ends of the composite plate is equal to the sum of the temperature differences of each layer, as shown in Figure 10. Equation (8) can be derived using Equations (6) and (7). The relationship between the thermal conductivity of the composite plate along the thickness direction and the thickness of the matrix is calculated as shown in Figure 11 and Table A2.

$$\dot{H} = \lambda \frac{\Delta T}{t} = \lambda_1 \frac{\Delta T_1}{t_1} = \lambda_2 \frac{\Delta T_2}{t_2} = \lambda_3 \frac{\Delta T_3}{t_3} \qquad (6)$$

$$\Delta T = \Delta T_1 + \Delta T_2 + \Delta T_3 \qquad (7)$$

$$\lambda = \frac{\frac{t_1}{\lambda_1} + \frac{t_2}{\lambda_2} + \frac{t_3}{\lambda_3}}{t_1 + t_2 + t_3} \qquad (8)$$

where $\dot{H}$ is the load-rated heat flow, W; $\Delta T, \Delta T_1, \Delta T_2, \Delta T_3$ are the temperature differences between the beginning and end of the composite plate, base material, copper mesh layer and carbon fiber layer in the thickness direction, respectively, °C; $\lambda, \lambda_1, \lambda_2, \lambda_3$ are the thermal conductivities of the composite plate, base material, copper mesh layer and carbon fiber layer in the thickness direction, respectively, W/(m·K); and t, t_1, t_2, t_3 are the thicknesses of the composite plate, base material, copper mesh layer and carbon fiber layer, respectively, m.

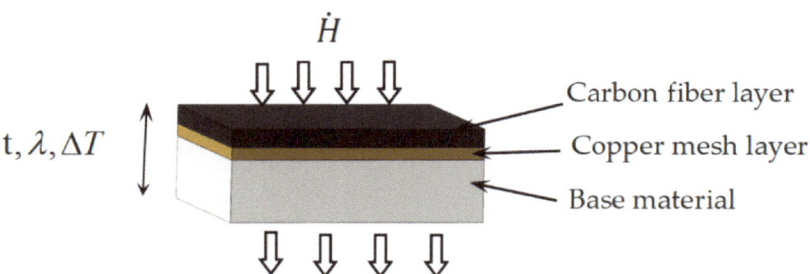

Figure 10. Schematic diagram of heat conduction in series mode.

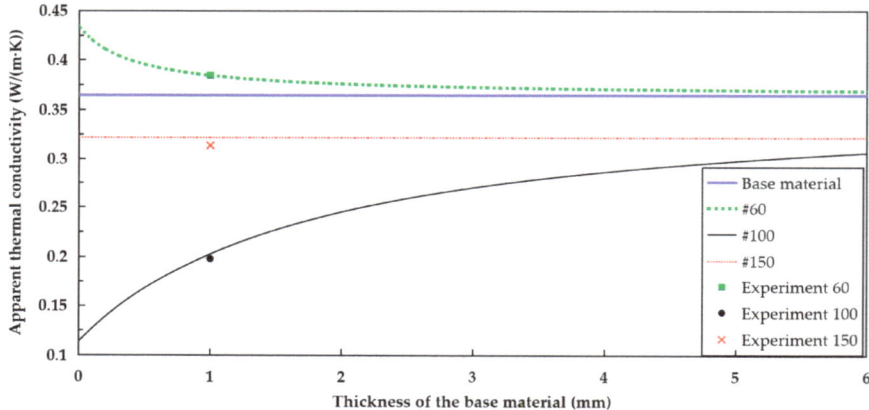

Figure 11. Thermal conductivity comparison between the calculated and experimental results in series mode.

Only the thermal conductivity of the #60 composite plate is slightly improved, and the improvement effect is negligible as the thickness of the base material increases. It is worth mentioning that the calculated values and experimental results are in excellent agreement, meaning that this numerical model is a useful tool for guiding the structure design of coated sCFRPs.

4.2.2. Length Direction (Parallel Mode)

According to the parallel heat transfer, theoretically, the total heat flow is equal to the sum of the heat flows of each layer, and the temperature difference between the beginning and the end of each layer is the same, as shown in Figure 12. Equation (11) can be derived by using Equations (9) and (10). The relationship between the thermal conductivity of the composite plate and the thickness of the base material at 70 °C is calculated as shown in Figure 13 and Table A2.

$$\dot{H} = \lambda t \frac{\Delta T}{L} = \lambda_1 t_1 \frac{\Delta T_1}{L} + \lambda_2 t_2 \frac{\Delta T_2}{L} + \lambda_3 t_3 \frac{\Delta T_3}{L} \qquad (9)$$

$$\Delta T = \Delta T_1 = \Delta T_2 = \Delta T_3 \qquad (10)$$

$$\lambda = \frac{\lambda_1 t_1 + \lambda_2 t_2 + \lambda_3 t_3}{t_1 + t_2 + t_3} \tag{11}$$

where $\dot{H}$ is the load-rated heat flow, W; L is the length of the composite plate; $\Delta T, \Delta T_1, \Delta T_2, \Delta T_3$ are the temperature differences between the left and right ends of the composite plate, base material, copper mesh layer and carbon fiber layer in the length direction, respectively, m; $\lambda, \lambda_1, \lambda_2, \lambda_3$ are the thermal conductivities of the composite plate, base material, copper mesh layer and carbon fiber layer in the length direction, respectively, W/(m·K); and t, t_1, t_2, t_3 are the thicknesses of the composite plate, base material, copper mesh layer and carbon fiber layer, respectively, m.

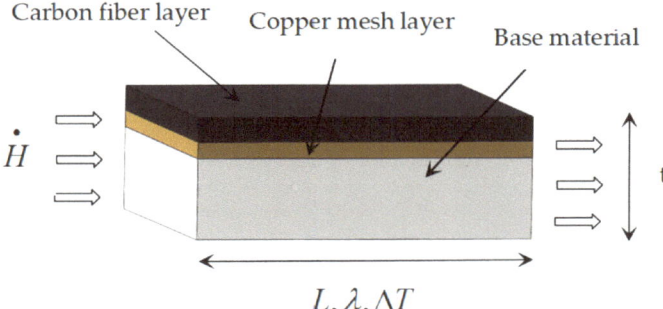

Figure 12. Schematic diagram of heat conduction in parallel mode.

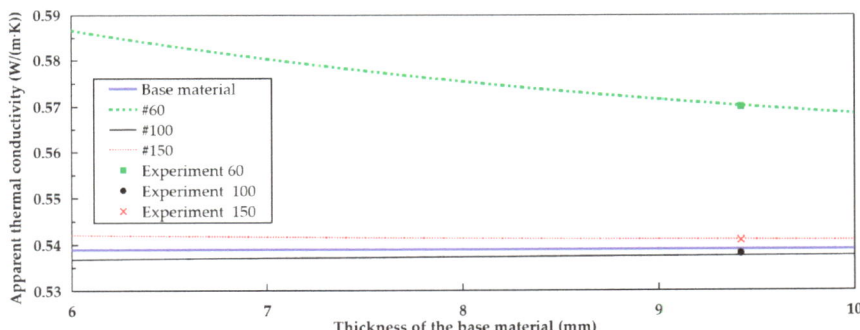

Figure 13. Thermal conductivity comparison between the calculated and experimental results in parallel mode.

Only the thermal conductivity of the #60 composite plate is significantly improved, and this improvement slowly decreases as the thickness of the base material increases. When the thickness of the base material is 1–10 mm, the thermal conductivity of the #60 composite plate can be increased by 5–38% compared with that of the base material. It is worth mentioning that the calculated values and experimental results are in excellent agreement, meaning that this numerical model is a useful tool for guiding the design of coated sCFRPs.

5. Conclusions

The mechanical and thermal properties of a coated carbon fiber reinforced composite were investigated using experimental measurement and numerical analysis. The effects of various copper meshes and thicknesses of the base material on the performance were discussed. The results show that the addition of different copper meshes had a significant influence on the mechanical and thermal properties of the composite plate, which contained carbon fiber fabric, copper mesh and the base material. This is because of the differences in the diameter and porosity of the copper mesh; the coated sCFRPs is mixed to different degrees at the interface. It is the existence of this composite material that directly affects the mechanical and thermal properties of the composite plate. The adhesive layer of the base adhesive on the base material directly affects the mechanical properties of the composite plate.

Among the plates, the mechanical and thermal properties of the composite plate with a 60-mesh copper mesh were significantly improved, while the improvement slowly declined with the increase in the thickness of the base material. The composite plate with 100-mesh and 150-mesh copper meshes had improved mechanical properties, whereas the influence on thermal conductivity was limited. For a thermal conductivity calculation, both the thickness and length directions of the heat transfer were considered. The comparative analysis indicated that the calculated values and experimental results are in excellent agreement, meaning that this numerical model is a useful tool for guiding the structure design of coated sCFRPs.

Author Contributions: Investigation, data curation, visualization, validation, formal analysis and writing—original draft preparation, H.C.; writing—review and editing, K.W. and Y.C.; conceptualization, methodology and writing—review and editing, H.L. All authors have read and agreed to the published version of the manuscript.

Funding: This research was funded by Tsinghua University-Foshan Advanced Manufacturing Research Institute and Foshan Longshen Robot Company through a joint research grant (2020THFS0307) and the Ministry of Science and Technology of China through a special grant (G2021102010L).

Institutional Review Board Statement: Not applicable.

Informed Consent Statement: Not applicable.

Data Availability Statement: Not applicable.

Acknowledgments: The financial support of Tsinghua University-Foshan Advanced Manufacturing Research Institute and Foshan Longshen Robot Company is acknowledged. The authors are grateful for the support of colleagues at the Future Materials & Design Research Center, The Future Lab, Tsinghua University. The assistance of the technical staff at the Micro-Nano Heat Transfer Laboratory of the School of Aeronautics, Tsinghua University, is also acknowledged.

Conflicts of Interest: The authors declare no conflict of interest.

Appendix A

Table A1. Apparent bending modulus (GPa) of the samples.

Thickness of the base material (mm)	Base Material	Carbon Fiber Fabric	#60	#100	#150
2.1	4.85 ± 0.29	9.11 ± 0.61	16.31 ± 0.29	14.19 ± 0.79	14.06 ± 2.45
3.15	4.80 ± 0.35	8.45 ± 0.45	11.51 ± 0.54	12.6 ± 0.06	11.97 ± 3.87
4.15	4.78 ± 0.23	9.37 ± 0.48	9.37 ± 0.95	9.68 ± 0.91	9.76 ± 0.92

Table A2. Thermal conductivities of the sample.

Apparent Thermal Conductivity (W/m·K)	Base Material	#60	#100	#150
Thickness direction	0.365 ± 0.03	0.385 ± 0.07	0.338 ± 0.06	0.314 ± 0.09
Length direction	0.539 ± 0.04	0.579 ± 0.05	0.198 ± 0.04	0.541 ± 0.06

References

1. Fan, X. Application status and development trend of carbon fiber composites. *Chem. Ind.* **2019**, *37*, 12–16+25.
2. Zheng, X.B. Research on the application of carbon fiber composites in aircraft structures. *China Plant Eng.* **2023**, *521*, 92–94.

3. Mou, S.X.; Chen, C.; Qiu, G.J.; Jia, Z.Y. Application of carbon fiber composite materials in wind power blades. *Adv. Mater. Ind.* **2012**, *219*, 25–29.
4. Duan, W.; Kong, X.X. Progress in the application of carbon fiber composites in the field of automotive lightweight. *Automob. Parts* **2023**, *178*, 84–87.
5. Chen, W.; Bai, Y.; Zhu, J.Q.; Meng, L.H. Application of carbon fiber composite material in sports equipment. *Tech. Text.* **2011**, *29*, 35–37+43.
6. Pang, S.X. Application of carbon fiber composite materials in water conservancy and hydropower reinforcement engineering. *Synth. Mater. Aging Appl.* **2023**, *52*, 120–122.
7. Zheng, H.; Zhang, W.J.; Li, B.; Zhu, J.J.; Wang, C.H.; Song, G.J.; Ma, L.C. Recent advances of interphases in carbon fiber-reinforced polymer composites: A review. *Compos. Part B Eng.* **2022**, *233*, 109639. [CrossRef]
8. Ning, F.D.; Cong, W.L.; Qiu, J.J.; Wang, S.R. Additive manufacturing of carbon fiber reinforced thermoplastic composites using fused deposition modeling. *Compos. Part B* **2015**, *80*, 369–378. [CrossRef]
9. Zhang, H.Q.; Zhang, K.; Aonan, L.; Wan, L.; Robert, C.; Brádaigh, C.M.Ó.; Yang, D.M. 3D printing of continuous carbon fibre reinforced powder-based epoxy composites. *Compos. Commun.* **2022**, *33*, 101239. [CrossRef]
10. Wang, C.L. Development and Application of Carbon Fiber Composites for 3D Printing. Master's Thesis, General Research Institute of Mechanical Sciences, Beijing, China, 2012.
11. Chen, X.M.; Yao, L.J.; Guo, L.C.; Sun, Y. A review of the research status of 3D printing continuous fiber reinforced composites. *Hangkong Xuebao* **2021**, *42*, 174–198.
12. Blok, L.G.; Longana, M.L.; Yu, H.; Woods, B.K.S. An investigation into 3D printing of fibre reinforced thermoplastic composites. *Addit. Manuf.* **2018**, *04*, 039. [CrossRef]
13. Yavas, D.Z.; Zhang, Z.Y.; Liu, Q.Y.; Wu, D.Z. Fracture Behavior of 3D Printed Carbon Fiber-Reinforced Polymer Composites. *Compos. Sci. Technol.* **2021**, *208*, 108741. [CrossRef]
14. Masahito, U.; Shun, K.; Masao, Y.; Antoine, L.D. 3D compaction printing of a continuous carbon fiber reinforced thermoplastic. *Compos. Part A* **2020**, *139*, 105985.
15. Song, X.D.; Pang, L.S. Research progress of carbon fiber resin matrix composites and their forming technology and application. *Packag. Eng.* **2021**, *42*, 81–91.
16. Dang, L.; Zhang, M.Y.; Cheng, Y.N.; Yan, C. Application and development of 3D printing technology in composite materials. *Technol. Innov. Appl.* **2022**, *12*, 166–169.
17. Yu, C.T.; Zhang, J.Y.; Wu, Y.B. Research progress in the application of lightweight materials for robots. *Adv. Mater. Ind.* **2019**, *313*, 41–45.
18. Chen, H.W.; Liu, G.; Wang, X.W.; Le, H.R. Application and prospect of lightweight composite materials and 3D printing technology in service robots. *J. Eng. Stud.* **2022**, *14*, 30–39. [CrossRef]
19. Li, D.D.; Shi, X.M.; Deng, S.P.; Qi, Y.M.; Chen, W.; Zhou, Y.Y.; Zhou, W.F. A review of the technology and development trend of collaborative robot industry. *Equip. Manuf. Technol.* **2021**, *320*, 73–76. [CrossRef]
20. Research Progress in 3D Printing of Fiber-Reinforced Thermoplastic Composites. Available online: http://kns.cnki.net/kcms/detail/10.1683.TU.20230425.1709.002.html (accessed on 25 May 2023).
21. Xu, Y.; Chi, Y.B. Effect of surface modification methods of carbon fiber on properties of reinforced nylon composites. *China Sci. Technol. Inf.* **2015**, *511*, 15–16.
22. Zhang, D.D.; Zhang, F.H.; Yang, J.X.; Li, X.F.; Li, Y.X.; Zeng, Y. Progress in surface modification of carbon fiber and its application in nylon composites. *Eng. Plast. Appl.* **2019**, *47*, 141–146.
23. Chu, C.X. Surface Modification of Carbon Fiber and Properties of Resin Matrix Composites. Master's Thesis, Jinan University, Jinan, China, 2020.
24. Wu, S.H. Study on Surface Metallization and Membrane Base Interface of Resin Matrix Composites. Master's Thesis, Lanzhou University, Lanzhou, China, 2013.
25. 3D Printing Continuous Carbon Fiber/Polyether Ketone Ketone Composite Process and Its Performance Control. Available online: https://doi.org/10.13801/j.cnki.fhclxb.20221215.002 (accessed on 26 May 2023).
26. Liu, Y.F. Surface treatment methods for mechanical engineering materials. *Coal Technol.* **2008**, *170*, 22–24.
27. Dong, X.Y.; Guo, J.H. Progress in surface metallization of fiber reinforced resin matrix composites. *Compos. Sci. Eng.* **2017**, *277*, 93–99.
28. Li, H.H.; Chen, H.X.; Deng, C.L. Engineering practice of repairing underwater cracks in concrete piles with carbon fiber. *Build. Struct.* **2021**, *51*, 2119–2121.
29. Wang, P.; Wang, X.; Hu, J.Y. Application of high strength carbon fiber fabric in the reinforcement technology of pressure steel pipe in hydropower plant. *Appl. IC* **2023**, *40*, 38–43.
30. Huang, Q.G.; Huang, Z.; Jiang, J.; Xue, X.D. Performances of Carbon Fiber Cloth Reinforced Bamboos. *Appl. Mech. Mater.* **2012**, *1801*, 174–177.
31. Al-Mahfooz, M.J.; Mahdi, E. Bending behavior of glass fiber reinforced composite overwrapping pvc plastic pipes. *Compos. Struct.* **2020**, *251*, 112656. [CrossRef]
32. Xiao, Y.; Li, S.L.; Yin, J.J.; Yao, X.L.; Zhang, X.H. Experimental study on direct effect of lightning current on copper mesh composites. *J. Aeronaut. Mater.* **2018**, *04*, 109–114.

33. Tian, M.H.; Liu, X.Y.; Wu, T.; Shan, Z.Z.; Lu, X. Analysis of the influence of the protective layer structure of copper mesh on the ablation damage of composite laminates by lightning strike. *Sci. Technol. Eng.* **2022**, *21*, 9071–9080.
34. Lu, W.B. Preparation and Properties of Layered Pantograph Slide Plate. Ph.D. Thesis, Shandong University, Jinan, China, 2012.
35. Li, P.M. Preparation and Properties of Epoxy Resin Based Thermal Conductivity Composites. Master's Thesis, University of Electronic Science and Technology of China, Chengdu, China, 2012.

Disclaimer/Publisher's Note: The statements, opinions and data contained in all publications are solely those of the individual author(s) and contributor(s) and not of MDPI and/or the editor(s). MDPI and/or the editor(s) disclaim responsibility for any injury to people or property resulting from any ideas, methods, instructions or products referred to in the content.

Article

Notched Behaviors of Carbon Fiber-Reinforced Epoxy Matrix Composite Laminates: Predictions and Experiments

Shaoyong Cao [1], Yan Zhu [2] and Yunpeng Jiang [2,*]

[1] School of Industrial Automation, Beijing Institute of Technology, Zhuhai 519088, China; lemonsayong2005@163.com
[2] College of Aerospace Engineering, Nanjing University of Aeronautics and Astronautics, Nanjing 210016, China
* Correspondence: ypjiang@nuaa.edu cn; Tel.: +86-25-84893240

Abstract: This paper experimentally studied the influence of the notch shape and size on the damage evolution and failure strength (tension and torsion) of carbon fiber-reinforced epoxy matrix (CFRP) laminates. Hashin's damage criteria were utilized to monitor the evolution of multi-damage modes, and FEM simulations were also performed by using the ABAQUS code to clarify the specific damage modes in detail as an instructive complement. The failure characteristics of all the notched samples were analyzed and compared with those without notches. The measured results presented that the existence of a variety of notches significantly impaired the load carrying capacity of CFRP laminates. The tensile strengths of C-notch and U-notch increase with an increasing notch radius, while the ultimate torques of C-notch and V-notch decrease with an increasing notch size and angle. The variation in notched properties was explained by different notch shapes and sizes, and the failure characteristics were also presented and compared among notched CFRP laminates with varied notches.

Keywords: carbon fiber-reinforced polymers (CFRPs); notch effect; finite element method (FEM); damage propagation; mechanical behaviors

1. Introduction

CFRP laminates have been widely applied in the aerospace, aviation and automobile industries due to their high specific stiffness, high specific strength and high design freedom. These composite laminate structures usually include discontinuities such as cut-outs for access and fastener holes for joining and inevitably become vulnerable regions under thermo-mechanical loading [1–3]. Understanding their notched behaviors is necessary for designing these complex structures, in which various parts are mostly connected with bolts and rivets [4]. The effect of these discontinuities on the mechanical performances is an important issue because it causes a relatively large reduction in strength compared to the unnotched laminates. Koricho et al. [5] proposed an innovative solution of using the 'tailored placement' of fibers around the holes/notch to make fabric laminates without the need for the drilling and machining of holes, thereby eliminating the sources of delamination. After drilling a hole, the tensile strength sharply decreased from 1012 MPa to 385 MPa, while the strength by adopting the developed processing could maintain 86% of the intact specimens.

A great number of experiments have been performed in studying the failure strength of notched CFRP laminates containing various shapes and sizes of notches under tensile, compressive and multi-axial loadings. Belgacem et al. [6] tested the center-notched CFRP laminates with radii of 2 mm, 6 mm and 10 mm and found that the ultimate strength is reduced by the order of 23%, 26% and 45% compared with those without notches. Torabi et al. [7] investigated the load-carrying capacity of glass/epoxy laminates with central U-shaped notches of various tip radii by using the virtual isotropic material concept without the need for ply-by-ply failure analysis. Xu et al. [8] studied the size effect in center-notched CFRP laminates under compression and found that the compressive strength of the small center-notched specimen is similar to that of the open-hole specimens. As the in-plane sizes increase, the center notches are weaker than the open holes. Serra et al. [9] carried out numerical and experimental studies on the size effect in notched CFRP laminates using the discrete ply modeling method [10] and showed that their strength reduces with an increasing specimen size. Lee et al. [11] considered the notch size, ply and laminate thickness to be the most important variables of scaling effects on the strength and found that the strength reduction is due to the hole size effect rather than the specimen thickness or volume increase. Wan et al. [12] studied the notch effect on the strength and fatigue life of double edge notched laminates and declared that the notch effect is strongly dependent on the fiber type, notch depth, load type and load sequence. Torabi et al. [13] also studied the E-glass/epoxy composites weakened by blunt V-notches with different notch angles and tip radii. Ghezzo et al. [14] analyzed the interaction between two holes set in different configurations with respect to the load direction to seek the minimum distance at which there is no superposition of notch effects within the area between two near holes.

Llobet et al. [15] measured the strength of CFRP notched laminates under static, tension–tension fatigue and residual strength tests. Lagattu et al. [16] characterized over-stress accommodation which develops near the notches. Vieille et al. [17] studied the influence of the matrix nature on the tensile thermo-mechanical behavior of notched laminates and indicated that the hole is an open access through the thickness for the heat flux causing thermal degradation and decreasing the laminate tensile properties. The tensile strength decreases from 472 MPa to 247 MPa after introducing a central hole, as for C/PPS laminates. Ye et al. [18] measured the residual strength of notched cross-ply CFRP laminates with different fiber/matrix adhesion and indicated that laminates with poor interfacial adhesion exhibit a higher residual strength than those with strong adhesion. Furthermore, the effect of interfacial adhesion on the fatigue residual strength of circular notched laminate was studied [19]; the strengths reduce from 1000 MPa to only 300 MPa with an increasing hole diameter. Czél et al. [20] prepared CFRP laminates with hybrid modulus carbon fibers and showed that reduced notch sensitivity was demonstrated for open holes and sharp notches. Qiao et al. [21] studied the failure behavior of notched laminates under multiaxial quasi-static and fatigue loading, and a significant non-linearity in the stress-strain curves was exhibited, becoming more and more significant with increasing shear load components.

The progressive damage analyses are usually employed on the notched behaviors of CFRP laminates. Liu [22,23] introduced the nonlocal integral theory into the damage model to solve the localization problem of composites and derived an FEM model on the nonlocal intra-laminar damage and interlaminar delamination of laminates. Hu et al. [24] studied the layer-by-layer stress components and the damage propagation and failure in notched laminates by the theoretical method and FEM. Divse et al. [25] investigated the stress concentrations factor, damage progression and tensile notched strength of CFRP laminates and presented that the FEM plane model slightly underestimated the extent of damage propagation and failure load when compared with a 3D progressive damage model. Ng et al. [26,27] proposed a progressive failure analysis together with a newly developed maximum notched strength method and confirmed that the location of failure initiation for laminates with large hole sizes is different from that for laminates with smaller holes. Riccio et al. [28] presented that the ABAQUS plane stress gradual degradation model overestimates the damage accumulation leading to a premature ultimate load, and the 3D degradation model could correctly predict the mechanical response and ultimate load. Maa et al. [29] combined the generalized standard material model with the principal damage concept of composite materials. Laurin et al. [30] presented a simplified strength analysis method for perforated plates with open-holes, ensuring design office requirements in terms of precision and computational time. Chen et al. [31] predicted the tensile and compressive strengths of notched laminates by extending Whitney and Nuismer's average stress failure criterion. Morgan et al. [4] developed a more complete picture of notch effects and analyzed the accuracy of the theory of critical distances.

In the light of the above literature survey, the purpose of this work is to reveal the effect of the notch size and shape on the mechanical behaviors of CFRP laminates under tension and torsion; especially, the torsion testing on the notched CFRP seems very limited at the present time and insufficient for structural engineering. A joint method of an experiment and FEM is used to understand the damage mechanism in these notched samples. The variation in tension and torsion properties is tested by using CFRP laminates containing varied notch shapes and sizes.

In Section 2, we present the preparation of samples for tensile and torsion testing. The user subroutine and FEM modeling are then presented in Section 3. The results of both measured and simulated failure mechanisms under tension are discussed in Section 4. The results of both tested and simulated failure mechanisms under torsion are discussed in Section 5. The conclusion is finally given in Section 5.

2. Materials and Methods

2.1. Materials

T700 carbon plain fabric (areal weight: 300 g/m^2, and thickness: 0.125 mm) is supplied from Japan Toray Industries, Inc. The epoxy resin used is E51 (Nanya epoxy Co., Ltd., Kunshan, China) with a low viscosity of 600–800 MPa·s at room temperature, and Polyamide 650 is used as the curing agent with a ratio of 30:10 parts in weight. The technical data of the Resin are listed in Table 1.

Table 1. Specification of epoxy resin.

Category	Parameters
Product categories	Bisphenol A epoxy vinyl ester resin
Appearance	Colorless
Viscosity	600–800 MPa·s, 25 °C
Tensile strength	85 MPa
Hardness (ShoreD)	88
Density	1.05 g/cm^3
Compress strength	300 MPa
Elastic modulus	1.0 GPa
Poisson's ratio	0.38

2.2. Manufacturing of Composites

The epoxy system is synthesized by mixing the resin with a hardener at the weight ratio of 30:10. The cross-ply $[0/90/0/90]_s$ laminates are stacked by hand lay-up, fabricated using the vacuum resin infusion process (ECVP425 vacuum pump, Easy Composites Asia Ltd., Beijing, China) and finally cured at room temperature for the duration of 24 h. Figure 1 illustrates the production process for CFRP laminate panels. The resulting volume fraction of the carbon fiber is 60% for all composites. The nominal thickness of the samples is measured as 1.0 ± 0.1 mm, The diamond tip water-cooled saw blade (ONEJET50-G30 × 15 Waterjet cutting machine, OneJet Co., Ltd., Shenzhen, China) is used to cut these CFRP panels into flat coupons with the specific dimensions and shapes of the testing samples shown in Figure 2, which also illustrates the photographs of various test samples. In order to study the notched behaviors of CFRP laminates, tension and torsion tests are conducted using specimens that are cut from the $[0/90/0/90]_s$ panels using a water jet cutter.

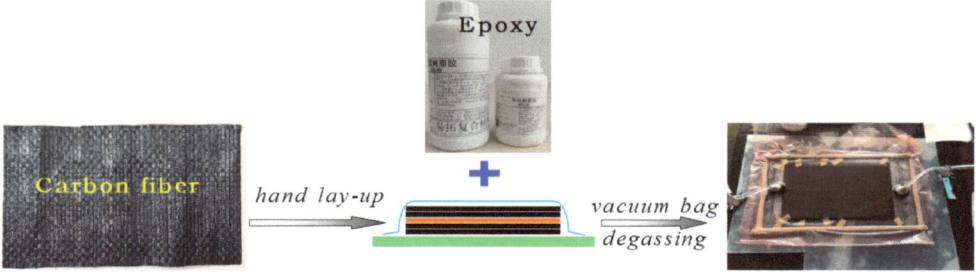

Figure 1. Production processing of composite samples.

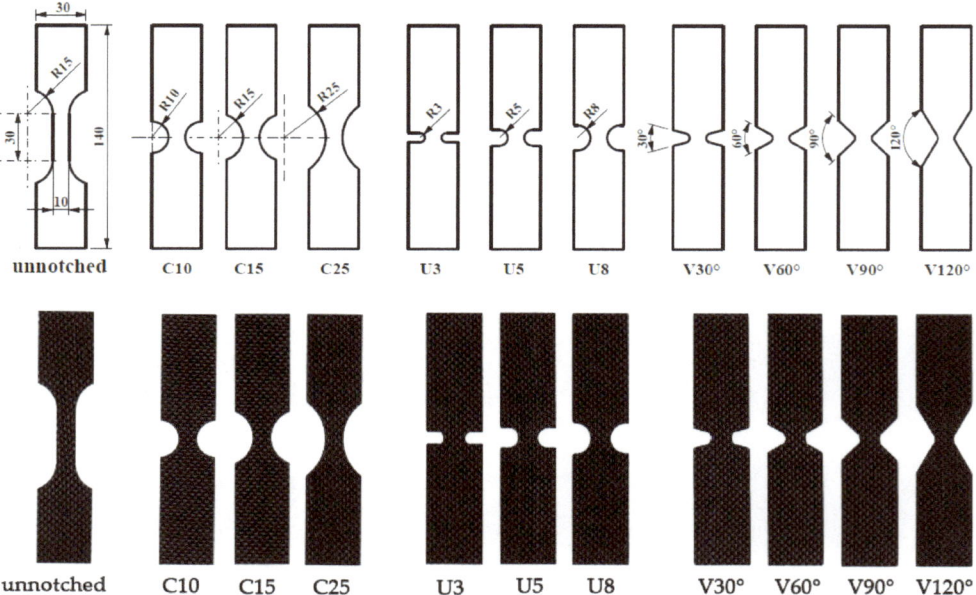

Figure 2. Specimen design of unnotched and various notched specimens, and photographs before testing.

2.3. Tension Tests

Tension tests are performed according to ASTM D638 for testing tensile properties of plastics, using a hydraulic-driven Instron testing machine 3365 series (Boston, MA, USA) and data acquisition card AD12 (Contec Corporation Ltd., Osaka, Japan). Tension tests are carried out by using a hydraulic-driven Instron tension tester at a constant crosshead displacement speed of 0.5 mm/min at room temperature, according to ASTM D3039 used for testing the tensile properties of polymer matrix composite materials. An extensometer with a gauge length of 50 mm is attached to the specimen to measure the average longitudinal strain. Three specimens of each composition are tested, and the average value is reported.

2.4. Torsion Tests

The specimens for the torsion test are similar to those of the tension test and measured using the electronic torsion testing machine NJ-S series (Beijing Timesun measurement and control technology Co., Ltd., Beijing, China). The cross-head speed for torsion tests is 10°/min, and Figure 3 shows a schematic diagram for the torsion test frame and the associated sample position in the frame.

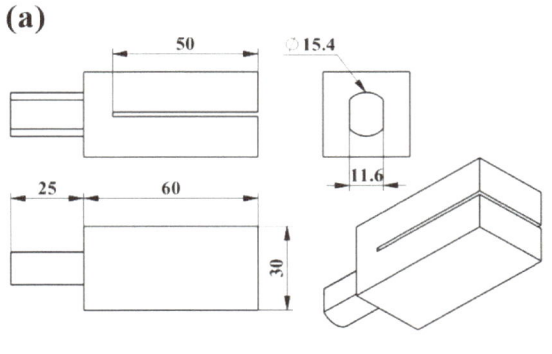 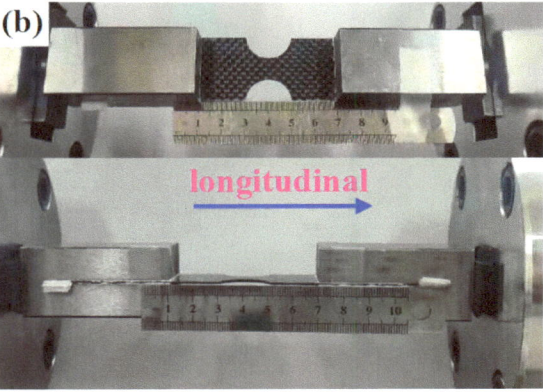

Figure 3. Schematic illustration of the torsion test frame in (**a**) and the sample position in the frame in (**b**); here, the upper picture is the top view, and the lower picture is the front view.

3. FEM Simulation

Eight-node quadrilateral, linear, thick-shell elements with six degrees of freedom per node are used. The user material subroutine (UMAT) is incorporated into the ABAQUS [32], and the geometric non-linearity is considered in the damage analysis. The geometric non-linearity and large deformation are accomplished by using the incremental loading and the NLGEOM parameter on the "STEP" option in ABAQUS. Damage modes in laminate structures strongly rely on the ply orientation, loading direction and panel geometry. There are four basic modes of failure that occur in a laminate structure. These failure modes are: matrix cracking, fiber–matrix shear failure, fiber failure and delamination. De-lamination failure is not considered here due to the high complication in modeling the interaction between plies. Mesh convergence is tested to ensure computational accuracy. To simulate the failure modes (matrix failure in tension or compression, fiber–matrix shear failure and fiber crack), the elastic properties are made linearly dependent on four field variables. The first field variable represents the matrix failure index, the second represents the fiber–matrix shear failure index and the third represents the fiber crack. All the elastic coefficients are dependent on the field variables to reflect the development of damage in the composites.

The finite element implementation of this progressive failure analysis is developed for the ABAQUS structural analysis program using the user subroutine USDFLD. ABAQUS calls the USDFLD subroutine at all material points of elements that have material properties defined in terms of the field variables. The subroutine provides access points to a number of variables, such as stresses, strains, material orientation, current load step and material name, all of which can be used to compute the field variables. Stresses and strains are computed at each incremental load step and evaluated by the failure criteria to determine the occurrence of failure and the mode of failure.

In this work, the subscript '−1' stands for the fiber direction, and subscript '2' is the direction perpendicular to the fiber direction. The Hashin's damage criteria [33] are adopted in order to simulate damage growth in each ply, and the failure criteria are written as

$$\text{Fiber failure in tension } (\sigma_1 \geq 0): e_f^2 = \left(\frac{\sigma_1}{S_{1T}}\right)^2 + \frac{2\tau_{12}^2/G_{12}^0 + 3\alpha\tau_{12}^4}{2S_{12}^2/G_{12}^0 + 3\alpha S_{12}^4} \quad (1)$$

$$\text{Fiber failure in compression } (\sigma_1 < 0): e_m^2 = \left(\frac{\sigma_1}{S_{1C}}\right)^2 + \frac{2\tau_{12}^2/G_{12}^0 + 3\alpha\tau_{12}^4}{2S_{12}^2/G_{12}^0 + 3\alpha S_{12}^4} \quad (2)$$

$$\text{Matrix failure in tension } (\sigma_2 \geq 0): e_m^2 = \left(\frac{\sigma_2}{S_{2T}}\right)^2 + \frac{2\tau_{12}^2/G_{12}^0 + 3\alpha\tau_{12}^4}{2S_{12}^2/G_{12}^0 + 3\alpha S_{12}^4} \quad (3)$$

Matrix failure in compression $(\sigma_2 < 0)$: $e_m^2 = \left(\dfrac{\sigma_2}{S_{2C}}\right)^2 + \dfrac{2\tau_{12}^2/G_{12}^0 + 3\alpha\tau_{12}^4}{2S_{12}^2/G_{12}^0 + 3\alpha S_{12}^4}$ (4)

Fiber/matrix shear failure : $e_{fm}^2 = \left(\dfrac{\sigma_1}{S_{1C}}\right)^2 + \dfrac{2\tau_{12}^2/G_{12}^0 + 3\alpha\tau_{12}^4}{2S_{12}^2/G_{12}^0 + 3\alpha S_{12}^4}$ (5)

where the factor $\alpha = 0.8 \times 10^{-14}$; these equations are used to monitor the evolutions of damage modes in each lamina, and the adhesion between laminas is assumed to be perfect, with no interface separation. The values of the field variables are set to be equal to zero in the undamaged state. After the failure index has exceeded 1.0, the associated user-defined field variable is set to be equal to 1. The mechanical properties in the damaged area are reduced appropriately, according to the property degradation model. The corresponding modulus and Poisson's ratio are assigned with a minimal value after the damage happens during the deformation. These damage modes are characterized by some internal state variables (SDV) as the output parameters.

To simulate the above failure modes, the elastic properties are made to be dependent on three field variables, $FV1$, $FV2$ and $FV3$. Four variables represent the matrix failure, fiber/matrix shearing failure and fiber failure, respectively. The values of the field variables are set to be equal to zero in the undamaged state. After a failure index has exceeded 1.0, the associated user-defined field variable is set to be equal to 1.0. The mechanical properties in the damaged area are reduced appropriately, according to the property degradation model defined in Table 2. For example, if the matrix failure criterion is satisfied, namely, $\sigma_2 \geq 0, e_m \geq 1$, then only $E_{22} \to 0$ and $\nu_{12} \to 0$ are degraded, while E_{11}, G_{12}, G_{13} and G_{23} equal the initial value. It is noted that after a damage happens, the iteration increment will be automatically selected and decreased gradually while increasing the damage degree. As the iteration increment is less than the predetermined minimum increment size, the computation process will be terminated correspondingly.

Table 2. The stiffness degradation rules based on the failure state (1—fiber direction, 2—transverse direction, FV1—matrix failure, FV2—fiber/matrix shear failure, FV3—fiber failure).

Intact	Matrix Failure	Fiber–Matrix Shear	Fiber Failure	All Failure Mode
E_{11}	E_{11}	E_{11}	$E_{11} \to 0$	$E_{11} \to 0$
E_{22}	$E_{22} \to 0$	E_{22}	E_{22}	$E_{22} \to 0$
ν_{12}	$\nu_{12} \to 0$	$\nu_{12} \to 0$	$\nu_{12} \to 0$	$\nu_{12} \to 0$
G_{12}	G_{12}	$G_{12} \to 0$	$G_{12} \to 0$	$G_{12} \to 0$
G_{13}	G_{13}	$G_{13} \to 0$	G_{13}	$G_{13} \to 0$
G_{23}	G_{23}	G_{23}	$G_{23} \to 0$	$G_{23} \to 0$
$FV1 = 0$	$FV1 = 1$	$FV1 = 0$	$FV1 = 0$	$FV1 = 1$
$FV2 = 0$	$FV2 = 0$	$FV2 = 1$	$FV2 = 0$	$FV2 = 1$
$FV3 = 0$	$FV3 = 0$	$FV3 = 0$	$FV3 = 1$	$FV3 = 1$

4. Results of Tension and Discussion

4.1. Tension Testing Results

Figure 4 shows the stress–strain curves of unnotched samples, where the used material properties in the present simulations are quoted from the other work [34], which are given as: $E_1 = 140$ GPa, $E_2 = 10$ GPa, $\nu_{12} = 0.3$, $G_{12} = G_{13} = 4.0$ GPa, $G_{23} = 4.32$ GPa, $S_{1T} = 2180$ MPa, $S_{2T} = 87.1$ MPa and $S_{12} = 165$ MPa. The stiffness degradation method allows us to obtain the stress softening stage after the failure point in Figure 4. Moreover, these above material properties are verified by comparing the present numerical simulations with the measured stress–strain relations.

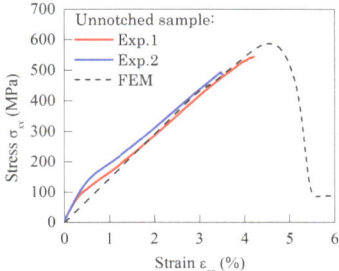

Figure 4. Stress–strain curves of unnotched samples under tensile loading (Exp.1 and Exp.2 denote two samples with the same sizes).

The mechanical characterization of the test samples in terms of tensile properties is conducted for different shapes and sizes of notches. The stress–strain relations are displayed in Figure 5, and it is noted that all samples show the brittle behaviors, exhibiting variations in both modulus and strength among these notched samples. Both numerical simulations and testing results present that all the circular notched samples exhibit similar stress–strain relations, and the C25 sample possesses the highest tensile strength, which is nearly similar to that of the unnotched samples. An increasing trend is found for tensile strength, suggesting that composite laminate is unaffected while increasing the notch diameter in such type of notched samples. The present predictions are in agreement with the measured results for the C-notched and U-notched samples, while the numerical calculations overestimate the tensile stiffness for the V-notched laminates and present lower predictions to their failure strengths. Furthermore, the stress–strain relations in the beginning stage of stretching are nonlinear, accompanying the marked transition from high stiffness to low stiffness. Such a change may originate from the inaccuracy in the ply orientation, off-centering clamp and sample cutting. The prediction errors in the ultimate strength between the numerical and measured results are in the range of 1~5% for the C-notch and range from 2% to 29.4% for the U-notch, while there is almost 14% underestimation for the V-notch.

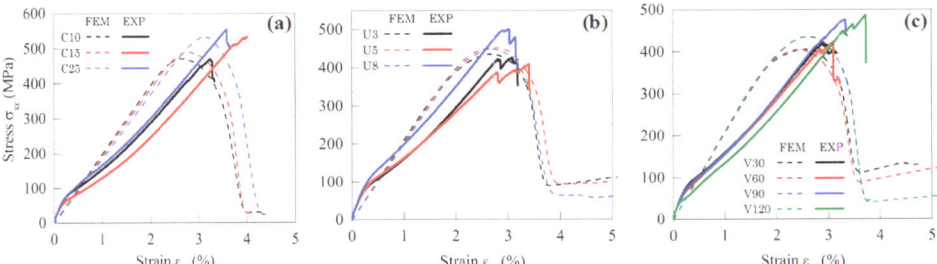

Figure 5. Stress–strain relations of C-notched samples in (**a**), U-notched samples in (**b**) and V-notched samples in (**c**).

Figure 6 shows the failure strength of C-notched, U-notched and V-notched samples; here, the errors are determined by measuring at least three specimens with the same size of notches. An increasing trend is observed for the tensile strength of the C-notch and U-notch specimens while increasing the notch radius, indicating that laminate structures become less sensitive while increasing the notch size. For the U-notched samples, it is noted that the changing tendency of strength with the notch radius is in accordance with what is found in Morgan's work [4]. In Morgan's research, the root radius of U-notched specimens changes from 0.5 mm to 20 mm, and the corresponding failure force increases from 4.19 kN up to 4.82 kN. Therefore, the dependence of the tensile strength on the root radius is the same as in our results. On the other hand, the tensile strength changes slightly with the notch angle for the V-notched samples, implying that the V-notched laminate structures have high tolerance to the V-notched shape. Based on the comparisons between the numerical results and testing datum, the present simulations could predict the tested failure strengths for the C-notch and U-notch samples, while the adopted numerical method for the V-notch specimens under-estimate the measured results.

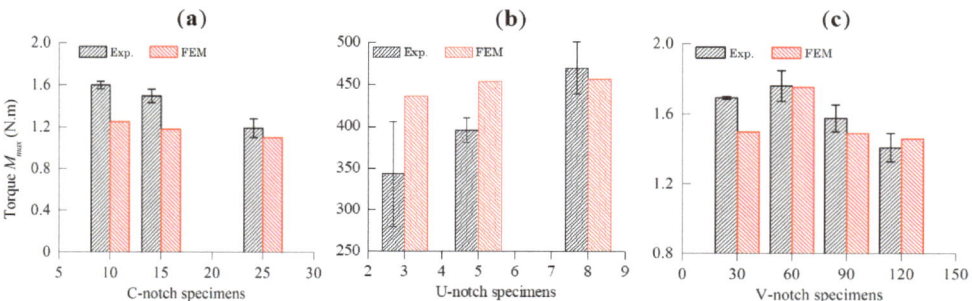

Figure 6. Dependence of the failure strength of the C-notched samples in (**a**), U-notched samples in (**b**) and V-notched samples in (**c**) with varied shapes.

4.2. Modeling of Tension Failure

Figure 7 shows the predicted damage modes in the C-notched, U-notched and V-notched specimens with varied sizes and angles. Here, fiber crack and shear damage are the two main failure modes in the laminate structures under tensile loading. It is observed that both the fiber crack and shear damage areas gradually become more and more wide along the C-shape notch while increasing the notch radius from 10 mm to 25 mm, and the severity of damage is relatively reduced; therefore, their corresponding tensile strengths increase with the decrease in these damage modes.

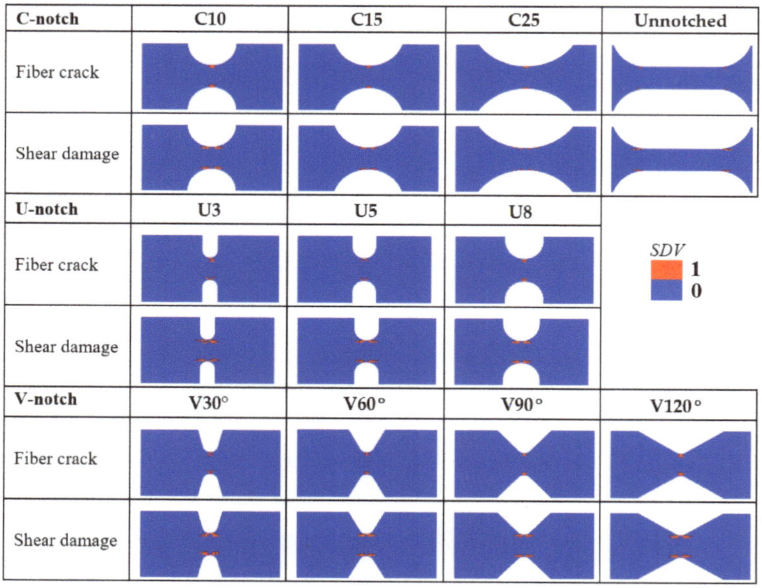

Figure 7. Damage modes in unnotched, C-notched, U-notched and V-notched laminates.

After examining the damage modes in the U-notched laminates in Figure 7, it is also found that both the fiber crack and shear damage areas are gradually reduced while increasing the notch radius from 3 mm to 8 mm, accompanied by an increment in their tensile strengths; therefore, the failure strength of the U-notched samples becomes closer to that of the unnotched sample.

Based on the damage contours of the V-notched samples in Figure 7, both the fiber crack and shear damage areas do not have an evident change in the samples with different notch angles from 30° to 120°, indicating that these notched laminates are insensitive to the notch angles, and, thus, their tensile strengths nearly remain unchanged.

4.3. The Damage Mechanism under Tension

The fractured specimens after tensile failure are demonstrated in Figure 8, and these failed samples display a brittle failure mode, which is attributed to the brittle nature of carbon fiber and epoxy resin. Long splitting, multi-mode fiber pull-out and delamination are observed. For the U8 and V120 samples, shear damage modes appear remarkably, and some carbon fibers along the transverse direction pull out, contributing to the higher loading capacity. A further increase in the notch radius results in the detriment of the laminates in terms of tensile strength. It is observed from failure modes that delamination formed in some local regions. These behaviors could be attributed to the brittle nature of the epoxy resin and the imperfect interphase cohesion between the carbon fiber and epoxy matrix. Failure mechanisms over the fracture surface indicate that some carbon fibers pull out, exhibiting the weak interfacial interaction induced by the vacuum infusion, and some micro-voids are still in existence. Moreover, the adhesion between the fiber and epoxy mainly stems from the mechanical friction between them, without any chemical bonding.

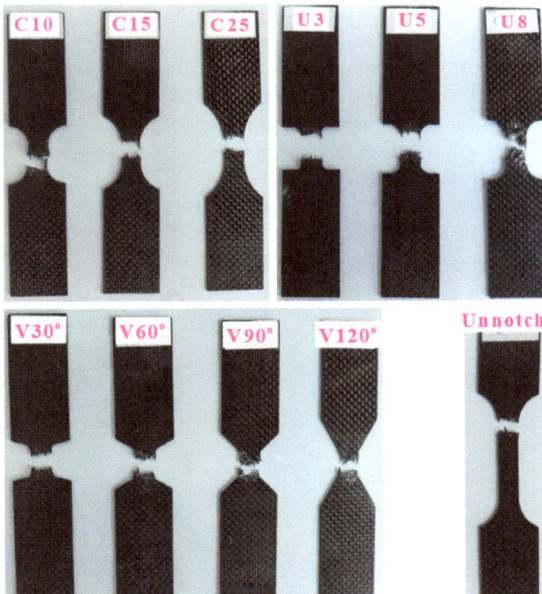

Figure 8. Failure surfaces of tension test samples.

5. Results of Torsion and Discussion

5.1. The Damage Mechanism under Tension

The mechanical characterization of torsion samples in terms of torsion properties is conducted for different shapes and sizes of notches. The torque–rotation angle relations are displayed in Figure 9; it is noted that all samples show the nonlinear behaviors, exhibiting progressive damage evolution among these notched samples. All the circular notched samples show the same stress–strain relations; the tensile strength of the C25 sample exhibited the highest value, which is nearly similar to that of unnotched samples. The present predictions are in good agreement with the measured results for all the notched laminates. Especially, the staged degradation associating the progressive multi-damage modes in the laminates is well captured in the predicted torque–rotation angle curves. These predicted torque–rotation angle curves present marked oscillations in the torque at the failure stage, which lie in the damage accumulations during the torsion deformation.

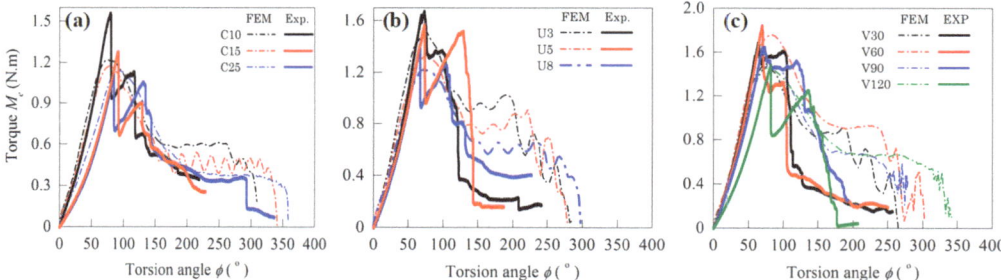

Figure 9. Torque–rotation angle (Me-φ) relations of the C-notched samples in (**a**), U-notched samples in (**b**) and V-notched samples in (**c**).

All the failure torques of these notched samples are summarized in Figure 10. The maximum torques of the C-notch and V-notch samples decrease with the increase in the notch size and angle, and the maximum torque becomes insensitive to the notch radius of the U-notched samples. The maximum torque decreases from 1.6 N·m to 1.2 N·m while increasing the notch radius in the C-notch specimens. Moreover, the tendency of the U-notched samples exhibits the 'U-shaped' variation, and the U5 sample processes the lowest torque among these specimens. Moreover, it should be mentioned that the changing tendency in the failure torque of the U-notched samples with the notch radius is also in agreement with that in Morgan's work [4]. In Morgan's research, the root radius of the U-notched specimens changes from 0.5 mm to 20 mm, and the corresponding failure torque increases from 2.76 N·m up to 3.66 N·m. Therefore, the dependence of the torsion strength on the root radius is the same as in our results.

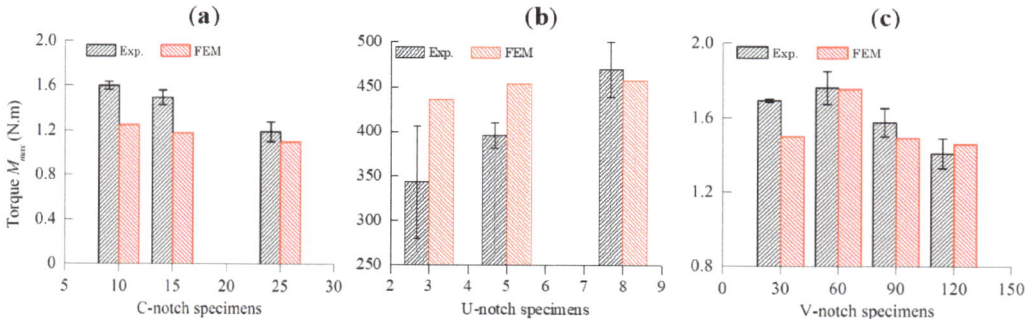

Figure 10. The measured and predicted failure torques of the C-notched samples in (**a**), U-notched samples in (**b**) and V-notched samples in (**c**).

Based on Hashin's damage criterion, the in-plane multi-damage modes in CFRP laminates can be well assessed by directly programing a user-subroutine in the ABAQUS code. However, the out-of-plane damage mode cannot be considered by this model, and thus, an additional cohesive zone with traction–separation laws should be involved to consider delamination growth, especially for the complicated multiple failure progress in the torsion loading.

5.2. Modeling of Torsion Failure

Figure 11 shows the damage modes in C-notched, U-notched and V-notched specimens under torsion. The present samples under torsion loading deform in the form of a very complicated shape and usually experience a large rotation angle before the final failure owing to their low torsion stiffness $G \times I_p$ (G is the shear modulus, and $I_p = \oint_A \rho^2 dA$ is the polar moment of inertia over the minimum transaction of the laminates, in which ρ is the distance from the shaft center to the element of the infinitesimal area dA). The highly twisted deformation constitutes the main part of the torsion damage evolution, in which shear damage and matrix failure happen during the torsion deformation, and some difference in the notch zones among these samples could be observed. For instance, there is nearly no discrepancy in the damage zones for the V-notched samples, and correspondingly, their maximum torques are also similar.

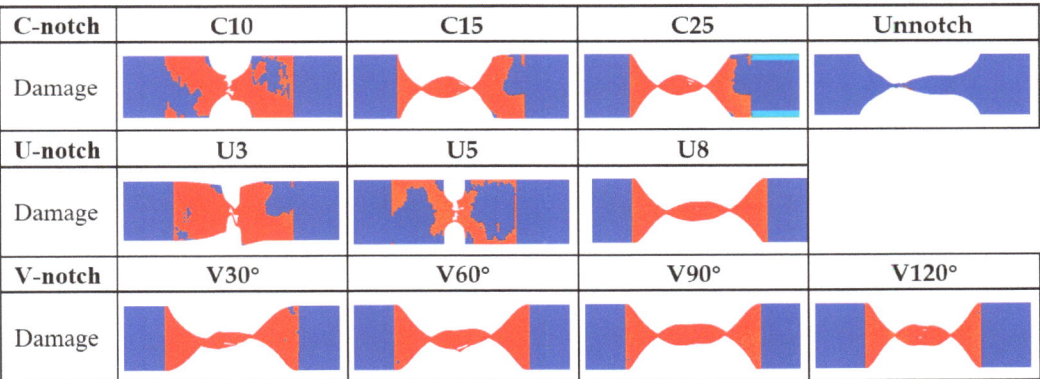

Figure 11. Damage mode by FEM simulations; red designates $SDV = 1$, and blue designates $SDV = 0$ in these contours.

5.3. The Torsion Mechanism

Figure 12 shows the failure surfaces of the torsion test samples, and it is observed that all the samples that fail in the form of fiber cracking follow a 45° spiral plane. The failure could be explained by the classic stress analysis, and the ±45° section plane is the principal plane in which the first principal stress σ_1 and third principal stress σ_3 lie. σ_1 would increase rapidly up to the critical strength of fibers under the torsion loading and then finally break the longitudinal fibers along this direction. The shear stress along the layers of the laminate structures under torsion also leads to the delamination damage between the plies due to their limited inter-laminar adhesive strength, which could be found from these failure surfaces in Figure 12. These behaviors could be attributed to the brittle nature of the sample, indicating imperfect interphase adhesion between fiber–epoxy interactions. Additionally, a striking permanent unrecoverable plastic deformation in the out-of-plane direction after the torsion loading is observed in all the samples, stemming from the progressive damage in the plies with the increase in the rotation angle.

Torsion provides a more complete picture of composite deformation for specific service scenarios as a complex deformation mode. However, the analysis of multi-damage progression under torsion should be challenging for CFRP laminates, which demands a comprehensive understanding in the next work.

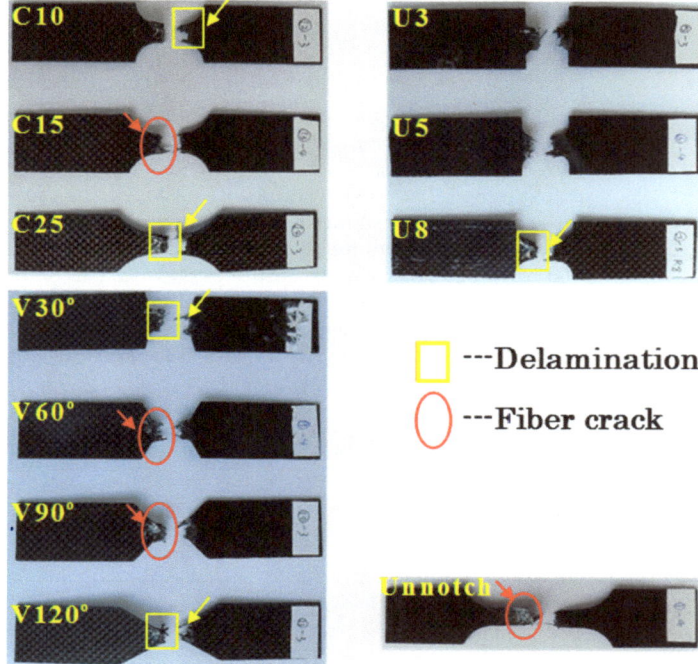

Figure 12. Failure surfaces of torsion test samples.

6. Conclusions

In this study, the mechanical characterization of the notched carbon fiber/epoxy composite laminates is investigated by incorporating notches with different shapes and sizes. The results indicate that the incorporation of notches significantly decreased the mechanical properties of carbon fiber/epoxy composites due to the fact that the stress concentration effect as a result of local stress changes with different sizes and shapes of notches. Several main conclusions are reached:

1. The tensile strengths of the C-notch and U-notch specimens increase while increasing the notch radius, and the tensile strength changes slightly with the notch angle for the V-notch samples.
2. The maximum torques of the C-notch and V-notch samples decrease while increasing the notch size and angle, and the maximum torque becomes insensitive to the notch radius of the U-notched samples.
3. All the notched samples displayed brittle failure modes, long splitting, multimode fiber pull-out and delamination. Shear damage modes appear remarkably, and some carbon fibers along the transverse direction pull out, contributing to the higher loading capacity.

Author Contributions: Conceptualization, Y.J. and Y.Z.; methodology, Y.Z.; validation, S.C.; formal analysis, S.C. and Y.Z.; writing—original draft preparation, Y.J.; writing—review and editing, S.C.; supervision, Y.J.; funding acquisition, Y.J. All authors have read and agreed to the published version of the manuscript.

Funding: This research was funded by the Fundamental Research Funds for the Central Universities, grant number NS2022012.

Data Availability Statement: The data that support the findings of this study are available from YPJ upon reasonable request.

Conflicts of Interest: The authors declare no conflict of interest.

References

1. Caminero, M.A.; Lopez-Pedrosa, M.; Pinna, C.; Soutis, C. Damage Monitoring and Analysis of Composite Laminates with an Open Hole and Adhesively Bonded Repairs Using Digital Image Correlation. *Compos. Part B Eng.* **2013**, *53*, 76–91. [CrossRef]
2. Olsson, R.; Iwarsson, J.; Melin, L.G.; Sjögren, A.; Solti, J. Experiments and Analysis of Laminates with Artificial Damage. *Compos. Sci. Technol.* **2003**, *63*, 199–209. [CrossRef]
3. Awerbuch, J.; Madhukar, M.S. Notched Strength of Composite Laminates: Predictions and Experiments—A Review. *J. Reinf. Plast. Compos.* **1985**, *4*, 3–159. [CrossRef]
4. Morgan, D.; Quinlan, S.; Taylor, D. Using the Theory of Critical Distances to Predict Notch Effects in Fibre Composites. *Theor. Appl. Fract. Mech.* **2022**, *118*, 103285. [CrossRef]
5. Koricho, E.G.; Khomenko, A.; Fristedt, T.; Haq, M. Innovative Tailored Fiber Placement Technique for Enhanced Damage Resistance in Notched Composite Laminate. *Compos. Struct.* **2015**, *120*, 378–385. [CrossRef]
6. Belgacem, L.; Ouinas, D.; Viña Olay, J.A.; Amado, A.A. Experimental Investigation of Notch Effect and Ply Number on Mechanical Behavior of Interply Hybrid Laminates (Glass/Carbon/Epoxy). *Compos. Part B Eng.* **2018**, *145*, 189–196. [CrossRef]
7. Torabi, A.R.; Pirhadi, E. Notch Failure in Laminated Composites under Opening Mode: The Virtual Isotropic Material Concept. *Compos. Part Engin.* **2019**, *172*, 61–75. [CrossRef]
8. Xu, X.; Paul, A.; Sun, X.; Wisnom, M.R. An Experimental Study of Scaling Effects in Notched Quasi-Isotropic Carbon/Epoxy Laminates under Compressive Loads. *Compos. Part Appl. Sci. Manuf.* **2020**, *137*, 106029. [CrossRef]
9. Serra, J.; Bouvet, C.; Castanié, B.; Petiot, C. Experimental and Numerical Analysis of Carbon Fiber Reinforced Polymer Notched Coupons under Tensile Loading. *Compos. Struct.* **2017**, *181*, 145–157. [CrossRef]
10. Serra, J.; Bouvet, C.; Castanié, B.; Petiot, C. Scaling Effect in Notched Composites: The Discrete Ply Model Approach. *Compos. Struct.* **2016**, *148*, 127–143. [CrossRef]
11. Lee, J.; Soutis, C. Measuring the Notched Compressive Strength of Composite Laminates: Specimen Size Effects. *Compos. Sci. Technol.* **2008**, *68*, 2359–2366. [CrossRef]
12. Wan, A.-S.; Xu, Y.; Xiong, J.-J. Notch Effect on Strength and Fatigue Life of Woven Composite Laminates. *Int. J. Fatigue* **2019**, *127*, 275–290. [CrossRef]
13. Torabi, A.R.; Pirhadi, E. On the Ability of the Notch Fracture Mechanics in Predicting the Last-Ply-Failure of Blunt V-Notched Laminated Composite Specimens: A Hard Problem Can Be Easily Solved by Conventional Methods. *Eng. Fract. Mech.* **2019**, *217*, 106534. [CrossRef]
14. Ghezzo, F.; Giannini, G.; Cesari, F.; Caligiana, G. Numerical and Experimental Analysis of the Interaction between Two Notches in Carbon Fibre Laminates. *Compos. Sci. Technol.* **2008**, *68*, 1057–1072. [CrossRef]
15. Llobet, J.; Maimí, P.; Turon, A.; Bak, B.L.V.; Lindgaard, E.; Carreras, L.; Essa, Y.; Martin de la Escalera, F. A Continuum Damage Model for Composite Laminates: Part IV-Experimental and Numerical Tests. *Mech. Mater.* **2021**, *154*, 103686. [CrossRef]
16. Lagattu, F.; Lafarie-Frenot, M.C.; Lam, T.Q.; Brillaud, J. Experimental Characterisation of Overstress Accommodation in Notched CFRP Composite Laminates. *Compos. Struct.* **2005**, *67*, 347–357. [CrossRef]
17. Vieille, B.; Coppalle, A.; Carpier, Y.; Maaroufi, M.A.; Barbe, F. Influence of Matrix Nature on the Post-Fire Mechanical Behaviour of Notched Polymer-Based Composite Structures for High Temperature Applications. *Compos. Part B Eng.* **2016**, *100*, 114–124. [CrossRef]
18. Ye, L.; Afaghi-Khatibi, A.; Lawcock, G.; Mai, Y.-W. Effect of Fibre/Matrix Adhesion on Residual Strength of Notched Composite Laminates. *Compos. Part Appl. Sci. Manuf.* **1998**, *29*, 1525–1533. [CrossRef]
19. Afaghi-Khatibi, A.; Ye, L.; Mai, Y.-W. An Experimental Study of the Influence of Fibre–Matrix Interface on Fatigue Tensile Strength of Notched Composite Laminates. *Compos. Part B Eng.* **2001**, *32*, 371–377. [CrossRef]
20. Czél, G.; Rev, T.; Jalalvand, M.; Fotouhi, M.; Longana, M.L.; Nixon-Pearson, O.J.; Wisnom, M.R. Pseudo-Ductility and Reduced Notch Sensitivity in Multi-Directional All-Carbon/Epoxy Thin-Ply Hybrid Composites. *Compos. Part Appl. Sci. Manuf.* **2018**, *104*, 151–164. [CrossRef]
21. Qiao, Y.; Deleo, A.A.; Salviato, M. A Study on the Multi-Axial Fatigue Failure Behavior of Notched Composite Laminates. *Compos. Part Appl. Sci. Manuf.* **2019**, *127*, 105640. [CrossRef]
22. Liu, P. *Localized Damage Models and Implicit Finite Element Analysis of Notched Carbon Fiber/Epoxy Composite Laminates under Tension*; Elsevier: Amsterdam, The Netherlands, 2021; pp. 69–107.
23. Liu, P. *Implicit Finite Element Analysis of Progressive Failure and Strain Localization of Notched Carbon Fiber/Epoxy Composite Laminates under Tension*; Elsevier: Amsterdam, The Netherlands, 2021; pp. 49–67.
24. Hu, J.; Zhang, K.; Cheng, H.; Liu, P.; Zou, P.; Song, D. Stress Analysis and Damage Evolution in Individual Plies of Notched Composite Laminates Subjected to In-Plane Loads. *Chin. J. Aeronaut.* **2017**, *30*, 447–460. [CrossRef]
25. Divse, V.; Marla, D.; Joshi, S.S. Finite Element Analysis of Tensile Notched Strength of Composite Laminates. *Compos. Struct.* **2021**, *255*, 112880. [CrossRef]
26. Ng, S.-P.; Tse, P.-C.; Lau, K.-J. Progressive Failure Analysis of 2/2 Twill Weave Fabric Composites with Moulded-in Circular Hole. *Compos. Part B Eng.* **2001**, *32*, 139–152. [CrossRef]

27. Ng, S.-P.; Lau, K.J.; Tse, P.C. 3D Finite Element Analysis of Tensile Notched Strength of 2/2 Twill Weave Fabric Composites with Drilled Circular Hole. *Compos. Part B Eng.* **2000**, *31*, 113–132. [CrossRef]
28. Riccio, A.; Di Costanzo, C.; Di Gennaro, P.; Sellitto, A.; Raimondo, A. Intra-Laminar Progressive Failure Analysis of Composite Laminates with a Large Notch Damage. *Eng. Fail. Anal.* **2017**, *73*, 97–112. [CrossRef]
29. Maa, R.-H.; Cheng, J.-H. A CDM-Based Failure Model for Predicting Strength of Notched Composite Laminates. *Compos. Part B Eng.* **2002**, *33*, 479–489. [CrossRef]
30. Laurin, F.; Carrere, N.; Maire, J.-F.; Mahdi, S. Enhanced Strength Analysis Method for Composite Open-Hole Plates Ensuring Design Office Requirements. *Compos. Part B Eng.* **2014**, *62*, 5–11. [CrossRef]
31. Chen, P.; Shen, Z.; Wang, J.Y. Prediction of the Strength of Notched Fiber-Dominated Composite Laminates. *Compos. Sci. Technol.* **2001**, *61*, 1311–1321. [CrossRef]
32. *ABAQUS Theory Manual*; HKS Inc.: Pawtucket, RI, USA, 2010.
33. Hashin, Z. Failure Criteria for Unidirectional Fiber Composites. *J. Appl. Mech.* **1980**, *47*, 329–334. [CrossRef]
34. Duan, M.M.; Shi, W.Y.; Zhang, X.Y. Numerical Analysis and Tests of Composite Laminates under Low-velocity Impact. *Struct. Environ. Engin.* **2020**, *2*, 26–31.

Disclaimer/Publisher's Note: The statements, opinions and data contained in all publications are solely those of the individual author(s) and contributor(s) and not of MDPI and/or the editor(s). MDPI and/or the editor(s) disclaim responsibility for any injury to people or property resulting from any ideas, methods, instructions or products referred to in the content.

Article

The Effect of Pulse Current on Electrolytically Plating Nickel as a Catalyst for Grafting Carbon Nanotubes onto Carbon Fibers via the Chemical Vapor Deposition Method

Kazuto Tanaka * and Shuhei Kyoyama

Department of Biomedical Engineering, Doshisha University, Kyotanabe 610-03948580, Japan
* Correspondence: ktanaka@mail.doshisha.ac.jp

Abstract: Carbon nanotubes (CNTs) can be directly grafted onto the surface of carbon fibers using the chemical vapor deposition method, in which nanometer-order nickel (Ni) particles, serving as catalysts, are plated onto the surface of carbon fibers via electrolytic plating. In our previous studies, in which a direct current (DC) was used to electrolytically plate Ni onto carbon fibers as a catalyst, the site densities and diameters of Ni particles increased simultaneously with the plating time, making it difficult to independently control the site densities and diameters of the particles. On the other hand, pulse current (PC) plating is attracting attention as a plating technique that can control the deposition morphology of nuclei. In this study, we clarify the effect of the parameters of the PC on the particle number per unit area (site density) and the particle diameters of Ni particles plated onto the surface of carbon fibers, using the PC to electrolytically plate Ni. Electrolytically plating Ni onto carbon fibers (via PC) after the removal of the sizing agent enable Ni particles with sparser site densities and larger diameters to be plated than those plated via DC. Using Ni particles with sparse site densities, it is shown that CNTs with sparse site densities can be grafted.

Keywords: carbon nanotubes (CNTs); carbon fibers; chemical vapor deposition method; electrolytically plating nickel; pulse current

Citation: Tanaka, K.; Kyoyama, S. The Effect of Pulse Current on Electrolytically Plating Nickel as a Catalyst for Grafting Carbon Nanotubes onto Carbon Fibers via the Chemical Vapor Deposition Method. *J. Compos. Sci.* **2023**, *7*, 88. https://doi.org/10.3390/jcs7020088

Academic Editor: Jiadeng Zhu

Received: 9 December 2022
Revised: 21 January 2023
Accepted: 13 February 2023
Published: 19 February 2023

Copyright: © 2023 by the authors. Licensee MDPI, Basel, Switzerland. This article is an open access article distributed under the terms and conditions of the Creative Commons Attribution (CC BY) license (https://creativecommons.org/licenses/by/4.0/).

1. Introduction

Carbon nanotubes (CNTs) are cylinders with diameters in the order of nanometers wound with graphene, which is a sheet-like structure consisting of single carbon atoms interconnected in a hexagonal lattice structure. CNTs can be generated via several methods, such as the arc discharge method, the laser ablation method and the chemical vapor deposition (CVD) method. The arc discharge method, in which carbon electrodes containing metal catalysts are evaporated by arc discharge, and the laser ablation method, in which graphite mixed with metal catalysts is evaporated by YAG laser irradiation, require temperatures above 1000 °C and produce only a few CNTs [1]. On the other hand, the CVD method, in which CNTs are generated on a metal catalyst by supplying gas containing carbon atoms in an inert atmosphere, can produce CNTs at relatively low temperatures and in large amounts [1]. CNTs have attracted attention as additives for fiber-reinforced plastics (FRPs) because of their advantages derived from graphene, such as nanoscale, high strength, high stiffness, and excellent electrical conductivity. Recently, one method of adding CNTs to carbon-fiber-reinforced plastics (CFRPs), a technique of grafting CNTs onto the surface of carbon fibers, has been developed to improve various properties of CFRPs [2–8]. This method has the advantage that agglomeration of CNTs is less likely to occur compared to the method in which CNTs are dispersed in the matrix. There are various metal catalysts that can be used in CNT grafting via the CVD method. Particularly, nickel (Ni) can be attached to the surface of carbon fibers more easily than iron (Fe) and cobalt (Co) via electrolytic plating; thus, CNTs can be selectively grafted onto the surface of carbon fibers in the CVD method. Sputtering is sometimes used to attach metal catalysts on substrates, but its drawbacks include the difficulty of attaching the catalyst on the inner surface of the carbon fibers in the fiber bundle and the high cost of the equipment. The disadvantage of Fe plating and Co plating is that the plating bath is easily oxidized and unstable. For the CVD method using ferrocene ($Fe(C_5H_5)_2$), which contains Fe as a metal catalyst element, it is difficult to control the location of CNT grafting, and this method requires a relatively high temperature of 750–1000 °C [9,10]. Ethanol is a carbon source used in CNT grafting via the CVD method because it is easier to handle than using methane and acetylene, and the grafting temperature can be lowered. Due to the lower processing temperature when using ethanol as a carbon source, the thermal degradation of carbon fibers can be reduced. In previous studies, CNTs were grafted onto carbon fibers at over 700 °C and the tensile strength of CNT-grafted carbon fibers was decreased due to degradation by oxidation [11,12]. We have developed a method to directly graft CNTs onto the surface of carbon fibers at a relatively lower temperature of 600 °C via the CVD method using ethanol (C_2H_5OH) as a carbon source, in which Ni particles from a nanometer order to a sub-micrometer order, serving as catalysts, are plated onto the surface of carbon fibers via electrolytic plating [13–15]. In studies regarding the grafting of CNTs onto carbon fibers via the CVD method using various catalysts, not only Ni, it has been reported that grafting CNTs onto the surface of carbon fibers can provide not only high fiber/matrix interfacial shear strength (IFSS) [16–20] due to the anchoring effect, but also a high level of resin impregnation into the fiber bundle [21,22] due to the capillary force of CNTs. In other words, it is expected that carbon-fiber-reinforced thermoplastics (CFRTPs) with superior mechanical properties can be obtained using CNT-grafted carbon fibers as they can improve both the fiber/resin IFSS and resin impregnation properties. On the other hand, excessive grafting of CNTs may cause agglomerated CNTs to be the crack initiation of a fracture [23] and inhibit resin impregnation into the fiber bundle [22], meaning the CNT grafting conditions need to be optimized.

Zhang et al. proposed two conditions regarding CNT grafting to provide CFRPs with superior mechanical properties with CNT-grafted carbon fibers as follows: a sparser site density of CNTs to reduce the number of stress concentrations at the ends of CNTs; and a longer length of CNTs to strengthen the resin layer [24]. However, a method to reduce the site density of CNTs on carbon fibers has not been yet developed. Using finite element analyses, Jia et al. suggested that there is a correlation between the diameter of CNTs grafted onto carbon fibers and interfacial shear strength [25]. This showed that it is necessary to develop a method to graft CNTs with a sparse site density and a large diameter onto carbon fibers. Meanwhile, past studies, in which silicon was used as a substrate, have shown that CNTs were grafted using catalyst particles with controlled numbers and diameters. It has been reported that the site densities and diameters of catalyst particles and the site densities and diameters of CNTs have correlations [26–29].

In our previous studies, in which a direct current (DC) was used for to electrolytically plate Ni onto carbon fibers as a catalyst, the site densities and diameters of Ni particles increased simultaneously with the plating time [30], making it difficult to independently control the site densities and diameters of these particles, i.e., the site densities and diameters of CNTs. In other words, if the plating time is shortened in order to reduce the site densities of particles to graft CNTs at a sparse site density, the particles' diameters also decrease. On the other hand, pulse current (PC) plating is attracting attention as a plating technique that can control the deposition morphology of nuclei. PC plating is generally used to obtain a plating film with a higher site density and a smaller diameter of nuclei by applying a larger current compared to DC plating [31,32]. Conversely, PC plating with a smaller current can promote nuclei growth while suppressing nucleation [33]. This can be applied to control the site density and diameter of Ni particles on carbon fibers. Pulse plating includes parameters such as current value (current, I), current density (J), plating time per pulse cycle (ON-time, t_{ON}), off time per pulse cycle (OFF-time, t_{OFF}) and the sum of the ON-time (actual total plating time, t), and it is necessary to clarify the effect of each of these parameters on the deposition morphology of Ni particles.

In this study, we clarify the effects of the parameters of the PC on the particle number per unit area (site density) and the particle diameter of Ni particles plated onto the surface of carbon fibers, using the PC to electrolytically plate Ni. Moreover, CNTs grafted onto carbon fibers via Ni particles with sparse site densities were observed using FE-SEM.

2. Materials and Methods

2.1. Materials and Sizing Agent Removal Method

Spread PAN-based carbon fibers (24 K, 23 mm width, 100 mm length, Nippon Tokushu Fabric, Japan) were used in this study. Generally, the sizing agent applied to the surface of carbon fibers serves as an insulating material for electrolytic plating. In our previous study, we reported that for carbon fibers with a sizing agent, Ni particles were less likely to be plated in areas where the sizing agent adhered, and Ni particles were concentrated in areas where the sizing agent did not adhere, resulting in an uneven distribution of Ni particles. Meanwhile, for unsized carbon fiber, Ni particles were plated evenly in all areas [16]. In this study, in order to reduce the effect of the sizing agent on Ni deposition, the sizing agent was removed. The sizing agent on the surface of the carbon fiber was removed via heat treatment at a temperature of 350 °C and at a holding time of 20 min in an argon (Ar) atmosphere using a chemical vapor deposition system (CVD system, MPCVD-70, Microphase, Japan). When removing sizing agents by heat treatment, it is important to not cause unnecessary chemical reactions. In a conventional oven, air oxidation of the fiber surface and sizing agent is a concern during heating. Therefore, heat treatment was performed in a CVD system in which the atmosphere can be replaced by Ar. In our previous study, the sizing agent on the surface of the carbon fiber was removed by the same methodology used in this study. We reported that the surface of the treated carbon fiber had stripe patterns, although the surface of the untreated carbon fiber had less stripes. The surface of the carbon fiber was also analyzed by an X-ray photoelectron spectrometer (ESCA, KRATOS ULTRA2, Shimadzu Corp. Japan). Graphite was not observed on the untreated carbon fibers, whereas it was detected in the treated carbon fibers [34]. The unsized carbon fibers are referred to as "Unsized CF".

2.2. Electrolytic Ni Plating onto Carbon Fibers

Ni particles, as a catalyst for CNT grafting via the chemical vapor deposition (CVD) method, were plated onto the carbon fiber surface using an electrolytic Ni plating method. The components of the plating bath, namely a "Watts bath", consisted of nickel sulfate hexahydrate ($NiSO_4 \cdot 6H_2O$, 240 g/L), nickel chloride hexahydrate ($NiCl_2 \cdot 6H_2O$, 45 g/L) and boric acid (H_3BO_3, 30 g/L). The temperature of the plating bath was 21 °C. The anode was connected to a Ni plate, and the cathode was connected to an Unsized CF, as shown in Figure 1. Two types of currents were used for electrolytic Ni plating, namely a direct current (DC) and a pulse current (PC). A schematic drawing of a PC wave is shown in Figure 2. For the DC, I = 200 Ma and t = 5 s were used as the standard conditions and for the PC, I = 200 Ma, t_{ON} = 10 ms, t_{OFF} = 100 ms and t = 5 s were used as the standard conditions. Electrolytic Ni plating was performed under various conditions as shown in Table 1. When assigning one parameter to the various sets of numerical values, other parameters remained constant at the standard condition shown in bold in Table 1. Carbon fibers plated via DC are referred to as "D-Ni-CF", and carbon fibers plated via the PC are referred to as "P-Ni-CF". They are collectively referred to as "Ni-CF". Ni particles on the surface of the carbon fiber were observed using a field-emission scanning electron microscope (FE-SEM, SU8020, Hitachi High-Technologies, Tokyo, Japan).

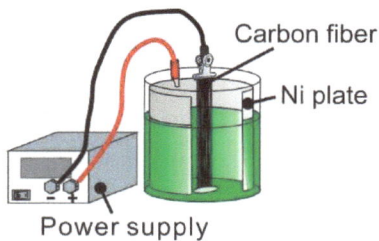

Figure 1. Schematic drawing of electrolytically plating Ni method.

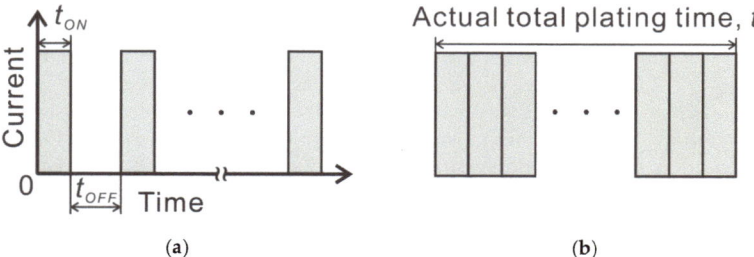

Figure 2. (**a**) Schematic drawing of a pulse current wave and (**b**) the definition of actual total plating time, sum of ON-time for PC plating.

Table 1. Parameters of plating condition.

Current (mA)	ON-Time (ms)	OFF-Time (ms)	Actual Total Plating Time (s)
100	10	10	5
200	50	100	10
300	100	1000	15

2.3. CNT Grafting onto Ni-CF via the CVD Method

CNTs were grafted onto the surface of Ni-CF via the CVD method using the CVD system, which is the same system used to remove the sizing agent. The grafting conditions of the CNTs were a grafting temperature of 600 °C, a grafting time of 10 min and 30 min, under a vacuum of −0.08 MPa and in an Ar atmosphere, and ethanol (C_2H_5OH) was supplied as a carbon source at 2 mL/min. CNT-grafted D-Ni-CF are referred to as "D-CNT_{10}-CF" and "D-CNT_{30}-CF", and CNT-grafted P-Ni-CF are referred to as "P-CNT_{10}-CF" and "P-CNT_{30}-CF". They are collectively referred to as "CNT-CF". CNT-CFs were observed using FE-SEM. The outer diameters of CNT-CFs were measured via image-processing program ImageJ.

3. Results and Discussion

3.1. Electrolytic Ni Plating onto Carbon Fibers

Figure 3 shows the FE-SEM observations of the Ni-CF and the elemental analyses performed by the EDS of the FE-SEM. The results show that Ni particles were successfully plated onto the carbon fibers. The number of Ni particles on the center of a carbon fiber was counted in a 5 μm × 5 μm square of FE-SEM images (Figure 4) via image-processing program, ImageJ. The counted number was divided by 25 and converted to the particle number per 1 $μm^2$ to obtain the site density. Figure 5 shows the enlarged FE-SEM images of Ni-CF. The diameters of the Ni particles were measured via ImageJ using these FE-SEM images. Figure 6 shows the particle site density and particle diameter of Ni-CF. It was found that P-Ni-CF had a sparser site density and a larger particle diameter than D-Ni-CF, although the same amounts of electrical current were supplied. It has been reported that when plating using pulse currents, most or part of the current is consumed to charge the electrical double-layer capacitance, resulting in less metal deposition [35]. However, in DC plating, the electrical double-layer capacitance is only charged once at the beginning of plating, while charging occurs during every pulse cycle in PC plating. In this study, the particle site density of Ni particles in P-Ni-CF is considered to be sparser than that in D-Ni-CF due to the same mechanism.

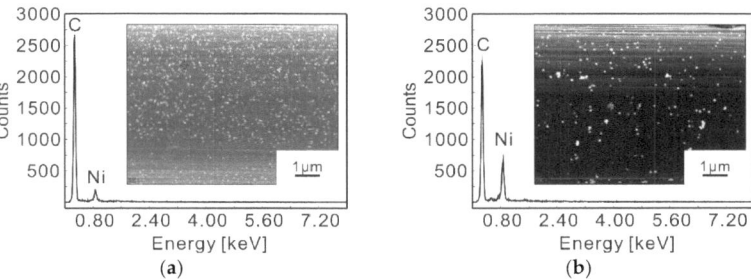

Figure 3. Elemental analyses of (**a**) D-Ni-CF (I = 200 mA, J = 3.79 A/m^2, t = 5 s) and (**b**) P-Ni-CF (I = 200 mA, J = 3.79 A/m^2, t_{ON} = 10 ms, t_{OFF} = 100 ms, t = 5 s) performed by EDS.

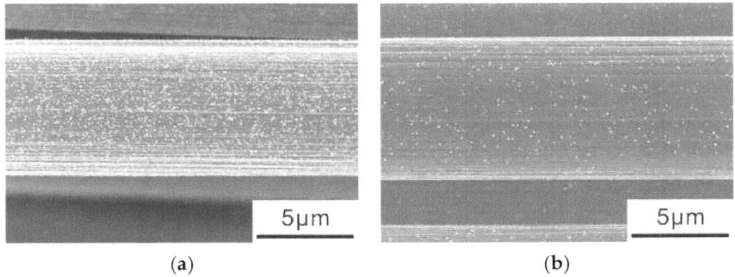

Figure 4. FE-SEM images of (**a**) D-Ni-CF (I = 200 mA, J = 3.79 A/m^2, t = 5 s) and (**b**) P-Ni-CF (I = 200 mA, J = 3.79 A/m^2, t_{ON} = 10 ms, t_{OFF} = 100 ms, t = 5 s).

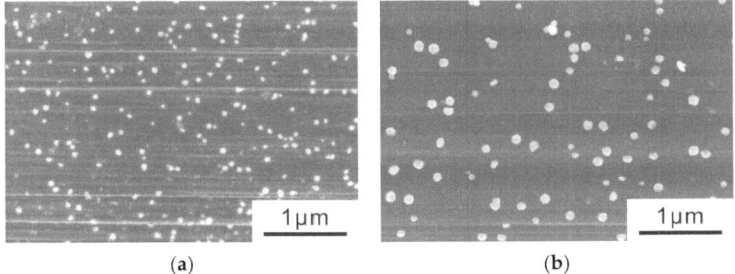

Figure 5. Enlarged FE-SEM images of (**a**) D-Ni-CF (I = 200 mA, J = 3.79 A/m^2, t = 5 s) and (**b**) P-Ni-CF (I = 200 mA, J = 3.79 A/m^2, t_{ON} = 10 ms, t_{OFF} = 100 ms, t = 5 s).

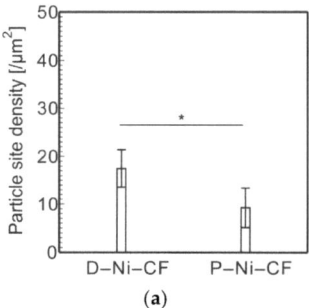 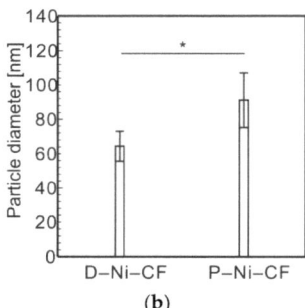

Figure 6. Measurement results of (**a**) particle site density (n = 27, mean ± S.D., * $p < 0.05$, Welch's t test) and (**b**) particle diameter (n = 45, mean ± S.D., * $p < 0.05$, Welch's t test) (I = 200 mA, J = 3.79 A/m^2, t_{ON} = 10 ms, t_{OFF} = 100 ms, t = 5 s).

The equations of ionic reaction on the cathode side are shown in Equations (1) and (2).

$$Ni^{2+} + 2e^- = Ni \qquad (1)$$

$$2H^+ + 2e^- = H_2 \qquad (2)$$

The cations involved in the reaction during plating are H$^+$ and Ni^{2+}. Figure 7 shows the possible mechanisms of (a) PC plating and (b) DC plating after the generation of Ni nuclei. The difference in hydrogen overvoltage [36] between nickel and carbon may have caused the following phenomena. Specifically, in the case of PC plating, when plating begins, H$^+$, which has a higher mobility than metal ions [37], first attaches to the cathode surface (i); H$^+$ on the surface of the Ni particle, which has a smaller hydrogen overvoltage than carbon, is hydrogenated (ii); and the hydrogen (H$_2$) leaves the surface of the Ni particle (iii). Secondly, when Ni^{2+} fills the space vacated by the release of H$_2$ (iv), Ni^{2+} is reduced and the diameter of the Ni particle increases (v), which is considered to be a cycle that results in the plating of Ni particles with a sparse particle site density and a large particle diameter. Meanwhile, in the case of DC plating, although H$^+$ initially adheres to the cathode surface (i), it is considered that, due to the long plating time, all H$^+$ on the cathode surface is hydrogenated regardless of the difference in hydrogen overvoltage (ii), and H$_2$ leaves the surface of the cathode (iii). Next, Ni^{2+} is reduced on the cathode surface (iv, v), and then, not only do the nuclei grow, but also, new nuclei are generated. As a result, Ni particles with thick site densities and small diameters are considered to be plated.

Figure 8 shows the particle site density and the particle diameter of Ni-CF plated using different currents. The particle site densities and particle diameters of both D-Ni-CF and P-Ni-CF tend to increase with the current.

Figure 9 shows the particle site density and the particle diameter of Ni-CF plated using different actual total plating times. The particle site density increases with the actual total plating time in both D-Ni-CF and P-Ni-CF. The particle diameter is proportional to the increase in the actual total plating time for DC, while for PC, the particle size is independent of the actual total plating time within the range of 5 s to 15 s. This result is confirmed in the FE-SEM images shown in Figure 10.

Figure 11 shows the particle site density and the particle diameter of P-Ni-CF plated using different ON-times. The particle site density tends to increase with the ON-time. However, the particle diameter is large at 10 ms, but small at 50 ms and above. Particularly, in order to increase the particle diameter, the ON-time should be set to a value as small as 10 ms.

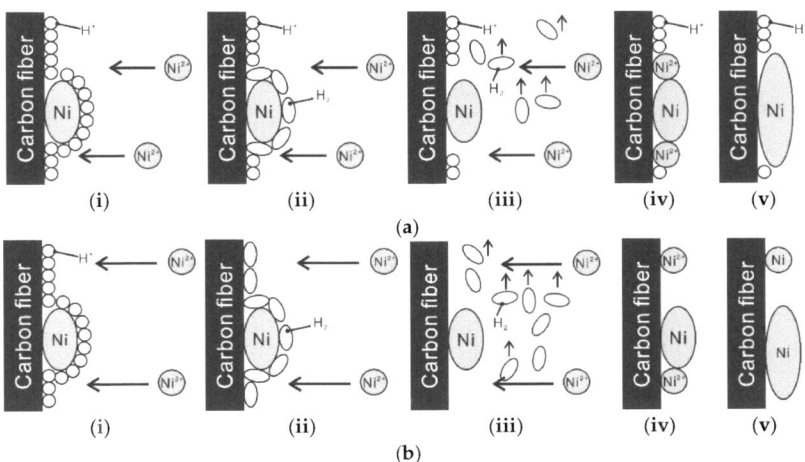

Figure 7. Schematic drawings of mechanisms of (**a**) PC plating and (**b**) DC plating.

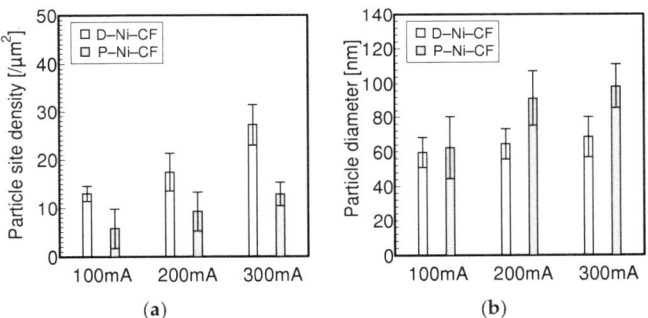

Figure 8. Measurement results of (**a**) particle site density (n = 27, mean ± S.D.) and (**b**) particle diameter (n = 45, mean ± S.D.) of Ni-CF plated using different currents (t_{ON} = 10 ms, t_{OFF} = 100 ms, t = 5 s).

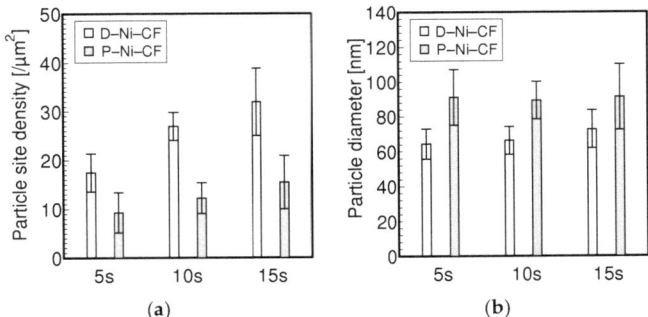

Figure 9. Measurement results of (**a**) particle site density (n = 27, mean ± S.D.) and (**b**) particle diameter (n = 45, mean ± S.D.) of Ni-CF plated using different actual total plating times (I = 200 mA, J = 3.79 A/m^2, t_{ON} = 10 ms, t_{OFF} = 100 ms).

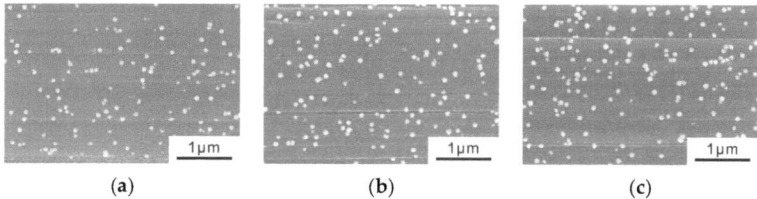

Figure 10. FE-SEM images of actual total plating time: (**a**) 5 s, (**b**) 10 s and (**c**) 15 s of P-Ni-CF (I = 200 mA, J = 3.79 A/m^2, t_{ON} = 10 ms, t_{OFF} = 100 ms, t = 5 s).

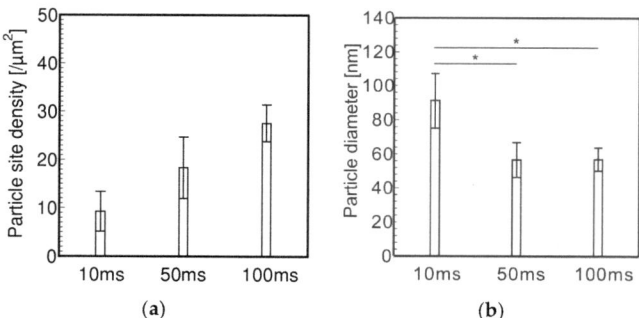

Figure 11. Measurement results of (**a**) particle site density (n = 27, mean ± S.D.) and (**b**) particle diameter (n = 45, mean ± S.D., * $p < 0.05$, Welch's t test with Bonferroni correction) of P-Ni-CF plated using different ON-times (I = 200 mA, J = 3.79 A/m^2, t_{OFF} = 100 ms, t = 5 s).

Figure 12 shows the particle site density and the particle diameter of the P-Ni-CF plate using different OFF-times. There is no difference in the particle site density and the particle diameter between OFF-times of 100 ms and 1000 ms. On the other hand, the particle site density is thicker and the particle diameters are smaller at an OFF-time of 10 ms. This behavior can be explained by the difference between the charging and discharging of the electrical double layer as shown in Figure 13. When the OFF-time of the PC is sufficiently longer, the charged electrical double layer is completely discharged and returns to its original uncharged state. Additionally, more current is consumed to charge the electrical double layer from the original uncharged state in the next ON-time. On the other hand, when the OFF-time is shorter, the next ON-time occurs before the electrical double layer is completely discharged and the electrical double layer is charged. In this case, the current consumed to charge the electrical double layer is considered to be smaller than the condition when the electrical double layer is charged from the original uncharged state. As shown in Figure 14, when the current is cut off at the OFF-time, the cations attracted near the cathode during the ON-time diffuse, and if the OFF-time is shorter, the cation diffusion distance is shorter and Ni^{2+} remains near the cathode until the next ON-time. H$^+$ diffuses faster than metal ions, and its direction is considered to be random, meaning the location of H$^+$ adsorption is considered to change each time during each pulse cycle. Therefore, Ni^{2+} near the cathode is considered to be adsorbed and reduced on the surface of the carbon fibers where H$^+$ is not adsorbed. As a result of this mechanism, when the OFF-time is relatively longer (above 100 ms), Ni particles with sparse site densities and large diameters are plated. In order to use PC to plate Ni particles with sparse site densities and large diameters, the OFF-time should be set to a value as large as 100 ms.

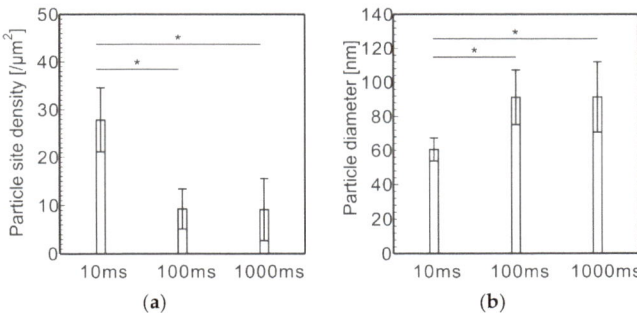

Figure 12. Measurement results of (**a**) particle site density (n = 27, mean ± S.D., * $p < 0.05$, Welch's t test with Bonferroni correction) and (**b**) particle diameter (n = 45, mean ± S.D., * $p < 0.05$, Welch's t test with Bonferroni correction) of P-Ni-CF plated using different OFF-times ($I = 200$ mA, $J = 3.79$ A/m^2, $t_{ON} = 10$ ms, $t = 5$ s).

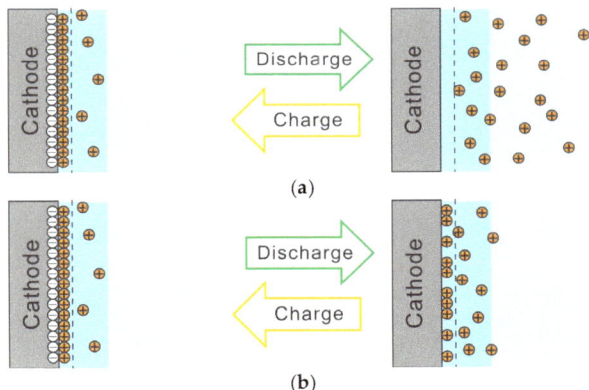

Figure 13. Schematic drawings of charge and discharge in (**a**) sufficiently longer OFF-time and (**b**) shorter OFF-time.

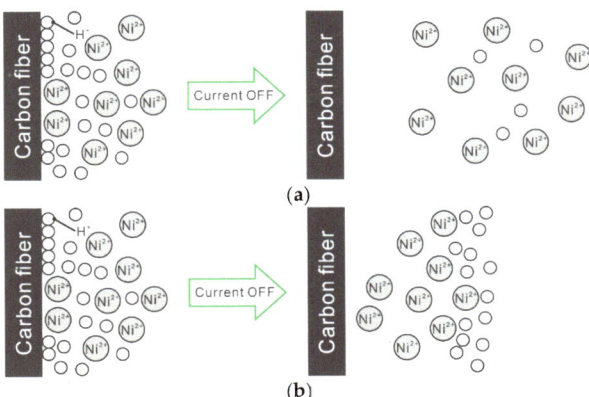

Figure 14. Schematic drawings of diffusion distance of cations in (**a**) sufficiently longer OFF-time and (**b**) shorter OFF-time.

3.2. CNT Grafting onto Ni-CF via the CVD Method

Figures 15 and 16 show the FE-SEM images of CNT-CF. They show that P-Ni-CF, which has a sparser site density of Ni particles, obtained CNTs at a sparser site density than D-Ni-CF. Figure 17 shows the magnified FE-SEM images of CNTs. The white particles are the Ni particles. Thick CNTs tend to be grafted from larger Ni particles. Figure 18 shows the outer diameter of the CNT-CF, indicating that the longer the CNT grafting time is, the larger the outer diameters of the CNTs that are grafted onto the carbon fibers. This result indicates that longer CNTs are grafted within longer grafting times.

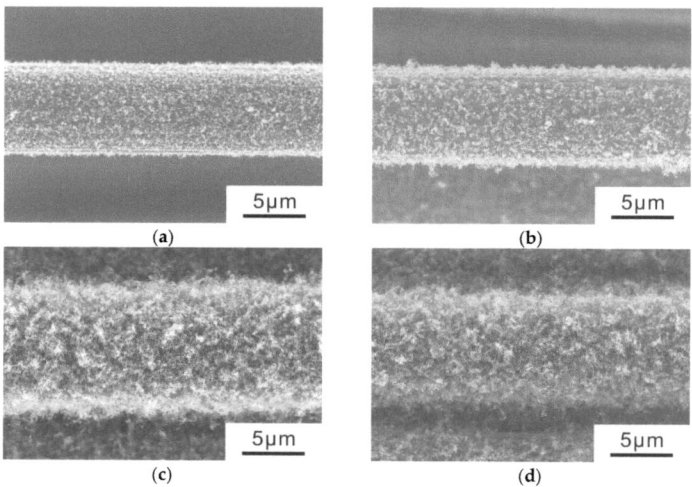

Figure 15. FE-SEM images of (**a**) D-CNT$_{10}$-CF, (**b**) P-CNT$_{10}$-CF, (**c**) D-CNT$_{30}$-CF and (**d**) P-CNT$_{30}$-CF.

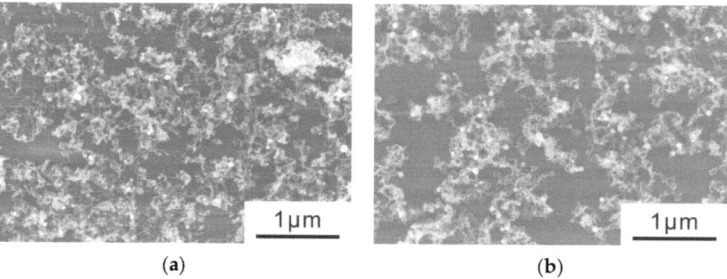

Figure 16. Enlarged FE-SEM images of (**a**) D-CNT$_{10}$-CF and (**b**) P-CNT$_{10}$-CF.

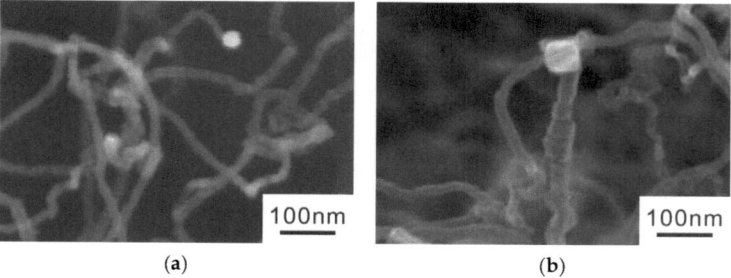

Figure 17. Magnified FE-SEM images of (**a**) thin CNT grafted from thin Ni particle and (**b**) thick CNT grafted from thick Ni particle.

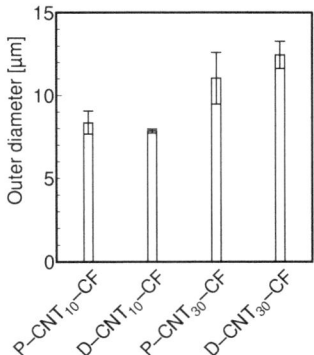

Figure 18. Outer diameter of CNT-grafted carbon fibers (n = 3, mean ± S.D.).

4. Conclusions

In this study, we have clarified the effects of the parameters of the pulse current (PC), such as the current, plating time per pulse cycle, off time per pulse cycle and sum of the plating time per pulse cycle, on the particle number per unit area (site density) and the particle diameters of nickel (Ni) particles plated onto the surface of carbon fibers using the PC to electrolytically plate Ni. Moreover, CNTs grafted via the chemical vapor deposition method (CVD method) were observed using FE-SEM. The investigation yielded the following conclusions:

1. Electrolytically plating Ni onto carbon fibers via the PC after the removal of the sizing agent attached to the surface of carbon fiber enabled Ni particles with sparser site densities and larger diameters to be plated rather than those plated via a direct current.
2. To electrolytically plate Ni with PC, it is necessary to shorten the plating time per pulse cycle to about 10 ms and lengthen the off time per pulse cycle to about 100 ms in order to plate Ni particles with sparse site densities and large diameters.
3. Using Ni particles with sparse site densities, CNTs with sparse site densities can be grafted.

Author Contributions: Conceptualization, K.T.; Data curation, S.K.; Funding acquisition, K.T.; Investigation, S.K.; Methodology, K.T.; Project administration, K.T.; Resources, K.T.; Supervision, K.T.; Validation, K.T. and S.K.; Visualization, S.K.; Writing—original draft, S.K.; Writing—review and editing, K.T. All authors have read and agreed to the published version of the manuscript.

Funding: This research was partially supported by JSPS KAKENHI (Japan Society for the Promotion of Science, Grant-in-Aid for Scientific Research (B), Grant Number JP19H02031).

Data Availability Statement: Not applicable.

Conflicts of Interest: The authors declare no conflict of interest.

References

1. Das, R.; Shahnavaz, Z.; Ali, M.E.; Islam, M.M.; Abd Hamid, S.B. Can We Optimize Arc Discharge and Laser Ablation for Well-Controlled Carbon Nanotube Synthesis? *Nanoscale Res. Lett.* **2016**, *11*, 510. [CrossRef]
2. Termine, S.; Trompeta, A.-F.A.; Dragatogiannis, D.A.; Charitidis, C.A. Novel CNTs grafting on carbon fibres through CVD: Investigation of epoxy matrix/fibre interface via nanoindentation. *MATEC Web Conf.* **2019**, *304*, 8. [CrossRef]
3. Pozegic, T.R.; Jayawardena, K.D.G.I.; Chen, J.-S.; Anguita, J.V.; Ballocchi, P.; Stolojan, V.; Silva, S.R.P.; Hamerton, I. Development of sizing-free multi-functional carbon fiber nanocomposites. *Compos. Part A Appl. Sci. Manuf.* **2016**, *90*, 306–319. [CrossRef]
4. Chen, J.; Xu, H.; Liu, C.; Mi, L.; Shen, C. The effect of double grafted interface layer on the properties of carbon fiber reinforced polyamide 66 composites. *Compos. Sci. Technol.* **2018**, *168*, 20–27. [CrossRef]
5. Naito, K. Development of high-performance and multi-functional carbon fibers. *JSPP* **2012**, *24*, 127–134. [CrossRef]

6. Wang, C.; Wang, Y.; Su, S. Optimization of Process Conditions for Continuous Growth of CNTs on the Surface of Carbon Fibers. *J. Compos. Sci.* **2021**, *5*, 111. [CrossRef]
7. Badakhsh, A.; An, K.-H.; Kim, B.-J. Enhanced Surface Energetics of CNT-Grafted Carbon Fibers for Superior Electrical and Mechanical Properties in CFRPs. *Polymers* **2020**, *12*, 1432. [CrossRef]
8. Lee, G.; Ko, K.D.; Yu, Y.C.; Lee, J.; Yu, W.-R.; Youk, J.H. A facile method for preparing CNT-grafted carbon fibers and improved tensile strength of their composites. *Compos. Part A Appl. Sci. Manuf.* **2015**, *69*, 132–1389. [CrossRef]
9. Naito, K.; Nagai, C. Effect of carbon nanotube surface modification on tensile properties of carbon fiber epoxy impregnated bundle composites. *Polym. Polym. Compos.* **2022**, *10*, 1176–1179. [CrossRef]
10. Komissarov, I.; Shaman, Y.; Fedotova, J.; Shulitski, B.; Zavadsky, S.; Kasiuk, J.; Karoza, A.; Pyatlitski, A.; Zhigulin, D.; Aleshkevych, P.; et al. Structural and magnetic investigation of single wall carbon nanotube films with iron based nanoparticles inclusions synthesized by CVD technique from ferrocene/ethanol solution. *Phys. Status Solidi C* **2013**, *10*, 1176–1179. [CrossRef]
11. Li, W.Z.; Wang, D.Z.; Yang, S.X.; Wen, J.G.; Ren, Z.F. Controlled growth of carbon nanotubes on graphite foil by chemical vapor deposition. *Chem. Phys. Lett.* **2001**, *335*, 141–149. [CrossRef]
12. Zhu, S.; Su, C.-H.; Lehoczky, S.L.; Muntele, I.; Ila, D. Carbon nanotube growth on carbon fibers. *Diam. Relat. Mater.* **2003**, *12*, 1825–1828. [CrossRef]
13. Tanaka, K.; Okuda, S.; Hinoue, Y.; Katayama, T. Effects of Water Absorption on Fiber Matrix Interfacial Shear Strength of Carbon Nanotube Grafted Carbon Fiber Reinforced Polyamide Resin. *J. Compos. Sci.* **2019**, *3*, 4. [CrossRef]
14. Tanaka, K.; Nishikawa, T.; Aoto, K.; Katayama, T. Effect of Carbon Nanotube Deposition Time to the Surface of Carbon Fibres on Flexural Strength of Resistance Welded Carbon Fibre Reinforced Thermoplastics Using Carbon Nanotube Grafted Carbon Fibre as Heating Element. *J. Compos. Sci.* **2019**, *3*, 9. [CrossRef]
15. Tanaka, K.; Habe, R.; Tanaka, M.; Katayama, T. Carbon Fiber Reinforced Thermoplastics Molding by Using Direct Resistance Heating to Carbon Nanofilaments Grafted Carbon Fiber. *J. Compos. Sci.* **2019**, *3*, 14. [CrossRef]
16. Tanaka, K.; Yamada, K.; Hinoue, Y.; Katayama, T. Influence of unsizing and carbon nanotube grafting of carbon fibre on fibre matrix interfacial shear strength of carbon fibre and polyamide 6. *Key Eng. Mater.* **2020**, *827*, 178–183. [CrossRef]
17. Lee, G.; Sung, M.; Youk, J.H.; Lee, J.; Yu, W.R. Improved tensile strength of carbon nanotube-grafted carbon fiber reinforced composites. *Compos. Struct.* **2019**, *220*, 580–591. [CrossRef]
18. Xiao, C.; Tan, Y.; Wang, X.; Gao, L.; Wang, L.; Qi, Z. Study on interfacial and mechanical improvement of carbon fiber/epoxy composites by depositing multi-walled carbon nanotubes on fibers. *Chem. Phys. Lett.* **2018**, *703*, 8–16. [CrossRef]
19. Qian, H.; Bismarck, A.; Greenhalgh, E.S.; Kalinka, G.; Shaffer, M.S.P. Hierarchical composites reinforced with carbon nanotube grafted fibers: The potential assessed at the single fiber level. *Chem. Mater.* **2008**, *20*, 1862–1869. [CrossRef]
20. Yao, Z.; Wang, C.; Qin, J.; Su, S.; Wang, Y.; Wang, Q.; Yu, M. Interfacial improvement of carbon fiber/epoxy composites using one-step method for grafting carbon nanotubes on the fibers at ultra-low temperatures. *Carbon* **2020**, *164*, 133–142. [CrossRef]
21. Lv, P.; Feng, Y.; Zhang, P.; Chen, H.; Zhao, N.; Feng, W. Increasing the interfacial strength in carbon fiber/epoxy composites by controlling the orientation and length of carbon nanotubes grown on the fibers. *Carbon* **2011**, *49*, 4665–4673. [CrossRef]
22. Tanaka, K.; Takenaka, T.; Katayama, T. Effect of grafted CNT on carbon fibers on impregnation properties of carbon fiber reinforced polyamide 6. *J. Soc. Mater. Sci.* **2020**, *70*, 670–677. [CrossRef]
23. Romanov, V.S.; Lomov, S.V.; Verpoest, I.; Gorbatikh, L. Modelling evidence of stress concentration mitigation at the micro-scale in polymer composites by the addition of carbon nanotubes. *Carbon* **2015**, *82*, 184–194. [CrossRef]
24. Zhang, L.; Greef, N.D.; Kalinka, G.; Bilzen, B.V.; Locquet, J.-P.; Verpoest, I.; Seo, J.W. Carbon nanotube-grafted carbon fiber polymer composites: Damage characterization on the micro-scale. *Compos. Part B-Eng.* **2017**, *126*, 202–210. [CrossRef]
25. Jia, Y.; Chen, Z.; Yan, W. A numerical study on carbon nanotube–hybridized carbon fibre pullout. *Compos. Sci. Technol.* **2014**, *91*, 38–44. [CrossRef]
26. Tu, Y.; Huang, Z.P.; Wang, D.Z.; Wen, J.G.; Ren, Z.F. Growth of aligned carbon nanotubes with controlled site density. *Appl. Phys. Lett.* **2002**, *80*, 4018–4020. [CrossRef]
27. Cheung, C.L.; Kurtz, A.; Park, H.; Lieber, C.M. Diameter-controlled synthesis of carbon nanotubes. *J. Phys. Chem. B* **2002**, *106*, 2429–2433. [CrossRef]
28. Gakis, G.P.; Termine, S.; Trompeta, A.-F.A.; Aviziotis, I.G.; Charitidis, C.A. Unraveling the mechanisms of carbon nanotube growth by chemical vapor deposition. *J. Chem. Eng.* **2022**, *445*, 1. [CrossRef]
29. Kato, T.; Hatakeyama, R. Growth of Single-Walled Carbon Nanotubes by Plasma CVD. *J. Nanotechnol.* **2010**, *2010*, 256906. [CrossRef]
30. Tanaka, K.; Okumura, Y.; Katayama, T. Effect of carbon nanotubes deposition form on carbon fiber and polyamide resin interfacial strength. *J. Soc. Mater. Sci.* **2016**, *65*, 586–591. [CrossRef]
31. Natter, H.; Hempelmann, R. Nanocrystalline copper by pulsed electrodeposition: The effects of organic additives, bath temperature, and pH. *J. Phys. Chem.* **1996**, *100*, 19525–19532. [CrossRef]
32. Puippe, J.-C.; SA, S.G. Qualitative Approach to Pulse Plating. *NASF Surf. Technol. White Pap.* **2021**, *85*, 6–14.
33. Lan, L.T.; Ohno, I.I.; Haruyama, S. Parusu mekki ni okeru kakuhassei oyobi seichou. (Generation and growth of nuclei in pulse current plating). *Denki Kagaku Oyobi Kogyo Butsuri Kagaku* **1983**, *51*, 167–168. (In Japanese) [CrossRef]
34. Tanaka, K.; Okuda, S.; Katayama, T. Effect of Air Oxidation of Carbon Fiber on Interfacial Shear Strength of Carbon Fiber Reinforced Thermoplastics. *J. Soc. Mater. Sci.* **2020**, *69*, 358–364. [CrossRef]

35. Ogata, M. Some problems in pulsed electrodeposition. *J. Met. Finish. Soc. Jpn.* **1988**, *39*, 180–184. [CrossRef]
36. En'yo, M. Recent research on hydrogen overvoltage. *Catalysis* **1956**, *13*, 155–177.
37. Shimao, K. Basic principles of electrophoresis. *Seibutsu Butsuri Kagaku* **1997**, *41*, 1–11. [CrossRef]

Disclaimer/Publisher's Note: The statements, opinions and data contained in all publications are solely those of the individual author(s) and contributor(s) and not of MDPI and/or the editor(s). MDPI and/or the editor(s) disclaim responsibility for any injury to people or property resulting from any ideas, methods, instructions or products referred to in the content.

Article

Self-Sensing Eco-Earth Composite with Carbon Microfibers for Sustainable Smart Buildings

Hasan Borke Birgin, Antonella D'Alessandro *, Andrea Meoni and Filippo Ubertini

Department of Civil and Environmental Engineering, University of Perugia, Via Goffredo Duranti 93, 06125 Perugia, Italy
* Correspondence: antonella.dalessandro@unipg.it; Tel.: +39-0755853910

Abstract: This paper proposes a new sustainable earth–cement building composite with multifunctional sensing features and investigates its properties through an experimental campaign. Earth and cement are proportioned as 2/7 in volume, while carbon microfibers are added in various amounts to achieve piezoresistivity, ranging from 0 to 1% with respect to the weight of the binder (i.e., earth + cement). The proposed material couples the construction performance with self-sensing properties in order to monitor the structural performance during the servile life of the building. The use of earth in the partial replacement of cement reduces the environmental footprint of the material while keeping sufficient mechanical properties, at least for applications that do not require a large load-bearing capacity (e.g., for plasters or for low-rise constructions). This paper analyzes the electrical and sensing behavior of cubic and beam samples through electrical and electromechanical tests. The results show that the samples with a filler percentage near the percolation zone, ranged between 0.025 and 0.25%, exhibit the best performance. From the cyclical compressive tests and linear developed models, it could be deduced that the filler content of 0.05% of carbon fibers, with respect to the binder weight, represents the best-performing smart composite for further investigation at higher scales. As demonstrated, the selected mix generated clear strain-sensing electrical signals, reaching gauge factors over 100.

Keywords: smart sustainable materials; multifunctional composites; load and strain sensing; piezoresistive self-sensing materials; structural health monitoring; carbon fillers

Citation: Birgin, H.B.; D'Alessandro, A.; Meoni, A.; Ubertini, F. Self-Sensing Eco-Earth Composite with Carbon Microfibers for Sustainable Smart Buildings. *J. Compos. Sci.* **2023**, *7*, 63. https://doi.org/10.3390/jcs7020063

Academic Editor: Jiadeng Zhu

Received: 31 December 2022
Revised: 21 January 2023
Accepted: 31 January 2023
Published: 6 February 2023

Copyright: © 2023 by the authors. Licensee MDPI, Basel, Switzerland. This article is an open access article distributed under the terms and conditions of the Creative Commons Attribution (CC BY) license (https://creativecommons.org/licenses/by/4.0/).

1. Introduction

Cement and cement-based composite materials are the most used structural materials in modern civil engineering [1] due to their mechanical properties, ease of production, and adaptability to different structural settings [2]. Nevertheless, their impact on the environment is quite high, considering the emissions of CO_2 and other greenhouse gasses, waste production, and high consumption of resources and energy during their production and use [3,4]. Recently, the growing interest in the development of structured approaches to safeguard human health and the environment has led to the diffusion of a multitude of strategies to reduce the impact of the concrete industry [5]. The use of less harmful components and inclusions in the design of mixtures, the optimization of production processes, and the enhancement of strength and durability are among the most frequently used approaches to mitigate the environmental impact of building materials [6,7]. Novel advanced materials, such as fiber-reinforced polymers (FRPs) [8,9] and composites [10,11], which exhibit enhanced properties such as a high strength-to-weight ratio, high durability, and an improved resistance to environmental effects [12,13], are now available to engineers and technicians as effective alternatives to traditional construction materials [14,15]. The use of natural or recycled components in the design of mixtures represents a further exemplification of strategies to reduce the environmental impact of cementitious materials [16,17].

Among the available natural materials, the earth stands forward as a promising alternative or partial substitute of cement. Earth–cement-based composites have been used for a long time as construction materials and are relatively widespread, especially thanks to their ecological sustainability [18]. They can be produced with relatively low energy consumption, do not require combustion processes, and are compatible with on-site production [19]. Moreover, earth concrete can be easily recycled, thus significantly reducing its ecological footprint [20]. Despite these benefits, the weakness of earth–cement-based composites lies in their mechanical properties. In particular, the tensile strength of cement mixtures containing earth is lower than that of normal concrete, making them even more susceptible to brittle collapse mechanisms [21,22]. The flexural, tensile, and compression strength of the earth concrete can be raised through additives to the binder matrix [23–25]. Some examples of reinforcing earth–cement-based matrices include the dispersion of polypropylene, steel fibers [26], and natural fibers [27,28]. The strengthening fibers can be obtained through recycling, thus further decreasing the environmental footprint of the resulting material [29,30]. Examples include the use of granite cutting residue [31], fibers from recycled tire and steel [32], as well as natural fibers [33]. The dispersion of fibers inside earth concrete can introduce multifunctionalities to the base material, as well as improve its mechanical properties [18,34]. Besides the strength and durability issues, the lack of adequate standards for designing earth-based constructions is another key aspect to be solved in order to achieve feasible practical applications of earth concrete [35]. Possible strategies for the safe application of this construction material are the use of stabilizers [36,37] or the integration of enhanced capabilities [38,39].

Nowadays, multifunctional structural materials are receiving significant interest [40,41]. Among the new functionalities that can be integrated within traditional construction materials through the addition of specific inclusions during their production, the integration of strain-sensing capabilities is a quite attractive strategy to obtain structures and infrastructures capable of self-monitoring their health state in a real-time fashion [34,42]. As demonstrated in the literature, structural materials with strain-sensing capabilities can be effective for the quick assessment of the structural integrity of constructions [43–45]. The literature works also indicate the importance of the optimal choice of the fillers, the design of the sensing material units [46], the tailoring of the electromechanical configuration, and of the post-processing of the signals [47,48] and the investigation of issues and strengths of self-sensing materials in real applications on constructions [49,50]. The analysis of the literature has also pointed out the importance of obtaining a homogeneous dispersion of the fillers within the matrix [51,52]. The authors have recently carried out several experimental campaigns in order to explore the potentials of smart cementitious materials [53,54] and composites [55,56] in applications to structural engineering. In this study, the authors build on their past experience to propose the development of a new sustainable earth–cement-based construction material with strain-monitoring capabilities. Following this approach, the present research explores and discusses key topics for the tailored production and multifunctional applications of innovative eco-friendly smart earth concretes with inclusions of carbon microfibers. The specific scopes can be pointed out in the following steps: (i) the development of a reproducible manufacturing process and (ii) the exploitation of the multifunctionalities of the proposed material using self-sensing properties.

The manuscript is organized as follows: Section 2 introduces the materials, the samples' characteristics, and the devices adopted for the various tests. Section 3 defines the tests carried out on the different types of samples, whose results are presented and discussed in Sections 4 and 5, respectively. Section 6 concludes the paper.

2. Materials and Setup

2.1. Components and Preparation of Samples

Small cubic samples with 5 cm sidelength and medium-sized beam samples with dimensions of $4 \times 4 \times 16$ cm were produced as standard samples for mortars. The doping of CMF for the cubes ranged from 0 to 1%, calculated as weight ratios to the earth–cement weight. Two sets were made for the cube samples, the first set containing one test sample out of each mixture with varying CMF doping, and the second set containing five samples per mixture with CMF doping levels of 0.025, 0.05, and 0.1%. Two additional beam samples were also prepared out of mixtures with CMF doping levels of 0.025, 0.05, and 0.1%.

The dry components of the earth composite of this study were a constitution of excavated earth, aggregates with particle sizes between 0 and 8 mm, and cement. The corresponding mixing proportions calculated per mass were 7/16, 7/16, and 2/16 for the earth, aggregates, and cement, respectively, according to the reference study [36]. The mixing to fabricate the earth composite of this study started with the manual mixing of the earth, aggregates, and cement, followed by the addition and mixing of CMF (Figure 1a). Lastly, the mixing was finished by adding water to the homogeneous mixture of dry components, creating the dough of material, providing sufficient workability for pouring the compound into the molds (Figure 1b). During the molding phase of the mixture, the material was poured into molds and manually compacted to reduce the presence of voids inside the dough. Copper wires were placed at a constant distance, in the central line of the sample, as electrodes for the electrical measurements. The samples were then cured for 28 days under laboratory conditions (Figure 1c).

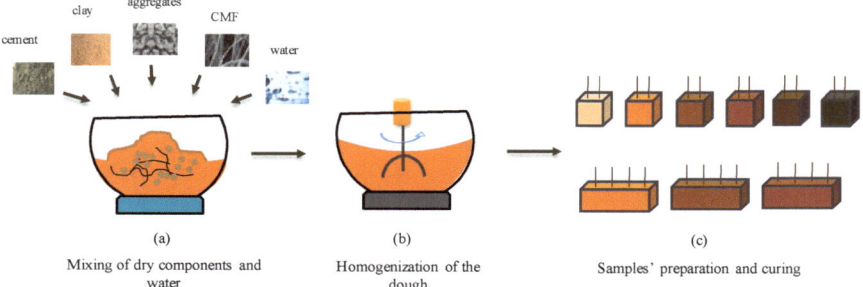

Figure 1. The manufacturing steps for smart earth samples: (**a**) addition of ingredients inside mixing bowl; (**b**) mixing of the compound until the homogeneity of the composite dough; (**c**) preparation of small cube and medium-size beam samples from the composite dough, together with the placement of copper electrodes.

Cement is portland cement 42.5R, earth is clay provided by a local regional quarry close to a brick factory in Umbria, Italy. CMF is from SGL Carbon, with the product type SIGRAFIL®, having 5 μm diameter and 6 mm fiber length. The electrodes are commercial copper wires with 0.8 mm diameter. A critical aspect of cementitious mixtures with carbon fillers results in the difficult dispersion of the fibers in the matrix for obtaining homogeneity, and in the integrity of the fibers after the preparation procedure. For this purpose, optical microscope and scanning electron microscope inspections were carried out on fragments of hardened material, as shown in Figure 2. Both types of micrographs demonstrate that the carbon fibers appear not damaged and are well dispersed in the composite.

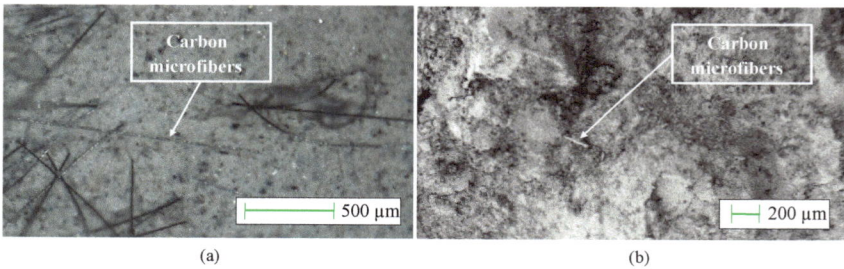

Figure 2. (**a**) Optical microscope and (**b**) scanning electron microscope inspections on fragments of hardened material.

2.2. Experimental Setups

The tests presented in this study were carried out to analyze the electrical resistivity of the developed earth material at different doping levels, as well as the piezoresistive capabilities of the proposed composite for strain sensing. The resistivity of the material samples was obtained through electrical readings taken from unloaded samples. The piezoresistivity was assessed through electromechanical tests performed on the test samples subjected to compressive cyclical loads. The electrical output, which measured the instantaneous electrical resistance of the test sample, was recorded by dedicated hardware capable of reading analog voltage. The recordings were later post-processed to observe the correlation of the test samples' electrical resistance to the induced load and strain. Figure 3 illustrates the experimental setups for resistance readings and electromechanical tests conducted on the different types of samples and setups, i.e., cubes with embedded electrodes placed at a distance of 2 cm and small beams with four aligned electrodes to obtain three subsequent measuring channels.

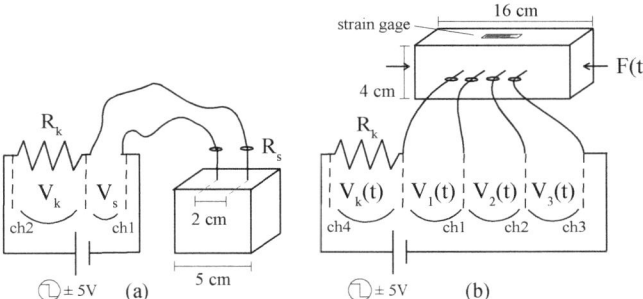

Figure 3. Experimental setups for electrical measurements on smart earth samples: (**a**) cubes for percolation investigation; (**b**) small-scale beams for electromechanical tests and piezoresistivity analysis.

To interpret the setups shown in Figure 3, the voltage applied to the samples was a 1 Hz biphasic square wave with a 10 V peak-to-peak difference, in order to reduce the polarization effect that may cause a constant positive drift in the resistance–time history and to avoid misleading results. For the data acquisition system during the tests, National Instruments devices and LABVIEW® environment were adopted for testing purposes. NI-PXIe 4138 module was used as voltage source apparatus. The analog voltage reader was a 32-channel NI-PXIe 4302 analog-to-digital converter module. The above-listed modules were mounted inside a NI-PXIe 1092 chassis operated in Windows system. The voltage–time history during the tests was sampled with 10 Hz, allowing the selection of 80% of the charge point on the positive phase of the measured square wave. The post-processing of recorded signals was carried out under LABVIEW® and MATLAB® environments.

For the mechanical equipment of the tests, a software-controlled dynamic compression machine Advantest 9 provided by Controls s.r.l. with a maximum load of 250 kN was used for load application. The induced strains to the material were measured through 2 cm monoaxial strain gauges by KYOWA with an internal resistance of 120 Ohms and a gauge factor of 2.1. The strain gauges attached to the samples were recorded by Advantest9 software simultaneously with the recording of the applied load–time history.

3. Electrical Characterization and Sensing Evaluation

The experimental investigation started with the percolation study for the determination of the optimal composite mixture with suitable electrical conductivity. As shown in Figure 3a, the percolation study was conducted by connecting the test sample cubes to an electrical circuit consisting of a biphasic voltage input and a shunt resistor of known value (i.e., R_k = 100 kΩ) placed in series. Ohm's law (Equation (1)) was employed to calculate the sample resistance–time history.

$$R_s(t) = R_k \frac{V_s(t)}{V_k(t)} \quad (1)$$

where resistance–time history of the sample in ohms ($R_s(t)$) derived from time recordings of voltages spent on the sample and resistor, denoted by $V_s(t)$ and $V_k(t)$ (in volts), through data acquisition channels ch1 and ch2. R_s was determined as the mean value of the readings with a duration of 10 seconds, which corresponded to a set of 10 samples after post-processing of the biphasic signal. Two data sets were investigated for the electrical resistance readings. As previously described, the first one was related to the cubes samples with CMF dopings at weight ratios of 0–0.025–0.050–0.100–0.250–0.500–1.000% of the earth–cement matrix weight. The second set was manufactured for a more detailed inspection of the percolation zone of CMF by containing six samples at the specific CMF weight ratios of 0.025–0.050–0.100% of the earth–cement matrix.

The load-sensing capabilities of the newly produced composite matrix were tested through electromechanical tests. As introduced in previous sections, an electromechanical test assessed and proved the strain-sensing power of the inspected sample by comparing the variation–time history of the electrical resistance with the induced strain–time history (ε, positive in compression). The tests for obtaining the strain-sensing performance of the composite material and manufactured sensors were structured following reference [57]. Accordingly, the strains were induced by cyclical triangular compression-applied loads with peak load magnitudes of 0.75–0.75–1.0–1.0 kN with a loading rate of 0.1 kN/s and a precompression load of 0.5 kN. A sample was instrumented during the electromechanical test, and the applied load–time history is plotted in Figure 4a. Referring to the illustration in Figure 3b, the small beams were instrumented by strain gauges, and the electrical readings were taken through three sequential electrode couples located in the central line of the samples (V_1, V_2, and V_3). In this way, the three sections between the three electrodes' couples behaved as three resistors connected in series. An additional voltage reading was taken through the shunt resistor (V_k) which served as the reference value for measurements from the composite sample. Considering the three monitored segments of the small beams, the related electrical resistances were calculated for all three segments by employing Equation (1) and using $V_i(t)\ \forall i \in \{1,2,3\}$ instead of V_s, resulting in $R_i(t)\ \forall i \in \{1,2,3\}$. The obtained electrical resistance time histories of the samples correlated with the strain time histories, with respect to the governing equation of the load sensing, which was adopted from the reference study [55] given by Equation (2), under the assumption uniform material properties could be post-processed to calculate the gauge factor λ_i:

$$\lambda_i = -\frac{\frac{\Delta R_i}{R_i}}{\varepsilon} = -\frac{\frac{\Delta \rho_i}{\rho_i}}{\varepsilon} + (1 + 2\nu) \quad \forall i \in \{1,2,3\} \quad (2)$$

where the unitless variable λ_i is the gauge factor of section i, ρ is the piezoresistivity of the material in ohm·meters, ν is the Poisson's ratio, ε is the induced strain in microstrains, and i is the segment index of measured electrode pairs. The gauge factor represents the measure of the sensitivity of the smart sensors, obtained with different levels of CMF doping. It is the coefficient of the linear model established between the measured variation in resistance and induced strain. Larger values for the gauge factor are desirable for minimizing the influence of signal noises, therefore improving the sensing quality. As shown by Equation (2), the value of the gauge factor is directly determined by the piezoresistivity of the material (ρ) and volumetric deformations of the sample geometry. An optimized composite material design determines a larger contribution of the piezoresistivity than that of the body deformations.

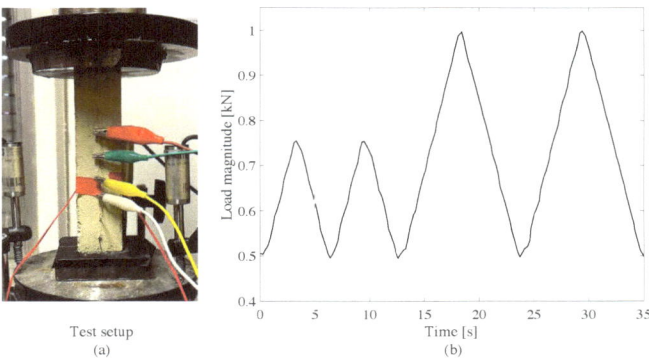

Figure 4. Electromechanical tests: (**a**) experimental setup of a smart earth composite sample, and (**b**) applied load–time history.

4. Results

The percolation curves obtained through the electrical readings revealed the transition of the electrical characteristics of the composite material, showing the percolation threshold which divided the behavior of the material from an electrical insulator to that of one as a conductor. Figure 5 plots the obtained curves out of the experimental readings. Figure 5a depicts the curve of the electrical resistance values, which were obtained by readings conducted on a full sample set. Subsequently, Figure 5b plots the resistance readings of multiple samples of selected CMF doping levels near the percolation threshold. The electrical conductivity transition of the composites started around the CMF doping level of 0.025%, causing a significant decrease in the resistance of the samples until it leveled around 0.25%. After that point, the material's conductivity increases further with the increasing doping level, with a less sloping trend. The data points plotted in Figure 5a show that the electrical resistance of the samples decreased by an order magnitude of four at the doping level of 1%. The percolation zone of the CMF fillers, according to the given curve in Figure 5a, appears to be around the CMF dopings of 0.025–0.050–0.100%, where it is expected to have an enhanced strain sensitivity due to the piezoresistivity. A further inspection of these levels, considering a higher number of samples, is plotted in Figure 5b. Accordingly, the doping level of 0.050% produced more consistent readings than the other doping levels and exhibited lower noise levels during the tests.

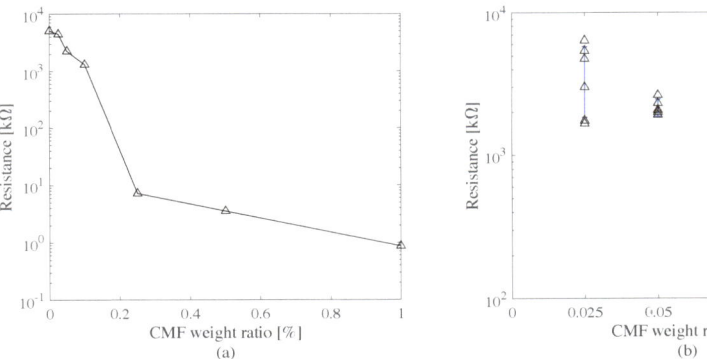

Figure 5. Electrical resistance readings of cube samples: (**a**) electrical resistance values of the sample set ranging 0–1% CMF doping ratios; (**b**) electrical resistance values of additional samples of critical CMF doping levels of 0.025–0.05–0.1% doping ratios, plotted with mean value and standard deviation.

The linear models of the correlation between the variation in the resistance and induced strain–time histories provide more information about the sensing behavior of the proposed smart earth composites. Figure 6 summarizes the results obtained through the electromechanical tests conducted on the beam samples. In Figure 6, column (a) plots together the recorded time histories of the strain and the fractional change in the sample resistance for all the samples, while column (b) plots the established linear model between the change in the resistance and strain based on the readings. These readings come from the whole body of the sensors. Accordingly, the samples with the 0.025% doping level produced the best performance among all the samples. Especially, the second sample produced a gauge factor of 80, with a very reliable linear model of strain sensing by holding an R^2 value of 0.95. The samples with 0.100% CMF did not provide meaningful results and outputted an excessive noise level. Despite the adequate performance of the first sample of the 0.05% CMF content, the second sample exhibited noise and produced a contradicting performance.

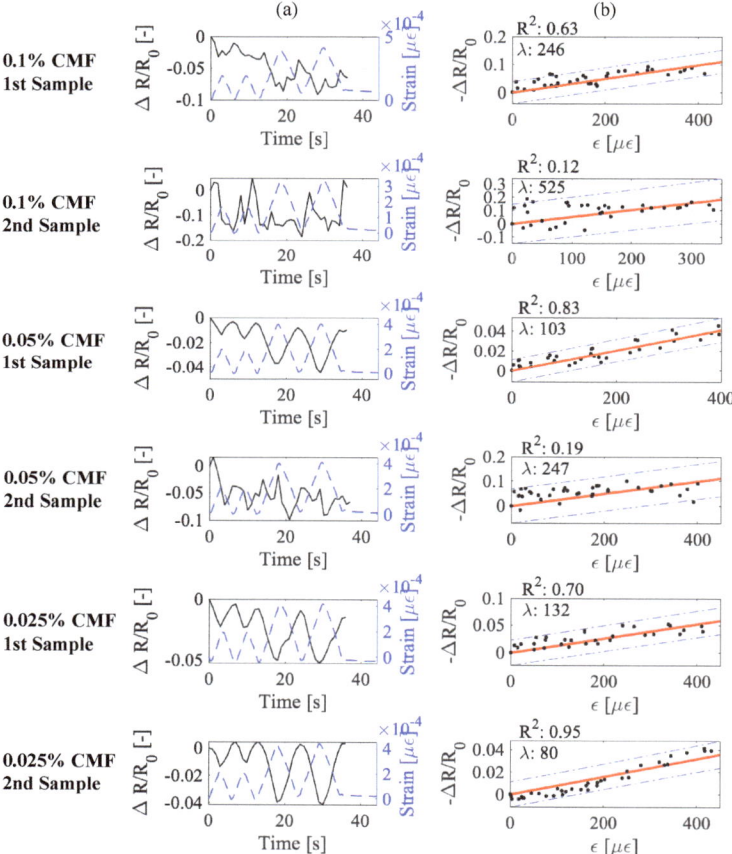

Figure 6. The measured signals from the set of beam samples when considering whole sensor volume as the sensor: (**a**) the time histories of induced strain and the fractional change in sample resistance; (**b**) the linear models of strain sensing established based on time histories.

The obtained results from the electromechanical tests have been expanded by the addition of multichannel sensing data. Figure 7 plots the fractional change in the resistance data obtained through all the channels indicated by Figure 3b. From the outputs, it was proven that also the second sample made from the 0.05% CMF composite is capable of sensing the induced strain. The poor performance of part of its volume could be related to the inhomogeneities that likely occur during the production phase of the sample. Even though the samples made from the 0.025% CMF composite performed well considering their full volume, internally they exhibited some irregularities. Unlike the composite material with 0.05%, these irregularities are observed to be consistent for both of the samples. Therefore, the behaviors of the 0.05 and 0.025% samples should have different explanations. Including the findings from Figure 5 into the discussion, it is reasonable to conclude that the composite material with 0.025% of the doping level is affected by the noise and conductivity issues related to poor dispersion of the microfibers. Such issues have played a role in obtaining irregular signals from different volume sections of the 0.025% samples.

The established linear models of the multichannels' strain sensing are shown in Figure 8. In the models, the obtained gauge factors are higher than the ones of the traditional strain sensors. Based on the findings, it can be concluded that the 0.05% composite material has the potential for reliable and reproducible strain sensing. The first sample of this composite material can be selected as the most reliable sensor in the set. However, it is evident that irregularities such as pores or agglomerations exist inside the second sample of this composite, pointing out the importance of the production phase which needs to be carefully controlled. As expected from the above discussion, the strain-sensing performance of the 0.025% CMF material varies through the channels, demonstrating the issues related to under-percolation. Therefore, it is concluded that the optimal CMF content is higher than 0.025% CMF and that the 0.05% CMF content is more promising for further development with the proposed type of material for advanced applications.

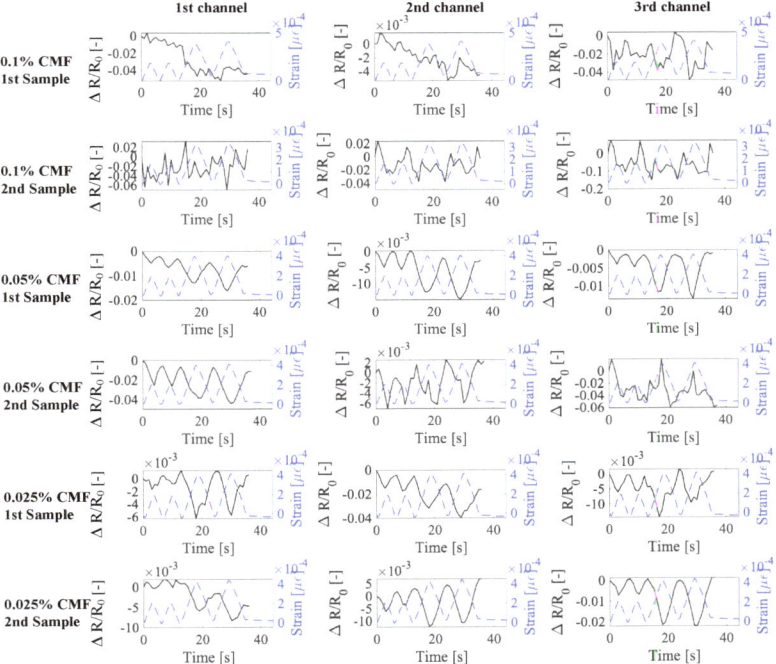

Figure 7. The measured signals from the beam samples; rows contain signals from each sample, and columns contain the data obtained through measurement channels 1, 2, and 3.

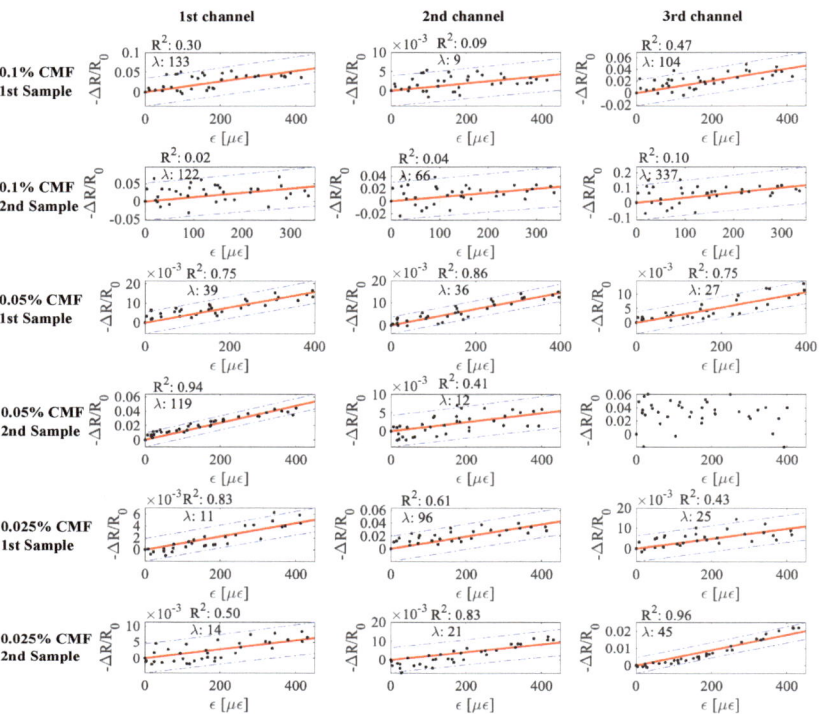

Figure 8. The established linear models of load sensing; rows contain signals from each sample, and columns contain the data obtained through measurement channels 1, 2, and 3.

5. Discussion

Multifunctional structural materials have huge application potential. In particular, self-monitoring ones could be particularly useful for enhancing the safety of structures and infrastructures, thus stimulating a lot of ongoing research efforts; however, it is still pivotal to explore different types of materials and investigate their potentialities. This research has proposed and investigated a smart earth–cement composite incorporating carbon microfibers to provide strain-sensing properties. The material has a reduced ecological footprint due to its earth-based matrix and reduced cement content. Different mixtures with varying carbon microfiber contents were studied, particularly considering the following weight percentages of carbon microfibers with respect to the binder matrix weight: 0.025, 0.050, 0.100, 0.250, 0.500, and 1.000%. The characterization tests consisted of electrical resistance readings and highlighted that the critical electrical conductivity state transition zone is between 0.025 and 0.25% of the microfiber content. Further experiments considering additional samples falling within this range pointed out 0.05% as the most promising microfiber content to achieve an effective electrical conductivity.

The second part of the experimental campaign consisted of electromechanical tests to explore the sensing capabilities of the newly produced material. The produced samples are made of mixtures with 0.025, 0.05, and 0.1% microfibers, covering the critical transition zone for the percolation of carbon microfibers. The samples were produced with a tailored design for multichannel readings. The multichannel reading allowed to verify the stability of the 0.05% sample. Although the sample with 0.025% microfibers exhibited a better sensing performance when compared to the 0.05% one when considering the whole sample volume, i.e., in a single-channel reading, the multichannel readings highlighted an improved stability and repeatability of the 0.05% mixture, which is conceivably related to more homogeneous conductive paths within the sample. These findings also agree with the conclusions made from the percolation curve. It was therefore concluded that despite its good performance, the 0.025% mixture corresponded to the under-percolation of the carbon microfibers, while the 0.05% mixture was closer to the optimum doping amount, and thus more promising for future advances.

Overall, the proposed smart earth–cement material exhibits a very high gauge factor compared to traditional strain sensors, such as resistive strain gauges, thus being suitable for designing sensing networks with reduced instrumentation complexity and costs. This material is also attractive for its sustainable properties related to the limited use of cement in comparison to traditional concrete.

6. Conclusions

This paper presents the results of an experimental campaign about the self-sensing capabilities of a novel smart composite composed of an earth–cement matrix doped with carbon microfibers. This composite was investigated for its prospective adoption as a multifunctional construction material with self-monitoring properties. In order to identify an optimal material formulation, conductive fillers were added in relative amounts ranging from 0 to 1% with respect to the weight of the binder. The results of the research show that the optimal mix design to enhance the sensitivity of the material can be identified through electrical and electromechanical tests.

The main remarks of the study are listed as follows. (i) The proposed earth–cement-based smart composite was a building material with sustainable characteristics that was found to possess self-monitoring capabilities. (ii) The electrical resistance readings have revealed that percolation occurred at doping levels from 0.025 to 0.25%. Within this range, the doping percentage of 0.05% of carbon microfibers was pointed out as the one leading to the best linearity and signal quality in the electrical resistance versus strain plots. (iii) Electromechanical tests were also repeated in a multichannel configuration, demonstrating the feasibility of distributed sensing, which is crucial for localizing cracks and internal defects. (iv) The obtained gauge factors were found much larger than those of traditional strain gauges. (v) The varying sizes of the test samples together with the employed manufacturing technique highlighted the scalability of the proposed material for more advanced field applications.

Future steps of the research will concern the sensing capabilities of higher-scale samples, mechanical improvements to the mixture for extensive usability, and the analysis of the different applications where the smart material could be adopted, including full-scale tests.

Author Contributions: The following contributions are made by the authors: conceptualization, A.D. and F.U.; methodology, H.B.B., A.D., and A.M.; software, H.B.B.; validation, A.D. and F.U.; resources, F.U.; writing—original draft preparation, H.B.B.; writing—review and editing, A.D. and F.U.; visualization, H.B.B. and A.D.; supervision, A.D. and F.U. All authors have read and agreed to the published version of the manuscript.

Funding: The authors acknowledge the support of the Italian Ministry of University and Research (MUR) through Project FISR 2019: "Eco Earth" (code 00245).

Institutional Review Board Statement: Not applicable.

Informed Consent Statement: Not applicable.

Data Availability Statement: The data presented in this study are available on request from the corresponding author.

Acknowledgments: The authors would like to acknowledge FBM Marsciano for providing dry clay.

Conflicts of Interest: The authors declare no conflict of interest.

References

1. Collepardi, M. *The New Concrete*; Tintoretto: Treviso, Italy, 2010; p. 426. ISBN 9788890377723.
2. Shah, S.P.; Hou, P.; Konsta-Gdoutos, M.S. Nano-modification of cementitious material: Toward a stronger and durable concrete. *J. Sustain.-Cem.-Based Mater.* **2016**, *5*, 1–22. [CrossRef]
3. Van den Heede, P.; De Belie, N. Environmental impact and life cycle assessment (LCA) of traditional and 'green' concretes: Literature review and theoretical calculations. *Cem. Concr. Compos.* **2012**, *34*, 431–442. [CrossRef]
4. Flower, D.; Sanjayan, J. Green house gas emissions due to concrete manufacture. *Int. J. Life Cycle Assess.* **2007**, *12*, 282–288. [CrossRef]
5. Müller, H.S.; Haist, M.; Vogel, M. Assessment of the sustainability potential of concrete and concrete structures considering their environmental impact, performance and lifetime. *Constr. Build. Mater.* **2014**, *67*, 321–337. [CrossRef]
6. Habert, G.; Roussel, N. Study of two concrete mix-design strategies to reach carbon mitigation objectives. *Cem. Concr. Compos.* **2009**, *31*, 397–402. [CrossRef]
7. Cabeza, L.F.; Barreneche, C.; Miró, L.; Morera, J.M.; Bartolí, E.; Inés Fernández, A. Low carbon and low embodied energy materials in buildings: A review. *Renew. Sustain. Energy Rev.* **2013**, *23*, 536–542. [CrossRef]
8. Tucci, F.; Vedernikov, A. Design Criteria for Pultruded Structural Elements. In *Encyclopedia of Materials: Composites*; Brabazon, D., Ed.; Elsevier: Oxford, UK, 2021; pp. 51–68. [CrossRef]
9. Onuralp, O.; Lokman, G.;Emrah, M.;Ceyhun, A.; İlker, K. Effect of the GFRP wrapping on the shear and bending Behavior of RC beams with GFRP encasement. *Steel Compos. Struct.* **2022**, *45*, 193–204.
10. Cassese, P.; Rainieri, C.; Occhiuzzi, A. Applications of Cement-Based Smart Composites to Civil Structural Health Monitoring: A Review. *Appl. Sci.* **2021**, *11*, 8530. [CrossRef]
11. Yu, K.; Lin, M.; Tian, L.; Ding, Y. Long-term stable and sustainable high-strength engineered cementitious composite incorporating limestone powder. *Structures* **2023**, *47*, 530–543. [CrossRef]
12. Vedernikov, A.; Safonov, A.; Tucci, F.; Carlone, P.; Akhatov, I. Analysis of Spring-in Deformation in L-shaped Profiles Pultruded at Different Pulling Speeds: Mathematical Simulation and Experimental Results. Available online: https://popups.uliege.be/esaform21/index.php?id=4743 (accessed on 30 December 2022).
13. Minchenkov, K.; Vedernikov, A.; Kuzminova, Y.; Gusev, S.; Sulimov, A.; Gulyaev, A.; Kreslavskaya, A.; Prosyanoy, I.; Xian, G.; Akhatov, I.; et al. Effects of the quality of pre-consolidated materials on the mechanical properties and morphology of thermoplastic pultruded flat laminates. *Compos. Commun.* **2022**, *35*, 101281. [CrossRef]
14. Zhou, P.; Li, C.; Bai, Y.; Dong, S.; Xian, G.; Vedernikov, A.; Akhatov, I.; Safonov, A.; Yue, Q. Durability study on the interlaminar shear behavior of glass-fibre reinforced polypropylene (GFRPP) bars for marine applications. *Constr. Build. Mater.* **2022**, *349*, 128694. . [CrossRef]
15. Madenci, E.; Özkılıç, Y.O.; Aksoylu, C.; Safonov, A. The Effects of Eccentric Web Openings on the Compressive Performance of Pultruded GFRP Boxes Wrapped with GFRP and CFRP Sheets. *Polymers* **2022**, *14*, 4567. [CrossRef]
16. Revilla-Cuesta, V.; Faleschini, F.; Pellegrino, C.; Skaf, M.; Ortega-López, V. Simultaneous addition of slag binder, recycled concrete aggregate and sustainable powders to self-compacting concrete: A synergistic mechanical-property approach. *J. Mater. Res. Technol.* **2022**, *18*, 1886–1908. [CrossRef]
17. Leng, Y.; Rui, Y.; Zhonghe, S.; Dingqiang, F.; Jinnan, W.; Yonghuan, Y.; Qiqing, L.; Xiang, H. Development of an environmental Ultra-High Performance Concrete (UHPC) incorporating carbonated recycled coarse aggregate. *Constr. Build. Mater.* **2023**, *362*, 129657. [CrossRef]
18. Van Damme, H.; Houben, H. Earth concrete. Stabilization revisited. *Cem. Concr. Res.* **2018**, *114*, 90–102. [CrossRef]
19. Pacheco-Torgal, F.; Jalali, S. Earth construction: Lessons from the past for future eco-efficient construction. *Constr. Build. Mater.* **2012**, *29*, 512–519. [CrossRef]
20. D'Alessandro, A.; Fabiani, C.; Pisello, A.L.; Ubertini, F.; Materazzi, A.L.; Cotana, F. Innovative concretes for low-carbon constructions: A review. *Int. J.-Low-Carbon Technol.* **2017**, *12*, 289–309. [CrossRef]
21. Fardoun, H.; Saliba, J.; Coureau, J.L.; Cointe, A.; Saiyouri, N. Long-Term Deformations and Mechanical Properties of Fine Recycled Aggregate Earth Concrete. *Appl. Sci.* **2022**, *12*, 11489. [CrossRef]
22. Yun, K.K.; Hossain, M.S.; Han, S.; Seunghak, C. Rheological, mechanical properties, and statistical significance analysis of shotcrete with various natural fibers and mixing ratios. *Case Stud. Constr. Mater.* **2022**, *16*, e00833. [CrossRef]
23. Rocha, J.H.A.; Galarza, F.P.; Chileno, N.G.C.; Rosas, M.H.; Peñaranda, S.P.; Diaz, L.L.; Abasto, R.P. Compressive Strength Assessment of Soil-Cement Blocks Incorporated with Waste Tire Steel Fiber. *Materials* **2022**, *15*, 1777. [CrossRef]
24. Venda Oliveira, P.; Correia, A.; Teles, J.; Custódio, D. Effect of fibre type on the compressive and tensile strength of a soft soil chemically stabilised. *Geosynth. Int.* **2016**, *23*, 171–182. [CrossRef]
25. Correia, A.A.; Oliveira, P.J.V.; Custódio, D.G. Effect of polypropylene fibres on the compressive and tensile strength of a soft soil, artificially stabilised with binders. *Geotext. Geomembranes* **2015**, *43*, 97–106. [CrossRef]
26. Sukontasukkul, P.; Jamsawang, P. Use of steel and polypropylene fibers to improve flexural performance of deep soil–cement column. *Constr. Build. Mater.* **2012**, *29*, 201–205. [CrossRef]
27. Tajdini, M.; Hajialilue Bonab, M.; Golmohamadi, S. An experimental investigation on effect of adding natural and synthetic fibres on mechanical and behavioural parameters of soil–cement materials. *Int. J. Civ. Eng.* **2018**, *16*, 353–370. [CrossRef]

28. Ren, G.; Yao, B.; Ren, M.; Gao, X. Utilization of natural sisal fibers to manufacture eco-friendly ultra-high performance concrete with low autogenous shrinkage. *J. Clean. Prod.* **2022**, *332*, 130105. [CrossRef]
29. da Silva Segantini, A.A.; Wada, P.H. An evaluation of the composition of soil cement bricks with construction and demolition waste/Estudo de dosagem de tijolos de solo-cimento com adicac de residuos de construcao e demolicao. *Acta Sci. Technol.* **2011**, *33*, 179–184.
30. Kouta, N.; Saliba, J.; Saiyouri, N. Fracture behavior of flax fibers reinforced earth concrete. *Eng. Fract. Mech.* **2021**, *241*, 107378. [CrossRef]
31. Nascimento, E.S.S.; de Souza, P.C.; de Oliveira, H.A.; Júnior, C.M.M.; de Oliveira Almeida, V.G.; de Melo, F.M.C. Soil-cement brick with granite cutting residue reuse. *J. Clean. Prod.* **2021**, *321*, 129002. [CrossRef]
32. Eko, R.M.; Offa, E.D.; Ngatcha, T.Y.; Minsili, L.S. Potential of salvaged steel fibers for reinforcement of unfired earth blocks. *Constr. Build. Mater.* **2012**, *35*, 340–346.
33. Koutous, A.; Hilali, E. Reinforcing rammed earth with plant fibers: A case study. *Case Stud. Constr. Mater.* **2021**, *14*, e00514. [CrossRef]
34. Gong, Z.; Han, L.; An, Z.; Yang, L.; Ding, S.; Xiang, Y. Empowering smart buildings with self-sensing concrete for structural health monitoring. In Proceedings of the ACM SIGCOMM 2022 Conference, Renton, WA, USA, 4–6 April 2022; pp. 560–575.
35. Thompson, D.; Augarde, C.; Osorio, J.P. A review of current construction guidelines to inform the design of rammed earth houses in seismically active zones. *J. Build. Eng.* **2022**, *54*, 104666. [CrossRef]
36. Curto, A.; Lanzoni, L.; Tarantino, A.M.; Viviani, M. Shot-earth for sustainable constructions. *Constr. Build. Mater.* **2020**, *239*, 117775. [CrossRef]
37. Bacciocchi, M.; Savino, V.; Lanzoni, L.; Tarantino, A.; Viviani, M. Multi-phase homogenization procedure for estimating the mechanical properties of shot-earth materials. *Compos. Struct.* **2022**, *295*, 115799. [CrossRef]
38. Bekzhanova, Z.; Memon, S.A.; Kim, J.R. Self-sensing cementitious composites: review and perspective. *Nanomaterials* **2021**, *11*, 2355. [CrossRef]
39. Azhari, F.; Banthia, N. Cement-based sensors with carbon fibers and carbon nanotubes for piezoresistive sensing. *Cem. Concr. Compos.* **2012**, *34*, 866–873. [CrossRef]
40. Taheri, S.; Georgaklis, J.; Ams, M.; Patabendigedara, S.; Belford, A.; Wu, S. Smart self-sensing concrete: the use of multiscale carbon fillers. *J. Mater. Sci.* **2022**, *57*, 2667–2682. [CrossRef]
41. Ramachandran, K.; Vijayan, P.; Murali, G.; Vatin, N.I. A Review on Principles, Theories and Materials for Self Sensing Concrete for Structural Applications. *Materials* **2022**, *15*, 3831. [CrossRef]
42. Qiu, L.; Dong, S.; Yu, X.; Han, B. Self-sensing ultra-high performance concrete for in-situ monitoring. *Sens. Actuators A Phys.* **2021**, *331*, 113049. [CrossRef]
43. Meoni, A.; D'Alessandro, A.; Cavalagli, N.; Gioffré, M.; Ubertini, F. Shaking table tests on a masonry building monitored using smart bricks: Damage detection and localization. *Earthq. Eng. Struct. Dyn.* **2019**, *48*, 910–928. [CrossRef]
44. Gautam, A.; Mo, Y.; Chen, Y.; Chen, J.; Joshi, B. Carbon nanofiber aggregate sensors for sustaining resilience of civil infrastructures to multi-hazards. *Adv. Civ. Eng. Technol.* **2019**, *3*, 263–269.
45. Castañeda-Saldarriaga, D.L.; Alvarez-Montoya, J.; Martínez-Tejada, V.; Sierra-Pérez, J. Toward structural health monitoring of civil structures based on self-sensing concrete nanocomposites: a validation in a reinforced-concrete beam. *Int. J. Concr. Struct. Mater.* **2021**, *15*, 1–18. [CrossRef]
46. Galao, O.; Baeza, F.J.; Zornoza, E.; Garcés, P. Carbon nanofiber cement sensors to detect strain and damage of concrete specimens under compression. *Nanomaterials* **2017**, *7*, 413. [CrossRef] [PubMed]
47. Chung, D. Self-sensing concrete: from resistance-based sensing to capacitance-based sensing. *Int. J. Smart Nano Mater.* **2021**, *12*, 1–19. [CrossRef]
48. Chung, D. A critical review of electrical-resistance-based self-sensing in conductive cement-based materials. *Carbon* **2023**, *203*, 311–325. [CrossRef]
49. Han, B.; Yu, X.; Ou, J. *Self-Sensing Concrete in Smart Structures*; Butterworth-Heineman: Oxford, UK, 2014.
50. Han, B.; Yu, X.; Ou, J. Multifunctional and Smart Carbon Nanotube Reinforced Cement-Based Materials. In *Nanotechnology in Civil Infrastructure: A Paradigm Shift*; Gopalakrishnan, K., Birgisson, B., Taylor, P., Attoh-Okine, N.O., Eds.; Springer: Berlin/Heidelberg, Germany, 2011; pp. 1–47. [CrossRef]
51. Reddy, P.N.; Kavyateja, B.V.; Jindal, B.B. Structural health monitoring methods, dispersion of fibers, micro and macro structural properties, sensing, and mechanical properties of self-sensing concrete—A review. *Struct. Concr.* **2021**, *22*, 793–805. [CrossRef]
52. Konsta-Gdoutos, M.S.; Metaxa, Z.S.; Shah, S.P. Highly dispersed carbon nanotube reinforced cement based materials. *Cem. Concr. Res.* **2010**, *40*, 1052–1059. [CrossRef]
53. D'Alessandro, A.; Ubertini, F.; Materazzi, A.; Laflamme, S.; Cancelli, A.; Micheli, L. Carbon cement-based sensors for dynamic monitoring of structures. In Proceedings of the 2016 IEEE 16th International Conference on Environment and Electrical Engineering (EEEIC), Florence, Italy, 7–10 June 2016; pp. 1–4.
54. D'Alessandro, A.; Birgin, H.B.; Ubertini, F. Carbon Microfiber-Doped Smart Concrete Sensors for Strain Monitoring in Reinforced Concrete Structures: An Experimental Study at Various Scales. *Sensors* **2022**, *22*, 6083. [CrossRef]
55. Birgin, H.B.; Laflamme, S.; D'Alessandro, A.; Garcia-Macias, E.; Ubertini, F. A weigh-in-motion characterization algorithm for smart pavements based on conductive cementitious materials. *Sensors* **2020**, *20*, 659. [CrossRef]

56. Birgin, H.B.; D'Alessandro, A.; Laflamme, S.; Ubertini, F. Hybrid carbon microfibers-graphite fillers for piezoresistive cementitious composites. *Sensors* **2021**, *21*, 518. [CrossRef]
57. Meoni, A.; D'Alessandro, A.; Downey, A.; García-Macías, E.; Rallini, M.; Materazzi, A.L.; Torre, L.; Laflamme, S.; Castro-Triguero, R.; Ubertini, F. An experimental study on static and dynamic strain sensitivity of embeddable smart concrete sensors doped with carbon nanotubes for SHM of large structures. *Sensors* **2018**, *18*, 831. [CrossRef]

Disclaimer/Publisher's Note: The statements, opinions and data contained in all publications are solely those of the individual author(s) and contributor(s) and not of MDPI and/or the editor(s). MDPI and/or the editor(s) disclaim responsibility for any injury to people or property resulting from any ideas, methods, instructions or products referred to in the content.

Article

Microstructural Analysis of the Transverse and Shear Behavior of Additively Manufactured CFRP Composite RVEs Based on the Phase-Field Fracture Theory

Matej Gljuščić [1], Domagoj Lanc [1,*], Marina Franulović [1] and Andrej Žerovnik [2]

[1] Faculty of Engineering, University of Rijeka, Vukovarska 58, 51000 Rijeka, Croatia
[2] Faculty of Mechanical Engineering, University of Ljubljana, Aškerčeva c. 6, 1000 Ljubljana, Slovenia
* Correspondence: dlanc@riteh.hr

Abstract: Due to the versatility of its implementation, additive manufacturing has become the enabling technology in the research and development of innovative engineering components. However, many experimental studies have shown inconsistent results and have highlighted multiple defects in the materials' structure thus bringing the adoption of the additive manufacturing method in practical engineering applications into question, yet limited work has been carried out in the material modelling of such cases. In order to account for the effects of the accumulated defects, a micromechanical analysis based on the representative volume element has been considered, and phase-field modelling has been adopted to model the effects of inter-fiber cracking. The 3D models of representative volume elements were developed in the Abaqus environment based on the fiber dimensions and content acquired using machine learning algorithms, while fulfilling both geometric and material periodicity. Furthermore, the periodic boundary conditions were assumed for each of the representative volume elements in transversal and in-plane shear test cases,. The analysis was conducted by adopting an open-source UMAT subroutine, where the phase-field balance equation was related to the readily available heat transfer equation from Abaqus, avoiding the necessity for a dedicated user-defined element thus enabling the adoption of the standard elements and features available in the Abaqus CAE environment. The model was tested on three representative volume element sizes and the interface properties were calibrated according to the experimentally acquired results for continuous carbon-fiber-reinforced composites subjected to transverse tensile and shear loads. This investigation confirmed the consistency between the experimental results and the numerical solutions acquired using a phase-field fracture approach for the transverse tensile and shear behavior of additively manufactured continuous-fiber-reinforced composites, while showing dependence on the representative volume element type for distinctive load cases.

Keywords: additive manufacturing; carbon-fiber-reinforced composites; phase field modelling; micromechanics; representative volume element

Citation: Gljuščić, M.; Lanc, D.; Franulović, M.; Žerovnik, A. Microstructural Analysis of the Transverse and Shear Behavior of Additively Manufactured CFRP Composite RVEs Based on the Phase-Field Fracture Theory. *J. Compos. Sci.* **2023**, *7*, 38. https://doi.org/10.3390/jcs7010038

Academic Editor: Jiadeng Zhu

Received: 1 December 2022
Revised: 1 January 2023
Accepted: 5 January 2023
Published: 12 January 2023

Copyright: © 2023 by the authors. Licensee MDPI, Basel, Switzerland. This article is an open access article distributed under the terms and conditions of the Creative Commons Attribution (CC BY) license (https://creativecommons.org/licenses/by/4.0/).

1. Introduction

Additive manufacturing (AM) technology of polymeric materials has more frequently been adopted in engineering applications [1,2], especially for prototyping in research and development [2,3], with prospects in mold tooling and tool jigs, limited-run production parts, service parts, maintenance and repair, and lightweight solutions in automotive industries [4–6], as well as in medical applications [7–9]. A comprehensive review aiming to synthesize the most relevant studies on carbon-based nano-composites was conducted in [10]. The authors discuss the mechanical and electrical properties of materials reinforced with carbonaceous nanofillers, such as graphene nanoplatelets and carbon nanotubes, and their compatibility with the requirements for biomedical applications, emphasizing the potential of the specific electrical, thermal, and mechanical properties of CNTs fillers in multifunctional, hight-strength, lightweight, and self-lubricating wear resistant composites. In addition, a comprehensive review on graphene nanofillers in metal matrix composites for biomedical applications was conducted in [11], highlighting the influences of nano-additive phases on cytotoxicity and the tailored mechanical properties achieved in implants by adopting scaffold structures. All thing considered, the application of AM technology was also further extended by the introduction of particle or fiber fillers to achieve the beneficial properties of these reinforcing materials. The prospects of embedding nanofillers in additively manufactured composite materials were investigated in [12], highlighting the increased strength and stiffness achieved when the components are reinforced with fibers and fillers, which led to a significant amount of research being concentrated on material selection and enhancing the characteristics of cost-effective printed components [12]. However, the experimental observations on the behavior of these enhanced materials show inconsistency with the performance of their conventionally manufactured counterparts [13–15], which has also been found as one of the major obstacles to the adoption of AM in the current industries [16]. A comprehensive review of the application of AM composites has been revied in [17], highlighting its potential in medical applications. A comprehensive review of the application of AM composites has been revied in [17], while the enhancement of the dielectric and thermal properties of polymeric composites by the addition of fillers was investigated in [18,19] respectively. The mechanical properties of carbon-fiber-reinforced specimens containing various LSSs were studied in [20]. While these reinforced materials exhibit heterogeneity, the AM components manifest in manufacturing inherent deposition deficiencies which are often attributed to the microscopic voids and inclusion [15,21,22]. Among these defects, the AM parts also depend on various controllable manufacturing parameters as well as the mechanical properties of the constituents, which often results in data inconsistencies and issues uncertainty in the prediction of the material behavior. These FDM-induced flaws were further studied in [21–23], indicating the necessity for microscopic analysis and NDT component examinations, as well as quantification of the deficiencies which affect the mechanical behavior of AM CFRP composites; however, the modelling approaches which account for the effects of the AM process are limited.

In order to calculate the resulting material properties in novel composite AM extrusion mixtures, microstructural analysis of these heterogenic materials is essential. This can be achieved by analyzing a representative volume element (RVE) designed according to the available microstructural information on the constituents' geometries and behaviors, as shown in [24–27], as well as being proven valid in multiple studies [28–33]. The constituents' geometry can be acquired from 2D microscopy [13,34] and then prepared using AI-thresholding [35] for statistical evaluations, based on which, and some assumptions, the RVE is built or it can be analyzed from the 3D μCT scans [36,37] using the voxel approach [38]. Such models are prepared in a CAE environment where each of the constituents adopts its constitutive behavior. Since these material models are usually acquired from macroscale analyses, additional assumptions on the intermaterial contact behavior [13,28,39,40] have to be adopted. In commercial FEA software, these interactions are implemented using cohesive elements [28] or alternatively using phase-field modelling [39–41].

The phase field approach was initially developed to overcome the problems of moving interface boundaries and topology adjustments when simulating the morphologies of evolving interfaces in cases of merging or division [42,43]. Based on the initial formulations presented in [44,45], the model uses a set of field variables Φ which are described using partial differential equations (PDEs) and they distinguish the two phases (0 and 1), between which a smooth transition is assumed when calculated in the proximity of the interface [46]. These sets of variables are continuous functions of time and spatial coordinates and can be integrated without explicit treatment of interface conditions [42,43]. By adopting the thermodynamic free energy approach, the phase field evolution can be defined in terms of temperature, strain, or concentration to predict various changes in microstructural features, leading to frequent application in simulations of microstructural evolutions [47] and the modelling of fracture mechanics [48].

Since complex crack trajectories and crack branching [49] can be achieved with arbitrary geometries and dimensions without tempering with the FE mesh, the application of this approach has been adopted for modelling various complex phenomena [49] including stress and temperature-induced transformations in shape memory alloys [50], functionally graded materials [51,52], hyper-elastic materials [53,54], compressive shear fracture in basalt, cement mortar and sandstone [55], chemo-mechanically assisted fractures [56], hydrogen embrittled alloys [57], piezoelectric [58], and fiber-reinforced composites [59,60]. However, despite still being limited in engineering applications, there are several reported applications in commercial FEA software including MATLAB [61–63], Python [64,65], Abaqus [46,50,57,66–71], and COMSOL [55,72], many of which are published in open-source archives. The constitutive relations used in these phase-field implementations are based on [45,73] as the AT2 and the AT1 model, respectively, and some of the presented implementations include the regularized cohesive zone model PF-CZM based on [74–77] while adopting newton monolithic [57], one-pass AM/staggered [57], or iterative AM/staggered solver algorithms [71] for acquiring solutions [49].

Furthermore, the development of solution algorithms for the coupled deformation-fracture problem was presented in [49,57,78,79]. The total potential energy is minimized according to the displacement field u and phase field Φ, while Φ is solved as a damage variable [80] at the finite element node instead of at the integration point as in cases of local damage models [46]. This is achieved using a user-defined element (UEL) prior to the implementation of the user defined material model (UMAT), which limits the modelling and visualization capabilities of commercial software [46]. This has been accounted for in [46,80] by adopting the similarity between the phase field balance and heat transfer equations which enables the calculation of phase-field damage Φ variable as an additional degree of freedom without the necessity for the UEL mesh definition. To retain all of the the Abaqus built-in modelling and meshing features, this approach has been implemented for the micromechanical damage modelling of AM carbon fiber-reinforced thermoplastic polymer (CFRP) composite RVE and validated on experimentally acquired data. All things considered, this work is focused on the application of the phase field approach in modelling the microstructural failure of AM CFRP composite RVE subjected to tensile and shear loads. Therefore, as an extension to the previously published work where the RVEs were modelled based on continuum mechanics and cohesive finer/matrix interfaces [13,81], the behavior of the AM CFRP composite was modelled in this work using a phase field modelling approach, introducing microscopic damage and showing better scalability of the RVE based on microscopic imaging in contrast to the previously acquired results in [13,81].

2. Materials and Methods

The consistency of an AM material may vary due to the ratios and microscopic properties of the constituents, the chemical properties of the sizing agents, layer thickness, nozzle diameter, humidity, filling type, and density, as well as the extrusion speed and temperature [82]. Since many of these parameters could be controlled, an iterative design process in novel material development is essential. Therefore, a micromechanical modelling approach based on the representative volume element has been proposed in this work.

The CFRP composite specimens for microscopic analysis were designed as 12 layered 25×25 mm laminated plates with a cross-ply stacking sequence equal to $[0/90_2/0/90_2]_s$, as shown in Figure 1, and additively manufactured using a Markforged X7 3D printer. This LSS was adopted after a preliminary microstructural examination of similar AM composites. A similar LSS is used in acquiring the multiaxial response through uniaxial tensile tests on rectangular specimens [83], which was also studied in [14]. This sequence also enables the measurements of unidirectionally reinforced layer composition and gives a better insight into laminar dimensions, layer-wise defect accumulation, and repeatability issues between single, double, and multiple adjacent layers. However, these specimens did not account for the lengthwise defect accumulation due to uneven material solidification.

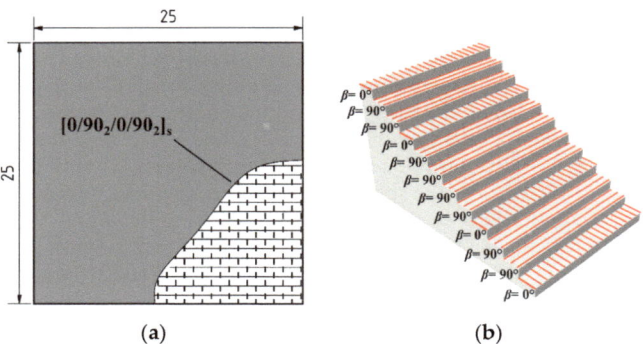

Figure 1. CFRP laminate sample [81]: (**a**) sketch in the x-y plane, and (**b**) LSS details.

The specimens were then cut along and perpendicular to the dominant fiber direction in order to acquire x-z and y-z cross-sections, while the x-y cross-section is acquired by grinding the polymeric top layer of the specimens. The cross-sections were polished and prepared for observation using the FEI-QUANTA-250-FEG scanning electron microscope. The acquired images were inspected and, due to the ununiform matrix background, the images were segmented based on pixel classification using Trainable WEKA segmentation machine learning algorithms which are readily available in FIJI [35] open-source software. The acquired results were statistically evaluated and compared with the data available in the relevant literature. Following the comparison, the RVEs were designed according to [13,31,84], while adopting the microscopic properties acquired from the four middle layers. The first case (RVE-1) was designed as a single fiber unit cell; the rectangular and hexagonal fiber arrangements have been adopted for the second (RVE-2) and the third (RVE-3), respectively, while keeping the smallest possible RVE size for the measured fiber volume ratio to increase the computational efficiency. However, to acquire more accurate results from the analysis of small-size RVEs, the fibers were placed at the RVE boundary edges [84], while ensuring geometrical, material, and mesh periodicity [13,31]. This was adopted for the cases of rectangular and hexagonal fiber arrays, while for the unit-cell case, both conditions could not be fulfilled.

In order to distinguish between the fiber, the matrix, and the fiber/matrix interface behaviors, each RVE domain was divided into three subdomains. The model was discretized using coupled temperature-displacement hexahedral elements (C3D8T), with the predefined temperature field indicating the initial state of damage within the material. Furthermore, the linear elastic behavior was assumed for the phase field damage framework [33], while the fiber and matrix's mechanical properties were adopted from [85] and [86], respectively; as shown in Table 1. The constituents' properties were adopted from the literature since the commercially available data are available only for the UD composite reinforced in the direction of the layers with the elastic modulus of 60 GPa and tensile strength of 800 MPa [87], while the matrix data are available for the AM polymer used in the secondary nozzle, and not in the analyzed CFRP material. The toughness K_{1C} was adopted for polyamide according to [88] and recalculated to acquire the G_C values. In contrast to the fiber and the matrix, the values for the interface properties were assumed and calibrated according to the experimentally observed material behavior.

Table 1. Constituents' mechanical properties.

	Fiber	Matrix	Interface
Elastic modulus, E_{11} [MPa]	191,000	3000	100
Poisson ratio, v [/]	0.2	0.3	0.3
Toughness G_c, [kJ/m^2]	0.763	1	0.3
Phase field length, l [mm]	1×10^{-4}	1×10^{-3}	1×10^{-3}

In order to appropriately simulate the behavior of the surrounding material, periodic boundary conditions (PBC) have been enforced on the RVEs boundary corners, edges, and surfaces [28,30–33], where the PBCs and the node constrain equations were defined using the automated procedure presented in [31,32]. However, instead of focusing on extracting the elastic properties, the procedure was modified to calculate the material degradation and damage using user-defined subroutines and simulate the complete material response until failure.

The mesh sensitivity was studied on three distinctive cases of RVEs with assumed hexahedral fiber placement. The curvature maximal deviation factor h/L was assumed as 0.01 for each of the tested cases, while the approximate element sizes were adopted as $2.5 \times 10^{-4}, 3 \times 10^{-4}, 4 \times 10^{-4}$, and 5×10^{-4} which resulted in 1,115,403, 60,860, 53,876, and 33,781 elements, respectively. Returning consistent results within the elastic region, the models were discretized using coupled temperature-displacement hexahedral elements (C3D8T) assuming the approximate element size equal to 4×10^{-4}, resulting in mesh sizes of 23,163, 57,300, and 53,876 elements for RVE-1, RVE-2, and RVE-3, respectively. Finally, the models were subjected to transversal and shear loads and analyzed using a phase field model.

In a standard phase field model formulation (AT2) presented in Equation (1), the damage is described using the phase field variable Φ which takes the values between zero for the undamaged and one for the fully damaged material, while the damage evolution depends on the balance between the energy stored within the material and the energy released during fracture [46,80].

$$\nabla^2 \phi = \frac{\phi}{l^2} - \frac{2 \cdot (1-\phi)}{G_c \cdot l} \cdot \psi \qquad (1)$$

In this expression, the fracture toughness of the material and the strain energy density are represented by the variables G_c, and ψ, respectively, while the variable l stands for the phase field length which is used to define the size of the damaged region. In Abaqus, this differential equation is solved by introducing the user-defined element (UEL) before the material model subroutine is called, thus losing the Abaqus built-in modelling, meshing, and visualization features in the process. To account for this, a solution based on the similarities between the phase field damage evolution and the heat transfer problem under steady-state conditions has been presented in [46,80]. In this framework, the change in the temperature T for a material with thermal conductivity k which is subjected to a heat source r is calculated according to Equation (2), and since this scheme is readily available as an Abaqus built-in feature, it was utilized for solving the differential equation for the phase field damage evolution [46,80].

$$k\nabla^2 T = -r \qquad (2)$$

The complete mathematical formulation of the phase field theory is presented in [46,80,89] as well as in Appendix A, and within this framework, the initial temperature is given as a predefined field variable, the material thermal conductivity is equal to one, while the variable r is redefined according to the phase field fracture theory and introduced using the UMAT subroutine.

Following these assumptions, the initial temperature is usually given equal to zero and it describes the initial state of damage, which evolves according to the phase field damage formulation presented as r until the fully damaged state is reached. Within the presented study, the constituents were adopted as isotropic, while presuming linear elastic behavior based on four mechanical properties for each of the constituents including the modulus of elasticity E, the Poisson ratio ν, the characteristic phase field length scale l, and the fracture toughness G_c. Besides these variables, the model variations are also presented in [46,80] including AT1 [73], AT2 [45], and the phase field cohesive approach PF-CZM [74], which also takes the material tensile strength into account. Furthermore, the monolithic and staggered solution schemes are also presented in [46,80], as well as a hybrid [90] or anisotropic [78] approach for the volumetric–deviatoric [91], and the spectral [78] strain energy split schemes. Within the framework of the presented research, the standard AT2 model has been adopted for each of the constituents in all of the RVE cases. While using the staggered solution scheme without strain decomposition in order to achieve convergence with a larger number of iterations, the Abaqus general solution controls I_0, I_R, I_P, I_C, I_L, and I_G have been set to 5000 [46,80]. The RVEs were analyzed using the adopted UMAT subroutine to be compared with the experimental results.

Therefore, two sets of continuous fiber-reinforced specimens with a polyamide matrix have been designed as flat rectangular laminates, as shown in Figure 2.

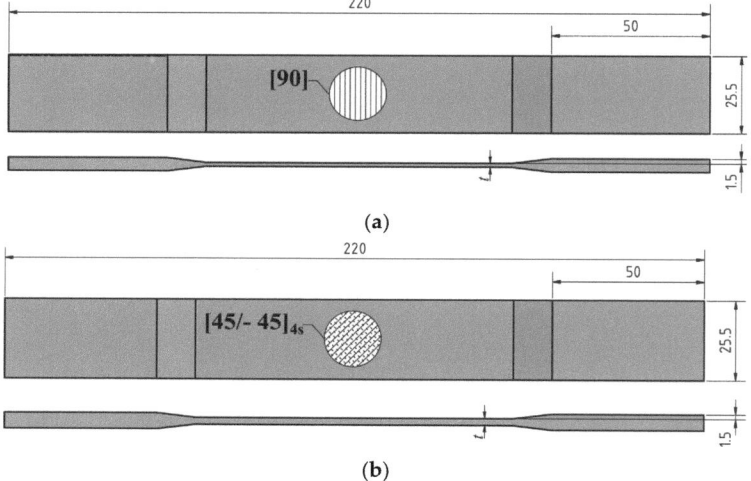

Figure 2. CAE dimensions: (**a**) UD-90 specimen dimensions, and (**b**) SH-45 specimen dimensions; recreated according to [81].

The specimens in the first set UD-90 were designed according to the ASTM-D3039 [92] standard as transversely reinforced, unidirectionally reinforced laminates with an LSS equal to $[90]_8$, while for the second set SH-45, according to the ASTM-D3518 [93] standard, they were designed as multidirectionally reinforced laminates with an LSS equal to $[45/-45]_{4S}$ which resolves the uniaxially applied tensile loads into in-plane shear loads; the visual representation of the adopted LSS is for each specimen set presented in Figure 3.

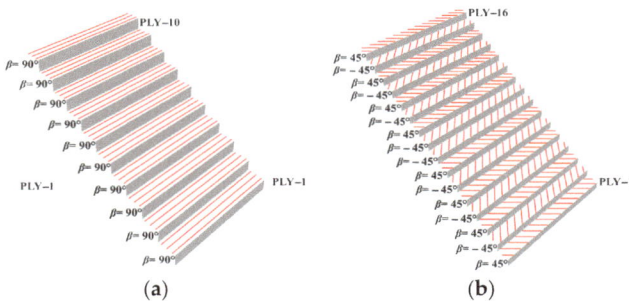

Figure 3. (**a**) Layer stacking sequence for the UD-90 set, and (**b**) layer stacking sequence for the SH-45 set; recreated according to [81].

Both sets of specimens were tested uniaxially in quasistatic conditions with $v = 0.01$ mm/s using a hydraulic testing system and monitored with both contact and optical extensometers. The measurements have been taken on the opposite sides of the specimens, monitoring the strains of one surface using the epsilon-tech axial extensometer Model-3542, and the opposite one with a GOM Aramis 5M (GigE) adjustable base system with 35 mm lenses. The image resolution was 2448 × 2050 pixels, while the cameras were positioned at 560 mm from the specimen, with the distance between the cameras being equal to 265 mm, closing the angle of 26°. The sensor calibration was conducted using the software-guided protocol at 22.5 °C resulting in an average deviation of 0.048 pixels in a measuring volume of 130 × 110 mm × 90 mm. Such a measurement protocol was selected to check the consistency of the measurements through the specimen thickness and to detect excessive delamination within the gauge length. The measurements have been systematized for comparison with the RVE results and they are presented in the following chapter.

3. Results

The results are given as the microstructural observations and macroscopic experimental results, following by design of the RVEs and numerical analysis.

3.1. Microstructural Inspection

In order to obtain the necessary input for the RVE design, the material microstructure was examined on multiple cross-sections in the y-z, x-z, and x-y planes. The cross-sections were polished and scanned using an FEI-QUANTA-250-FEG scanning electron microscope; the results are presented in Figure 4.

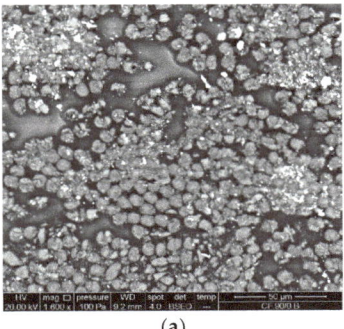
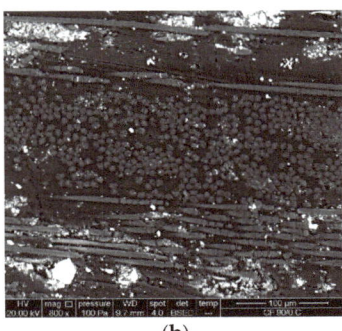
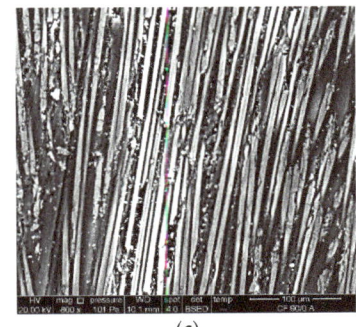

Figure 4. CFRP SEM images: (**a**) y-z cross-section 1600× magnification, (**b**) x-z cross-section 800× magnification, and (**c**) x-y cross-section 800× magnification.

The acquired images were inspected and, due to the ununiform matrix background, the images were segmented based on pixel classification using Trainable WEKA segmentation machine learning algorithms in FIJI as shown in Figure 5. In the classification process, three sets of categories have been defined: the fibers, matrix, and voids with debris, each associated with the appropriate pixels. This procedure was conducted on ten randomly selected areas, while the number of included layers depended on the particular cross-section.

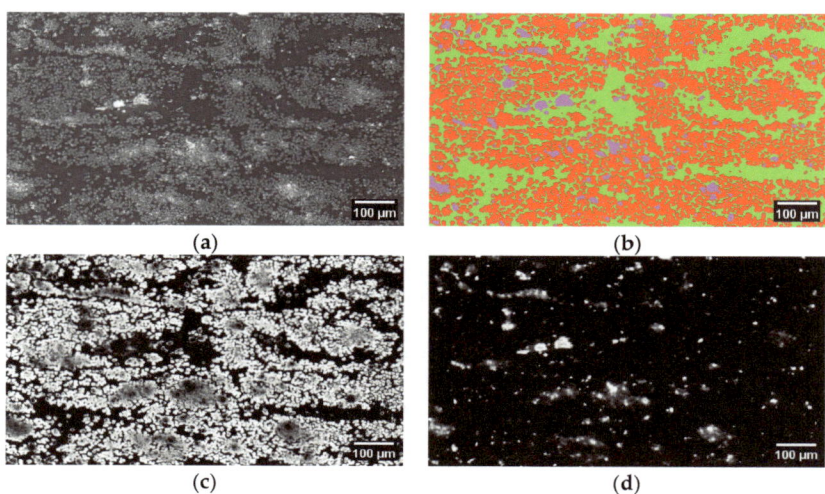

Figure 5. CFRP laminate's four middle unidirectional layers: (**a**) SEM image, (**b**) WEKA classification, (**c**) the fiber vs. matrix probability map, and (**d**) voids and debris probability map.

The Figure 5b shows the classified microstructure after being analyzed based on pixel recognition using the machine learning algorithms. The coloured regions define the detected phases after the training sessions and correspond to the constituents: the fibers, matrix, and voids and debris present within the analyzed material. Furthermore, the constituent's volume ratios were identified using the colour threshold adjustment on multiple randomly selected images from each of the analyzed cross-sections, where the major contribution to the result was the images containing four middle layers in the y-z cross-section, as shown in Figure 5. Following the classification algorithm, the fibers and the ratios of the fibers, matrix, and debris and voids were identified for each of the selected images from which the average value was acquired and is presented in Table 2. However, due to the irregularity of the fiber arrangement, the local fiber volume ratio was measured on ten randomly selected 100 μm × 100 μm zones within the middle four-layer area, showing consistency with the measured average, while being inconsistent with the constituents' volume ratios found in material agglomerations. In contrast to the volume ratios, the geometrical variable such as the fiber diameter and layer height could have been measured directly from the acquired images, and additional image refinements were not necessary for those datasets. As presented in Figure 6, the statistical analysis of the fiber diameters showed consistency with the normal distribution, while larger data discrepancies from the normal distribution could be observed in the layer height distribution which is also indicated by the Anderson–Darling value lower than 0.005.

Table 2. Volume fraction comparison.

	Laminate [13]	Filament [85]
Fiber volume ratio	0.536 ± 0.026	0.34 ± 0.002
Matrix volume ratio	0.41 ± 0.02	0.66 ± 0.002
Fiber local ratio	0.568 ± 0.028 [1]	0.90 [2]
Fiber diameter, μm	7.00 ± 0.41	7.2 ± 0.30
Layer height, μm	138.22 ± 5.11	/

[1] Measurements acquired in multiple randomly selected 100 μm × 100 μm zones; [2] measurements based on selected fiber agglomerations [85].

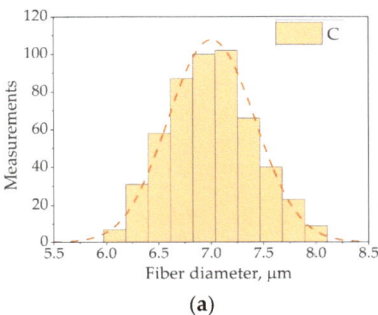 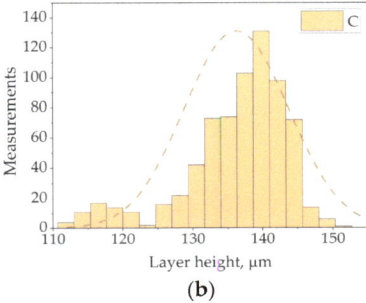

(a) (b)

Figure 6. (a) Carbon fiber diameter measurements, and (b) the specimen's layer height.

The results of the statistical evaluations have been compared with the values published in the relevant literature, and the differences between the volume ratio acquired from the sample using microscopy and from the filament using pyrolysis are highlighted in Table 2. The comparison yielded consistency for the values of fiber diameters, while significant discrepancies were documented for the total and local fiber volume ratios since different machines and parameters were used in manufacturing.

Conclusively, despite the microstructural inspection having been proven valid for the identification of the constituents' properties, it also disclosed the inconsistent repeatability of layer thickness in the AM process, which is seldom discussed, but essential for the performance of fiber-reinforced composite laminates. Despite the comparison with the relevant data in the literature yielding consistency for the values of the fiber diameters, significant discrepancies were observed for the total and local fiber volume ratios. These discrepancies could be influenced by the distinctive machines and parameters used in the manufacturing process but they were also expected since the microstructural images were extracted from the laminated specimen [13], instead of the filament presented in [85].

3.2. Experimental Acquisition of Lamina Properties

In order to identify the mechanical properties of AM continuous-fiber-reinforced composites, experimental studies on transversal and shear behavior have been conducted, following the guidelines presented in ASTM-D3039 [92] and ASTM-D3518 [93] standards. The dimensions of each specimen were assessed in multiple cross-sections within the gauge length using a digital micrometer, and the acquired values for the widths and thicknesses were averaged and are reported in Table 3. The acquired values were analyzed, showing inconsistencies in comparison with the CAE data, as presented in Figure 2, caused by the removal of the support material. Therefore, to represent the manufactured material more accurately, the measured values were adopted in the numerical analysis.

Table 3. Dimensions of the additively manufactured specimens.

	Length (*L*), mm	Width (*W*), mm	Thickness (*t*), mm
UD-90	220	26.903 ± 0.154	1.33 ± 0.017
SH-45	220	27.250 ± 0.017	2.30 ± 0.017

The specimens were subjected to uniaxial tensile loads and monitored with both contact and non-contact strain measurement techniques. During the experimental protocol, it was confirmed that the behavior of AM UD-90 specimens is highly influenced by the matrix response and also the fact that these specimens exhibit inferior mechanical properties in comparison to their injection-moulded counterparts [86]. The material bonding defects influenced by the fiber/matrix interface and the raster-induced defects of the additive manufacturing process act as stress concentrators which can be observed in Figure 7 as localized strains between the deposition paths. Consequently, these effects diminish the strength and stiffness of the AM materials, and due to their stochastic nature, they cause material inconsistencies, which is also confirmed by the scatter in the experimental data, and thus had to be considered in material modelling.

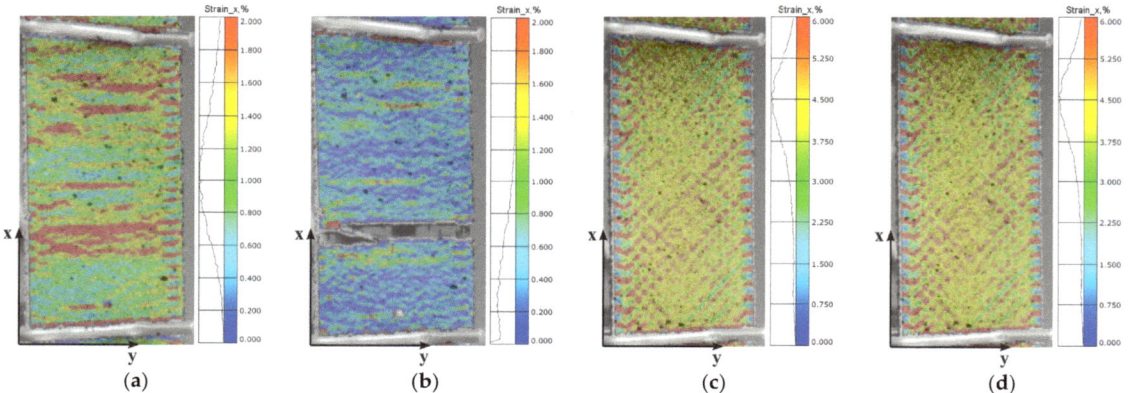

Figure 7. DIC images of the 50 mm gauge length during the experimental procedure: (**a**) UD-90 pre-failure, (**b**) UD-90 at failure, (**c**) SH-45 at 5% of shear strain, and (**d**) SH-45 at 5% of axial strain.

Additionally, the SH-45 specimens represent a specific case of multidirectionally reinforced laminates where the adopted LSS induces multiaxial in-plane shear stress during uniaxial tension, while, similar to the case of UD-90, multiple strain localizations can be observed, as shown in Figure 7. In contrast to the UD-90 case, the SH-45 specimens demonstrated more consistent results up to 2% of shear strain, which is followed by an increase in data scatter up to 5% of shear strain. The anticipated ductile behavior for the SH-45 specimens resolves in the fiber rearrangement phenomenon which can lead to 25% of strain before failure. Since these large strains are not applicable in composite structures, the ASTM-D3518 standard [93] recommends limiting the observations of the experimental results at 5% of shear strain, in case the failure does not occur before. The shear strength in SH-45 specimens was also acquired according to the ASTM standard as the intersection between the shear stress–strain curve and the 0.2% offset of its shear modulus. The experimentally acquired data were systematized and prepared for comparison with the numerical results for each of the RVEs.

3.3. RVE Design

According to the acquired microscopic constituents' properties, three sets of RVE models have been developed as presented in Figure 8, where RVE-1 represents a unit cell containing only one central fiber, while both RVE-2 and RVE-3 contain another fiber which is divided into four quarters placed in the RVE corners to ensure better model predictions using a smaller RVE. Furthermore, to account for the material bonding deficiencies, both the fiber/matrix and the raster-induced bonding inconsistencies have to be accounted for. However, since the scale of raster-induced defects exceeded the minimal RVE dimensions, both the matrix/fiber and the deposition contact deficiencies have been introduced together at the fiber/matrix interface as a distinctive subdomain.

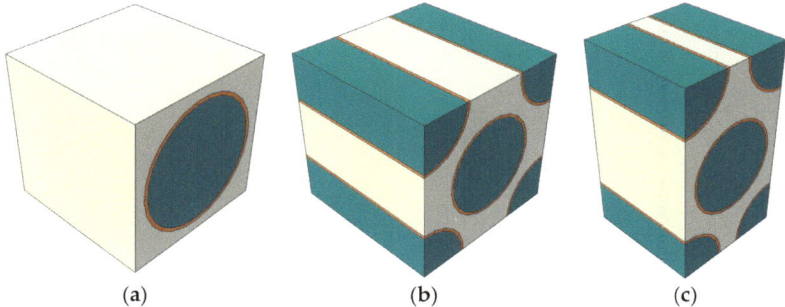

Figure 8. (**a**) Single-fiber unit cell (RVE-1), (**b**) RVE with rectangular fiber placement (RVE-2), and (**c**) RVE with hexagonal fiber placement (RVE-3).

In order to check the significance of the discretization on the numerical results, a mesh sensitivity analysis was conducted on the RVE-3 case. The models were meshed using hexahedral elements (C3D8T) with characteristic lengths of 2.5×10^{-4}, 3×10^{-4}, 4×10^{-4}, and 5×10^{-4}, resulting in 1,115,403, 60,860, 53,876, and 33,781 elements, respectively, returning consistent results within the elastic region, while deviating from the average values for the post-yielding region in the case of 2.5×10^{-4}, as presented in Figure 9.

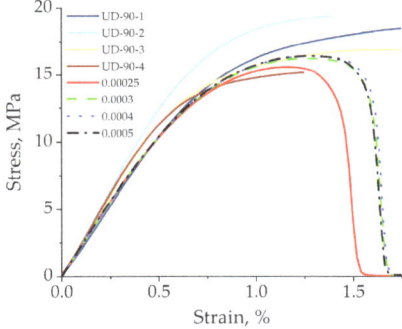

Figure 9. Comparison between the experimental results and the RVE-3 response for characteristic element length in the range from 2.5×10^{-4} to 5×10^{-4}.

Since the presented discrepancies were less significant than the scatter within the experimental dataset, the characteristic element length of 4×10^{-4} was adopted for all of the RVE cases. With the mesh size adopted, and the domains discretized, an automated procedure [31] was adopted for enforcing the periodic boundary conditions for transverse and in-plane shear cases. In order to include the failure analysis of heterogenic materials, the procedure was modified to use the phase-field model subroutines presented in [46,80]. The necessary mechanical properties for the fiber and matrix materials were adopted from the literature, while the properties for the interface were fitted according to the experimental results. The RVEs were analyzed in the ABAQUS CAE environment, with them reaching convergence in each tested case, and the results are compared with the experimentally acquired data as presented in Figure 10, thus showing consistency for the transverse case and deviations from the experiments in the in-plane shear.

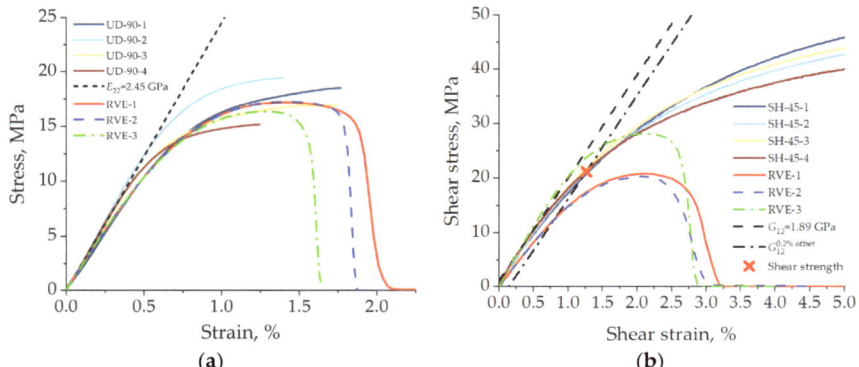

Figure 10. Comparison between the experimental and the numerical results: (**a**) transversally loaded unidirectionally reinforced case, and (**b**) the in-plane shear case.

According to the data presented in Figure 10, there is a consistency between the modelled behaviors for the transverse tensile cases regardless of the RVE type. Furthermore, the model predictions correspond to the experimentally acquired data returning values within the scatter range. However, when subjected to in-plane shear loads, the RVE-1 and RVE-2 diverge from both the experimental results and the results acquired for the RVE-3, underestimating the shear strength and modulus, which in contrast are overestimated by the RVE-3 response. Furthermore, the resulting contour plots of the damage index are presented in Figure 11, thus supporting the interfiber damage initiation.

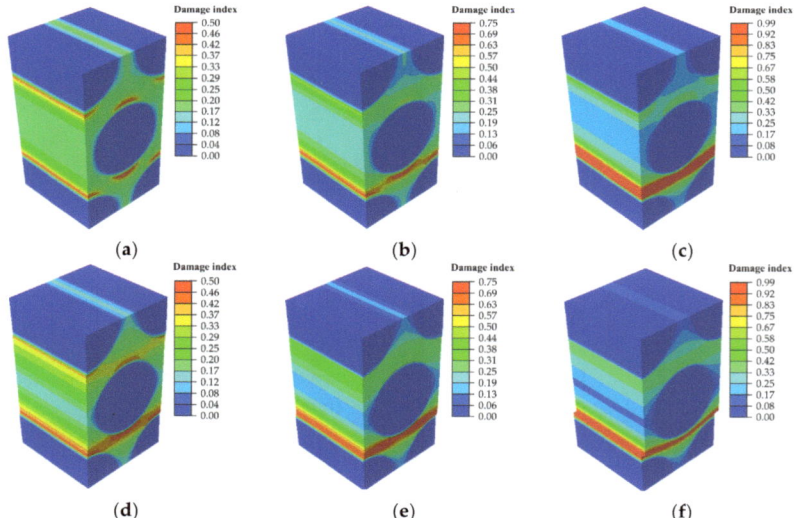

Figure 11. Damage evolution in the transverse tensile and shear cases: (**a**) damage index of 0.5 in transverse tension, (**b**) damage index of 0.75 in transverse tension, (**c**) failure in transverse tension, (**d**) damage index of 0.5 in shear, (**e**) damage index of 0.75 in shear, and (**f**) failure in shear.

As shown in Figure 11, both RVEs subjected to transverse tensile and shear loads manifest a damage initiation at the fiber/matrix interface, which is then propagated in the matrix region, resulting in fracture. Since the fibers within the RVEs are evenly distributed by the assumption of the hexagonal array, the crack path is significantly influenced by their placements. Therefore, to investigate the crack path evolution, a randomly distributed fiber array should be adopted in further studies. All things considered, it has to be emphasized that despite the transverse micromechanical behavior being well represented, the in-plane shear cases do not follow the experimental results after 1% of shear strain and predict a complete loss of bearing capabilities after exceeding the shear strength calculated according to the ASTM guidelines, which is not the case in these laminates. Therefore, different techniques for the experimental identification and modelling of shear behavior in AM-fiber-reinforced composite materials should be considered.

4. Discussion

In order to predict the behavior of additively manufactured composite materials based on the material microstructural parameters and the constituents' arrangements, a micromechanical modelling approach based on the representative volume element has been proposed in this work. The microscopic analysis was conducted on the representative (25 × 25 mm) cross-ply specimens through multiple cross-sections. Despite acquiring a lot of information on the material microstructure which was validated with the relevant data from the literature and then implemented in the RVE design, it was not possible to recognize any of the deposition paths in either of the cross-sections. This could be caused by the fact that the small specimen size, which is manufactured faster than the previously deposited layer, could have been completely cooled, resulting in a more homogenous microstructure without the characteristic triangular void patterns found in large-scale prints. Therefore, a scale-oriented microstructural inspection should be considered in further studies to address this issue. Furthermore, the 2D microscopic inspection in multiple cross-sections did not offer a complete insight into the materials' microstructure. The reference coordinate system, the printing direction, and misalignments were difficult to establish without assumptions. Therefore, a µCT approach should be considered in the future both for the microstructure and the inspection of failure mechanisms.

In this study, the RVEs were designed assuming various geometrical fiber arrangements with consideration given to the measured fiber volume ratio, while keeping the RVE surfaces symmetrical to enable periodic mesh generation and the implementation of the periodic boundary conditions. Such RVEs fulfill both the geometrical and material periodicity conditions; hence, the fibers complete each other if RVEs' segments are arranged together. However, the geometrical arrangement in these RVEs does not reflect the proper nature of the fiber arrangement found in these materials. Therefore, a statistically random fiber distribution or an input from the µCT scan should be considered in future studies. Three distinctive cases of RVEs were developed in the Abaqus CAE environment based on the results from microscopic inspections. The analysis was conducted using the phase-field fracture model available as an open-source UMAT subroutine, where the phase-field balance equation was related to the heat transfer equation readily available in the Abaqus CAE environment, hence avoiding the necessity for a dedicated UEL. This ensured the adoption of standard elements and features available in Abaqus and therefore the application of automatized procedures for enforcing the periodic boundary conditions for transversal and in-plane shear cases. In this analysis, the linear elastic constituents' behavior was assumed, while the fracture properties were calculated according to the available properties for similar materials. Thus, for better insight into the microstructural behavior, these properties should be measured.

In order to acquire the transversal tensile and in-plane shear mechanical properties of AM CFRP composites, macroscopic evaluations were conducted on unidirectional, transversely reinforced UD-90 and multidirectionally reinforced SH-45 laminates, respectively. Both cases were tested uniaxially, where for SH-45, the adopted LSS resolved the uniaxial tensile stresses into in-plane shear. As anticipated, the transverse behavior was confirmed to be dominated by the matrix's mechanical properties; however, the overall material response was inferior to that of the injection-moulded counterparts found in the literature, thus it was attributed to the fiber/matrix interface and the raster-induced defects of the additive manufacturing process. These effects act as stress concentrators and reduce the strength and stiffness of the AM materials, which was also confirmed in the scattering of the experimentally acquired data. These effects were also documented in the SH-45 specimens, which showed 2% divergence from the average before reaching 5% of shear strain. Therefore, both the fiber/matrix and the raster-induced bonding inconsistencies had to be accounted for in the material modelling. However, as the scale of raster-induced defects exceeded the size of the adopted RVEs, both the matrix/fiber and deposition contact deficiencies were introduced uniformly at the fiber/matrix interface. Since this approach is not physically consistent, yet based on an assumption, a multiscale representation to distinguish these effects should be considered in further studies. However, the properties of the fiber/matrix interface or the material deposition contact need to be acquired experimentally.

5. Conclusions

Numerical calculations of the transverse tensile and in-plane shear micromechanical behavior of AM CFRP materials based on the material microscopic analysis and the macroscopic response were presented in this study. The procedures used in material manufacturing and preparation for microscopic inspections were discussed, the positive and the negative aspects of the adopted approach were presented, and the key features for acquiring better results in further studies have been highlighted. The RVE design based on geometrical assumptions and supported by the microstructural measurements was also discussed, where many benefits of using μCT over microscopy have been presented. Furthermore, the presented microstructural analysis is based on the proposed analogy between phase-field fracture and the thermal conductivity differential equations since it supports the ABAQUS built-in features, thus enabling the development and analysis of the periodic RVEs using the phase-field fracture theory. This approach simplified the domain divisions based on the various materials and their interfaces, as well as the load application and finally the visualization of the acquired result. Furthermore, both the data scatter and the DIC images confirmed that the inconsistencies in AM CFRP composites are caused in the contact zones of the deposited material. However, since the scale of these deficiencies exceeded the size of the proposed RVEs, the effects were introduced together with the fiber/matrix influence at the constituents' interface. The negative aspects of this approach were discussed, and a multiscale approach with experimentally acquired interface properties was proposed for further study.

All things considered, this study confirmed consistency between the experimental results and the numerical prediction using the phase-field fracture approach for the transverse tensile behavior of AM CFRP composites. The numerical results were independent of the RVE type, returning values within the range of the experimental scatter. However, when analyzing the in-plane shear behavior, RVE-1 and RVE-2 revealed divergence from both the experimental results and the results acquired for the RVE-3, thus underestimating the shear strength and modulus of the tested materials. In contrast, RVE-3 overestimated the mechanical properties in the case of in-plane shear. It was shown that despite the transverse micromechanical behavior being well characterized using this phase-field modelling approach, the in-plane shear cases do not follow the experimental results after 1% of shear strain. Furthermore, after exceeding the shear strength calculated according to the ASTM guidelines, the model predicts a complete specimen failure which does not occur in the experiments. Therefore, different techniques for experimental identification and modelling of shear behavior in AM-fiber-reinforced composite materials should be considered in future work.

Author Contributions: M.G.: writing–original draft, visualization, software, methodology, investigation, and formal analysis. D.L.: writing—review and editing, supervision, project administration, and funding acquisition. M.F.: writing—review and editing, supervision, project administration, and funding acquisition. A.Ž.: writing—review and editing, validation, resources, methodology, and investigation. All authors have read and agreed to the published version of the manuscript.

Funding: The research presented in this paper was made possible by the financial support of Croatian Science Foundation (project No. IP-2019-04-8615) and (project No. IP-2019-04-3607), and the University of Rijeka (uniri-tehnic-18-139) and (uniri-tehnic-18-34).

Data Availability Statement: Data will be available on request.

Acknowledgments: Special thanks to IB-CADDY d.o.o., the Institute of Metals and Technology (IMT) in Ljubljana, Slovenia, and the Center for Advanced Computing and Modelling (CNRM) at the University of Rijeka for their technical assistance in the manufacturing and experimental procedures.

Reprints: Figure 7 was reprinted as detail from Additive Manufacturing, 56, M. Gljušćić, M. Franulović, D. Lanc, A. Žerovnik, Representative volume element for microscale analysis of additively manufactured composites/ Results and discussion, Page 9., Copyright (2022), with permission from Elsevier.

Conflicts of Interest: The authors declare that they have no known competing financial interests or personal relationships that could have appeared to have influence the work reported in this paper.

Appendix A

A formulation of the phase-field fracture theory reproduced according to [46,80,89].

The strain tensor is calculated based on the displacement field vector **u**, under the assumption of small strain sand isothermal conditions, as shown in Equation (A1) [46,80,89].

$$\varepsilon = \frac{1}{2}\left(\nabla \mathbf{u}^T + \nabla \mathbf{u}\right) \tag{A1}$$

The degree of damage is described with a continuous phase field scalar ϕ which takes values between zero and one, regarded as undamaged and fully damaged material, respectively [46,80,89]. The smeared crack approach is used to represent the discrete nature of the cracks, and it is regulated by a length scale variable $l > 0$, under the approximation of the fracture energy over a discontinuous surface Γ; where γ is the crack surface density functional, while the material toughness is given by G_c, as shown in Equation (A2) [46,80,89].

$$\phi = \int_\Gamma G_c dS \approx \int_\Omega G_c \gamma(\phi, \nabla \phi) dV \tag{A2}$$

Under the assumption that the phase field depends only on the solution of the displacement problem, the external traction forces are not included in the calculation of the principle of virtual work, as shown in Equation (A3).

$$\int_\Omega \{\sigma : \delta\varepsilon + \omega\delta\phi + \xi \cdot \delta\nabla\phi\}dV \approx \int_{\partial\Omega} (\mathbf{T}\cdot\delta\mathbf{u})dS \tag{A3}$$

In the Equation (A3), the Cauchy stress tensor is represented by σ, which is work-conjugate to the strains represented by ε. The traction **T** is defined for the outward unit normal **n** at boundary $\partial\Omega$, and it is work-conjugate to the displacements **u** [46,80,89]. The work-conjugate to the phase field ϕ is given by ω, while ξ is the micro-stress vector, a work-conjugate to the $\nabla\phi$, as the virtual quantity is given by δ [46,80,89].

In the material domain Ω, the local force balance is calculated according to the expressions given in Equation (A4), assuming $\sigma \cdot \mathbf{n} = \mathbf{T}$ and $\xi \cdot \mathbf{n} = 0$ boundary conditions on $\partial\Omega$ [46,80,89].

$$\nabla \cdot \sigma = 0; \quad \nabla \cdot \xi - \omega = 0 \tag{A4}$$

The generalized expression for the potential energy is presented in Equation (A5) [46,80,89], where the elastic strain and the facture energy densities are given by ψ and φ, respectively, and $g(\phi)$ represents the degradation function [46,80,89].

$$W(\varepsilon(\mathbf{u}), \phi, \nabla\phi) = \psi(\varepsilon(\mathbf{u}), g(\phi) + \varphi(\phi, \nabla\phi)) \tag{A5}$$

By adopting the degradation function $g(\phi)$ and assuming the criteria presented in Equation (A6) [46,80,89], the value of the potential energy is decreased with the increase in damage in the material [46,80,89].

$$g(0) = 1$$
$$g(1) = 0 \qquad (A6)$$
$$g'(\phi) \leq 0, \quad \text{for } \phi \in [0,1]$$

According to the presented assumptions, the expression for the fracture energy density is given in Equation (A7) [46,80,89], where the length scale of the phase field is represented by l, c_w is a scaling constant, and $w(\phi)$ is a geometric crack function, as the damage growth is realized with the stored elastic energy which is characterized by the elastic strain energy density ψ_0 in the undamaged state [46,80,89].

$$\varphi(\phi, \nabla\phi) = G_c \gamma(\phi, \nabla\phi) = G_c \frac{1}{4c_w l}\left(w(\phi) + l^2|\nabla\phi|^2\right) \qquad (A7)$$

In order to prevent the crack development in compression, the fracture driving force is decomposed to the active and inactive parts, represented by ψ_0^+ and ψ_0^-, respectively [46,80,89], which leads to the definition of the elastic strain energy density presented in Equation (A8) [46,80,89].

$$\psi(\varepsilon(\mathbf{u}), g(\phi)) = \psi_0^+(\varepsilon(\mathbf{u}), \phi) + \psi_0^-(\varepsilon(\mathbf{u})) = g(\phi)\psi_0^+(\varepsilon(\mathbf{u})) + \psi_0^-(\varepsilon(\mathbf{u})) \qquad (A8)$$

The damage irreversibility is represented as $\dot{\phi} \geq 0$, and it is achieved using the history field variable $\mathcal{H}$ which fulfills the Karush–Kuhn–Tucker (KKT) conditions presented in Equation (A8) [46,80,89].

$$\psi_0^+ - \mathcal{H} \leq 0, \quad \dot{\mathcal{H}} \geq 0, \quad \mathcal{H}(\psi_0^+ - \mathcal{H}) = 0 \qquad (A9)$$

This leads to a history field definition for a specific time referred to as t over a total time given by τ, as shown in Equation (A10), and the reformulation of the potential energy is shown in Equation (A11) [46,80,89].

$$\mathcal{H} = \max_{t \in [0,\tau]} \psi_0^+(t) \qquad (A10)$$

$$W = g(\phi)\mathcal{H} + G_c \frac{1}{4c_w}\left(\frac{1}{l}w(\phi) + l|\nabla\phi|^2\right) \qquad (A11)$$

Scalar micro-stress scalar ω and the micro-stress vector ξ are calculated according to the expressions presented in Equations (A12) and (A13), respectively [46,80,89], which were incorporated into Equation (A14), and this leads to the phase field evolution law presented in Equation (A15) [46,80,89].

$$\omega = \frac{\partial W}{\partial \nabla\phi} = g'(\phi)\mathcal{H} + \frac{G_c}{4c_w}w'(\phi) \qquad (A12)$$

$$\xi = \frac{\partial W}{\partial \nabla\phi} = \frac{l}{2c_w}G_c \nabla\phi \qquad (A13)$$

$$\frac{G_c}{2c_w}\left(\frac{w'(\phi)}{2l} - l\nabla^2\phi\right) + g'(\phi)\mathcal{H} = 0 \qquad (A14)$$

The heat transfer analogy is based on the evolution of the temperature field T in a specific time t as shown in Equation (A15) [46,80,89], where the thermal conductivity of the material is represented by k, the specific heat is represented by c_p, and the density is represented by ρ [46,80,89].

$$k\nabla^2 T - \rho c_p \frac{\partial T}{\partial t} = -r \qquad (A15)$$

If steady-state conditions are assumed, the $\frac{\partial T}{\partial t}$ reduces to zero, and the expression is simplified as shown in Equation (A16) [46,80,89].

$$k\nabla^2 T = -r \qquad (A16)$$

In order to correspond with Equation (A16), the phase field evolution law presented in Equation (A14) can be rearranged as shown in Equation (A17) [46,80,89].

$$\nabla^2 \phi = \frac{g'(\phi)\mathcal{H}}{lG_c} + \frac{w'(\phi)}{2l^2} \qquad (A17)$$

According to [46,80,89], this analogy can be adopted considering the equivalence between the temperature and the phase field ($T = \phi$) assuming the thermal conductivity is equal to one ($k = 1$) and by defining the heat flux according to the internal heat generation, as shown in Equation (A18), while in order to compute the Jacobian matrix, the heat flux rate of change is defined according to Equation (A19).

$$r = \frac{g'(\phi)2\mathcal{H}c_w}{lG_c} - \frac{w'(\phi)}{2l^2} \tag{A18}$$

$$\frac{\partial r}{\partial \phi} = \frac{g''(\phi)2\mathcal{H}c_w}{lG_c} - \frac{w''(\phi)}{2l^2} \tag{A19}$$

Since the total potential energy is minimized according to the displacement field u and phase field Φ, while Φ is being solved at the finite element nodes such as a damage variable at the finite element node instead of at the integration point as in the case of local damage models, the calculation would require the user-defined element (UEL) before the implementation of the user-defined material model (UMAT). However, since this approach uses the heat transfer analogy, the UMAT subroutine can be run without the necessity for the user-defined element.

Steps for the implementation of the subroutine [89]:

- Define the user material with five material properties including the Young's modulus E, Poisson's ratio ν, phase field length scale l, fracture toughness G_c, and the tensile strength f_t which is applicable for the phase field cohesive zone models, while otherwise neglected
- Set a solution-dependent state variable (SDV)
- Define the material conductivity equal to one
- State the analysis step as coupled temperature-displacement with steady-state or transient options, a constant increment size, a separated solution technique, and symmetric equation solver matrix storage
- Change the values of the solution controls parameters I_0, I_R, I_P, I_C, I_L, and I_G to 5000, to avoid convergence problems due to large number of iterations
- Define the initial temperature condition equal to zero to describe the undamaged material in the initial step
- Adopt the element type as coupled temperature–displacement

References

1. Alghamdi, S.S.; John, S.; Choudhury, N.R.; Dutta, N.K. Additive Manufacturing of Polymer Materials: Progress, Promise and Challenges. *Polymers* **2021**, *13*, 753. [CrossRef]
2. Ferreira, I.; Machado, M.; Alves, F.; Torres Marques, A. A Review on Fibre Reinforced Composite Printing via FFF. *Rapid Prototyp. J.* **2019**, *25*, 972–988. [CrossRef]
3. Ngo, T.D.; Kashani, A.; Imbalzano, G.; Nguyen, K.T.Q.; Hui, D. Additive Manufacturing (3D Printing): A Review of Materials, Methods, Applications and Challenges. *Compos. Part B* **2018**, *143*, 172–196. [CrossRef]
4. van de Werken, N.; Tekinalp, H.; Khanbolouki, P.; Ozcan, S.; Williams, A.; Tehrani, M. Additively Manufactured Carbon Fiber-Reinforced Composites: State of the Art and Perspective. *Addit. Manuf.* **2020**, *31*, 100962. [CrossRef]
5. van de Werken, N. Additively Manufactured Continuous Carbon Fiber Thermoplastic Composites for High-Performance Applications. Doctoral Dissertation, The University of New Mexico, Albuquerque, NM, USA, 2019.
6. Karaş, B.; Smith, P.J.; Fairclough, J.P.A.; Mumtaz, K. Additive Manufacturing of High Density Carbon Fibre Reinforced Polymer Composites. *Addit. Manuf.* **2022**, *58*, 103044. [CrossRef]
7. Rengier, F.; Mehndiratta, A.; Von Tengg-Kobligk, H.; Zechmann, C.M.; Unterhinninghofen, R.; Kauczor, H.U.; Giesel, F.L. 3D Printing Based on Imaging Data: Review of Medical Applications. *Int. J. Comput. Assist. Radiol. Surg.* **2010**, *5*, 335–341. [CrossRef]
8. Schubert, C.; Van Langeveld, M.C.; Donoso, L.A. Innovations in 3D Printing: A 3D Overview from Optics to Organs. *Br. J. Ophthalmol.* **2014**, *98*, 159–161. [CrossRef]
9. Diegel, O.; Nordin, A.; Motte, D. *Additive Manufacturing Technologies*; Springer: Berlin/Heidelberg, Germany, 2019; ISBN 9781493921126.
10. Abazari, S.; Shamsipur, A.; Bakhsheshi-Rad, H.R.; Ismail, A.F. Carbon Nanotubes (CNTs)-Reinforced Magnesium-Based Matrix Composites: A Comprehensive Review. *J. Higher Educ.* **2020**, *13*, 38. [CrossRef]
11. Abazari, S.; Shamsipur, A.; Bakhsheshi-Rad, H.R.; Ramakrishna, S.; Berto, F. Graphene Family Nanomaterial Reinforced Magnesium-Based Matrix Composites for Biomedical Application: A Comprehensive Review. *Metals* **2020**, *10*, 1002. [CrossRef]
12. Monfared, V.; Bakhsheshi-Rad, H.R.; Ramakrishna, S.; Razzaghi, M.; Berto, F. A Brief Review on Additive Manufacturing of Polymeric Composites and Nanocomposites. *Micromachines* **2021**, *12*, 24. [CrossRef]
13. Gljušćić, M.; Franulović, M.; Lanc, D.; Žerovnik, A. Representative Volume Element for Microscale Analysis of Additively Manufactured Composites. *Addit. Manuf.* **2022**, *56*, 102902. [CrossRef]
14. Gljušćić, M.; Franulović, M.; Žužek, B.; Žerovnik, A. Experimental Validation of Progressive Damage Modeling in Additively Manufactured Continuous Fiber Composites. *Compos. Struct.* **2022**, *295*, 115869. [CrossRef]

15. Iragi, M.; Pascual-González, C.; Esnaola, A.; Lopes, C.S.; Aretxabaleta, L. Ply and Interlaminar Behaviours of 3D Printed Continuous Carbon Fibre-Reinforced Thermoplastic Laminates; Effects of Processing Conditions and Microstructure. *Addit. Manuf.* **2019**, *30*, 100884. [CrossRef]
16. Carlota, V. Essentium's Latest Survey: What Is the Future of Industrial 3D Printing? Available online: https://www.3dnatives.com/en/essentium-190320195/ (accessed on 15 October 2022).
17. Gide, K.M.; Islam, S.; Bagheri, Z.S. Polymer-Based Materials Built with Additive Manufacturing Methods for Orthopedic Applications: A Review. *J. Compos. Sci.* **2022**, *6*, 262. [CrossRef]
18. Deeba, F.; Shrivastava, K.; Bafna, M.; Jain, A. Tuning of Dielectric Properties of Polymers by Composite Formation: The Effect of Inorganic Fillers Addition. *J. Compos. Sci.* **2022**, *6*, 355. [CrossRef]
19. Tran, C.C.; Nguyen, Q.K. An Efficient Method to Determine the Thermal Behavior of Composite Material with Loading High Thermal Conductivity Fillers. *J. Compos. Sci.* **2022**, *6*, 214. [CrossRef]
20. Kabir, S.M.F.; Mathur, K.; Seyam, A.F.M. Maximizing the Performance of 3d Printed Fiber-Reinforced Composites. *J. Compos. Sci.* **2021**, *5*, 136. [CrossRef]
21. Melenka, G.W.; Cheung, B.K.O.; Schofield, J.S.; Dawson, M.R.; Carey, J.P. Evaluation and Prediction of the Tensile Properties of Continuous Fiber-Reinforced 3D Printed Structures. *Compos. Struct.* **2016**, *153*, 866–875. [CrossRef]
22. He, Q.; Wang, H.; Fu, K.; Ye, L. 3D Printed Continuous CF/PA6 Composites: Effect of Microscopic Voids on Mechanical Performance. *Compos. Sci. Technol.* **2020**, *191*, 108077. [CrossRef]
23. Baechle-Clayton, M.; Loos, E.; Taheri, M.; Taheri, H. Failures and Flaws in Fused Deposition Modeling (FDM) Additively Manufactured Polymers and Composites. *J. Compos. Sci.* **2022**, *6*, 202. [CrossRef]
24. Suquet, P. *Continuum Micromechanics*; Kaliszky, S., Sayir, M., Schneider, W., Bianchi, G., Tasso, C., Eds.; Springer: Berlin/Heidelberg, Germany, 1997; ISBN 9783211829028.
25. Michel, J.C.; Moulinec, H.; Suquet, P. Effective Properties of Composite Materials with Periodic Microstructure: A Computational Approach. *Comput. Methods Appl. Mech. Eng.* **1999**, *172*, 109–143. [CrossRef]
26. Rémond, Y.; Ahzi, S. *Applied RVE Reconstruction and Homogenization of Heterogeneous Materials*; John Wiley & Sons: Hoboken, NJ, USA, 2016; ISBN 9781848219014.
27. Soutis, C.; Beaumont, P.W.R. *Multi-Scale Modelling of Composite Material Systems: The Art of Predictive Damage Modelling*; Woodhead Publishing Limited: Sawston, UK; CRC Press LLC: Boca Raton, FL, USA, 2005; Volume 3, ISBN 978-1-85573-936-9.
28. Múgica, J.I.; Lopes, C.S.; Naya, F.; Herráez, M.; Martínez, V.; González, C. Multiscale Modelling of Thermoplastic Woven Fabric Composites: From Micromechanics to Mesomechanics. *Compos. Struct.* **2019**, *228*, 111340. [CrossRef]
29. Riaño, L.; Joliff, Y. An ABAQUS™ Plug-in for the Geometry Generation of Representative Volume Elements with Randomly Distributed Fibers and Interphases. *Compos. Struct.* **2019**, *209*, 644–651. [CrossRef]
30. Raju, B.; Hiremath, S.R.; Roy Mahapatra, D. A Review of Micromechanics Based Models for Effective Elastic Properties of Reinforced Polymer Matrix Composites. *Compos. Struct.* **2018**, *204*, 607–619. [CrossRef]
31. Omairey, S.L.; Dunning, P.D.; Sriramula, S. Development of an ABAQUS Plugin Tool for Periodic RVE Homogenisation. *Eng. Comput.* **2019**, *35*, 567–577. [CrossRef]
32. Akpoyomare, A.I.; Okereke, M.I.; Bingley, M.S. Virtual Testing of Composites: Imposing Periodic Boundary Conditions on General Finite Element Meshes. *Compos. Struct.* **2017**, *160*, 983–994. [CrossRef]
33. Okereke, M.I.; Akpoyomare, A.I. A Virtual Framework for Prediction of Full-Field Elastic Response of Unidirectional Composites. *Comput. Mater. Sci.* **2013**, *70*, 82–99. [CrossRef]
34. Kempesis, D.; Iannucci, L.; Ramesh, K.T.; Del Rosso, S.; Curtis, P.T.; Pope, D.; Duke, P.W. Micromechanical Analysis of High Fibre Volume Fraction Polymeric Laminates Using Micrograph-Based Representative Volume Element Models. *Compos. Sci. Technol.* **2022**, *229*, 109680. [CrossRef]
35. Schindelin, J.; Arganda-Carreras, I.; Frise, E.; Kaynig, V.; Longair, M.; Pietzsch, T.; Preibisch, S.; Rueden, C.; Saalfeld, S.; Schmid, B.; et al. Fiji: An Open-Source Platform for Biological-Image Analysis. *Nat. Methods* **2012**, *9*, 676–682. [CrossRef]
36. Breite, C.; Melnikov, A.; Turon, A.; de Morais, A.B.; Le Bourlot, C.; Maire, E.; Schöberl, E.; Otero, F.; Mesquita, F.; Sinclair, I.; et al. Detailed Experimental Validation and Benchmarking of Six Models for Longitudinal Tensile Failure of Unidirectional Composites. *Compos. Struct.* **2022**, *279*, 114828. [CrossRef]
37. Liu, Y.; Straumit, I.; Vasiukov, D.; Lomov, S.V.; Panier, S. Multi-Scale Material Model for 3D Composite Using Micro CT Images Geometry Reconstruction. In Proceedings of the 7th European Conference on Composite Materials (ECCM), Munich, Germany, 26–30 June 2016.
38. Straumit, I.; Lomov, S.V.; Wevers, M. Quantification of the Internal Structure and Automatic Generation of Voxel Models of Textile Composites from X-Ray Computed Tomography Data. *Compos. Part A Appl. Sci. Manuf.* **2015**, *69*, 150–158. [CrossRef]
39. Dean, A.; Asur Vijaya Kumar, P.K.; Reinoso, J.; Gerendt, C.; Paggi, M.; Mahdi, E.; Rolfes, R. A Multi Phase-Field Fracture Model for Long Fiber Reinforced Composites Based on the Puck Theory of Failure. *Compos. Struct.* **2020**, *251*, 112446. [CrossRef]
40. Guillén-Hernández, T.; García, I.G.; Reinoso, J.; Paggi, M. A Micromechanical Analysis of Inter-Fiber Failure in Long Reinforced Composites Based on the Phase Field Approach of Fracture Combined with the Cohesive Zone Model. *Int. J. Fract.* **2019**, *220*, 181–203. [CrossRef]
41. Tan, W.; Martínez-Pañeda, E. Phase Field Fracture Predictions of Microscopic Bridging Behaviour of Composite Materials. *Compos. Struct.* **2022**, *286*, 115242. [CrossRef]

42. Griffith, A.A., VI. The phenomena of rupture and flow in solids. *Philos. Trans. R. Soc. Lond.* **1921**, *221*, 163–198. [CrossRef]
43. Biner, S.B. *Programming Phase-Field Modeling*; Springer: Cham, Switzerland, 2017; ISBN 9783319411965.
44. Francfort, G.A.; Marigo, J.J. Revisiting Brittle Fracture as an Energy Minimization Problem. *J. Mech. Phys. Solids* **1998**, *46*, 1319–1342. [CrossRef]
45. Bourdin, B.; Francfort, G.A.; Marigo, J. Numerical Experiments in Revisited Brittle Fracture. *J. Mech. Phys. Solids* **2000**, *48*, 797–826. [CrossRef]
46. Navidtehrani, Y.; Betegón, C.; Martínez-Pañeda, E. A Unified Abaqus Implementation of the Phase Field Fracture Method Using Only a User Material Subroutine. *Materials* **2021**, *14*, 1913. [CrossRef]
47. Tourret, D.; Liu, H.; LLorca, J. Phase-Field Modeling of Microstructure Evolution: Recent Applications, Perspectives and Challenges. *Prog. Mater. Sci.* **2022**, *123*, 100810. [CrossRef]
48. Egger, A.; Pillai, U.; Agathos, K.; Kakouris, E.; Chatzi, E.; Aschroft, I.A.; Triantafyllou, S.P. Discrete and Phase Field Methods for Linear Elastic Fracture Mechanics: A Comparative Study and State-of-the-Art Review. *Appl. Sci* **2019**, *9*, 2436. [CrossRef]
49. Wu, J.; Huang, Y. Comprehensive Implementations of Phase-Field Damage Models in Abaqus. *Theor. Appl. Fract. Mech.* **2020**, *106*, 102440. [CrossRef]
50. Simoes, M.; Martínez-pañeda, E. Phase Field Modelling of Fracture and Fatigue in Shape Memory Alloys. *Comput. Methods Appl. Mech. Eng.* **2021**, *373*, 113504. [CrossRef]
51. Natarajan, S.; Annabattula, R.K.; Martínez-pañeda, E. Phase Field Modelling of Crack Propagation in Functionally Graded Materials. *Compos. Part B* **2019**, *169*, 239–248. [CrossRef]
52. Asur Vijaya Kumar, P.K.; Dean, A.; Reinoso, J.; Lenarda, P.; Paggi, M. Phase Field Modeling of Fracture in Functionally Graded Materials: Γ-Convergence and Mechanical Insight on the Effect of Grading. *Thin-Walled Struct.* **2021**, *159*, 107234. [CrossRef]
53. Miehe, C.; Schänzel, L. Phase Field Modeling of Fracture in Rubbery Polymers. Part I: Finite Elasticity Coupled with Brittle Failure. *J. Mech. Phys. Solids* **2014**, *65*, 93–113. [CrossRef]
54. Loew, P.J.; Peters, B.; Beex, L.A.A. Rate-Dependent Phase-Field Damage Modeling of Rubber and Its Experimental Parameter Identification. *J. Mech. Phys. Solids* **2019**, *127*, 266–294. [CrossRef]
55. Zhou, S.; Zhuang, X.; Rabczuk, T. Phase Field Modeling of Brittle Compressive-Shear Fractures in Rock-like Materials: A New Driving Force and a Hybrid Formulation. *Comput. Methods Appl. Mech. Eng.* **2019**, *355*, 729–752. [CrossRef]
56. Schuler, L.; Ilgen, A.G.; Newell, P. Chemo-Mechanical Phase-Field Modeling of Dissolution-Assisted Fracture. *Comput. Methods Appl. Mech. Eng.* **2020**, *362*, 112838. [CrossRef]
57. Kristensen, P.K.; Martínez-pañeda, E.; Engineering, M.; Lyngby, D.-K. Phase Field Fracture Modelling Using Quasi-Newton Methods and a New Adaptive Step Scheme. *Theor. Appl. Fract. Mech.* **2020**, *107*, 102446. [CrossRef]
58. Abdollahi, A.; Arias, I. Phase-Field Modeling of Crack Propagation in Piezoelectric and Ferroelectric Materials with Different Electromechanical Crack Conditions. *J. Mech. Phys. Solids* **2012**, *60*, 2100–2126. [CrossRef]
59. Pillai, U. Damage Modelling in Fibre-Reinforced Composite Laminates Using Phase Field Approach. Doctoral Dissertation, University of Nottingham, Nottingham, UK, 2021.
60. Quintanas-corominas, A.; Reinoso, J.; Casoni, E.; Turon, A.; Mayugo, J.A. A Phase Field Approach to Simulate Intralaminar and Translaminar Fracture in Long Fiber Composite Materials. *Compos. Struct.* **2019**, *220*, 899–911. [CrossRef]
61. Ahmadi, M. A Hybrid Phase Field Model for Fracture Induced by Lithium Diffusion in Electrode Particles of Li-Ion Batteries. *Comput. Mater. Sci.* **2020**, *184*, 109879. [CrossRef]
62. Nguyen, T.T.; Yvonnet, J.; Zhu, Q.Z.; Bornert, M.; Chateau, C. A Phase Field Method to Simulate Crack Nucleation and Propagation in Strongly Heterogeneous Materials from Direct Imaging of Their Microstructure. *Eng. Fract. Mech.* **2015**, *139*, 18–39. [CrossRef]
63. Goswami, S.; Anitescu, C.; Rabczuk, T. Adaptive Phase Field Analysis with Dual Hierarchical Meshes for Brittle Fracture. *Eng. Fract. Mech.* **2019**, *218*, 106608. [CrossRef]
64. Goswami, S.; Anitescu, C.; Rabczuk, T. Adaptive Fourth-Order Phase Field Analysis Using Deep Energy Minimization. *Theor. Appl. Fract. Mech.* **2020**, *107*, 102527. [CrossRef]
65. Huynh, G.D.; Zhuang, X.; Nguyen-Xuan, H. Implementation Aspects of a Phase-Field Approach for Brittle Fracture. *Front. Struct. Civ. Eng.* **2019**, *13*, 417–428. [CrossRef]
66. Msekh, M.A.; Sargado, J.M.; Jamshidian, M.; Areias, P.M.; Rabczuk, T. Abaqus Implementation of Phase-Field Model for Brittle Fracture. *Comput. Mater. Sci.* **2015**, *96*, 472–484. [CrossRef]
67. Liu, G.; Li, Q.; Msekh, M.A.; Zuo, Z. Abaqus Implementation of Monolithic and Staggered Schemes for Quasi-Static and Dynamic Fracture Phase-Field Model. *Comput. Mater. Sci.* **2016**, *121*, 35–47. [CrossRef]
68. Molnár, G.; Gravouil, A. 2D and 3D Abaqus Implementation of a Robust Staggered Phase-Field Solution for Modeling Brittle Fracture. *Finite Elem. Anal. Des.* **2017**, *130*, 27–38. [CrossRef]
69. Pillai, U.; Heider, Y.; Markert, B. A Diffusive Dynamic Brittle Fracture Model for Heterogeneous Solids and Porous Materials with Implementation Using a User-Element Subroutine. *Comput. Mater. Sci.* **2018**, *153*, 36–47. [CrossRef]
70. Bhowmick, S.; Liu, G.R. A Phase-Field Modeling for Brittle Fracture and Crack Propagation Based on the Cell-Based Smoothed Finite Element Method. *Eng. Fract. Mech.* **2018**, *204*, 369–387. [CrossRef]
71. Seleš, K.; Lesičar, T.; Tonković, Z.; Sorić, J. A Residual Control Staggered Solution Scheme for the Phase-Field Modeling of Brittle Fracture. *Eng. Fract. Mech.* **2019**, *205*, 370–386. [CrossRef]

72. Zhou, S.; Rabczuk, T.; Zhuang, X. Phase Field Modeling of Quasi-Static and Dynamic Crack Propagation: COMSOL Implementation and Case Studies. *Adv. Eng. Softw.* **2018**, *122*, 31–49. [CrossRef]
73. Pham, K.; Amor, H.; Marigo, J.J.; Maurini, C. Gradient Damage Models and Their Use to Approximate Brittle Fracture. *Int. J. Damage Mech.* **2011**, *20*, 618–652. [CrossRef]
74. Wu, J.Y. A Unified Phase-Field Theory for the Mechanics of Damage and Quasi-Brittle Failure. *J. Mech. Phys. Solids* **2017**, *103*, 72–99. [CrossRef]
75. Wu, J.Y. A Geometrically Regularized Gradient-Damage Model with Energetic Equivalence. *Comput. Methods Appl. Mech. Eng.* **2018**, *328*, 612–637. [CrossRef]
76. Wu, J.Y.; Nguyen, V.P. A Length Scale Insensitive Phase-Field Damage Model for Brittle Fracture. *J. Mech. Phys. Solids* **2018**, *119*, 20–42. [CrossRef]
77. Zhang, P.; Hu, X.; Wang, X.; Yao, W. An Iteration Scheme for Phase Field Model for Cohesive Fracture and Its Implementation in Abaqus. *Eng. Fract. Mech.* **2018**, *204*, 268–287. [CrossRef]
78. Miehe, C.; Hofacker, M.; Welschinger, F. A Phase Field Model for Rate-Independent Crack Propagation: Robust Algorithmic Implementation Based on Operator Splits. *Comput. Methods Appl. Mech. Eng.* **2010**, *199*, 2765–2778. [CrossRef]
79. Gerasimov, T.; De Lorenzis, L. A Line Search Assisted Monolithic Approach for Phase-Field Computing of Brittle Fracture. *Comput. Methods Appl. Mech. Eng.* **2016**, *312*, 276–303. [CrossRef]
80. Navidtehrani, Y.; Betegón, C.; Martínez-Pañeda, E. A Simple and Robust Abaqus Implementation of the Phase Field Fracture Method. *Appl. Eng. Sci.* **2021**, *6*, 100050. [CrossRef]
81. Gljuščić, M. Multiscale Modelling of Additively Manufactured Composite Material Behaviour. Doctoral Dissertation, University of Rijeka, Rijeka, Croatia, 2022.
82. Iragi, M.; Pascual-Gonzalez, C.; Esnaola, A.; Aurrekoetxea, J.; Lopes, C.S.; Aretxabaleta, L. Characterization of Elastic and Resistance Behaviours of 3D Printed Continuous Carbon Fibre Reinforced Thermoplastics. In Proceedings of the ECCM18—18th European Conference on Composite Materials, Athens, Greece, 24–28 June 2018; pp. 24–28.
83. Carraro, P.A.; Quaresimin, M. A Stiffness Degradation Model for Cracked Multidirectional Laminates with Cracks in Multiple Layers. *Int. J. Solids Struct.* **2014**, *58*, 34–51. [CrossRef]
84. Okereke, M.; Keates, S. *Finite Element Applications: A Practical Guide to the FEM Process*; Seung-Bok, C., Habinin, D., Fu, Y., Guardiola, C., Sun, J.-Q., Eds.; Springer International Publishing: Cham, Switzerland, 2018; ISBN 978-3-319-67124-6.
85. Pascual-González, C.; Iragi, M.; Fernández, A.; Fernández-Blázquez, J.P.; Aretxabaleta, L.; Lopes, C.S. An Approach to Analyse the Factors behind the Micromechanical Response of 3D-Printed Composites. *Compos. Part B Eng.* **2020**, *186*, 107820. [CrossRef]
86. VDI/VDE 2479; Materials for Precision Engineering; Polyamide Moulding Materials Unreinforced. VDI-Verlag GmbH: Dusseldorf, Germany, 1978; Volume 1.
87. MarkForged. *Material Datasheet Composites*; Markforged: Watertown, MA, USA, 2018.
88. Nabavi, A.; Goroshin, S.; Frost, D.L.; Barthelat, F. Mechanical Properties of Chromium–Chromium Sulfide Cermets Fabricated by Self-Propagating High-Temperature Synthesis. *J. Mater. Sci.* **2015**, *50*, 3434–3446. [CrossRef]
89. Navidtehrani, Y.; Martinez-Paneda, E. A Simple yet General ABAQUS Phase Field Fracture Implementation Using a UMAT Subroutine. *Eng. Sci.* **2021**, *6*, 100390.
90. Ambati, M.; Gerasimov, T.; De Lorenzis, L. A Review on Phase-Field Models of Brittle Fracture and a New Fast Hybrid Formulation. *Comput. Mech.* **2015**, *55*, 383–405. [CrossRef]
91. Amor, H.; Marigo, J.J.; Maurini, C. Regularized Formulation of the Variational Brittle Fracture with Unilateral Contact: Numerical Experiments. *J. Mech. Phys. Solids* **2009**, *57*, 1209–1229. [CrossRef]
92. *ASTM D3039/D3039M-17*; Standard Test Method for Tensile Properties of Polymer Matrix Composite Materials. ASTM International: West Conshohocken, PA, USA, 2017.
93. *ASTM D3518/D3518M*; 18 Standard Test Method for In-Plane Shear Response of Polymer Matrix Composite Materials by Tensile Test of a +/−45° Laminate. ASTM International: West Conshohocken, PA, USA, 2001; pp. 1–7.

Disclaimer/Publisher's Note: The statements, opinions and data contained in all publications are solely those of the individual author(s) and contributor(s) and not of MDPI and/or the editor(s). MDPI and/or the editor(s) disclaim responsibility for any injury to people or property resulting from any ideas, methods, instructions or products referred to in the content.

Article

Optimization of Electrical Intensity for Electrochemical Anodic Oxidation to Modify the Surface of Carbon Fibers and Preparation of Carbon Nanotubes/Carbon Fiber Multi-Scale Reinforcements

Mengfan Li [1,2], Yanxiang Wang [1,2,*], Bowen Cui [1,2], Chengjuan Wang [1,2], Hongxue Tan [1,2], Haotian Jiang [1,2], Zhenhao Xu [1,2], Chengguo Wang [1,2] and Guangshan Zhuang [1,2,*]

[1] Key Laboratory of Liquid-Solid Structural Evolution and Processing of Materials of Ministry of Education, Shandong University, Jinan 250061, China
[2] Carbon Fiber Engineering Research Center, School of Material Science and Engineering, Shandong University, Jinan 250061, China
* Correspondence: wangyanxiang079@163.com (Y.W.); zhuangguangshan01@163.com (G.Z.)

Abstract: Carbon fiber (CF) reinforced composites are widely used due to their excellent properties. However, the smooth surface and few functional groups of CFs can lead to fiber fractures and pullout, which reduce the service life of the composites. The overall performance of composites can be improved by growing carbon nanotubes (CNTs) on the CF surface. Before this, CF surface should be modified to enhance the loading amount of catalyst particles and thus make the CNTs more uniform. In this paper, CNTs were grown on a CF surface by one-step chemical vapor deposition to prepare multi-scale CNTs/CF reinforcements, and the effects of different methods on the CF surface modification were explored. After setting four intensities of electrochemical anodic oxidation, i.e., 50 C/g, 100 C/g, 150 C/g and 200 C/g, it was found that the distribution and quantity of CNTs were improved under both the 100 C/g and 150 C/g conditions. Considering the influence of electrical intensity on the (002) interplanar spacing of CFs, which affects the mechanical properties of the samples, 100 C/g was finally selected as the optimal electrochemical treatment intensity. This finding provides a reference for continuous and large-scale modification of CF surfaces to prepare CNTs/CF multi-scale reinforcements.

Keywords: carbon nanotube; carbon fiber; surface modification; electrochemical anodic oxidation

1. Introduction

Carbon fiber (CF) has excellent properties, including high tensile strength, high tensile modulus, good compressive strength, low density, and excellent electrical and thermal/chemical resistance [1]. CF reinforced polymer composites have been widely used in aerospace, military industry, transportation, leisure and sports, and other fields [2–4]. These composite materials are made of resin and CFs by laminating and pressing under high temperature and pressure. Resin, as a continuous phase, and CFs, as a discontinuous phase, both contribute to the overall performance of the composite materials [5]. However, the contribution of the resin matrix to the overall performance of the composite materials is limited and cannot be easily improved. Therefore, the improvement of the strength of composite materials mainly depends on the performance of the interface between the resin and fibers, and enhancements of interface performance are usually achieved by the surface modification of CFs [6–8]. Common CF surface modification treatment methods include the chemical grafting method [9,10], chemical vapor deposition (CVD) method [11–13], etc. Among them, the method of in situ growth of CNTs on the surface of CFs by CVD has received a great deal of attention. This method has the advantages of adjustable process parameters, controllable growth of CNTs, economical-growth equipment, etc. [14–17]. Moreover, as a

nanomaterial, CNTs have a special one-dimensional tubular structure, and their mechanical properties, electrical and thermal conductivity, are marvelous [18,19]. Therefore, growing CNTs on the CF surface to prepare CNTs/CF multi-scale reinforcements can introduce the excellent properties of CNTs into CF reinforced composites, resulting in a good reinforcement effect [20–23]. Ideal CNTs/CF multi-scale reinforcements need to uniformly grow a layer of CNTs on the surface of CFs. The question of whether catalyst particles, as the "seeds" for growing CNTs, can achieve firm attachment and uniform distribution is crucial in obtaining high-performance samples [24]. Nevertheless, as an industrialized product, CF has undergone pre-oxidation and medium-high temperature carbonization treatment. The surface is composed of an ordered, six-membered ring graphite sheet structure with almost no active functional groups. The surface is smooth and chemically stable [25,26], which prevents the attachment of a catalyst precursor and gives rise to the phenomenon of uneven distribution or agglomeration and deactivation of catalyst particles. Hence, the surface of CFs must be modified to make it easier to attach a catalyst precursor in order to ensure the uniform growth of CNTs. Common surface treatment methods for CFs include electrochemical anodic oxidation (EAO) [27–29], plasma treatment [30,31], liquid oxidation [32], etc. Among these, EAO and the liquid oxidation methods have simple operation processes and low economic cost and are worthy of in-depth research and development [27–29,32]. In particular, since CF has good chemical stability in various aqueous electrolytes and excellent electric conductivity, EAO can be used as an ideal surface modification approach to treat continuous fibers [33]. Additionally, many parameters can be adjusted during EAO, such as current, potential, and electrolyte composition [33,34]. Moreover, the treatment process of EAO is relatively mild and will not cause significant damage to the fiber surface [35]. Generally speaking, EAO includes two processes: anodic oxidation and cathodic reduction. The input electric energy is converted into chemical energy, and water is decomposed into strong oxidants at the anode to etch the surface of CFs, causing changes in surface functional groups and roughness [36,37].

In this study, the effects of EAO and liquid oxidation on the surface modification of CFs were compared. The EAO process was determined to be the most suitable method for the production of CNTs/CF multi-scale reinforcements. The process was then optimized by adjusting the electrical intensity.

2. Materials and Methods

2.1. Materials

PAN-based CFs (T-700-12k) were provided by Toray Co. Ltd., Tokyo, Japan. Nitric acid (HNO_3, 68%), hydrogen peroxide (H_2O_2, 30%), ethanol (C_2H_5OH, 99.7%), and ammonium dihydrogen phosphate ($NH_4H_2PO_4$, 99%) were purchased from Sinopharm Co. Ltd., Shanghai, China. Nickel nitrate ($Ni(NO_3)_2 \cdot 6H_2O$, 99.9%) was supplied by Aladdin Reagent Co. Ltd., Shanghai, China. N_2, H_2 and C_2H_2 were purchased from Jinan Gas Factory, Jinan, China. Deionized water was self-made in the laboratory.

2.2. Preparation of CNTs/CF Multi-Scale Reinforcements

The surface of the finished CFs adheres to the sizing agent. The CFs need to be placed in a vertical CVD furnace in advance and kept at 450 °C for 1.5 h in N_2 atmosphere. After removing the sizing agent on the surface of CFs, the desized CFs are obtained.

The surface modification treatments of desized CFs were carried out by HNO_3 treatment, H_2O_2 oxidation, and EAO, respectively.

- As for HNO_3 treatment, the desized CFs were fully immersed in a solution tank filled with HNO_3 aqueous solution (10 wt%), which kept in a 60 °C oven for 1 h.
- With regard to H_2O_2 oxidation, we fully submerged the desized CFs into a solution tank filled with H_2O_2 aqueous solution (30 wt%) and kept them in a 60 °C oven for 1.5 h.
- Regarding EAO, this study adopted a laboratory-made EAO treatment device, as shown in Figure 1a. The desized CFs entered the electrolytic tank through the guide

roller, and the electrolyte was an aqueous solution of $NH_4H_2PO_4$ (5 wt%). The two variables affecting the etching degree of the CF surface, i.e., the wire speed and the electrolytic intensity, were integrated into the electrical intensity per unit mass of CFs in order to quantify the intensity of the EAO treatment.

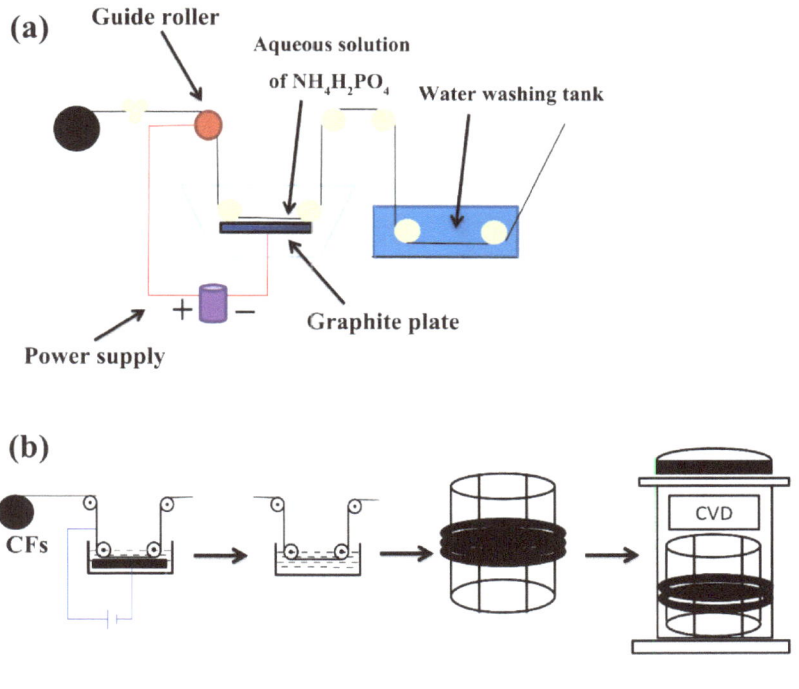

Figure 1. (**a**) EAO device, (**b**) Schematic diagram of CNTs grown on the CF surface.

Using the solution impregnation method, the catalyst solution was a 0.05 M $Ni(NO_3)_2$ alcohol solution. We then put the CFs in a drying oven at 70 °C. A one-step method was used to grow CNTs on the CF surface. The dried CFs were then placed into a vertical CVD furnace, which was heated to a reduction temperature of 450 °C, and H_2 was introduced to reduce the catalyst precursor to elemental Ni, before being adjusted to 550 °C. Additionally, 0.3 L/min H_2, 0.6 L/min N_2, and 0.3 L/min C_2H_2 were injected and kept for a long enough time to make CNTs grow on the surface of CFs. The CVD process in this device was kept in a quasi-vacuum state and a N_2 atmosphere. Figure 1b is a schematic diagram of the process, from CF surface modification treatment to the final growth of the CNTs.

2.3. Characterization

The morphological characteristics of CNTs/CF multi-scale reinforcement samples under different conditions were observed by scanning electron microscope (SEM, SU-70). After the surface modification treatment, the chemical composition and content of CF surface were analyzed by X-ray photoelectron spectroscopy (XPS, Thermo Fisher ESCALAB 250). Changes in the crystal structure of the samples were tested by X-ray diffraction (XRD, Rigaku D/max-RC).

The single-filament tensile strength of the prepared samples was tested according to ASTM D3822-07 standard. At least 40 CF filaments were prepared for each sample, and the average value was taken as the single-filament tensile strength of a single sample after testing in sequence.

3. Results and Discussion

3.1. Effect of Surface Modification on CF Surface Morphology

Figure 2 shows SEM images of the samples with different surface modification treatments, the surfaces after growing CNTs, and a structure diagram of CNTs/CF multi-scale reinforcements. It can be seen from Figure 2a,b that the surfaces of the desized CFs were very smooth and that CNTs had grown on the surface after the CVD process. However, the distribution was extremely uneven. Many CFs were not covered with CNTs to expose the smooth surface, and a large number of impurities and clustered CNTs appeared in the gaps between adjacent CFs. This was because the unmodified CF surface could not attract the catalyst precursor particles in solution effectively, meaning that it was left in the gaps. In this case, the catalyst precursor remaining after the volatilization of anhydrous ethanol was prone to agglomeration to form coarse catalyst particles during reduction. Additionally, local aggregations of CNTs formed during the catalysis process, and the catalyst particles that were too large lost their activity. In contrast, in Figure 2c–h, it can be seen that the surface of samples underwent slight changes after the modification treatment. This was due to the surface of CFs being oxidized and etched to make an unstable and disordered turbostratic graphite structure which fell off, leaving an indentation [38]. In comparison, the grooves on the surface of CFs treated with HNO_3 solution were too thin and shallow. Additionally, the inconspicuous grooves made a limited contribution to the adhesion of the catalyst particles, the distribution was not uniform, and there were point defects on the surface. The surface of EAO-treated samples had a relatively obvious groove structure, which was evenly distributed, and the groove morphology was better than those of the other two methods. The obvious groove morphology showed that its width and depth were large. Such a structure can effectively provide a "landing field" for precursor particles. The effect of H_2O_2 treatment was somewhere in between. The morphology images of the grown CNTs also effectively reflect the role of the grooves. Compared with the desized CFs, the growth quantity of CNTs in the H_2O_2-treated sample was not significantly increased, and the distribution uniformity was slightly improved. The number of CNTs in the HNO_3-treated sample was significantly increased, but the CNTs only grew on half of the sides of CF, presumably because the CFs were not under tension during the immersion modification treatment, and some CF surfaces were not fully etched by lamination. The effect of CNTs/CF obtained by EAO treatment was most satisfactory; the CNTs were evenly distributed and the density was appropriate.

The growth mechanism of CNTs on the CF surface is shown in Figure 2i. Firstly, the catalyst particles attached to the modified CF surface, and then the activated carbon atoms formed by the cracking of the carbon source gas diffused through the interior or surface of the catalyst particles. Finally, after being closely connected with the CF surface through chemical bonds, the remaining activated carbon atoms gathered together to grow hollow CNTs in a "top growth" mode. CFs, CNTs connected with them, and catalyst particles together constitute CNTs/CF multi-scale reinforcements.

3.2. Effect of Surface Modification on Single-Filament Tensile Strength

During the preparation of CNTs/CF multi-scale reinforcements, CFs will be damaged to different degrees, which will affect the mechanical properties of the material. The CF surface modification pretreatment process essentially destroys the surface structure of CFs and etches the trench structure. As such, the decrease in strength should be minimized while achieving a certain effect. The modified CFs obtained by different treatment methods were used for a single-filament tensile test; the results are shown in Figure 3.

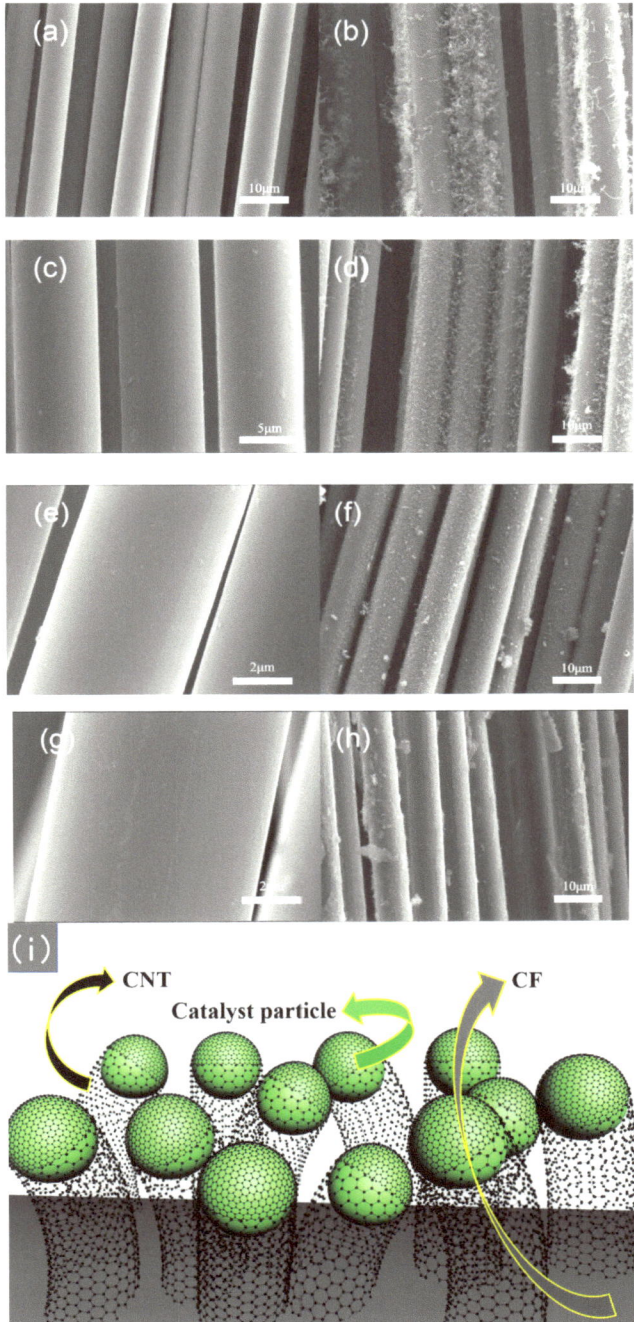

Figure 2. Surface morphology of (**a**,**b**) desized CFs and CFs pretreated with (**c**,**d**) HNO$_3$, (**e**,**f**) EAO, (**g**,**h**) H$_2$O$_2$ surface modification and CNTs grown on modified CFs, and (**i**) Structure diagram of CNTs/CF multi-scale reinforcements.

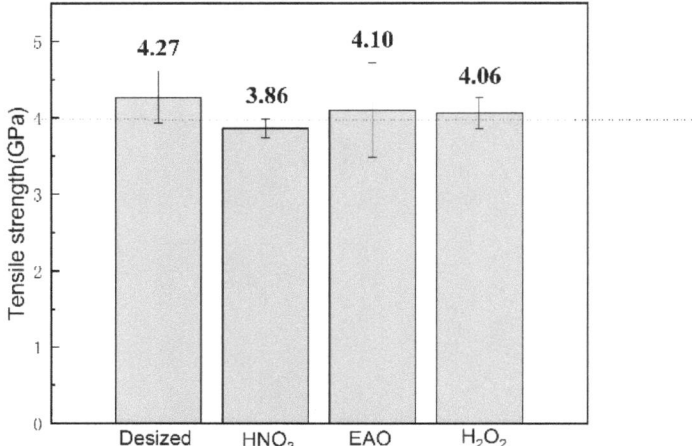

Figure 3. Single-filament tensile strength of CFs obtained using different surface pretreatment methods. (The dotted line in the figure indicates we set 4 GPa as the reference.)

The single-filament tensile strength of desized CFs was about 4.27 GPa. After modification, the single-filament tensile strength of each sample was reduced to varying degrees. The strength after HNO_3 treatment was 3.86 GPa, and that after H_2O_2 treatment was 4.06 GPa. Significantly, the strength after EAO treatment was 4.10 GPa, which shows that the etching on the CF surface after pretreatment with H_2O_2 and EAO treatment is relatively mild, while the oxidation reaction of HNO_3 treatment is severe, so the graphite microcrystals on the surface of CFs were significantly damaged, and the strength decreased by up to 9.6%.

By comprehensively comparing the three treatment methods, it was found that EAO treatment has great advantages. It has an obvious effect on promoting the growth of CNTs, the degree of strength reduction is acceptable, the improvement and optimization space is large, the treatment time is short, a large number of samples can be processed continuously, and the risk and economic cost are low. Therefore, EAO was selected as the pretreatment method of the CF surface and the optimization of its process conditions were explored.

3.3. Effect of EAO Treatment on Surface Chemical Composition of CFs

The electrochemical pretreatment method includes two processes: anodic oxidation and cathodic reduction. The input electrical energy is converted into chemical energy, and water is decomposed at the anode to generate strong oxidants to etch the CF surface, resulting in changes in chemical elements and functional groups on the surface of CFs. Qualitative and quantitative analyses were carried out by XPS tests. Table 1 shows the changes in the elemental composition of C, O, and N on the surface of CFs before and after EAO treatment.

Table 1. Main elemental contents of CFs before and after EAO treatment.

Main Element	Desized (At%)	EAO (At%)
C	92.36	63.52
O	7.64	33.83
N	0	2.65

Comparing the two samples, the elemental composition of the CF surface changed significantly after EAO treatment; notably, the oxygen content increased from 7.64% before EAO treatment to 33.83%. The desized CFs underwent high-temperature carbonization treatment, and the nitrogen-containing functional groups on the surface were removed. After electrolysis, 2.65% nitrogen appeared on the surface of the CFs. This was due to the reaction of $NH_4H_2PO_4$ with CFs under the action of an electrical field, introducing nitrogen-containing functional groups into the CF surface.

3.4. Optimization of Process Parameters of EAO

EAO treatment also has two sides. On the one hand, it provides the conditions for CFs to attach catalyst precursor particles, while on the other, it causes damage and reduces the mechanical strength of CFs. In order to achieve a more balanced effect, samples with four parameters, i.e., 50 C/g, 100 C/g, 150 C/g and 200 C/g, were created for a comparative analysis. Figure 4 shows SEM images of the sample surface treated with different electrochemical intensities. It can be seen that when the intensity was 50 C/g, heavy magnification was necessary to observe the fine grooves. With the increase of intensity, the number of grooves on the CF surface gradually increased, as did the depth and width of the groove structure and the adsorption capacity for catalyst precursor; however, the excessive etching intensity penetrated deep into the body structure of CFs to cause a loss of mechanical properties.

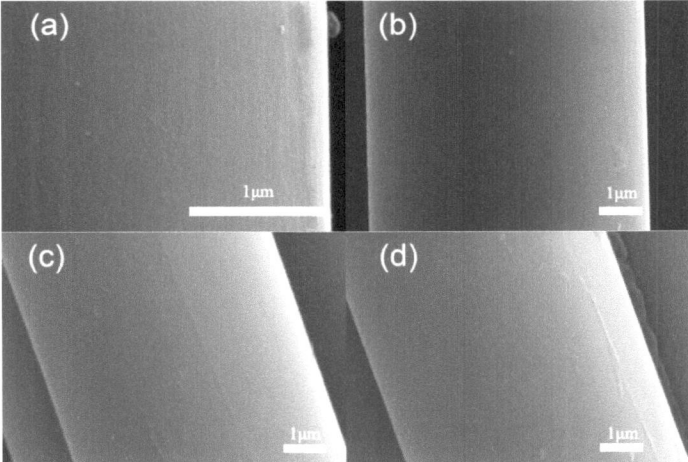

Figure 4. SEM images of CF surface treated with different EAO intensities. (**a**) 50 C/g, (**b**) 100 C/g, (**c**) 150 C/g, (**d**) 200 C/g.

Figure 5 presents images of CNTs grown on a CF surface treated with different EAO intensities at 550 °C. It can be seen that under the condition of 50 C/g, the number of CNTs was sparse and the distribution was uneven. Additionally, the surface of some CFs was smooth, without growing CNTs. The insufficient attracting ability of the precursor particles led to the aggregation of the precursor particles in the gaps between adjacent CFs to form larger-sized particles, so coarse carbon nanofibers emerged at the same position after the CVD process. Under the conditions of 100 C/g and 150 C/g electrochemical treatment intensity, the surface of CFs could be coated with a uniform layer of CNTs, and the surface morphology of the two was not much different. Under the condition of 200 C/g, the distribution of CNTs was too dense, and some parts demonstrated the phenomenon of clustering. This was due to the fact that the trenches obtained by high-intensity electrochemical oxidation etching were too deep, and the catalyst precursor easily aggregated, which was not conducive to the growth of CNTs. Therefore, when the EAO treatment intensities were 100 C/g and 150 C/g, the morphology of the CNTs/CF multi-scale reinforcements was optimal.

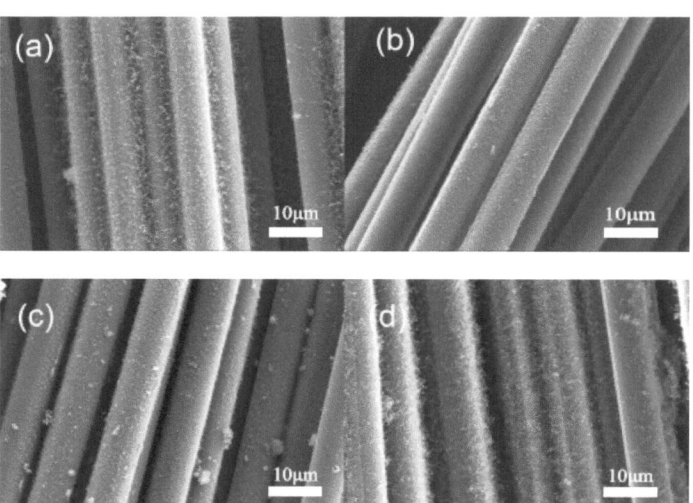

Figure 5. Surface morphology of CFs with CNTs treated with different EAO intensities. (**a**) 50 C/g, (**b**) 100 C/g, (**c**) 150 C/g, (**d**) 200 C/g.

Figure 6a shows the tensile strength of each sample after treatment with different electrical intensities. Compared with the value of 4.27 GPa for the desized sample, the tensile strength under the condition of 50 C/g was 4.16 GPa, i.e., a decrease of 2.57%. Additionally, corrosion occurred at this time. However, the oxidative damage to the CFs was not large. The tensile strength was 4.10 GPa at 100 C/g, a decrease of 3.98%. As the treatment intensity increased, the mechanical properties of the CFs continued to decline; the tensile strength at 150 C/g was 3.95 GPa. Under the condition of 200 C/g, the tensile strength was 3.86 GPa, marking a decrease of nearly 10%. At this time, the CFs were seriously damaged. During the electrochemical treatment process, the oxidation etching penetrated into the interior of the CFs, and the etching caused by the aggregation of catalyst particles exacerbated the strength loss.

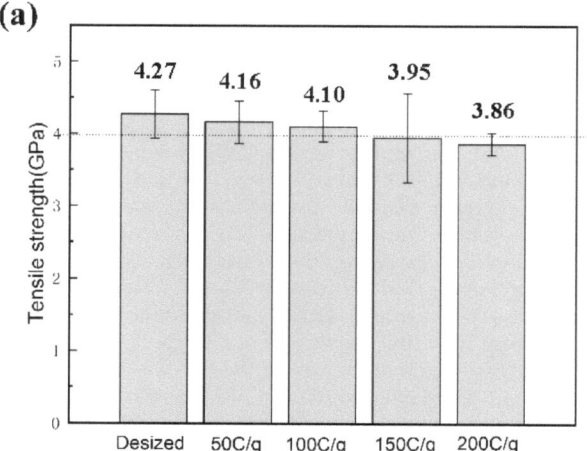

Figure 6. *Cont.*

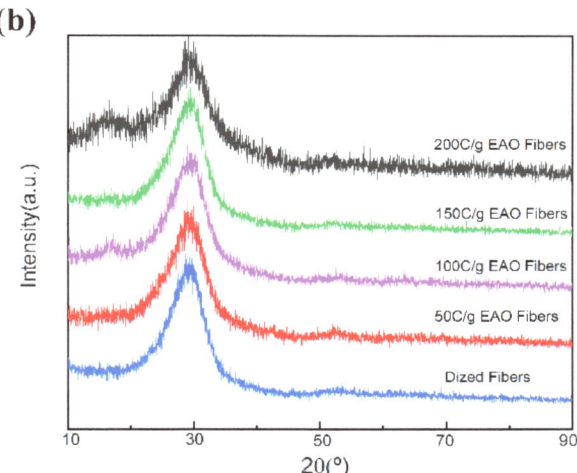

Figure 6. (**a**) Single-filament tensile strength (The dotted line in the figure indicates we set 4 GPa as the reference.) and (**b**) XRD curve of desized CFs and CFs treated with different EAO intensities.

It can be seen from the XPS test that the EAO treatment added functional groups containing O and N elements to the surface of CFs. These active functional groups contributed to the loading of catalyst precursor particles. These changes are reflected in the structure of the surface graphite crystallites. Figure 6b demonstrates the XRD patterns of the desized CFs and CFs with different electrochemical treatment intensities. It can be seen that each sample had a wide diffraction peak around $2\theta = 25.6°$, which is the characteristic diffraction peak of the (002) crystal plane of the CF surface through data comparison; the more complete the graphite turbostratic structure, the sharper the peak shape. Compared with the desized CFs, the peak shapes of the treated samples were broadened, indicating that the graphite structure was damaged to different degrees.

The microcrystalline structure of the (002) crystal plane of each sample was analyzed, and the parameters are illustrated in Table 2. Taking 3.354 Å as the ideal spacing of the standard graphite (002) crystal plane as the benchmark, the distance between the (002) crystal planes of each sample was closer to this value, indicating that the CF surface was less damaged and the degree of graphitization was high. The surface of CF itself had small defects, and the high temperature process of desizing damaged the surface structures of the CFs, which were different from the standard values. After electrochemical treatment, the interplanar spacing of all samples was larger than that of the desized fibers, indicating that the oxidation reaction changed the microcrystalline structure of the CF surface. Additionally, with the increase of the electrical intensity, the $d_{(002)}$ of the CF surface increased, the graphite sheet structure became increasingly loose, and the mechanical properties of the fibers decreased to a greater extent; this was also related to the tensile strength of the single-filament. The test results were consistent with these observations.

Table 2. Microstructure parameters of the (002) crystal plane of desized CFs and CFs treated with different EAO intensities.

Sample	2θ (°)	$d_{(002)}$ (Å)
Desized	25.693	3.465
50 C/g	25.537	3.490
100 C/g	25.236	3.504
150 C/g	25.429	3.496
200 C/g	25.512	3.533

4. Conclusions

CF surface pretreatment is an indispensable step before the CVD process. A H_2O_2 solution has a limited etching effect on the surface modification of CFs and cannot effectively increase the number of CNTs grown. A HNO_3 solution can effectively oxidize and etch graphite microcrystals, forming a groove structure on the surface of CFs; however, the reaction is violent, and the fiber is significantly damaged. Using the EAO processing equipment designed to oxidize and etch the surface of CFs can effectively increase the adhesion rate of the catalyst precursor, which increases the quantity of CNTs grown and significantly improves the surface morphology of CNTs/CF multi-scale reinforcements. After electrochemical treatment, the single-filament tensile strengths of 50 C/g, 100 C/g, 150 C/g, and 200 C/g samples decreased by 2.57%, 3.98%, 7.49%, and 9.60% respectively, compared with the desized CFs. The EAO method uses an oxidation reaction to etch the graphite microcrystalline structure on the surface of CFs; the higher the electrical intensity, the larger the (002) interplanar spacing of the CFs. EAO is appropriate for the pretreatment of CFs before the CVD process, which can effectively improve the loading quantity and distribution uniformity of catalyst precursor particles.

Author Contributions: Conceptualization, M.L., Y.W. and C.W. (Chengguo Wang); Data curation, M.L.; Investigation, M.L., Y.W., B.C. and C.W. (Chengjuan Wang); Methodology, M.L., B.C. and C.W. (Chengjuan Wang); Project administration, Y.W. and G.Z.; Software, H.T., H.J. and Z.X.; Supervision, Y.W.; Validation, M.L.; Visualization, M.L.; Writing—original draft, M.L., B.C. and C.W. (Chengjuan Wang); Writing—review and editing, M.L., Y.W., B.C., C.W. (Chengjuan Wang), H.T., H.J. and Z.X. All authors have read and agreed to the published version of the manuscript.

Funding: This research was funded by the Natural Science Foundation in Shandong Province (ZR2020ME134, ZR2020ME039, ZR2021ME194).

Institutional Review Board Statement: Not applicable.

Informed Consent Statement: Not applicable.

Data Availability Statement: All data which support the findings of this study are included within the article.

Acknowledgments: The authors thank the editor and the anonymous reviewers for their valuable comments on this manuscript. The authors also acknowledge the support of technical staff for assisting in preparing samples and analyzing them. This work was supported by the key research and development program of Shandong Province (2021ZLGX01).

Conflicts of Interest: The authors declare no conflict of interest.

References

1. Liu, Y.; Kumar, S. Recent Progress in Fabrication, Structure, and Properties of Carbon Fibers. *Polym. Rev.* **2012**, *52*, 234–258. [CrossRef]
2. Yao, Z.; Wang, C.; Wang, Y.; Qin, J.; Ma, Z.; Cui, X.; Wang, Q.; Wei, H. Effect of CNTs deposition on carbon fiber followed by amination on the interfacial properties of epoxy composites. *Compos. Struct.* **2022**, *292*, 115665. [CrossRef]
3. Gangineni, P.K.; Yandrapu, S.; Ghosh, S.K.; Anand, A.; Prusty, R.K.; Ray, B.C. Mechanical behavior of Graphene decorated carbon fiber reinforced polymer composites: An assessment of the influence of functional groups. *Compos. Part A Appl. Sci. Manuf.* **2019**, *122*, 36–44. [CrossRef]
4. Nie, H.-J.; Shen, X.-J.; Tang, B.-L.; Dang, C.-Y.; Yang, S.; Fu, S.-Y. Effectively enhanced interlaminar shear strength of carbon fiber fabric/epoxy composites by oxidized short carbon fibers at an extremely low content. *Compos. Sci. Technol.* **2019**, *183*, 107803. [CrossRef]
5. Zhang, Y.; Zhang, Y.; Liu, Y.; Wang, X.; Yang, B. A novel surface modification of carbon fiber for high-performance thermoplastic polyurethane composites. *Appl. Surf. Sci.* **2016**, *382*, 144–154. [CrossRef]
6. Dong, J.; Jia, C.; Wang, M.; Fang, X.; Wei, H.; Xie, H.; Zhang, T.; He, J.; Jiang, Z.; Huang, Y. Improved mechanical properties of carbon fiber-reinforced epoxy composites by growing carbon black on carbon fiber surface. *Compos. Sci. Technol.* **2017**, *149*, 75–80. [CrossRef]
7. Wu, Q.; Li, M.; Gu, Y.; Li, Y.; Zhang, Z. Nano-analysis on the structure and chemical composition of the interphase region in carbon fiber composite. *Compos. Part A Appl. Sci. Manuf.* **2014**, *56*, 143–149. [CrossRef]

8. Adstedt, K.; Stojcevski, F.; Newman, B.; Hayne, D.J.; Henderson, L.C.; Mollenhauer, D.; Nepal, D.; Tsukruk, V. Carbon Fiber Surface Functional Landscapes: Nanoscale Topography and Property Distribution. *ACS Appl. Mater. Interfaces* **2022**, *14*, 4699–4713. [CrossRef]
9. Hu, J.; Li, F.; Wang, B.; Zhang, H.; Ji, C.; Wang, S.; Zhou, Z. A two-step combination strategy for significantly enhancing the interfacial adhesion of CF/PPS composites: The liquid-phase oxidation followed by grafting of silane coupling agent. *Compos. Part B Eng.* **2020**, *191*, 107966. [CrossRef]
10. Yongqiang, L.; Chunzheng, P. Chemically grafting carbon nanotubes onto carbon fibers for enhancing interfacial strength in carbon fiber/HDPE composites. *Surf. Interface Anal.* **2018**, *50*, 552–557. [CrossRef]
11. Qin, J.; Wang, C.; Wang, Y.; Su, S.; Lu, R.; Yao, Z.; Wang, Q.; Gao, Q. Communication—A Technique for Online Continuous Manufacture of Carbon Nanotubes-Grown Carbon Fibers. *ECS J. Solid State Sci. Technol.* **2019**, *8*, M23–M25. [CrossRef]
12. Russello, M.; Diamanti, E.K.; Catalanotti, G.; Ohlsson, F.; Hawkins, S.C.; Falzon, B.G. Enhancing the electrical conductivity of carbon fibre thin-ply laminates with directly grown aligned carbon nanotubes. *Compos. Struct.* **2018**, *206*, 272–278. [CrossRef]
13. Wang, X.; Wang, C.; Wang, Z.; Wang, Y.; Lu, R.; Qin, J. Colossal permittivity of carbon nanotubes grafted carbon fiber-reinforced epoxy composites. *Mater. Lett.* **2018**, *211*, 273–276. [CrossRef]
14. Yao, Z.; Wang, C.; Qin, J.; Wang, Y.; Wang, Q.; Wei, H. Fex-Co1-x bimetallic catalysts for highly efficient growth of carbon nanotubes on carbon fibers. *Ceram. Int.* **2020**, *46*, 27158–27162. [CrossRef]
15. Su, S.; Wang, Y.; Qin, J.; Wang, C.; Yao, Z.; Lu, R.; Wang, Q. Continuous method for grafting CNTs on the surface of carbon fibers based on cobalt catalyst assisted by thiourea. *J. Mater. Sci.* **2019**, *54*, 12498–12508. [CrossRef]
16. Zheng, L.; Wang, Y.; Qin, J.; Wang, X.; Lu, R.; Qu, C.; Wang, C. Scalable manufacturing of carbon nanotubes on continuous carbon fibers surface from chemical vapor deposition. *Vacuum* **2018**, *152*, 84–90. [CrossRef]
17. Wang, C.; Wang, Y.; Su, S. Optimization of Process Conditions for Continuous Growth of CNTs on the Surface of Carbon Fibers. *J. Compos. Sci.* **2021**, *5*, 111. [CrossRef]
18. Guo, M.; Liu, J.; Yuan, Y.; Zhang, Z.; Yin, S.; Leng, J.; Huang, N. CNTs/Cf based counter electrode for highly efficient hole-transport-material-free perovskite solar cells. *J. Photochem. Photobiol. A Chem.* **2020**, *403*, 112843. [CrossRef]
19. Rong, H.; Dahmen, K.-H.; Garmestani, H.; Yu, M.; Jacob, K.I. Comparison of chemical vapor deposition and chemical grafting for improving the mechanical properties of carbon fiber/epoxy composites with multi-wall carbon nanotubes. *J. Mater. Sci.* **2013**, *48*, 4834–4842. [CrossRef]
20. Qin, J.; Wang, C.; Lu, R.; Su, S.; Yao, Z.; Zheng, L.; Gao, Q.; Wang, Y.; Wang, Q.; Wei, H. Uniform growth of carbon nanotubes on carbon fiber cloth after surface oxidation treatment to enhance interfacial strength of composites. *Compos. Sci. Technol.* **2020**, *195*, 108198. [CrossRef]
21. Gao, B.; Zhang, R.; He, M.; Sun, L.; Wang, C.; Liu, L.; Zhao, L.; Cui, H.; Cao, A. Effect of a multiscale reinforcement by carbon fiber surface treatment with graphene oxide/carbon nanotubes on the mechanical properties of reinforced carbon/carbon composites. *Compos. Part A Appl. Sci. Manuf.* **2016**, *90*, 433–440. [CrossRef]
22. Zhang, W.; Deng, X.; Sui, G.; Yang, X. Improving interfacial and mechanical properties of carbon nanotube-sized carbon fiber/epoxy composites. *Carbon* **2019**, *145*, 629–639. [CrossRef]
23. Wang, C.; Wang, Y.; Jiang, H.; Tan, H.; Liu, D. Continuous in-situ growth of carbon nanotubes on carbon fibers at various temperatures for efficient electromagnetic wave absorption. *Carbon* **2022**, *200*, 94–107. [CrossRef]
24. Fan, W.; Wang, Y.; Wang, C.; Chen, J.; Wang, Q.; Yuan, Y.; Niu, F. High efficient preparation of carbon nanotube-grafted carbon fibers with the improved tensile strength. *Appl. Surf. Sci.* **2016**, *364*, 539–551. [CrossRef]
25. Wang, B.; Fu, Q.; Sun, L.; Lu, Y.; Liu, Y. Improving the tribological performance of carbon fiber reinforced resin composite by grafting MWCNT and GNPs on fiber surface. *Mater. Lett.* **2022**, *306*, 130953. [CrossRef]
26. Xu, S.; Jiang, Q. Surface modification of carbon fiber support by ferrous oxalate for biofilm wastewater treatment system. *J. Clean. Prod.* **2018**, *194*, 416–424. [CrossRef]
27. Fan, W.; Wang, Y.; Chen, J.; Yuan, Y.; Li, A.; Wang, Q.; Wang, C. Controllable growth of uniform carbon nanotubes/carbon nanofibers on the surface of carbon fibers. *RSC Adv.* **2015**, *5*, 75735–75745. [CrossRef]
28. Qin, R.; Tian, Y.; Zheng, L.; Qin, J.; Tao, Y.; Song, C.; Fan, W.; Wang, Y. Studying the growth of carbon nanotubes on carbon fibers surface under different catalysts and electrochemical treatment conditions. *Fuller. Nanotub. Carbon Nanostruct.* **2017**, *25*, 156–162. [CrossRef]
29. Yue, Y.; Wang, Y.; Xu, X.; Cui, B.; Yao, Z.; Wang, Y.; Wang, C.; Wang, Y.; Wang, Y. Continuous growth of carbon nanotubes on the surface of carbon fibers for enhanced electromagnetic wave absorption properties. *Ceram. Int.* **2022**, *48*, 1869–1878. [CrossRef]
30. Chang, Q.; Zhao, H.; He, R. The mechanical properties of plasma-treated carbon fiber reinforced PA6 composites with CNT. *Surf. Interface Anal.* **2017**, *49*, 1244–1248. [CrossRef]
31. Zhao, L.; Liu, W.; Liu, P.; Tian, J.; Xu, M.; Sun, S.; Wang, Y. Study on atmospheric air glow discharge plasma generation and surface modification of carbon fiber fabric. *Plasma Process. Polym.* **2020**, *17*, 1900148. [CrossRef]
32. Liang, Y.; Tuo, Z.; Zhao, Q.; Lin, S.; Lin, Z.; Han, Z.; Ren, L. Study on preparation and mechanical properties of bionic carbon fiber reinforced epoxy resin composite with eagle feather structure. *Mater. Res. Express* **2021**, *8*, 065301. [CrossRef]
33. Li, H.; Liebscher, M.; Ranjbarian, M.; Hempel, S.; Tzounis, L.; Schröfl, C.; Mechtcherine, V. Electrochemical modification of carbon fiber yarns in cementitious pore solution for an enhanced interaction towards concrete matrices. *Appl. Surf. Sci.* **2019**, *487*, 52–58. [CrossRef]

34. Fu, Y.; Li, H.; Cao, W. Enhancing the interfacial properties of high-modulus carbon fiber reinforced polymer matrix composites via electrochemical surface oxidation and grafting. *Compos. Part A Appl. Sci. Manuf.* **2020**, *130*, 105719. [CrossRef]
35. Qian, X.; Zhong, J.; Zhi, J.; Heng, F.; Wang, X.; Zhang, Y.; Song, S. Electrochemical surface modification of polyacrylonitrile-based ultrahigh modulus carbon fibers and its effect on the interfacial properties of UHMCF/EP composites. *Compos. Part B Eng.* **2019**, *164*, 476–484. [CrossRef]
36. Jiang, H.; Wang, Y.; Wang, C.; Xu, X.; Li, M.; Xu, Z.; Tan, H.; Wang, Y. Effect of electrochemical anodization and growth time on continuous growth of carbon nanotubes on carbon fiber surface. *Ceram. Int.* **2022**, *48*, 29695–29704. [CrossRef]
37. Wen, Z.; Xu, C.; Qian, X.; Zhang, Y.; Wang, X.; Song, S.; Dai, M.; Zhang, C. A two-step carbon fiber surface treatment and its effect on the interfacial properties of CF/EP composites: The electrochemical oxidation followed by grafting of silane coupling agent. *Appl. Surf. Sci.* **2019**, *486*, 546–554. [CrossRef]
38. Ma, Z.; Wang, Y.; Qin, J.; Yao, Z.; Cui, X.; Cui, B.; Yue, Y.; Wang, Y.; Wang, C. Growth of carbon nanotubes on the surface of carbon fiber using Fe–Ni bimetallic catalyst at low temperature. *Ceram. Int.* **2021**, *47*, 1625–1631. [CrossRef]

Article

Surface Damage in Woven Carbon Composite Panels under Orthogonal and Inclined High-Velocity Impacts

Veronica Marchante Rodriguez [1], Marzio Grasso [1,*], Yifan Zhao [1], Haochen Liu [1], Kailun Deng [1], Andrew Roberts [2] and Gareth James Appleby-Thomas [2]

1. School of Aerospace, Transport and Manufacturing, Cranfield University, Cranfield MK43 0AL, UK
2. Centre for Defence Engineering, Cranfield University, Defence Academy of the United Kingdom, Shrivenham SN6 8LA, UK
* Correspondence: marzio.grasso@cranfield.ac.uk

Abstract: The present research is aimed at the study of the failure analysis of composite panels impacted orthogonally at a high velocity and with an angle. Woven carbon-fibre panels with and without external Kevlar layers were impacted at different energy levels between 1.2 and 39.9 J. Sharp and smooth gravels with a mass from 3.1 to 6.7 g were used to investigate the effects of the mass and the contact area on the damage. Optical microscopy and thermography analyses were carried out to identify internal and surface damage. It was identified that sharp impactors created more damage on the impacted face of the panels, while the presence of a Kevlar layer increased the penetration limit and reduced the damage level in the panel at a higher energy.

Keywords: impact tests; gas gun; damage analysis; optical microscopy; thermography

Citation: Rodriguez, V.M.; Grasso, M.; Zhao, Y.; Liu, H.; Deng, K.; Roberts, A.; Appleby-Thomas, G.J. Surface Damage in Woven Carbon Composite Panels under Orthogonal and Inclined High-Velocity Impacts. *J. Compos. Sci.* **2022**, *6*, 282. https://doi.org/10.3390/jcs6100282

Academic Editors: Jiadeng Zhu and Francesco Tornabene

Received: 1 August 2022
Accepted: 16 September 2022
Published: 26 September 2022

Publisher's Note: MDPI stays neutral with regard to jurisdictional claims in published maps and institutional affiliations.

Copyright: © 2022 by the authors. Licensee MDPI, Basel, Switzerland. This article is an open access article distributed under the terms and conditions of the Creative Commons Attribution (CC BY) license (https://creativecommons.org/licenses/by/4.0/).

1. Introduction

Composite materials are widely used in automotives and motorsports, as their design flexibility and mechanical properties enable the advanced light weighting needed to achieve high competitiveness. However, composites do not tolerate impact damage due to their limited ductility and significant sensitivity to the strain rate. Small flying objects can damage the aerodynamic surfaces of racing cars, thus greatly influencing their performance and the safety of the driver. The FIA regulates the size and shape of the gravel on every circuit, resulting, in most cases, in either spherical or river-washed stones (from 5 to 15 mm in diameter).

Four different types of impact can be defined depending on the velocity achieved by the impactor [1], namely: low-velocity impacts (LVIs) below 11 m/s, high-velocity impacts (HVIs) below 500 m/s, ballistic impacts below 2000 m/s, and hypervelocity impacts above 2000 m/s. The main difference in the structural response among these types of impacts is the deformation induced, since the higher the velocity is, the more localized the effect will be, with a high shear compared to bending for low-velocity impacts. LVIs are dominated by the global response of the panel, and HVIs are dominated by the local response (Figure 1) [2,3].

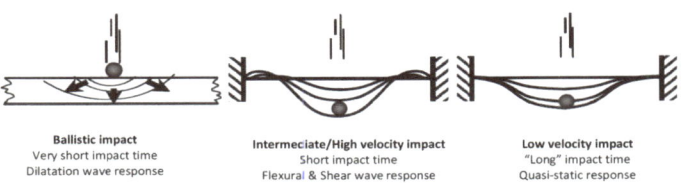

Figure 1. Comparison of high- and low-velocity impact responses (adapted from [4]).

The limit between low- and high-velocity impacts is very difficult to identify, as it will depend on the shape and geometry of the impactor, the target characteristics, and external factors, such as temperature [5]. However, in a low-velocity impact, the duration of the contact time allows elastic waves to propagate through a large portion of the target, resulting in a global response dominated by elastic strain energy that is mainly stored in the yarns [6]. These are also responsible for the propagation of elastic waves through impacted composite panels. The damage sequence is a combination of matrix cracking and fibre fracture with delamination that involves several layers depending on the energy level and the stiffness of the panel. Failure modes and energy absorption are also greatly affected by the target's size and boundary conditions [7].

Under a high-velocity impact, most of the available energy is dissipated over a small volume close to the impact point. Stress waves propagate through the thickness of the target within a short time, causing localized damages [8]. The damage sequence is characterized by transverse shear cracks that occur at an angle of approximately 45° and close to the impact area, as well as cracks due to bending that occur on the back layer [3]. With the increase in the plate thickness, the impact velocity, and the bluntness of the projectile, shear plugging becomes a likely failure mode of the final perforation of a plate. A hole is then created, and friction will be responsible for the dissipation of part of the energy while the projectile penetrates the plate [9,10] (Figure 1).

The factors affecting the size and morphology of the damage under low- and high-velocity impacts are tensile strength, strain at failure, and density, as these are related to the wave speed and the propagation of energy [11]. Woven fabrics provide better interlaminar fracture toughness than that of unidirectional ones, resulting in less damage, as the delamination is reduced. However, the in-plane properties of woven composites are affected by the fibre architecture, and 3D composites have thorough thickness reinforcement, which allows to them provide excellent in-plane and transverse properties.

The effect of the resin on the type and shape of damage depends on the toughness and strength of the resin. Thermoplastic resins possess a greater toughness and strain to failure than thermoset. Béland [12] studied the performance of thermoplastics and thermoset resins, focussing on their toughness, indentation due to fibre breakage, and matrix plasticisation, which improved the Mode-I fracture toughness and, thus, crack propagation. Thermoplastics provide a better stress distribution along the fibres, improving fibre bonding. Vieille et al. [13] compared polyether ether ketone (PEEK) and polyphenylene sulphide (PPS) with epoxy resins to study the effect of the matrix on the delamination of woven carbon-fibre laminates under impacts and observed that the two thermoplastic resins improved the delamination resistance thanks to the fibre bridging.

Lopes et al. [14] observed that limiting the orientation angle between two consecutive plies greatly improved damage tolerance. York [15] studied different lay-up configurations and observed that anti-symmetric laminates provided the best damage tolerance, and symmetric ones provide the worst behaviour. Dorey [16] reported that composites with $+/-45°$ surface layers provided higher impact resistance and compressive residual strength with respect to those with 0° surface plies. This was attributed to the higher flexibility of the composite, which improved its ability to absorb elastic energy [8].

Park and Jang [17] studied the influence of the stacking sequence on the low-velocity impact response of hybrid aramid–glass laminates. They reported that the placement of aramid plies at the outer surface of a carbon-fibre laminate increased the impact resistance because the fibres with a high strain to failure on the outer side could undergo greater deformation [8].

The thickness of the target and its in-plane dimensions dictate the bending stiffness, the magnitude of the maximum contact force, and the induced damages. Flexible targets are mainly influenced by bending, which causes tensile stress in the ply that is exposed to the impact, and matrix cracks and, thus, delamination are developed. This leads to the reverse-pine-tree damage. For stiffer targets, the high-stress region is located directly where the impact takes place (the pine tree) [18]. According to Gellert [19], the energy absorption in thin-glass-fibre-reinforced targets is largely independent of the geometry of the projectile nose, and thicker composites are more ballistically efficient, as they improve indentation, which is a significant mechanism of energy absorption. At a small span-to-thickness ratio and low velocity, a high stiffness, a large peak force, and a short duration occur; thus, delamination is more likely to occur. At a high velocity, the short contact duration of the event reduces the effect of the span-to-thickness ratio [12].

Icten et al. [20] found that decreasing the diameter of the impactor results in a decrease in the impact resistance at the same energy. The most common shapes are spherical, cylindrical, flat, and diamond-shaped. Sevkat et al. [21] found that the contact duration and peak force are modified by the shape of the impactor, with a large contact area creating the highest force.

Blunt projectiles cause failure through shear plugging, while conical projectiles cause petalling in thin plates. Ductile hole enlargement is observed in thick plates, and hemispherical projectiles cause tensile stretching after a huge indentation of the target plate. During high velocity impacts, ballistic limit of blunt projectile is lower than hemispherical and conical ones [22]. Conical projectiles induce a greater local and global energy [9]. Mitrevski et al. [5] concluded that blunt hemispherical impactors produce more damages than conical and ogival ones in low-velocity impacts, whilst Lee et al. [23] reported that flat and hemispherical impactors produce similar responses.

The mass of the projectile will influence the energy absorption mechanisms of the target. Small masses cause limited damage at low velocity and greater localised damages at high velocity because an increase in mass will induce a shift from global to local in the failure mechanisms. Small masses cause through-thickness waves, intermediate masses involve shear and flexural waves, and large masses involve a quasi-static response [1].

Fibre hybridisation has been reported to improve impact response when used in combination with a given lay-up. Hazell et al. [24] performed high-velocity tests and compared the effects of adding Kevlar layers in different places. When penetration occurred, placing the layer on the face that was directly impacted was the least effective solution. Gustin et al. [25] used Kevlar for low-velocity impacts and observed a 10% increase in absorbed energy.

The current literature reviewed for this work deals with the ballistic limit and internal damages produced during high-velocity impacts. There is very little related to the impact response of composite panels under inclined im-pact with incident angle close to 0° or 180° degree. The experimental data were collected at energy levels below those corresponding to the ballistic limit of the composite panel. This characterization is aimed at the definition of damage mechanisms in order to be able to devise systems (including coatings) that can be used to protect surfaces and reduce damages.

Thus, in this work, hybrid composite plates made with woven carbon and Kevlar fabrics were tested with a gas gun that used gravel with smooth and sharp surfaces at different incident angles. Then, the damaged surfaces were analysed with optical microscopy and thermal imaging to correlate the surface damages, energy levels, and gravel shapes.

2. Materials and Methods

In order to study the effects of high-velocity impacts on carbon-fibre-reinforced polymers (CFRPs) and a hybrid configuration with an external layer of Kevlar, the first step was to perform experimental tests with a wide range of impact velocities at two different impact angles of 0° and 88°. The composite laminates used in this study were manufactured at Cranfield University using pre-impregnated 5HS CFRP supplied by Cytec, which consisted of 5HS woven carbon 3K Toray T1000 fibres and MTM49 toughened epoxy [24] with a resin weight of 42%. The 5HS CFRP material had an areal density of 283 g/m^2, a nominal ply cure thickness of 0.35 mm, and the mechanical properties listed in Table 1. The curing cycle (pressure and temperature) followed the datasheet [24].

Table 1. Mechanical properties of MTM49-3-42%-3KFT300B40B-5H-283-1000.

Mechanical Properties	Results
0° Tensile strength (MPa)	1065
0° Tensile modulus (GPa)	44.6
90° Tensile strength (MPa)	1035
90° Tensile modulus (GPa)	42.8
0° Compressive strength (MPa)	640
0° Compressive modulus (GPa)	59
90° Compressive strength (MPa)	610
90° Compressive modulus (GPa)	57
In-plane shear strength (MPa)	108
In-plane shear modulus (GPa)	2.5
0° Interlaminar shear strength (MPa)	64.2

Ten samples were made from carbon fibre (CF) with three layers that were 0.35 mm thick each, and the other four samples were made with carbon fibre with three layers that were 0.35 mm thick each plus one layer of Kevlar that was 0.35 mm thick. All specimens were cured for 90 min at 135 °C in an autoclave oven. The test samples were cut into plates with dimensions of 150 × 100 mm with different thicknesses. Table 2 summarises details of the panels and lay-ups.

Table 2. Panel lay-up configuration.

Specimen Number	Panel Thickness (mm)	Number of Layers	Type of Fibre
1, 2, 3, 4 13, 14, 15, 16,17, 18	1.05	3	CF
5, 6, 7, 8 9, 10, 11, 12	1.40	4	CF + kevlar

2.1. Impact Testing

The impact testing was conducted using a single-stage light gas gun (Figure 2), which used compressed air to propel the projectiles. The gas gun consisted of a 1.75-m-long smooth-bore barrel with a calibre of 22 mm. Gravel was used as a projectile; this was saboted to enable the firing of sub-calibre sizes and loaded into the gun at the breech end. The breech was then sealed and the was gun charged with the correct pressure to produce the velocity required. The velocity was measured via a set of infrared LEDs and receivers, and the time between them was recorded on an oscilloscope. The gun was fired remotely via an electronic solenoid. A Phantom V1212 high-speed camera operating at 10,000 frames per second was used to record the impact and assess the residual velocity.

The composite panels were held in an aluminium-frame-type fixture, as shown in Figure 3. This fixture had a 130 × 80 mm rectangular opening. The two frames were held together using eight hex-head cap-screw alloy steel bolts to restrain out-of-plane motion and rotation. These held the panel in place without any gaps. All impacts were targeted toward the centre of the frame.

Two types of projectiles were used—projectiles with smooth and sharp surfaces—with different masses. Before the test, each projectile was weighed, and the major and minor dimensions were measured. The gravel was fired with the help of a sabot made of Styrofoam LB-X with a density of 33 kg/m^3. A summary of all of the tests conducted, including the panels, gravel weights, types, impact speeds, and energies, is presented in Table 3.

Figure 2. Picture of the setup for the gas gun and chamber.

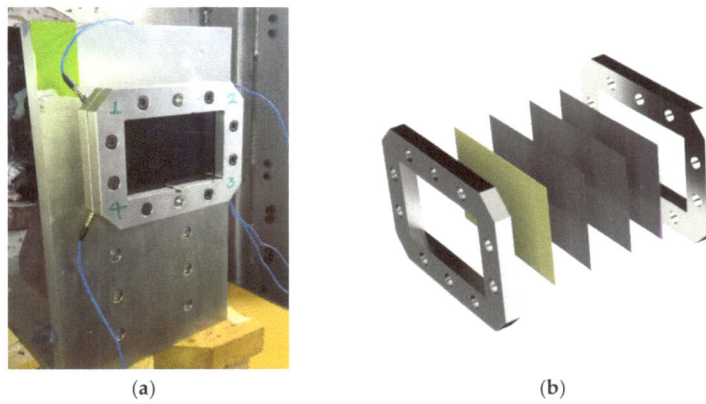

Figure 3. (a) Image of the holding of the sample for gas gun testing and (b) representation of the frame and panel layers.

Table 3. Summary of the impact tests conducted and their details ("p": full penetration of the panel).

Sample Number	Panel Materials	Projectile Type	Angle of Impact	Projectile Weight (g)	Impact Velocity (m/s)	Energy (J)
1	CF	Sharp	28.5°	5.12	38.4	3.1
2 (p)	CF	Sharp	28.5°	6.12	76.9	15.4
3	CF	Smooth	28.5°	5.32	35.0	2.7
4 (p)	CF	Smooth	28.5°	5.82	68.9	11.6
5	CF + Kevlar	Sharp	28.5°	6.40	29.4	2.8
6	CF + Kevlar	Sharp	28.5°	4.70	29.4	2.0

Table 3. Cont.

Sample Number	Panel Materials	Projectile Type	Angle of Impact	Projectile Weight (g)	Impact Velocity (m/s)	Energy (J)
6	CF + Kevlar	Sharp	28.5°	5.62	57.1	7.7
7	CF + Kevlar	Smooth	28.5°	4.80	37.7	3.4
8	CF + Kevlar	Smooth	28.5°	6.09	58.8	8.9
9	CF + Kevlar	Smooth	90°	4.20	24.2	1.2
10	CF + Kevlar	Smooth	90°	4.70	76.9	13.9
11	CF + Kevlar	Sharp	90°	5.60	18.3	0.9
11	CF + Kevlar	Sharp	90°	5.60	44.4	5.5
12	CF + Kevlar	Sharp	90°	4.70	74.0	12.9
13	CF	Sharp	90°	3.10	37.7	2.2
14	CF	Sharp	90°	4.10	71.0	10.3
14	CF	Sharp	90°	6.60	45.8	6.9
14(p)	CF	Sharp	90°	6.60	102.5	34.7
15	CF	Smooth	90°	2.90	37.4	2.0
16	CF	Smooth	90°	5.70	15	0.6
16	CF	Smooth	90°	5.70	25	1.8
16 (p)	CF	Smooth	90°	5.70	74	15.6
17	CF	Sharp	28.5°	5.18	27.4	1.9
17	CF	Sharp	28.5°	4.87	33.9	2.8
18	CF	Smooth	28.5°	5.12	58.8	7.3

2.2. Samples' Analysis

An optical microscope was used to observe and analyse the damage topographies on the top and bottom surfaces of the specimens. In order to quantify the extent of the damage accurately, the thermography technique was used to measure the damaged areas. However, this non-destructive inspection method could not distinguish the damage modes. Since the focus of the work was on the surface damage, the samples were also observed to identify the key features of their surface damage.

Pulsed thermography is a reliable non-destructive testing technique for detecting near- and sub-surface damage. It is a more robust and faster method compared to ultrasonic testing and X-radiography [26]. The typical thermographic setup is illustrated in Figure 4. A pulse was emitted by a flash lamp onto the specimen's surface. Heat conduction took place on the sample and led to a decrease from the surface to the interior. An infrared (IR) camera measured the temperature of the sample surface against time. Point 1 in Figure 4 represents an undamaged area of the sample, while Point 2 represents a damaged one. In the presence of a defect, a temperature deviation occurred, and the time when it occurred (t1 in Figure 4) allowed the estimation of the defect's depth. Initially, thermographic analysis was based on contrasts in the post-heating images. Most modern systems analyse each pixel individually and independently. This allows to not be dependent of a reference standard. Thermographic signal reconstruction (TSR) allows noise reduction, and its first and second derivatives are invariant to ambient conditions, surface preparation, and input energy, thus allowing the analysis of the sub-surface of a specimen. TSR is, then, a polynomial fitting, and its first and second derivatives allow the visualisation of heat flow and heat flow variation. In this study, the model order was chosen as 7. Considering the thickness of the samples and the low thermal diffusivity of the composites, a sampling rate of 20 Hz was used, and a total of 1000 frames (equivalent to 50 s) were captured after the flash.

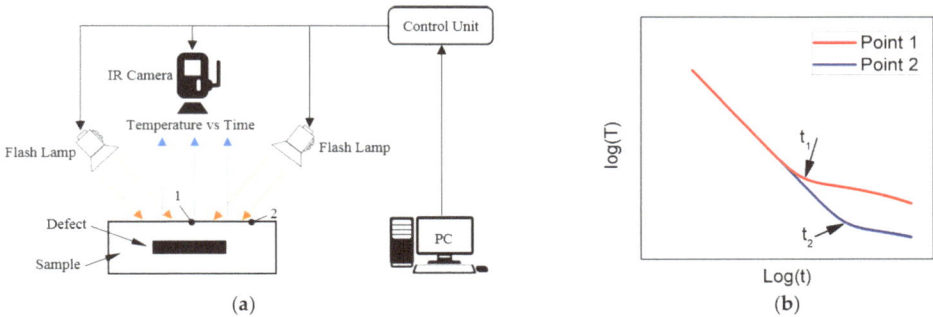

Figure 4. (**a**) Experimental configuration of the pulsed thermographic inspection under the reflection mode, where Point 1 denotes a surface location with a defect underneath and Point 2 denotes a location on the sample surface with no defect underneath. (**b**) Typical observed time–temperature decay curves in the logarithmic domain for Points 1 and 2, respectively, where the times of t1 and t2 are the key for measuring the thickness of the local materials.

3. Results and Discussion

The samples were tested with different energy levels, as described in Table 4. The results are reported by grouping together the orthogonal impact and inclined impact. The images from the optical microsccpy and thermography were combined to reconstruct the internal and surface damages.

Table 4. Summary of damage observed on the composite panels under orthogonal impacts.

Sample	Material	Projectile	Energy (J)	Front Face Damage	Back Face Damage
13	CF	Sharp	2.2	Matrix cracking	No visual damage
15	CF	Smooth	2.7	Matrix cracking and indentation	Matrix crack
9	CF + Kevlar	Smooth	0.9	None	No visual damage
11	CF + Kevlar	Sharp	1.2 & 6.5	Fibre peel-off on the first layer and indentation	Matrix crack
10	CF + Kevlar	Smooth	16.9	Matrix cracking, fibre peel-off, and indentation	Cross fracture and delamination
12	CF + Kevlar	Sharp	15.6	Matrix cracking, fibre pee-off, and indentation	Cross fracture and delamination

3.1. Orthogonal Impact

A summary of the damage observed on the panels that were impacted at a 90° angle is presented in Table 4. The samples were analysed according to the following parameters: effect of the projectile surface (sharp or smooth) and the presence of a Kevlar reinforcement layer.

3.1.1. Effect of the Shape of the Projectile

CF panels 13 and 15 were impacted at 2.2 J with sharp gravel and 2.0 J with smooth gravel, respectively. Visible surface damage was observed with optical microscopy, with matrix cracking developing perpendicularly to the fibre direction in both cases. However, an indentation with a round-shaped crater was observed on panel 15, which was impacted by smooth gravel, whilst on panel 13 (sharp projectile), the matrix cracking area was larger than that on panel 15. These damages can be observed in Figures 5 and 6.

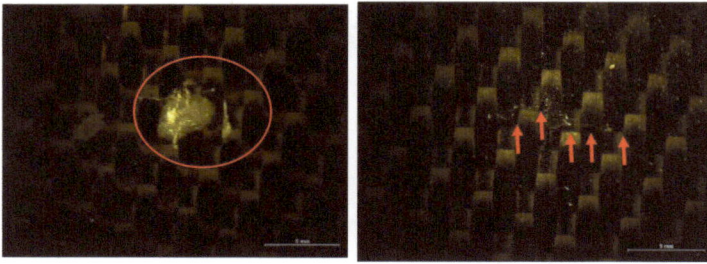

Figure 5. Optical images of the damages observed on the front faces of panels 13 (**left-hand side**) and 15 (**right-hand side**).

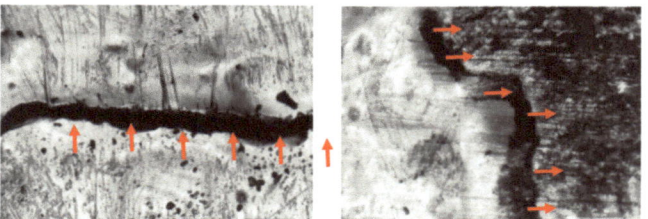

Figure 6. Optical images of the front faces of panels 13 (**left-hand side**) and 15 (**right-hand side**) detailing the matrix cracks perpendicular to the fibre direction (crack highlighted with red arrows).

The thermographic images of panels 13 and 15 allowed the identification of internal damages (Figure 7). Both panels presented similar internal damages, with indentations that were clearly identified on the front faces of the panels. On the back faces of these panels, a matrix crack in a line was noticeable on panel 13, while a different shape was observable on panel 15 (Figure 8).

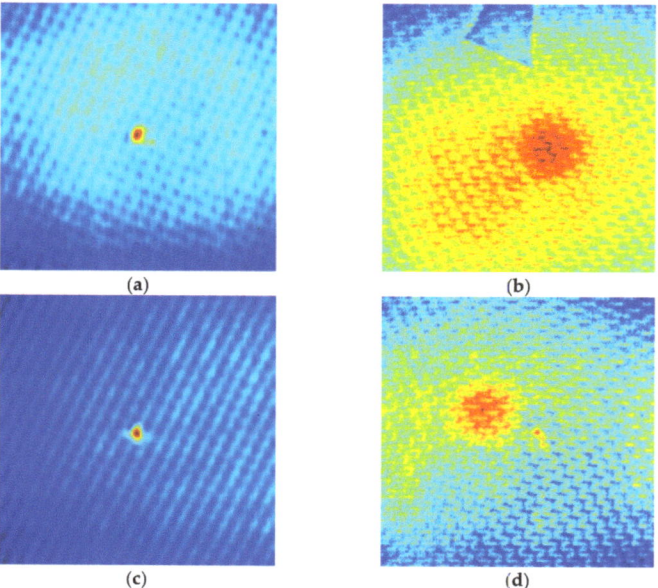

Figure 7. Thermographic images of the impact on the (**a**) front face and (**b**) back face of panel 13 and the (**c**) front face and (**d**) back face of panel 15.

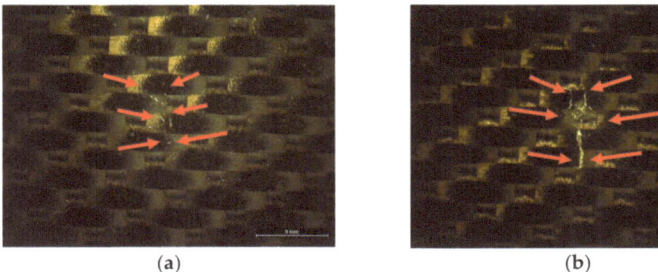

Figure 8. Microscopic images of the back faces of (**a**) panel 13 and (**b**) panel 15.

3.1.2. Effect of the Layer of Kevlar for Reinforcement

Concerning the Kevlar-reinforced panels, panel 9 was impacted at 1.2 J with smooth gravel with no visible damage. In the case of panel 11, a first impact was carried out at 0.9 J with a sharp projectile. This impact created a small damaged area (the red square in Figure 9) without matrix cracking., but with fibre peel-off and failure on the first layer (Figure 10). The other marks were caused by another impact at 6.5 J with a sharp impactor (the blue triangle in Figure 9). This created a larger damaged area than the previous impact, and it was likely caused by the rotation of the gravel. The point of impact was clearly noticeable in the thermographic images (Figure 11). However, internal damages did not seem to occur (Figure 11).

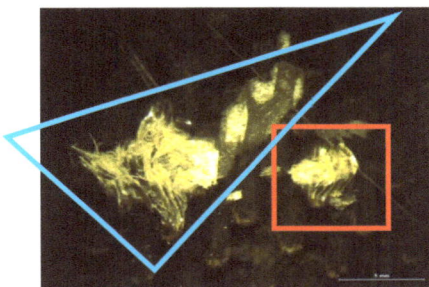

Figure 9. Optical image of the top face of panel 11 highlighting the damages cause by the first impact (red square) and the second impact (blue triangle).

Figure 10. Microscopic image of panel 11 showing fibre loss on the first layer of the panel.

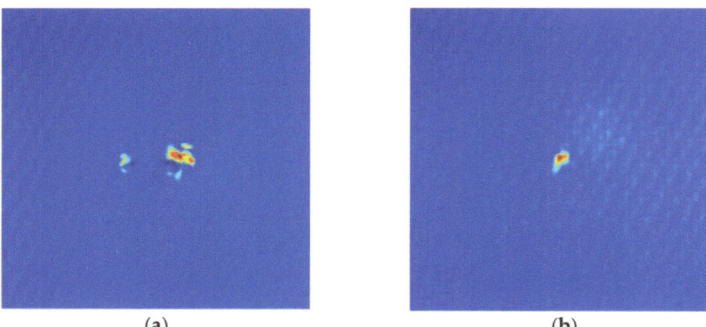

Figure 11. Thermographic images of the (**a**) front face and (**b**) back face of panel 11.

On the back face, the main damage that was identifiable with optical microscopy was a matrix crack (Figure 12).

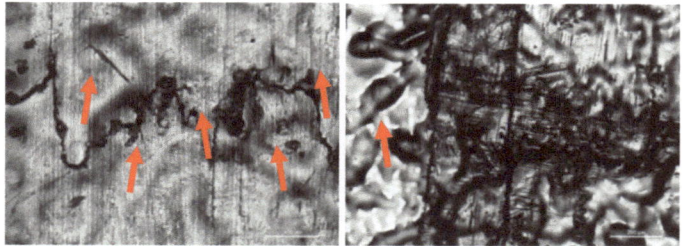

Figure 12. Microscopic images of the back face of panel 11 with a matrix crack (**left**) and localised damage (**right**).

Panels 10 and 12, which were made of CF and Kevlar, were impacted at 16.9 J with a smooth impactor and 15.6 J with a sharp one (a higher impact energy than that of panels 9 and 11), respectively. Similar damage to that of panel 11 was observed on the top face with fibre peel-off, fibre failure, and matrix damage. The thermographic and TSR images (Figures 13 and 14) illustrated a significant area of internal damages. The second derivative clearly enabled the identification of an intense but more restricted damage area close to the point of impact, which was likely delamination.

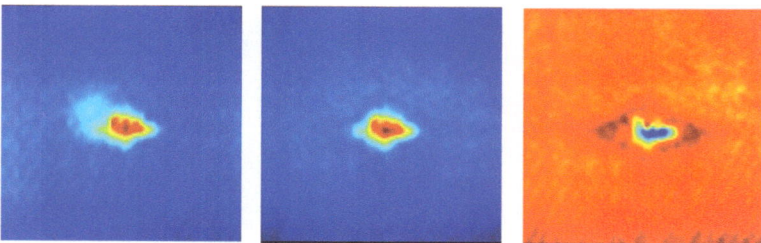

Figure 13. Thermographic images of the front face of panel 10: thermographic image (**left-hand side**), TSR image (**centre**), and second derivative (**right-hand side**).

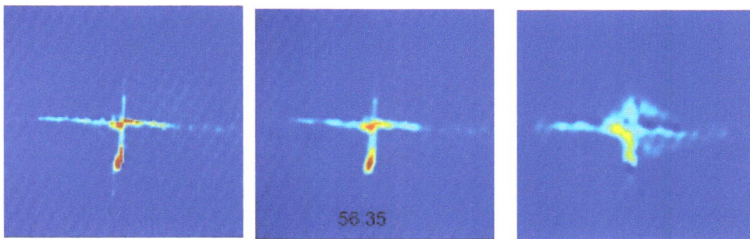

Figure 14. Thermographic images of the back face of panel 10: thermographic image (**left-hand side**), TSR image (**centre**), and second derivative (**right-hand side**).

Cross failures were observable on the back faces of the panels. The cross shape was clearly noticeable in the TSR and raw images (Figure 15). Damaged areas close to the cross were observable on the second derivative image, and they also seemed to indicate delamination.

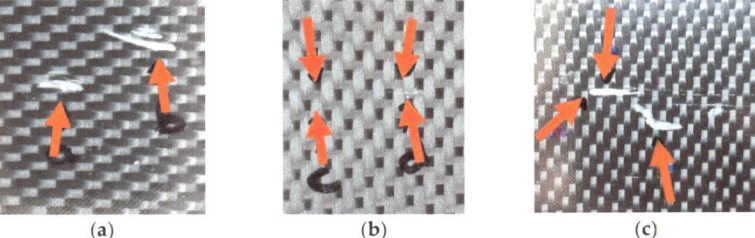

Figure 15. Images of damages on the (**a**) front face and (**b**) back face of panel 1 and on the (**c**) front face of panel 3.

3.2. Inclined Impact

A summary of the damage observed on the panels impacted at an angle of 28.5° with respect to the plane of the panel is presented in Table 5, followed by an analytical discussion on the effect of the shape of the projectile and the presence of a layer of Kevlar as reinforcement.

Table 5. Summary of the damages observed on composite panels under an inclined (28.5°) impact.

Sample	Material	Projectile	Energy	Type of Front Face Damage	Type of Back Face Damage
1	CF	Sharp	3.1 J	Matrix cracking & Fibres exposed	Matrix cracking
3	CF	Smooth	2.7 J	Fibres exposed	No visual damage
7	CF + Kevlar	Smooth	3.4 J	Scratch	No visual damage
6	CF + Kevlar	Sharp	7.7 J & 2 J	Matrix cracking & Fibre breakage	no visual damage
8	CF + Kevlar	Smooth	8.9 J	Concentration region	Matrix cracking
18	CF	Smooth	7.3 J	Matrix concentration & cracking (a,b)	Matrix cracking

3.2.1. Effect of the Shape of the Projectile

Panels 1 and 3 were impacted at 3.1 J with sharp gravel and 2.7 J with a smooth projectile, respectively, at an angle of 28.5°. Panel 1 presented damages on the front and back faces—matrix cracking and fibre peel-off—while panel 3 presented some damage only on the front face with fibre peel-off (Figure 15). The thermography images of panels 1 and 3 (Figure 16) highlighted the areas that were damaged on the front and back of the panels. The main difference between these samples and the ones that were impacted orthogonally (panels 13 and 15 in Figure 8) is the fact that, with the orthogonal impact, the damage was concentrated/localised at the contact point, while with the inclined impact, the internal damaged area is larger.

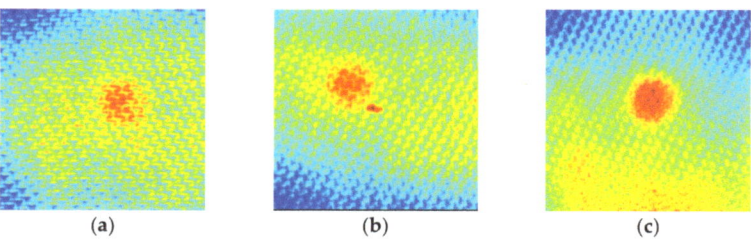

(a) (b) (c)

Figure 16. Thermographic images of the (**a**) front face and (**b**) back face of panel 1 and (**c**) the front face of panel 3.

3.2.2. Effect of the Layer of Kevlar for Reinforcement

To assess the effect of the Kevlar layer with an inclined impact, panels 3 and 7 were compared. These panels were impacted with smooth projectiles at 2.7 and 3.4 J, respectively; panel 7 was the one with the layer of Kevlar for reinforcement. On panel 7, the front face presented some scratches, while the back face had no visible damage (Figure 17). The levels of damage obtained were similar for both panels, suggesting that the reinforcement layer of Kevlar had little effect.

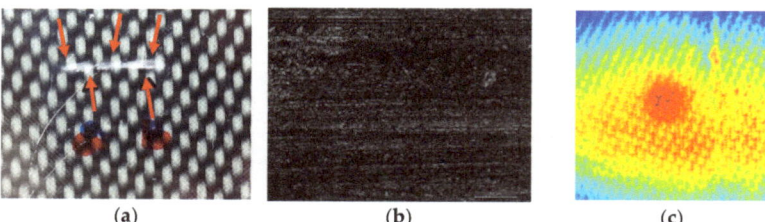

(a) (b) (c)

Figure 17. Images of the front face of panel 7: (**a**) microscopic image showing damage from the impact, (**b**) magnification showing matrix cracking, and (**c**) thermographic image showing internal damages.

Panels 6 and 18 were impacted at 7.7 J with a sharp projectile and 7.3 J with a smooth projectile, respectively, with both at an angle of 28.5°. Both panels presented scratches in the area of the impact. However, panel 6 did not present damage on its back face, while panel 18 had a matrix crack (Figure 18).

In this case, the smooth projectile caused more layer damage to the panel than the sharp one did. This could be related to the effect of the reinforcement of the Kevlar layer in the panel configuration. At low levels of impact energy, the effect of the Kevlar layer was not observed, but at higher levels, it reduced the damage on the back face of the plate. However, some internal damage was identified (Figure 19).

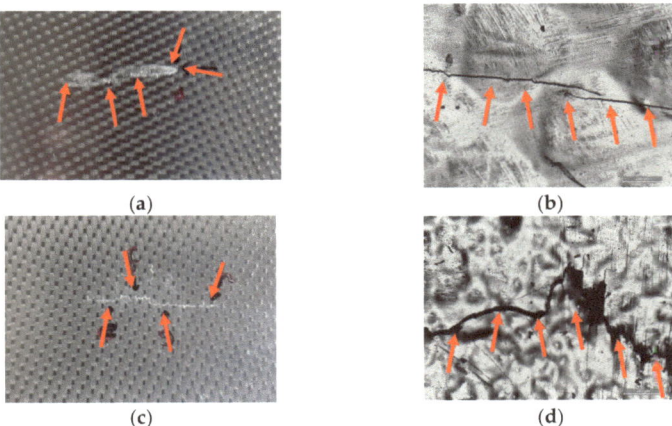

Figure 18. Microscopic images of panel 18: (**a**) front face, (**b**) magnification of the front face, (**c**) back face, and (**d**) magnification of the back face.

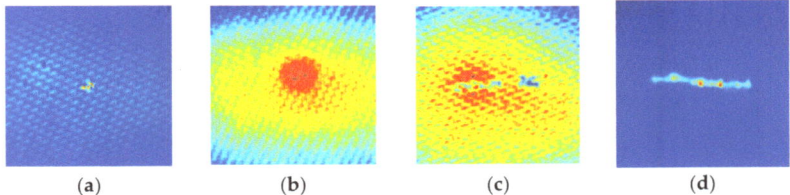

Figure 19. Thermographic images of the (**a**) front face and (**b**) back face of panel 6 and the (**c**) front face and (**d**) back face of panel 18.

In terms of the angle of impact, by comparing panels 3 and 15, it was noticed that inclined impacts caused less damage on the composite panels than perpendicular impacts did.

3.3. Effect of Impact Energy

For the plate thicknesses of 1.05 and 1.40 mm, several levels of damage were observed in relation to the impact energy. For a low level of energy (2–6 J), damage on the top surface was observed; for a medium level (6–14 J), there was visible damage on the front and back faces and through the thickness; for high levels (higher than 15 J), total penetration of the plates occurred. However, the effect of the reinforcement with the Kevlar layer was evident. For example, when impacted with a smooth projectile, plate 4 (CF lay-up) suffered penetration at an impact energy of 11.6 J, while plate 10 (CF + Kevlar lay-up) did not experience penetration at an impact energy of 13.9 J.

Further investigation of the impact through the images captured with a high-speed camera showed a contribution of the Kevlar layer to a reduction in the amount of damage produced on the front face. Figures 20–23 show sequences of three frames captured during the test to show the start of the impact, an intermediate position, and the final position of the gravel before it detached from the specimen. The four different sequences refer to orthogonal and inclined impacts with either sharp or smooth gravel. The gravel, which was either sharp or round, produced limited damages on the surfaces when the panels were reinforced with a Kevlar layer on the bottom face. The sharp gravel could cut the panel, with partial perforation and extensive damages, whilst the round gravel impacted the surface and rolled over it, as happened without the Kevlar reinforcement, producing a fibre cut and scratches in the resin. These impact conditions refer to energy levels that did not produce perforation.

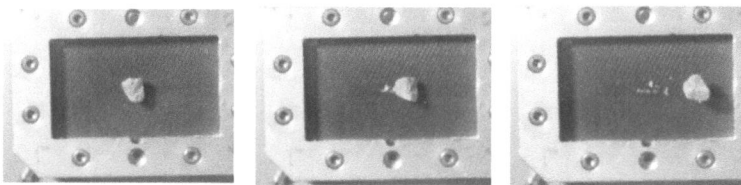

Figure 20. Sample 5, sharp gravel, CF + Kevlar, 2.8 J, showing the initial (**left**), intermediate (**centre**), and gravel detachment stages (**right**).

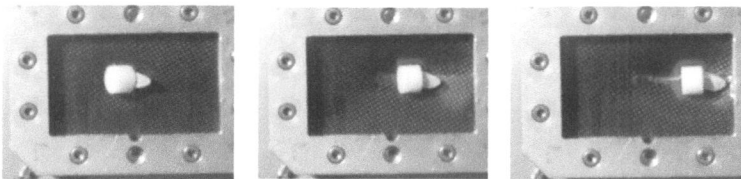

Figure 21. Sample 2 (CF, 28.5 degrees, 15.4 J), test n. 19, sharp gravel, showing the initial (**left**), intermediate (**centre**), and gravel detachment stages (**right**).

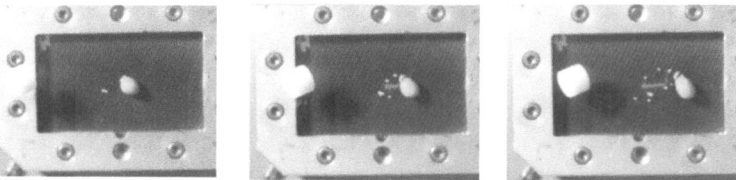

Figure 22. Sample 7, CF + Kevlar, 3.4 J, smooth gravel, showing the initial (**left**), intermediate (**centre**), and gravel detachment stages (**right**).

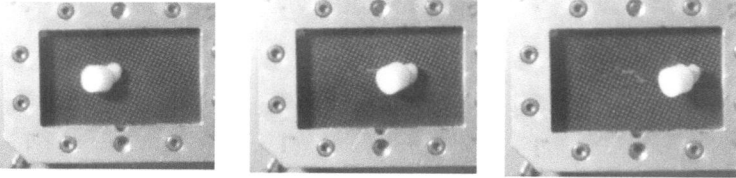

Figure 23. Sample 3, CF, smooth gravel, 2.7 J, 28.5 degrees, showing the initial (**left**), intermediate (**centre**), and gravel detachment stages (**right**).

If we compare our observations with the previous literature review, the same damages as those observed by Liu et al. [24] were noticed for panels 10, 12, and 14. The damages were caused by compressive stress waves generated and transmitted through the volume after impact. These waves were then reflected on the back face of the panel and formed tensile waves, thus explaining the fibre failure on the back face. It is then likely that delamination occurred. Compared with the results from Othman et al. [27], who assessed impacts on unidirectional laminates, the damages were different. The matrix cracks were in the direction of the fibres, while with these woven panels, the cracks were perpendicular to the fibres.

4. Conclusions

Gas gun testing for high-velocity impacts was used to characterize the responses of composite panels that were impacted under different energy levels and different angles by gravel of different shapes.

Sharp gravel produced more damage on the top face compared to smooth gravel, whilst the damage on the back face was not affected by the shape, but only by energy and angle. From the thermographic results, more internal damages were observed with sharp gravel, but this would need to be confirmed through further analyses.

Matrix damage was the main mechanism observed at a low impact energy (around 2 J), whilst at higher energies, Kevlar delayed the fibre damages by improving the ballistic limits of the panels. However, at a low level of energy, no major improvements were noticed when using Kevlar, as matrix cracking was dominant.

Further work on the through-thickness analysis of the panels is required, as thermography cannot provide a definitive answer.

Author Contributions: Data curation, V.M.R., M.G.; Formal analysis, V.M.R., M.G.; Experimental Data A.R., M.G., V.M.R., K.D., H.L.; Writing—original draft, M.G., V.M.R.; Review & Editing, M.G., V.M.R., Y.Z., A.R., G.J.A.-T. All authors have read and agreed to the published version of the manuscript.

Funding: This research received no external funding.

Institutional Review Board Statement: Not applicable.

Informed Consent Statement: Not applicable.

Data Availability Statement: The data that support the findings of this study are available on request from the corresponding author. The data are not publicly available due to privacy or ethical restrictions.

Acknowledgments: The Authors would like to thank Nick Smith and SHD Composites for supplying the composite materials.

Conflicts of Interest: The authors declare no conflict of interest.

References

1. Ursenbach, D.O. Penetration of CFRP Laminates by Cylindrical Indenters. Master's Thesis, The University of British Columbia, Vancouver, BC, Canada, October 1995.
2. Bhatnagar, A. *Lightweight Ballistic Composites*; Woodhead Publishing: Sawston, UK, 2006. [CrossRef]
3. Cantwell, W.J.; Morton, J. Comparison of the low and high velocity impact response of CFRP. *Composites* **1989**, *20*, 545–551. [CrossRef]
4. Vaidya, U.K. Impact response of laminated and sandwich composites. In *Impact Engineering of Composite Structures*; Abrate, S., Ed.; Springer: Vienna, Austria, 2011; pp. 97–191. [CrossRef]
5. Mitrevski, T.; Marshall, I.H.; Thomson, R. The influence of impactor shape on the damage to composite laminates. *Compos. Struct.* **2006**, *76*, 116–122. [CrossRef]
6. Cantwell, W.J.; Morton, J. Impact perforation of carbon fibre reinforced plastic. *Compos. Sci. Technol.* **1990**, *38*, 119–141. [CrossRef]
7. Shahkarami, A.; Cepus, E.; Vaziri, R.; Poursartip, A. Material responses to ballistic impact. *Lightweight Ballist. Compos. Mil. Law-Enforc. Appl.* **1995**, *2*, 72–100. [CrossRef]
8. Andrew, J.J.; Srinivasan, S.M.; Arockiarajan, A.; Dhakal, H.N. Parameters influencing the impact response of fiber-reinforced polymer matrix composite materials: A critical review. *Compos. Struct.* **2019**, *224*, 111007. [CrossRef]
9. Børvik, T.; Langseth, M.; Hopperstad, O.S.; Malo, K.A. Perforation of 12 mm thick steel plates by 20 mm diameter projectiles with flat, hemispherical and conical noses: Part I: Experimental study. *Int. J. Impact Eng.* **2002**, *27*, 19–35. [CrossRef]
10. Hoskin, B.C.; Baker, A.A. *Lectures on Composite Materials for Aircraft Structures*; Aeronautical Research Labs: Melbourne, Australia, 1982.
11. Tang, E.; Wang, J.; Han, Y.; Chen, C. Microscopic damage modes and physical mechanisms of CFRP laminates impacted by ice projectile at high velocity. *J. Mater. Res. Technol.* **2019**, *8*, 5671–5686. [CrossRef]
12. Beland, S. *High Performance Thermoplastic Resins and Their Composites*, 1st ed.; William Andrew: Norwich, NY, USA, 1990.
13. Vieille, B.; Casado, V.M.; Bouvet, C. About the impact behavior of woven-ply carbon fiber-reinforced thermoplastic-and thermosetting-composites: A comparative study. *Compos. Struct.* **2013**, *101*, 9–21. [CrossRef]
14. Lopes, C.; Seresta, O.; Abdalla, M.; Gurdal, Z.; Thuis, B.; Camanho, P. Stacking sequence dispersion and tow-placement for improved damage tolerance. In Proceedings of the 49th AIAA/ASME/ASCE/AHS/ASC Structures, Structural Dynamics, and Materials Conference, 16th AIAA/ASME/AHS Adaptive Structures Conference, 10th AIAA Non-Deterministic Approaches Conference, 9th AIAA Gossamer Spacecraft Forum, 4th AIAA Multidisciplinary Des., Schaumburg, IL, USA, 7–10 April 2008; p. 1735.

15. York, C.B. Unified approach to the characterization of coupled composite laminates: Benchmark configurations and special cases. *J. Aerosp. Eng.* **2010**, *23*, 219–242. [CrossRef]
16. Dorey, G. Failure Mode of Composite Materials with Organic Matrices and Their Consequences in Design. 1975. Available online: https://apps.dtic.mil/sti/citations/ADA018178 (accessed on 31 July 2022).
17. Park, R.; Jang, J. Impact behavior of aramid fiber/glass fiber hybrid composites: The effect of stacking sequence. *Polym. Compos.* **2001**, *22*, 80–89. [CrossRef]
18. Abrate, S. *Impact Engineering of Composite Structures*; Springer Science & Business Media: Berlin/Heidelberg, Germany, 2011; Volume 526.
19. Gellert, E.P.; Cimpoeru, S.J.; Woodward, R.L. A study of the effect of target thickness on the ballistic perforation of glass-fibre-reinforced plastic composites. *Int. J. Impact Eng.* **2000**, *24*, 445–456. [CrossRef]
20. Icten, B.M.; Kıral, B.G.; Deniz, M.E. Impactor diameter effect on low velocity impact response of woven glass epoxy composite plates. *Compos. Part. B Eng.* **2013**, *50*, 325–332. [CrossRef]
21. Sevkat, E.; Liaw, B.; Delale, F. Drop-weight impact response of hybrid composites impacted by impactor of various geometries. *Mater. Des. 1980–2015* **2013**, *52*, 67–77. [CrossRef]
22. Safri, S.N.A.; Sultan, M.T.H.; Yidris, N.; Mustapha, F. Low velocity and high velocity impact test on composite materials—A review. *Int. J. Eng. Sci* **2014**, *3*, 50–60.
23. Lee, S.-M.; Cheon, J.-S.; Im, Y.-T. Experimental and numerical study of the impact behavior of SMC plates. *Compos. Struct.* **1999**, *47*, 551–561. [CrossRef]
24. Liu, J.; Liu, H.; Kaboglu, C.; Kong, X.; Ding, Y.; Chai, H.; Blackman, B.R.; Kinloch, A.J.; Dear, J.P. The impact performance of woven-fabric thermoplastic and thermoset composites subjected to high-velocity soft- and hard-impact loading. *Appl. Compos. Mater.* **2019**, *26*, 1389–1410. [CrossRef]
25. Gustin, J.; Joneson, A.; Mahinfalah, M.; Stone, J. Low velocity impact of combination Kevlar/carbon fiber sandwich composites. *Compos. Struct.* **2005**, *69*, 396–406. [CrossRef]
26. Zhao, Y.; Addepalli, S.; Sirikham, A.; Roy, R. A confidence map based damage assessment approach using pulsed thermographic inspection. *NDT E Int.* **2018**, *93*, 86–97. [CrossRef]
27. Othman, R.; Ogi, K.; Yashiro, S. Characterization of microscopic damage due to low-velocity and high-velocity impact in CFRP with toughened interlayers. *Mech. Eng. J.* **2016**, *3*, 16-00151. [CrossRef]

Article

Mechanical Response and Processability of Wet-Laid Recycled Carbon Fiber PE, PA66 and PET Thermoplastic Composites

Uday Vaidya [1,2,3,*], Mark Janney [4], Keith Graham [4], Hicham Ghossein [1] and Merlin Theodore [2,5]

[1] Tickle College of Engineering, University of Tennessee, 1512 Middle Drive, Knoxville, TN 37996, USA; hghossei@utk.edu
[2] Manufacturing Sciences Division (MSD), Oak Ridge National Laboratory (ORNL), 2350 Cherahala Blvd, Oak Ridge, TN 37932, USA; theodorem@ornl.gov
[3] The Institute for Advanced Composites Manufacturing Innovation, IACMI-The Composites Institute, 2370 Cherahala Blvd, Knoxville, TN 37932, USA
[4] Carbon Conversions, Lake City, SC 29560, USA; mjanney52@gmail.com (M.J.); kgraham@carbonconversions.com (K.G.)
[5] Carbon Fiber Technology Facility, Oak Ridge National Laboratory, Oak Ridge, TN 37830, USA
* Correspondence: uvaidya@utk.edu

Abstract: The interest in recycled carbon fiber (rCF) is growing rapidly and the supply chain for these materials is gradually being established. However, the processing routes, material intermediates and properties of rCF composites are less understood for designers to adopt them into practice. This paper provides a practical pathway for rCFs in conjunction with low cost and, for the most part, commodity thermoplastic resins, namely polyethylene (PE), polyamide 66 (PA66) and polyethylene terephthalate (PET). Industrially relevant wet-laid (WL) process routes have been adopted to produce mats using two variants of WL mats, namely (a) high speed wet-laid inclined wire to produce broad good 'roll' forms and (b) 3DEP™ process patented by Materials Innovation Technologies (MIT)-recycled carbon fiber (RCF), now Carbon Conversions, which involves mixing fibers and water and depositing the fibers on a water-immersed mold. These are referred to as 'sheet' forms. The produced mats were evaluated for their processing into composites as 'fully consolidated mats' and 'non-consolidated' as-produced mats. Comprehensive mechanical data in terms of tensile strength, tensile modulus and impact toughness for rCF C/PE, C/PA66 and C/PET are presented. The work is of high value to sustainable composite designers and modelers.

Keywords: recycled carbon fiber; thermoplastics; wet-laid processing; compression molding

1. Introduction

Discontinuous carbon fibers have a number of advantages, such as (a) fiber aspect ratio can be greater than critical fiber length, hence, superior mechanical properties can be realized; (b) higher drapeability offered due to fiber movement during processes, such as compression and thermo-stamping; (c) ability to hybridize fiber lengths and types; and (d) lower cost, since secondary weaving and braiding are not necessary. Traditional processes, such as injection molding and extrusion, result in significant fiber length attrition due to friction and interaction with the screw with the material. Wet-laid (WL) processing offers a low-energy alternative to traditional processes, such as weaving and/or stitch bonding, producing mats in desired fiber-matrix weight fraction. Both reinforcing fibers and resin fibers (resin in fiber form) are mixed in desired weight proportion in water (with dispersant and flocculent) and mixed till the material assumes a homogenous form. The water is drained rapidly from the fiber bulk, resulting in a well-dispersed fiber-polymer mat.

Two types of industrially relevant WL mat process routes have been investigated in this work, namely—(a) high speed wet-laid line to produce broad good 'roll' forms and (b) 3DEP™ process patented by Carbon Conversions, mixing fibers and water, and depositing on a water-immersed mold. The underlying hypothesis is that the high-speed WL process would yield preferred fiber alignment in the 'roll' direction, while the 3DEP^{TM} would produce randomized fiber orientation. This work reports mechanical properties of WL-processed rCF mat composites in conjunction with commodity thermoplastics, namely polyethylene (PE) and polyethylene terephthalate (PET) and engineering thermoplastic polyamide (PA66). There is no work, to our knowledge, that quantifies the mechanical performance of rCF WL mat composites, while such information would be valuable to a designer and modeler(s).

2. Literature Review

With increased emphasis on circular economy, rCFs are finding use in applications, such as automotive, sporting goods and industrial parts [1–5]. The processes used to obtain rCF are pyrolysis and solvolysis of out-of-date prepregs and end-of-life CF intensive parts. Other sources include manufacturing scrap, edge trims and waste from textile processes. Carbon Conversions specializes in pyrolysis-based recovery of CFs [6], the primary focus of this effort. Several efforts have emphasized the importance of processing discontinuous carbon fiber thermoplastic composites [7–9].

Thomason [10] and Vaidya [11] illustrated the importance of fiber aspect ratio of discontinuous fibers. Polyamide (PA), polyethylene (PE) and polyethylene terephthalate (PE) are of continued interest as thermoplastic matrices in reinforced composites due to their recyclability and superior mechanical properties [12–14].

The WL process is promising in terms of fiber length retention. Hemamalini and Dev [15] discussed that WL is an emerging technique to produce nonwovens using short natural cellulosic fibers and synthetic fibers and their blends. The steps involved in wet laying are dispersion, deposition and consolidation. Uniform dispersion is the key to attain defect-free nonwovens in web laying. WL processing is like the papermaking process with differences in fiber length and density of the fibers [16,17]. The quality of the dispersion depends on material parameters, such as fiber length, surfactant, source of the fibers, linear density of the fibers and machine parameters, such as dispersion time and mechanical agitation.

There are only limited studies with WL and thermoplastic polymers in conjunction with high-performance CFs [17]. Product opportunities for automotive and aerospace can expand using WL intermediates. This work considers WL rCFs in conjunction with commodity thermoplastics, such as PP and PET, as well as engineering thermoplastics, such as PA66. Yan et al. [18] investigated process parameters of WL rCF-reinforced thermoplastic (CFRTP) nonwoven mats. They used response surface methodology to optimize the heat-molding compression parameters in terms of temperature, pressure and time, respectively. They also reported that CFRTP comprising 30 wt% CF fiber length of 6 mm provided the highest tensile strength. Ghossein et al. [19] evaluated the mechanical behavior of WL-CF mats in conjunction with the microstructure predicted through Object-Oriented Finite Element Analysis (OOF). The authors used novel mixing methods to reduce time to create optimal mats. Barnett et al. [20] created CF-PPS WL mats, similar to organosheets used in automotive production. Erland et al. [21] investigated the re-manufacture and repairability of thick-section poly(ether ether ketone) PEEK CFRTPs. They reported results on C/PEEK tested under three-point bend loaded to fracture before being re-heated, re-pressed and re-tested. Their study showed that C/PEEK composites could be repaired with minimal loss of mechanical performance, even when significant fracture occurs. They attained a flexural modulus of 80 GPa and a maximum bending stress of 900 MPa. Brahma et al. [22] investigated discontinuous WL CF mats and compared them to liquid-molded PA6. There was roughly a 10–13% increase in its tensile strength, modulus and impact strength properties at 30 and 40% weight fractions and almost a 120% increase at 50% weight fraction. Yeole et al. [23] studied the effects of dispersant and flocculent in glass fiber WL thermoplastic composites. Kore et al. [24] hybridized bamboo fibers with carbon fiber mats with the WL process and reported the property bounds.

There are presently no systematic guidelines of using rCF in composite products. This paper attempts to address this gap, and addresses commodity and engineering thermoplastics with rCFs to provide a comprehensive understanding of lower- and upper-bound properties with these materials. The work is of high relevance to sustainable composite designers and end users.

3. Materials and Methods

Two WL-processing approaches were considered in this study. Nonwoven rCF-thermoplastic mats were produced (a) in a WL machine capable of producing 'roll' forms; and (b) 3DEP^{TM} process where rCF mats were deposited as a 'sheet' in a water tank. Throughout this manuscript these variants are referred to as 'roll' and 'sheet' forms, respectively.

The 'roll' mats were produced in 1.2 m (48″) wide rolls, while the 'sheets' were produced as 3DEP^{TM} mats using a water-based deposition on a screen tool. A 'sheet' was typically 350 mm × 350 mm WL mat.

'Sheet': Carbon Conversions developed an innovative method for making WL fiber preforms [5]. The 3DEP^{TM} process lends itself to converting loose recycled fibers into nonwoven carbon fiber mats. The 3DEP^{TM} process uses advanced slurry molding process for creating nonwoven rCF preforms. 3DEP^{TM} produces homogeneous fiber distribution within the mat with consistent areal weight and acceptable dimensional tolerance. In this work 3DEP^{TM} was used to produce WL rCF mats. rCF obtained from pyrolysis of T800 prepreg were used. The recycled fiber had nominal 12.7 mm fiber length and 8–10 mm diameter.

'Roll': Carbon Conversions produces continuous, WL, nonwoven fabrics on a 1.2 m wide RotoFormer machine (Allimand Interweb, Inc., Glen Falls, NY, USA). Compositions include chopped carbon fiber and blends of carbon fiber with thermoplastic polymer staple fibers. Areal density can range from 100 to 500 g/m^2 (gsm). Areal density coefficient of variation (COV) is typically <3%. After forming, the web is sent through a continuous dryer and then bound onto 50–200 m rolls. rCF mats were processed via WL with three resin systems: PE, PA66 and PET, respectively. The molecular weights are as follows: PET—25,000 g/mol, PE—30,000 g/mol and PA66—25,000 g/mol. The tensile modulus of neat (unreinforced) PET is 2.8 GPa, PE is 0.9 GPa and PA66 is 3.2 GPa, respectively.

The C/PET and C/PA66 were processable at 500 F due to their higher melting point, while PE was processed at 250 F since PE melts at a lower temperature. The work was conducted in two batches referred to as *Batch 1* and *Batch 2*. The lessons from *Batch 1* were applied in producing *Batch 2* mats. Table 1 summarizes the rCF mats designed for the 'roll' and 'sheet' forms under *Batch 1*. *Batch 2* mats are discussed later. Composite panels were made from the WL mats using compression-molding process with the process conditions identified in the table.

Table 1. Sample variants, preform type and processing conditions for *Batch 1* mats.

Sample Variant **	Preform Type	Processing Notes ^
PA66/CF/68/30	Sheets	Tool at 500 F and 1000 psi
PA66/CF/78/20	Sheets	Tool at 500 F and 1000 psi
PA66/CF/88/10	Sheets	Tool at 500 F and 1000 psi
PA66/CF/77/20	Roll	Tool at 500 F and 1000 psi
PE/CF/78/20	Roll	Tool at 250 F and 1000 psi
PA66/CF/77/20	Roll	Tool at 509 F and 1000 psi
PA66/CF/78/20	Dried Sheets	Tool at 509 F and 1000 psi
PE/CF/77/20	Roll	Tool at 265 F and 1000 psi
PE/CF/77/20	Roll	
PET/CF/77/20	Sheets	Tool at 500 F and 1000 psi

^ All plates were compression molded; ** e.g., nomenclature PA66/CF/68/30 means, 68% resin, 30% carbon fiber by weight.

4. Results and Discussion

4.1. Partial and Fully Consolidated Panels

Compression molding was used to produce partially and fully consolidated panels as illustrated in Figure 1a–f. Five (5) layers of 300 mm × 300 mm preforms were compression molded in a matched metal tool. For C/PA66, the press platens were heated to a temperature of 500 °F at 6.895 MPa (1000 psi). In a few cases, the tool temperature was held at 250–265 °F. The hold time was approximately 20 min at temperature. The tool was cooled to room temperature. In some cases, slight discoloration was noted along the edges of the panel—12 mm wide band along the four edges.

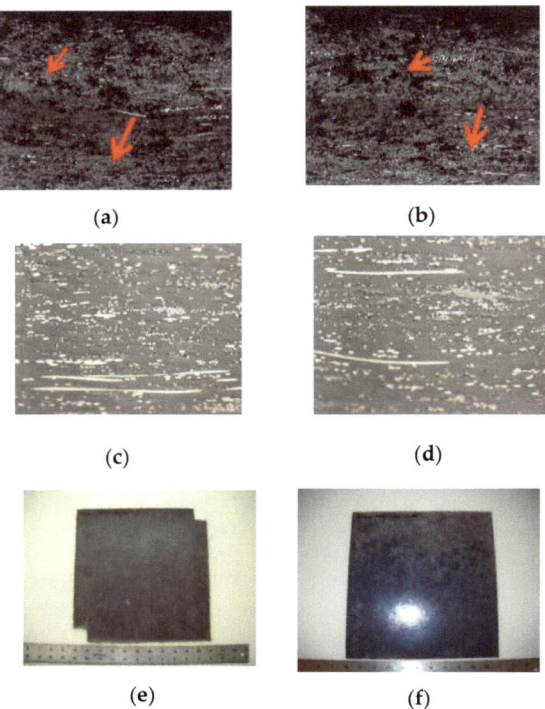

Figure 1. Effect of consolidation; (**a**) PA66/CF/68/30, 5 layers of preform (less-consolidated panel); (**b**) PA66/CF/88/10, 5 layers of preform (less-consolidated panel). Arrows point to representative voids in both (**a**,**b**); (**c**) PA66/CF/68/30, 5 layers of preform (well-consolidated panel); (**d**) PA66/CF/88/10, 5 layers of preform (well-consolidated panel); (**e**) PA66/CF/68/30 less-consolidated panel, the PA66/CF/88/10 was similar in look; (**f**) PA66/CF/68/30 well-consolidated panel, the PA66/CF/88/10 was similar in look; Panel size 275 mm × 275 mm.

The C/PE and C/PET panels were produced in a similar manner to C/PA66. Two panels of PE-CF-78-20 roll were processed as three layers of preforms were compression molded in a 300 × 300 mm matched metal tool. The tool was heated to a temperature of 250 °F (for C/PE) and 500 °F (for C/PET) at 6.895 MPa (1000 psi). The hold time was approximately 20 min at these temperatures. The tool was cooled to room temperature.

4.2. 'Roll' versus 'Sheet' Forms

Preform "sheets" and "roll" forms were evaluated in similar weight fraction and resin type(s). For example, 'sheet(s)' PA66-CF-68-30 and PA66-CF-78-20 and 'roll' PA66-CF-77-20 (e.g., PA66-CF-77-20 means 77 wt% PA66 and 20 wt% CF) were evaluated and compared. Qualitatively, the 'roll' form processed under similar conditions consolidated better (less voids) than 'sheets'. Figure 2 illustrates a representative 'roll' and 'sheet' form composite panel. Moisture analysis revealed that the 'roll' form had less moisture content. The material was dried before consolidation. Parallel edge coupons were tabbed and tested in two (2) directions 'along' and 'across' the machine direction. The direction was more relevant in the 'roll' due to preferential fiber orientation along the warp (machine) direction.

(a)

(b)

Figure 2. 'Roll' and 'Sheet' form panel; (**a**) 'Roll' form panel—optimal compression-molded C/PA66 panel from WL mats; (**b**) panel produced from WL PA66-CF mats in 'sheet' form. Panel size 300 × 300 mm^2.

4.3. Moisture Analysis

Moisture analysis was conducted to determine moisture content in the preform 'sheets and 'roll'. Percentage moisture was determined by weight analysis. The samples studied were PA66-CF-78-20 preform 'sheets' and PA66-CF-77-20 preform "roll". The materials were dried at 250 °F for 8 h. Table 2 illustrates the moisture percent in the 'sheet' versus 'roll' preform. The 'sheet' exhibited an average moisture content of 3% while the 'roll' exhibited average moisture of 1.68%, about 45% lower than the 'sheet'.

Table 2. Moisture analysis of WL PA66-CF 'sheet' and 'roll' forms.

Sample ID	Wet Sample	Dry Sample	Moisture Content	Moisture %
PA66/CF/78/20/Preform Sheets-1	2.6057	2.53	0.08	3.03
PA66/CF/78/20/Preform Sheets-2	2.8037	2.72	0.08	3.00
PA66/CF/77/20/Preform Roll-1	3.7176	3.65	0.06	1.70
PA66/CF/77/20/Preform Roll-2	3.5691	3.51	0.06	1.66

Tension samples were cut from the consolidated 300 × 300 mm^2 plate. Flat-wise tabs were used for the tension samples (25.4 mm wide and 200 mm length). Some dog bone samples were also tested in a couple of variants to observe the effect of sample shape and size of final properties. Strength and modulus were determined for three specimens each, 'along' and 'across' the fiber directions at a rate 2 mm/min. The modulus was determined with an extensometer (0.2% to 1 % strain).

4.4. Batch 1 Results

Tables 3 and 4 summarize the tensile modulus and strength for composites made with C/PA66 'sheets' versus 'roll', respectively. It is seen that the C/PA66 'roll' had 67% higher average tensile strength (108 MPa ('sheet') versus 187.73 MPa ('roll')) and 72% higher average modulus (10 GPa ('sheet') versus 16.5 GPa ('roll')). The high values for the 'roll' can be attributed to the preferential fiber alignment in the 'roll' direction while the 'sheet' exhibits quasi-isotropic/random orientation. There was no statistical difference in the tensile strength and modulus between the flat edge specimens compared to the dog bone specimen geometry as shown in Table 5.

The 30 wt% C/PE 'roll' specimens exhibited an average tensile strength of 45 MPa and tensile modulus of 6.5 GPa. These were approximately half that of the 30 wt% C/PA66 composites. The C/PE was only tested (available) in the 'roll' direction.

Table 3. Tensile modulus and strength of C/PA66 (preform 'Sheets').

Type	Sample ID	Direction	Average Modulus (GPa)	Average Modulus 10⁶ (psi)	Average Strength (MPa)	Average Strength 10³ (psi)	Density (g/cc)	Specific Strength	Specific Modulus
Flat Tension Coupons	PA66-CF-68-30	1	10.01	1.45	108.00	15.66	1.24	8.07	87.09
	PA66-CF-68-30	2	9.95	1.44	103.49	15.01	1.24	8.02	83.46
	PA66-CF-78-20	1	10.01	1.45	108.00	15.66	1.21	8.27	89.25
	PA66-CF-78-20	2	9.98	1.45	105.74	15.34	1.21	8.25	87.39
	PA66-CF-88-10	1	9.99	1.45	106.49	15.44	1.18	8.46	90.25
	PA66-CF-88-10	2	9.98	1.45	105.74	15.34	1.18	8.46	89.61
Dog Bone	PA66-CF-78-20	1	9.98	1.45	105.99	15.37	1.21	8.25	87.60

Table 4. Tensile modulus and strength of C/PA66 (preform 'Roll').

Type	Sample ID	Avg. Modulus (GPa)	Avg. Modulus 10⁶ (psi)	Avg. Strength (MPa)	Avg. Strength 10³ (psi)	Density (g/cc)	Specific Strength	Specific Modulus
Flat Samples	PA66-CF-77-20 *	15.32	2.22	187.72	27.23	1.21	12.66	155.14
Samples	PA66-CF-77-20 ^	17.07	2.48	178.16	25.84	1.21	14.11	147.24
Dog Bone	PA66-CF-77-20 ^	16.20	2.35	182.94	26.53	1.21	13.38	151.19

* 2 direction, ^ 1direction.

Table 5. Tensile modulus and strength—C/PE (preform Roll).

Sample Type	Sample ID	Avg. Modulus (GPa)	Avg. Modulus 10⁶ (psi)	Avg. Strength (MPa)	Avg. Strength 10³ (psi)	Density	Specific Modulus	Specific Strength
Flat Samples	PE-CF-78-20	5.19	0.75	40.39	5.86	1.03	5.04	39.21
Dog Bone	PE-CF-78-20	4.21	0.61	47.72	6.92	1.03	4.09	46.33

4.5. Batch 2 Results—Tensile Modulus and Tensile Strength

Based on the results from *Batch 1*, a controlled set of preforms was prepared with approximately 20 wt% CF for PA66, PE and PET, respectively. Composite plates were produced in two configurations, namely, 'no cross-stack' and 'cross-stack', respectively. The rationale for the two configurations was to evaluate if the preferential fiber orientation in the 'roll' influenced the stacking sequence.

Tables 6–8 summarize the results from these materials. The trend of the 'roll' form of higher values than the 'sheet' forms was similar to that in Batch 1. The 'roll' form had 88% higher strength and 137% higher modulus compared to the 'sheet' form. This indicates the influence of significant fiber orientation in the 'roll' form. The effect of drying the mats in *Batch 2* had a marked influence in the 'sheet' form. Drying improved the tensile strength and modulus by an average factor of two or greater.

Table 6. Tensile modulus and strength of PA66-CF (preform 'Roll').

Sample ID	Preform Type	Stacking Sequence	Direction	Tensile Modulus (GPa)	Tensile Strength (MPa)
PA66-CF-77-20-CS	Roll	Cross Stack	1	19.98	257.70
PA66-CF-77-20-CS	Roll	Cross Sack	2	15.16	217.34
PA66-CF-77-20-NCS	Roll	No Cross Stack	1	20.57	242.06
PA66-CF-77-20-NCS	Roll	No Cross Stack	2	16.92	248.29

Table 7. Tensile modulus and strength of PA66-CF (preform 'Sheet').

Sample ID	Preform Type	Stacking Sequence	Direction	Tensile Modulus (GPa)	Tensile Strength (MPa)
PA66-CF-78-20-Predried	Sheets	N/A	1	10.66	169.95
PA66-CF-78-20-Predried	Sheets	N/A	2	6.39	84.31

Table 8. Tensile modulus and strength—PE-CF (preform 'Roll').

Sample ID	Preform Type	Stacking Sequence	Direction	Tensile Modulus (GPa)	Tensile Strength (MPa)
PE-CF-77-20-CS	Roll	Cross Stack	1	5.50	53.74
PE-CF-77-20-CS	Roll	Cross Stack	2	5.57	52.43
PE-CF-77-20-NCS	Roll	No Cross Stack	1	4.20	39.41
PE-CF-77-20-NCS	Roll	No Cross Stack	2	6.97	66.48
PE-CF-77-20-MIT	Roll	N/A	1	5.80	53.27
PE-CF-77-20-MIT	Roll	N/A	2	7.57	75.11

Batch 2 of the PE/CF panels was processed at a higher temperature than Batch 1 (260 °F instead of 245 °F). Increasing the processing temperature increased the tensile strength marginally. The cross-stack panel exhibits similar properties in both the directions while the no-cross stack exhibits a difference in properties in the two directions as shown in Table 9.

Table 9. Tensile modulus and strength—C/PET (preform 'Roll').

Sample ID	Preform Type	Stacking Sequence	Direction	Tensile Modulus (GPa)	Tensile Strength (MPa)
PET-CF-77-20	Sheets	N/A	1	17.35	243.84
PET-CF-77-20	Sheets	N/A	2	17.41	271.19

C/PET sheet preforms were processed at 500 F and 100 psi. These exhibit excellent tensile modulus and strength and are comparable to the C/PA66 samples. Table 10 compares the density of the mats for different weight fractions for C/PA66, C/PET and C/PE, respectively. The C/PA66 composite density ranged from 1.18 and 1.21 to 1.24 g/cc for 10%, 20% and 30 wt%, respectively. The C/PE was 1.03 and C/PET 1.42 g/cc for 20 wt%, respectively. The densest of the materials was C/PET. Table 11 summarizes typical (standard) materials, such as aluminum, ABS and long glass fiber thermoplastics, for comparison to the carbon fiber mats in terms of the density, strength and modulus, respectively. Table 12 provides a detailed summary of all material variants studied in this work C/PA66, C/PE and C/PET for 'roll' and 'sheet' forms in no-stack and cross-stack configurations, where applicable. The data are summarized in terms of density, strength, modulus, specific strength and specific modulus.

Table 10. Density of the rCF thermoplastic variants.

Sample Variants	Fiber Weight %	Resin Weight %	Fiber Density g/cm^3	Resin Density g/cm^3	Fiber Volume Fraction	Resin Volume Fraction	Composite Density g/cm^3
Carbon/PA66	10	90	1.50	1.15	0.078	0.922	1.18
Carbon/PA66	20	80	1.50	1.15	0.161	0.839	1.21
Carbon/PA66	30	70	1.50	1.15	0.247	0.753	1.24
Carbon/Polyethylene	20	80	1.50	0.955	0.137	0.863	1.03
Carbon/PET	20	80	1.50	1.40	0.189	0.811	1.42

Table 11. Specific strength and specific modulus of other engineering materials.

Sample ID	Density (g/cm^3)	Young's Modulus (GPa)	Tensile Strength (MPa)	Specific Modulus (GPa/(g/cm^3))	Specific Strength (MPa/(g/cm^3))
Aluminum	2.70	70.00	570.00	25.93	211.11
ABS (Impact Grade) Min	1.02	1.40	28.00	1.37	27.45
ABS (Impact Grade) Max	1.20	2.80	138.00	2.33	115.00
Glass-PP-40-60	1.21	8.27	80.00	6.83	66.12

Table 12. Comprehensive summary of tensile strength, tensile modulus, specific strength and specific modulus for all rCF variants in this study. The effect of stacking sequence and 'roll' versus 'sheet' form are included.

Sample ID	Preform Type	Stacking Sequence	Direction	Tensile Modulus (GPa)	Tensile Strength (MPa)	Density g/cm^3	Specific Modulus ((GPa)/(g/cm^3))	Specific Strength ((MPa)/(g/cm^3))
PA66-CF-77-20-CS	Roll	Cross Stack	1	19.98	257.70	1.21	16.51	212.97
PA66-CF-77-20-CS	Roll	Cross Stack	2	15.16	217.34	1.21	12.53	179.62
PA66-CF-77-20-NCS	Roll	No Cross Stack	1	20.57	242.06	1.21	17.00	200.05
PA66-CF-77-20-NCS	Roll	No Cross Stack	2	16.92	248.29	1.21	13.98	205.20
PA66-CF-78-20-Predried	Sheets	N/A	1	10.66	169.95	1.21	8.81	140.45
PA66-CF-78-20-Predried	Sheets	N/A	2	6.39	84.31	1.21	5.28	69.68
PE-CF-77-20-CS	Roll	Cross Stack	1	5.50	53.74	1.03	5.34	52.17
PE-CF-77-20-CS	Roll	Cross Stack	2	5.57	52.43	1.03	5.40	50.90
PE-CF-77-20-NCS	Roll	No Cross Stack	1	4.20	39.41	1.03	4.08	38.27
PE-CF-77-20-NCS	Roll	No Cross Stack	2	6.97	66.48	1.03	6.76	64.54
PE-CF-77-20-MIT	Roll	N/A	1	5.80	53.74	1.03	5.63	51.74
PE-CF-77-20-MIT	Roll	N/A	2	7.57	75.11	1.03	7.35	72.92
PET-CF-77-20	Sheets	N/A	1	17.35	243.84	1.42	12.22	171.72
PET-CF-77-20	Sheets	N/A	2	17.41	271.19	1.42	12.26	190.98

4.6. Low-Velocity Impact Testing

The specimens were subjected to drop tower impact on a Dynatup 8250 under clamped plates 100 × 100 mm with drop height impact for two energy levels (5 J and 15 J) (or drop heights), referred to as 'low-energy 5 J' and 'high-energy 15 J' impact. Tables 13 and 14 summarized the impact data for all variants tested for drop weight impact. Figures 3 and 4 compares the normalized load and normalized energy for variants of 20 wt% carbon fiber in each of C/PA66, C/PET and C/PE for no-stack versus cross-stack, where applicable.

Table 13. Low-velocity impact results at low-impact energy (5 J).

Variant	Thickness (mm)	Max Load (kN)	Energy at Max Load (Joule)	Normalized Max Load (kN/mm)	Normalized Energy (Joule/mm)
MIT-C/PE/77/20	2.50	1.83	7.10	0.73	2.84
C/PE/77/20 Cross Stack	2.85	1.84	7.17	0.65	2.52
C/PE/77/20 No Cross Stack	2.69	1.59	6.89	0.59	2.56
C/PA66/77/20 Cross Stack	2.25	1.46	7.45	0.65	3.31
C/PA66/77/20 No Cross Stack	2.03	1.15	7.15	0.57	3.52
C/PET/77/20	2.78	2.08	7.26	0.75	2.61

Table 14. Low-velocity impact results at higher impact energy (15 J).

Sample Variant	Sample Thickness (mm)	Max Load (kN)	Energy at Max Load (Joule)	Normalized Max Load (kN/mm)	Normalized Energy (Joule/mm)
C/PE/77/20	2.46	1.84	11.64	0.75	6.33
C/PE/77/20 Cross Stack	2.77	2.03	12.41	0.73	6.11
C/PE/77/20 No Cross Stack	2.56	1.63	10.83	0.64	6.64
C/PA66/77/20 Cross Stack	1.81	1.66	3.30	0.92	1.99
C/PA66/77/20 No Cross Stack	2.13	1.46	4.91	0.69	3.36
C/PET/77/20	2.69	2.17	6.63	0.81	3.06

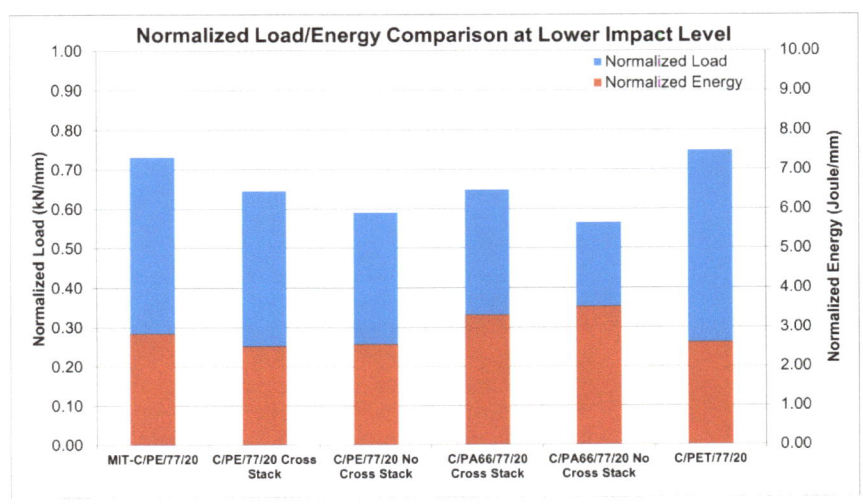

Figure 3. Comparison at normalized load and normalized energy for rCF variants at 5 J impact energy.

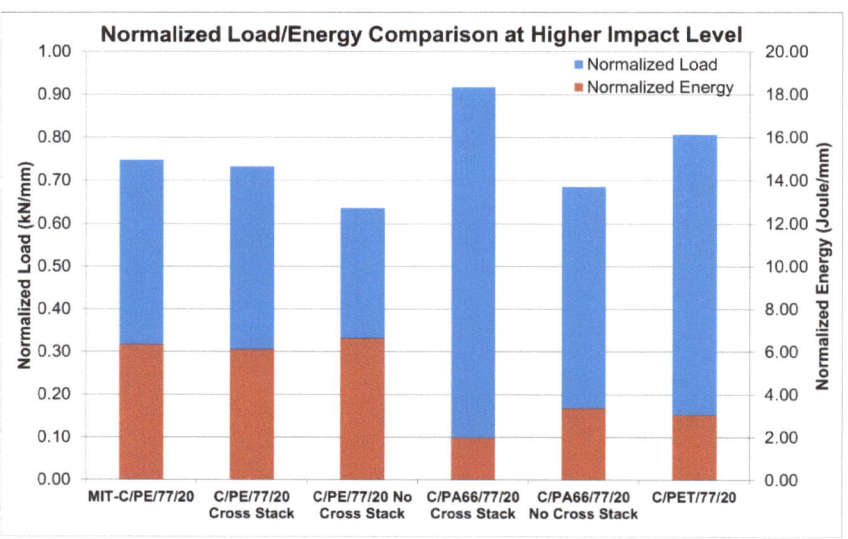

Figure 4. Comparison normalized load and normalized energy for rCF variants at 15 J impact energy.

At lower energy, the highest peak loads attained were from C/PE and C/PET, respectively. In both these systems, once the peak on the force–time curve was attained, there was penetration of the impactor through the thickness, and the unloading was, hence, sudden. While the normalized energy was highest for C/PA66, both no-stack and cross-stack compared to the rest. This suggests that C/PA66-exhibited-energy absorption occurs both in the loading and unloading phase. There is no penetration of the indenter for C/PA66.

For higher energy impact, C/PA66 and C/PET exhibited the highest normalized load bearing for the cross-stack. The highest energy absorbed was noted for all C/PE variants, regardless of no-stack or cross-stack.

The effect of stacking was less pronounced in all the impact tested samples. This may be due to localized transverse impact and only limited contact area between the impactor and the specimen. Cross-stack or no-stack is more of a function for in-plane loading. The peak load in case of drop weight impact is the onset at which the unloading phase begins. Energy absorption continues into the unloading phase for damage-tolerant materials. Overall, the PA66 offered higher damage tolerance in terms of energy absorption, for both low- and high-energy impact.

5. Discussion

Figures 5–7 provide a comprehensive visual of all tests conducted for C/PA66, C/PE and C/PET, respectively. Where applicable, the no-stack versus cross-stack has been reported. The overall tensile strength of 'roll' form of C/PA65 ranged from 217 to 248 MPa and tensile modulus of 15–20 GPa, respectively. The differences between cross-stack and no-stack are not very definitive, indicating fiber entanglement occurs in discontinuous fibers, masking the distinct effect of fiber orientation. In some cases, modulus and strength for cross-stack were lower by 12.5% compared to no-stack.

For 77 wt% C/PET (i.e., 23% CF), the highest values of strength ranged from 243 to 271 MPa and modulus 16–17 GPa for the 'sheet' form. For the 'roll' form, there was a distinct difference in the cross-stack versus no-stack, or high anisotropy. The values ranged from 128 to 170 MPa and modulus of 5–10 GPa, much lower than other variants. Further, the fiber content in these was only 15 wt%, unlike the others, which were >20 wt% carbon fiber. It may also be noted for the no-stack roll form, when a high degree of fiber orientation in the machine direction occurs, the strength and modulus are high, i.e., 256 MPa and 16 GPa, respectively.

The 77 wt% C/PE (i.e., ~20 wt% carbon fiber) exhibited the lowest values of all. For cross-stack, the average modulus was 53 MPa and average strength was 5.5 GPa, similar in both directions for cross-stack. For the no-cross stack, significant anisotropy was observed at 39 and 66 MPa and 4 and 6 GPa modulus.

Figure 8 considers a long fiber thermoplastic C/PPS with 40 wt% 25 mm (1") fiber length, which has strength of 175 MPa and modulus 25 GPa. Although this is not a one-to-one comparison, both C/PA66 and C/PET rCF mat composites have much higher modulus (by 37 %), higher than LFT C/PPS. The strength of C/PPS was 25% higher, the C/PA66 rCF mats providing the closest to the LFT values.

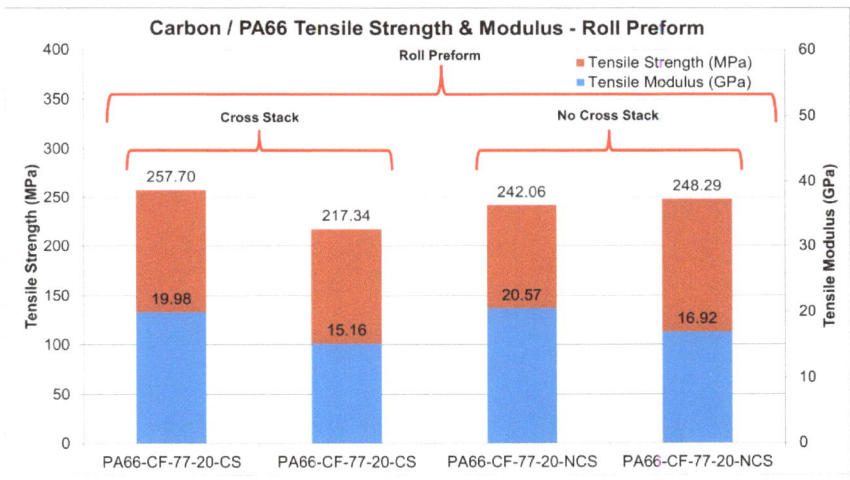

Figure 5. Comprehensive summary of tensile strength and tensile modulus for 'roll' form C/PA66.

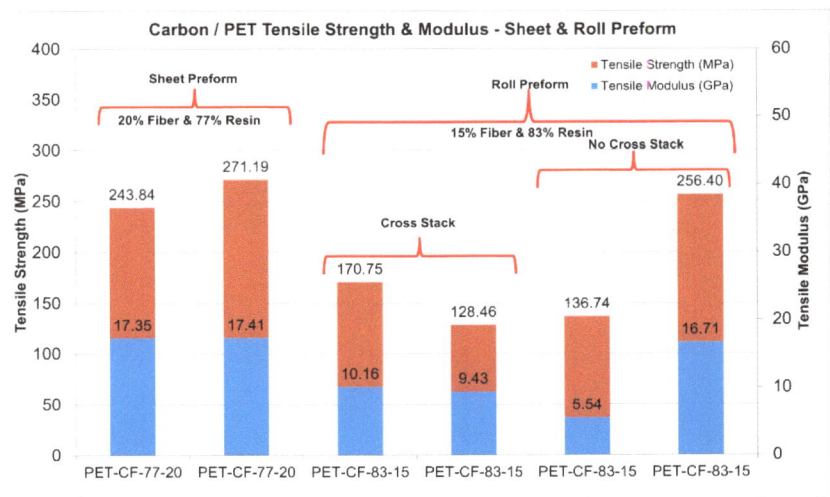

Figure 6. Comprehensive summary of tensile strength and tensile modulus for 'roll' and 'sheet' form C/PET.

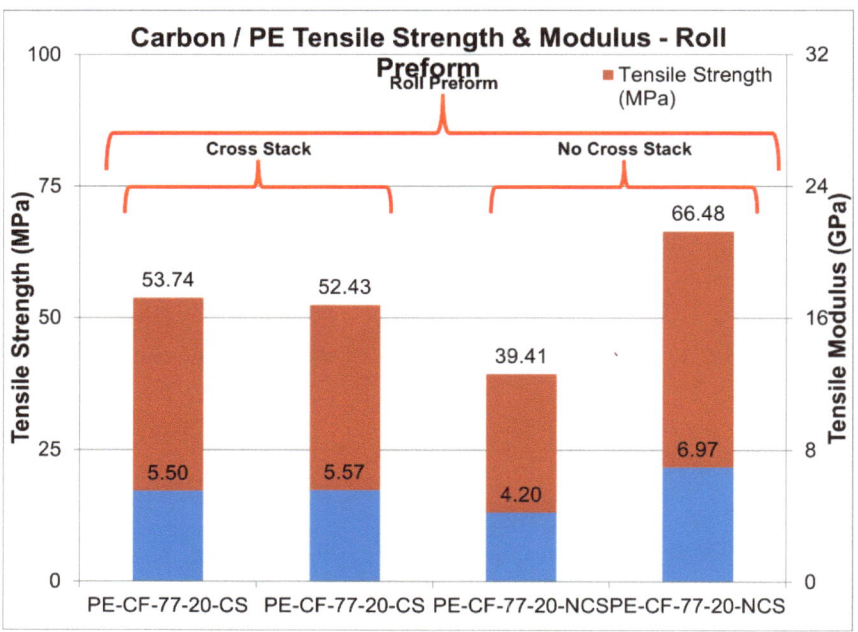

Figure 7. Comprehensive summary of tensile strength and tensile modulus for 'roll' form no stack and cross-stack—C/PE.

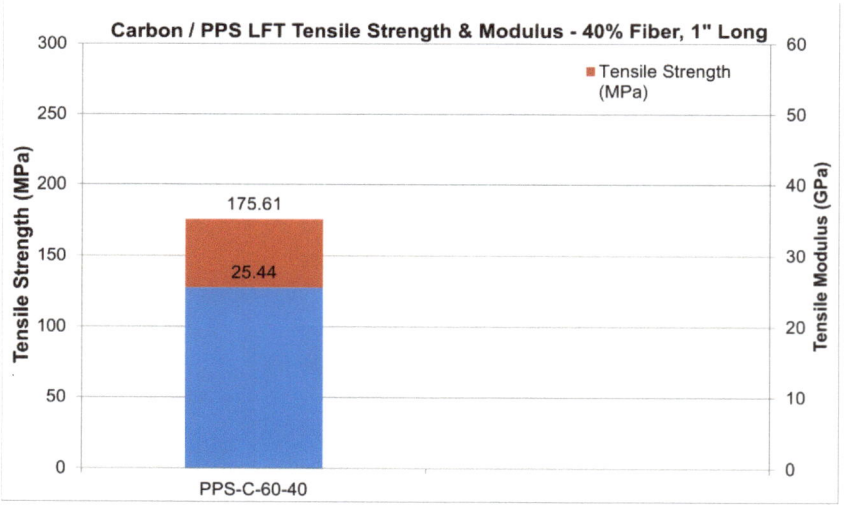

Figure 8. Benchmark tensile strength and tensile modulus for LFT C/PPS 40 wt% CF.

6. Processing Studies

Optimal processing results were obtained from panels produced with tool temperature at 500 °F. The panels processed at 250–265 F tool temperature exhibited voids, as seen in Figure 1a,b. All panels used for testing were, hence, processed at 500 °F tool temperature. Several processing routes were attempted from the rCF mats.

(a) Compression molding of the preform in matched metal tool produced composite plates. The compression molding of the PA66 rCF mats was conducted to different consolidation pressures. This helped understand process temperature–pressure–microstructure relationships. The fully consolidated panels were used for mechanical testing/data generation; (b) compression molding of C/PA66 panels followed by pre-heating the consolidated panel and subsequently subjecting the heated panel to single-diaphragm thermoform (SDF), and (c) pre-heating the C/PA66 mats without compression molding (hence, a less stiff mat) and subjecting it to SDF.

6.1. Single-Diaphragm Forming of Pre-Consolidated Panel

The purpose of this study was to evaluate the formability of the mat(s) in terms of draw. PA66/CF/78/20 was consolidated using the 300 × 300 mm^2 tool for a 2.5 mm thick panel at 500 °F and 6.895 MPa (1000 psi). The consolidated panel (blank) was then re-heated in a convection oven for approximately 5 min at 490–500 °F. There was very little sag (if any) evident. A toy car mold (250 × 100 × 125 mm^3) was used as a tool to thermoform the consolidated blank. The blank exhibited some discoloration inside the oven. The blank was unable to soften and did not reach the melt temperature without degradation. As such, one atmosphere vacuum was used to form the part. The consolidated blank failed catastrophically during forming, as seen in Figure 9.

(a)   (b)

Figure 9. Pre-consolidated blank (after heating and thermoforming). Sample—PA66/CF/78/20, 5 layers of preform; (**a**) exposed side in oven shows much yellowing; (**b**) non-exposed side shows less yellowing.

The pre-consolidated C/PA66 plate did not sag, hence, the plate was stiff when transferred from the oven to the forming station. Due to this, it appears that well-consolidated plates possess limited ability to form to shape, resulting in cracking in the C/PA66 resin. It appears that heating of the preforms must be done in a vacuum oven/inert condition to prevent discoloration (yellowing). Whether the yellowing is from moisture remains unclear. To find the cause for discoloration when the PA66/CF is heated in an oven, the preform was heated under vacuum or inert atmosphere to determine if discoloration occurs due to the presence of air, as shown in Table 15.

Table 15. Mats produced via different processing routes.

Sample Variants	Preform Type	Compression Molding	Single Diaphragm	Oven Compression Molding
Sample-PA66-CF-68-30	Sheets of 14″ × 14″	Yes	Yes	Yes
Sample-PA66-CF-78-20	Sheets of 14″ × 14″	Yes	Yes	No
Sample-PA66-CF-88-10	Sheets of 14″ × 14″	Yes	No	No
Sample-PA66-CF-77-20	Roll	Yes	No	
Sample-PE-CF-78-20	Roll	Yes	Yes	Yes

6.2. Compression Molding—External Heating (Heating the Preforms in a Convection Oven)

Two layers of PA66/CF/88/10 were placed in a convection oven at 500 °F for 5 min. A heated mold (with oil heating up to 350 °F) was used to compression mold the heated preforms. The blanks exhibited some discoloration inside the oven, see Figure 10a,b. Only the top layers of the preform became discolored due to heat. The bottom did not reach the processing temperature, nor did it discolor. Further, 1000 psi of positive pressure was applied on the tool.

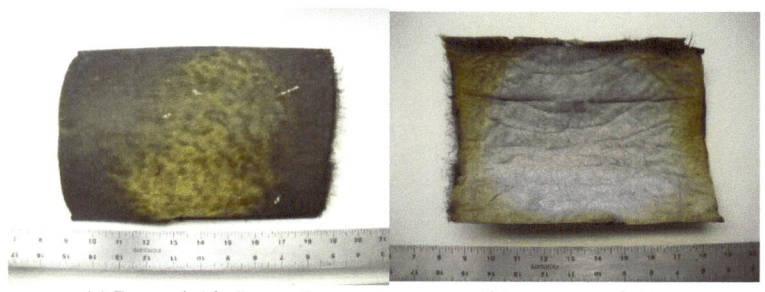

(**a**) Exposed side (in oven)   (**b**) non-exposed side

Figure 10. Heated preforms after compression molding.

6.3. Single-Diaphragm Forming—PE/CF

One layer of PE-CF-78-20 was heated in an oven at 350 F for 5–6 min. The heated preform was transferred to the mold and subjected to one atmosphere of vacuum. The material formed well without any discoloration, see Figure 11.

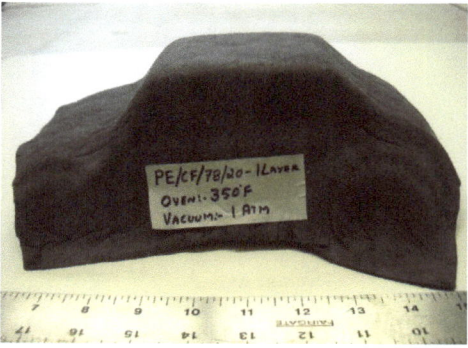

Figure 11. Forming via SDF exhibited optimal draw and consolidation.

6.4. Compression Molding—External Heating (Heating the Preforms in a Convection Oven)

Two layers of PE/CF/78/20 were heated in a convection oven at 400 °F for 10 min. An in-house heated mold (with oil heating at 250 °F) was used as a tool to compression mold the heated preforms. There was "no" discoloration of the blank inside the oven, see Figure 12a,b. Subsequently, 6.895 MPa (1000 psi) of positive pressure was applied on the tool.

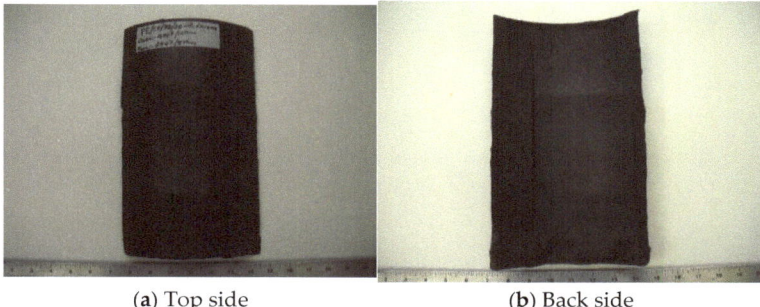

(a) Top side   (b) Back side

Figure 12. Forming of shell shape through external heating and compression molding.

6.5. Discussion on Heating the Mats

Since the mats have significant open porosity and air (before consolidation), getting the mats to attain their processing temperature is important. Hence, pre-heating brings the mats to a uniform temperature and assists with the processing. It was observed that prior to consolidation, uniform heating of the mats, either in infrared oven or via contact heating in the closed-cavity, brings the mats to a processable condition. While the mechanical properties are more a function of optimal temperature and consolidation pressure, the efficient way to get to these conditions is via pre-heating to minimize time in the press (hence, higher process efficiency).

7. Conclusions

rCF WL mats were successfully produced in three resin types—PA66, PE and PET. The processing method had significant influence on properties. The 'sheet' form exhibited random/quasi-isotropic properties while the properties in the 'roll' form were guided by the preferred fiber orientation. The tensile strength and modulus were 80–120% higher on average in the 'roll' form compared to the 'sheet' form.

The tensile strength and modulus of the 77 wt% resin, ~20 wt% fiber mats ranked as C/PA66 > C/PET > C/PE guided by the resin properties. Some variants, such as C/PE 20 wt% carbon fiber, had higher anisotropy, i.e., they were more sensitive for the cross-stack versus no-stack, while in some variants, the fiber entanglement seems to minimize the influence of fiber orientation, i.e., differences in properties in the no-stack versus cross-stack were less discernable.

The impact response of the rCF mats indicated the best performance came from C/PA66, while the energy absorbed by C/PE is assumed to be the highest, due to the weaker bonding between C and PE; as evident from the strength and modulus, this helps with energy absorption.

For PA66-CF, pre-drying was an important step as it influenced the properties by a factor of 2 or greater; pre-dried mats performed higher. The formability of pre-consolidated WL composites was poor due to high stiffness. Matched metal die provided the best forming of the WL mats for all the resin systems. The material loses heat rapidly, hence, the forming must be conducted immediately to pre-heating. The best formability was achieved in the PE-CF mats.

* Notice of Copyright: This manuscript has been authored by UT-Battelle, LLC under Contract No. DE-AC05-00OR22725 with the U.S. Department of Energy. The United States Government retains and the publisher, by accepting the article for publication, acknowledges that the United States Government retains a non-exclusive, paid-up, irrevocable, world-wide license to publish or reproduce the published form of this manuscript, or allow others to do so, for United States Government purposes. The Department of Energy will provide public access to these results of federally sponsored research in accordance

with the DOE Public Access Plan (http://energy.gov/downloads/doe-public-access-plan) (accessed on 15 June 2022).

Author Contributions: Conceptualization, M.J. and U.V.; methodology, U.V. and H.G.; validation, M.J., K.G., M.T. and U.V.; formal analysis, U.V.; investigation, U.V. and H.G.; resources, K.G., M.J. and M.T.; data curation, U.V.; writing—original draft preparation, M.J., K.G. and M.T.; writing—review and editing, supervision, U.V. and M.J.; project administration, M.J.; funding acquisition, M.J. All authors have read and agreed to the published version of the manuscript.

Funding: This study was supported by the Department of Energy for the project entitled Low Cost Carbon Fiber Composites for Lightweight Vehicle Parts, Materials Innovation Technologies, LLC, Contract #DOE-EE0004539, US DOE SBIR Phase III.

Institutional Review Board Statement: Not applicable.

Informed Consent Statement: Not applicable.

Data Availability Statement: Not applicable.

Acknowledgments: The discussions with the Institute for Advanced Composites Manufacturing Innovation (IACMI)—The Composites regarding light-weighting technologies and recycling carbon fiber materials is gratefully acknowledged. IACMI was funded in part by the Office of Energy Efficiency and Renewable Energy (EERE), U.S. Department of Energy, under Award Number DE-EE0006926.

Conflicts of Interest: The authors declare no conflict of interest.

References

1. Bledzki, A.K.; Seidlitz, H.; Goracy, K.; Urbaniak, M.; Rösch, J.J. Recycling of Carbon Fiber Reinforced Composite Polymers—Review—Part 1: Volume of Production, Recycling Technologies, Legislative Aspects. *Polymers* **2021**, *13*, 300. [CrossRef] [PubMed]
2. Bledzki, A.K.; Seidlitz, H.; Krenz, J.; Goracy, K.; Urbaniak, M.; Rösch, J.J. Recycling of Carbon Fiber Reinforced Composite Polymers—Review—Part 2: Recovery and Application of Recycled Carbon Fibers. *Polymers* **2020**, *12*, 3003. [CrossRef] [PubMed]
3. BMW i Production CFRP Wackersdorf: Carbon Fiber Recycling Material for the Use in the BMW i3, e.g., the Roof of the BMW i3. 2013. Available online: www.press.bmwgroup.com/global/photo/detail/P90125888 (accessed on 20 May 2020).
4. The State of Recycled Carbon Fiber. 2019. Available online: www.compositesworld.com/articles/the-state-of-recycled-carbon-fiber (accessed on 26 July 2020).
5. Technology Roadmap, IACMI-The Composites Institute. Available online: www.iacmi.org (accessed on 22 April 2022).
6. Closing the Loop, Carbon Conversions. Available online: https://carbonconversions.com/ (accessed on 17 June 2022).
7. Wöllinga, J.; Schmiega, M.; Manisa, F.; Drechslerb, K. Nonwovens from recycled carbon fibres—Comparison of processing technologies. In Proceedings of the 1st Cirp Conference on Composite Materials Parts Manufacturing, Procedia CIRP, Karlsruhe, Germany, 8–9 June 2017; Volume 66, pp. 271–276. [CrossRef]
8. Song, Y.; Gandhi, U.; Sekito, T.; Vaidya, U.K.; Hsu, J.; Yang, A.; Osswald, T. A Novel CAE Method for Compression Molding Simulation of Carbon Fiber-Reinforced Thermoplastic Composite Sheet Materials. *J. Compos. Sci.* **2018**, *2*, 33. [CrossRef]
9. Selezneva, M.; Lessard, L. Characterization of mechanical properties of randomly oriented strand thermoplastic composites. *J. Compos. Mater.* **2016**, *50*, 2833–2851. [CrossRef]
10. Thomason, J. The influence of fibre length and concentration on the properties of glass fibre reinforced polypropylene: 5. Injection moulded long and short fibre PP. *Compos. Part A Appl. Sci. Manuf.* **2002**, *33*, 1641–1652. [CrossRef]
11. Vaidya, U.K. *Composites for Automotive, Mass Transit and Heavy Truck*; DesTech Publishers: Lancaster, PA, USA, 2010.
12. Stoeffler, K.; Andjelic, S.; Legros, N.; Roberge, J.; Schougaard, S.B. Polyphenylene sulfide (PPS) composites reinforced with recycled carbon fiber. *Compos. Sci. Technol.* **2013**, *84*, 65–71. [CrossRef]
13. Borkar, A.; Hendlmeier, A.; Simon, Z.; Randall, J.D.; Stojcevski, F.; Henderson, L.C. A comparison of mechanical properties of recycled high-density polyethylene/waste carbon fiber via injection molding and 3D printing. *Polym. Compos.* **2022**, *43*, 2408–2418. [CrossRef]
14. Baek, Y.-M.; Shin, P.-S.; Kim, J.-H.; Park, H.-S.; Kwon, D.-J.; DeVries, K.L.; Park, J.-M. Investigation of Interfacial and Mechanical Properties of Various Thermally-Recycled Carbon Fibers/Recycled PET Composites. *Fibers Polym.* **2018**, *19*, 1767–1775. [CrossRef]
15. Hemamalini, T.; Dev, V.R.G. Wet Laying Nonwoven Using Natural Cellulosic Fibers and Their Blends: Process and Technical Applications. A Review. *J. Nat. Fibers* **2019**, *18*, 1823–1833. [CrossRef]
16. Melani, L.; Kim, H.-J. The surface softness and mechanical properties of wood pulp–lyocell wet-laid nonwoven fabric. *J. Text. Inst.* **2020**, *112*, 1191–1198. [CrossRef]
17. Ghossein, H.; Hassen, A.A.; Kim, S.; Ault, J.; Vaidya, U.K. Characterization of Mechanical Performance of Composites Fabricated Using Innovative Carbon Fiber Wet-laid Process. *J. Compos. Sci.* **2020**, *4*, 124. [CrossRef]

18. Yan, X.; Wang, X.; Yang, J.; Zhao, G. Optimization of process parameters of recycled carbon fiber-reinforced thermoplastic prepared by the wet-laid hybrid nonwoven process. *Text. Res. J.* **2021**, *91*, 1565–1577. [CrossRef]
19. Ghossein, H.K. Novel Wet-Laid Nonwoven Carbon Fiber Mats and Their Composites. Ph.D. Thesis, University of Tennessee, Knoxville, TN, USA, 2018.
20. Barnett, P.R.; Young, S.A.; Chawla, V.; Foster, D.M.; Penumadu, D. Thermo-mechanical characterization of discontinuous recycled/repurposed carbon fiber reinforced thermoplastic organosheet composites. *J. Compos. Mater.* **2021**, *55*, 3409–3423. [CrossRef]
21. Erland, S.; Stevens, H.; Savage, L. The re-manufacture and repairability of poly(ether ether ketone) discontinuous carbon fibre composites. *Polym. Int.* **2021**, *70*, 1118–1127. [CrossRef]
22. Brahma, S.; Pillay, S.; Ning, H. Comparison and characterization of discontinuous carbon fiber liquid-molded nylon to hydroentanglement/compression-molded composites. *J. Thermoplast. Compos. Mater.* **2020**, *33*, 1078–1093. [CrossRef]
23. Yeole, P.; Hassen, A.A.; Vaidya, U.K. The effect of flocculent and dispersants on wet–laid process for recycled glass fiber/PA6 composite. *Polym. Polym. Compos.* **2018**, *26*, 259–269. [CrossRef]
24. Kore, S.; Spencer, R.; Ghossein, H.; Slaven, L.; Knight, D.; Unser, J.; Vaidya, U. Performance of hybridized bamboo-carbon fiber reinforced polypropylene composites processed using wet-laid technique. *Compos. Part C Open Access* **2021**, *6*, 100185. [CrossRef]

Article

Static and Vibration Analyses of a Composite CFRP Robot Manipulator

Mohammad Amir Khozeimeh, Reza Fotouhi * and Reza Moazed

Department of Mechanical Engineering, University of Saskatchewan, Saskatoon, SK S7N 5A9, Canada; mok101@mail.usask.ca (M.A.K.); rem980@mail.usask.ca (R.M.)
* Correspondence: reza.fotouhi@usask.ca

Abstract: This paper reports analyses of a 5-degrees-of-freedom (5-DOF) carbon fiber-reinforced polymer (CFRP) robot manipulator, which has been developed for farm applications. The manipulator was made of aluminum alloy (AA) and steel materials. However, to check the effectiveness of CFRP materials on the static and free-vibration performance of the manipulator, the AA parts were replaced with CFRP. For this purpose, the effects of various cross-sections and layups on three design criteria—deflection, load-carrying capacity, and natural frequency—were investigated. Two types of thin-walled laminated sections, specifically the I section and rectangular tubular sections, were used for the composite parts. These parts were made from three hollow square section ("SSS" section) beams and three I section ("III" section) beams. These multi-cell beams were modeled using the finite element (FE) method. Three configurations were selected for analysis based on the manipulator's most common operating conditions. The results indicated that the use of CFRP increased the manipulator's natural frequencies, increased the load-carrying capacity, and decreased the manipulator's tip deflection when compared with its AA counterpart. An analysis showed that using CFRP in the manipulator's structure could improve static and vibrational performances. It was observed that the "SSS" section beams were 1.17 times stiffer, could carry a 1.20 times higher load, and were 1.40 times heavier than the "III" section beams. Also, decreasing the fiber direction in angle-ply layups from 90° to 0° and adding 0° plies, while keeping the total number of layers constant, decreased the manipulator's tip deflection and increased its natural frequencies.

Keywords: composite beams; finite element analysis; free vibration; glass carbon fiber

Citation: Khozeimeh, M.A.; Fotouhi, R.; Moazed, R. Static and Vibration Analyses of a Composite CFRP Robot Manipulator. *J. Compos. Sci.* **2022**, *6*, 196. https://doi.org/10.3390/jcs6070196

Academic Editor: Jiadeng Zhu

Received: 6 June 2022
Accepted: 29 June 2022
Published: 4 July 2022

Publisher's Note: MDPI stays neutral with regard to jurisdictional claims in published maps and institutional affiliations.

Copyright: © 2022 by the authors. Licensee MDPI, Basel, Switzerland. This article is an open access article distributed under the terms and conditions of the Creative Commons Attribution (CC BY) license (https://creativecommons.org/licenses/by/4.0/).

1. Introduction

Thin-walled structures can be made of steel, aluminum alloy (AA) [1,2], and composites [3–5]; AA and steel beams can also be reinforced by composite materials [6,7]. Among the wide area of potential applications of thin-walled structures, composite structures are used in the automotive [8,9], aerospace [10], and robotic industries [11]. Many factors, including the material properties, the cross-sectional shape, and the loading conditions, could affect the static and free-vibration performance of these structures [12,13]. Using composite materials with a lower density, higher specific stiffness, and higher specific strength in a structure over conventional materials could improve these parameters [14,15]. Samal et al. [16] studied the effect of fiber orientation on the free-vibration performance of glass fiber-reinforced polymer (GFRP) beams. The authors showed that, as the fiber orientation increased from 0° to 90°, the natural frequencies decreased while the damping ratios increased. Ding et al. [17] investigated the effect of a square cross-section and fiber orientation on the load-carrying capacity of carbon fiber-reinforced polymer (CFRP) beams. The results showed that the number of 0° and +45° plies could affect the ultimate load-carrying capacity of CFRP box girders when the total number of layers was the same. The load-carrying capacity of composite beams can be determined using two widely accepted failure criteria: Tsai–Wu and Tsai–Hill [18]. Gliszczyski et al. [19] used the Tsai–Wu criterion to estimate the load-carrying capacity of a GFRP beam under pure bending, while considering the effect of different layups. The results indicated that the ply scheme with 50 percent 0° plies (e.g., $[90/0/90/0]_S$ and $[0/90/0/90]_S$) had the highest failure load values. Debski et al. [20] experimentally estimated the failure load of CFRP channel columns with different ply schemes under compression loading. The study showed that columns had a higher failure load when the 0° plies were located in the outer surface of the channel.

In this paper, thin-walled laminated composite beams were used to modify the structure of a 5-DOF (degrees-of-freedom) robot manipulator [21]. This robot was intended to be used for a crop-monitoring application (phenotyping) and was made of AA and steel materials; as a result, it was relatively heavy and required powerful motors for its operation. In farm applications, the deflection and vibration of a manipulator's end effector are very important, as the manipulator carries several sensors that measure important crop traits. Using lightweight materials such as CFRP could reduce the manipulator's mass and possibly increase its overall stiffness; this would increase the manipulator's natural frequencies, which is important for its application. Parts made of CFRP could reduce the structural mass up to 70% compared with steel materials [14]. Prior research by Wyatt et al. [22] showed that a CFRP robot hand had a lower mass compared with steel and AA hands by about 71% and 41%, respectively. Hagenah et al. [23] used titanium and AA for a manipulator structure, which resulted in a balanced mass distribution and a desirable stiffness. Yin et al. [24] developed a hybrid manipulator structure using CFRP and AA. The study showed that the hybrid structure had better energy conservation, a better driving ability, and a higher natural frequency than AA and steel structures. Lee et al. [25] created a manipulator wrist from composite materials instead of AA to enhance its efficiency. In addition, several researchers have used CFRP to improve parameters such as deflection [26], load-carrying capacity, and natural frequencies [27,28] in robot structures.

In Section 2 of this paper, the 5-DOF manipulator is introduced. In Section 3, the results from [17] are used to evaluate the finite element analysis (FEA). Section 4 examines the composite's mechanical properties and defines the composite's structure using FEA software. Section 5 provides an analysis of the effects of different layups on the beam's deflection. In Section 6, the beam's load-carrying capacity is analyzed using the Tsai–Wu (TW) and Tsai–Hill (TH) [18] failure criteria. In Section 7, the effects of different layups on the composite beam's natural frequency are investigated. In Sections 8 and 9, the selected layup is used to model the composite robot parts and the robot's natural frequencies for AA and CFRP are compared.

2. Robot Manipulator's Composite Parts

The robot under consideration was a 5-DOF manipulator that was designed, built, and tested at the University of Saskatchewan for plant monitoring applications [21]. In Figure 1c, the robot is shown partially as it is being patented. This manipulator had four rotational joints and one prismatic joint. The robot's top part was created from two links connected with a prismatic joint. Each link was about 1.2 m long and the robot's fully extended configuration could reach up to about 3 m. Each of the two top links was made of three AA beams with a length of 1.2 m and a rectangular cross-section of 40 mm × 60 mm. The resulting multi-cell beam had dimensions of 60 mm × 120 mm. The robot's end effector carried several sensors with a total payload mass capacity of about 20 kg. The robot's vibration and tip (end effector) displacement affected the sensors' measurement performance in terms of the photo quality and the collected data. To obtain a desired tip deflection and acceptable natural frequencies for vibration, some of the robot's AA parts were replaced with composite parts. Two types of thin-walled laminated composite beams, specifically an I-beam and a rectangular tube, were used in the model. These parts were used to form three combined squared tubular sections and three combined I sections, referred to as "SSS" and "III", respectively. These multi-cell beams were modeled with FEA using eight layups with CFRP materials. Figure 1a,b show the schematics of the "SSS" and "III" beams.

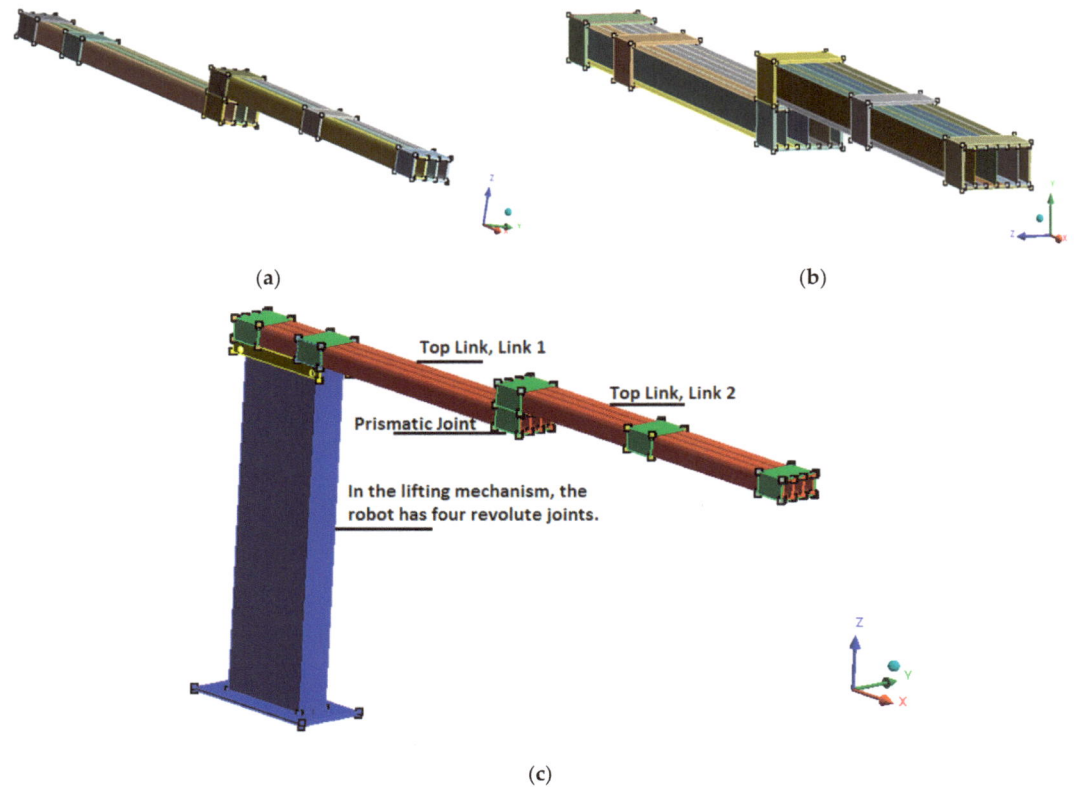

Figure 1. Schematic of the 5-DOF robot manipulator and its parts: (**a**) composite beams with three squared tubular sections (SSS) for the upper link of the manipulator, (**b**) composite beams with three I sections (III) for the upper link of the manipulator, and (**c**) the 5-DOF robot manipulator.

It should be noted that, while a nuts-and-bolts connection is ideal for connecting AA beams, for the assembly of composite CFRP beams this type of connection may degrade the beams' quality and cause structural damage. Therefore, glue and AA clamps were used to assemble the CFRP beams. For simplification and modeling, other components such as motors were not included in the FEA.

3. Verifying the FE Model

The FE model results were verified and compared with the experimental results reported in [17]. The authors of [17] did a very good job of testing several CFRP beams and reporting their results. Since we do not currently have the facilities to conduct similar tests, we decided to rely on the academically reported and peer-reviewed results from [17]. In the near future, we plan to procure experimental facilities and conduct similar static tests, and add vibrational tests as well. For the vibrational tests, we plan to use several IMUs (inertia measurement units) to measure the accelerations in different locations of the manipulator that are induced by an impact or by continuous excitation signals to the base of the manipulator. Such a vibration test would reveal the natural frequencies and mode shapes of the manipulator. The experimental results reported in [17] are related to a three-point bending test on a composite beam. The strain values were measured using a strain gauge on a specimen with a symmetric layup created from eight 0° plies and two 90° plies ($[0,0,0,0,90]_S$). To compare with the experimental results, a hollow square composite beam with dimensions of 40 × 60 mm and a length of 280 mm was modeled. The beam was created from ten layers with a total thickness of 2 mm. Figure 2 shows the three-point bending test schematic setup and the detailed mesh of the CFRP box with its supports.

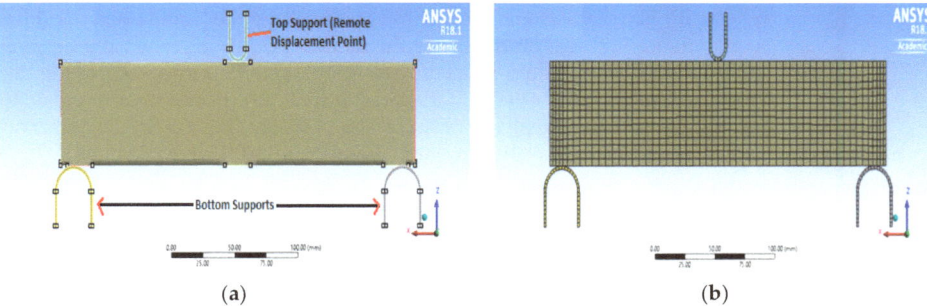

Figure 2. Three-point bending test schematic setup and FEA details: (**a**) the three-point bending test schematic setup, modeled in FEA, and (**b**) schematic of the meshed CFRP box beam in FEA.

Numerical modelling of the composite beams was performed using the commercial finite element analysis software ANSYS (FEA software, Canonsburg, PA, USA) [29]. Solid elements were used to mesh the composite beams. Specifically, Solid185 was selected for meshing the CFRP beam. This element is defined by eight nodes with three degrees of freedom at each node, specifically translations in the nodal x-, y-, and z-directions. The parts of the beam in contact with the supports were selected as the target surfaces and the top face of the supports was chosen as the contact surface. Conta174 and Targe170 were used for meshing the contact and target face. The connections between the supports and the beam were considered to be frictionless. A mesh convergence study was performed to verify the accuracy of the FEA, and the results are shown in Table A9. A remote point was defined at the top support to apply the load. Defining remote points in the selected parts allowed for the application of loading and displacement. The reference points inside the two bottom supports were fully constrained to limit their freedom in any direction. The load values in Table 1 were taken from the literature [17] and the material properties are given in Table A7. From Table 1, it can be seen that there is a good agreement between the FEA results and the experimental results from [17].

Table 1. Comparison of the FEA results with the experimental results [17].

Load (N)	Experimental Strain (10^{-6}) (Specimen D) [17]	FEA Strain (10^{-6})	Difference (%)
435	15	16	6.25
822	44	46	4.35
1225	74	75	1.33
2430	176	180	2.22
2825	181	188	3.72
3226	220	229	3.93
3625	248	235	−5.53
4024	269	285	5.61

4. Effect of Layup on Mechanical Properties of Composite Parts

As mentioned in the Introduction, eight layups were selected to evaluate the effectiveness of CFRP beams in terms of the robot static and free-vibration performance. For this purpose, the effects of the layup on the mechanical properties of the CFRP beams were investigated by following the approaches of Moazed et al. [11] and Ding et al. [17]. These references investigated the role of the layup in the design of CFRP beams to meet high bending resistance and lightweight design criteria. In these studies, the effect of increasing the fiber direction in the angle-ply layups (L1 to L3 layups) [11] and adding plies with fibers at 45° and 90° (L4 to L8 layups) on the beam's static and free-vibration performance [17] was studied. For the mentioned 5-DOF robot, the top link's bending deflection had the greatest effect on the sensors' stability; thus, calculating the equivalent Young's elastic modulus (E_x) in the longitudinal direction was of interest. To perform a more comprehensive analysis, the equivalent shear modulus (G_{xy}) was also calculated. To calculate the elastic and shear moduli, Equations (1)–(3) were used [18].

$$\begin{bmatrix} a & b \\ b^T & d \end{bmatrix} = \begin{bmatrix} A & B \\ B & D \end{bmatrix}^{-1} \quad (1)$$

$$E_x = \frac{\begin{vmatrix} A_{11} & A_{12} & A_{16} & B_{11} & B_{12} & B_{16} \\ A_{12} & A_{22} & A_{26} & B_{12} & B_{22} & B_{26} \\ A_{16} & A_{26} & A_{66} & B_{16} & B_{26} & B_{66} \\ B_{11} & B_{12} & B_{16} & D_{11} & D_{12} & D_{16} \\ B_{12} & B_{22} & B_{26} & D_{12} & D_{22} & D_{26} \\ B_{16} & B_{26} & B_{66} & D_{16} & D_{26} & D_{66} \end{vmatrix}}{\begin{vmatrix} A_{22} & A_{26} & B_{12} & B_{22} & B_{26} \\ A_{26} & A_{66} & B_{16} & B_{26} & B_{66} \\ B_{12} & B_{16} & D_{11} & D_{12} & D_{16} \\ B_{22} & B_{26} & D_{12} & D_{22} & D_{26} \\ B_{26} & B_{66} & D_{16} & D_{26} & D_{66} \end{vmatrix}} \frac{1}{n.t} \quad (2)$$

$$G_{xy} = \frac{\begin{vmatrix} A_{11} & A_{12} & A_{16} & B_{11} & B_{12} & B_{16} \\ A_{12} & A_{22} & A_{26} & B_{12} & B_{22} & B_{26} \\ A_{16} & A_{26} & A_{66} & B_{16} & B_{26} & B_{66} \\ B_{11} & B_{12} & B_{16} & D_{11} & D_{12} & D_{16} \\ B_{12} & B_{22} & B_{26} & D_{12} & D_{22} & D_{26} \\ B_{16} & B_{26} & B_{66} & D_{16} & D_{26} & D_{66} \end{vmatrix}}{\begin{vmatrix} A_{11} & A_{12} & B_{11} & B_{12} & B_{16} \\ A_{12} & A_{22} & B_{12} & B_{22} & B_{26} \\ B_{11} & B_{12} & D_{11} & D_{12} & D_{16} \\ B_{12} & B_{22} & D_{12} & D_{22} & D_{26} \\ B_{16} & B_{26} & D_{16} & D_{26} & D_{66} \end{vmatrix}} \frac{1}{n.t} \quad (3)$$

where [A], [B], and [D] are the laminate extensional, coupling, and bending stiffness matrices, respectively, while [a], [b], and [d] refer to the laminate extensional, coupling, and bending compliance matrices, respectively. E_x is the laminate equivalent of Young's modulus in the longitudinal direction, G_{xy} is the equivalent shear modulus, n is the number of layers, and t is the layer thickness. As mentioned in Section 2, square tubular sections and I sections were used to model the robot's top links. The beam's equivalent bending stiffness EI_{yy} (and torsional stiffnesses, GI_t) was calculated using Equations (4) and (5) for the square beam and I-beam, respectively [14].

$$EI_{yy} = \frac{b_f}{(a_{11})_f} \frac{d^2}{2} + \frac{2b_f}{(d_{11})_f} + \frac{2b_w^3}{12(a_{11})_w}, \quad GI_t = \frac{2d_f^2 d^2}{(a_{66})_f d_f + (a_{66})_w d} \quad (4)$$

$$EI_{yy} = \frac{b_f}{(a_{11})_f} \frac{d^2}{2} + \frac{2b_f}{(d_{11})_f} + \frac{2b_w^3}{12(a_{11})_w} \quad (5)$$

where EI_{yy} is the equivalent bending stiffness about the Y-axis, a_{11} is the laminate extensional compliance, a_{66} is the laminate shear compliance, and d_{11} is the laminate bending compliance, as shown in Equations (4) and (5). Figure 3 shows the geometrical details for the square and I-beams and a laminate schematic depicting the individual plies for L8 = $[0_2/\pm 45/90]_s$.

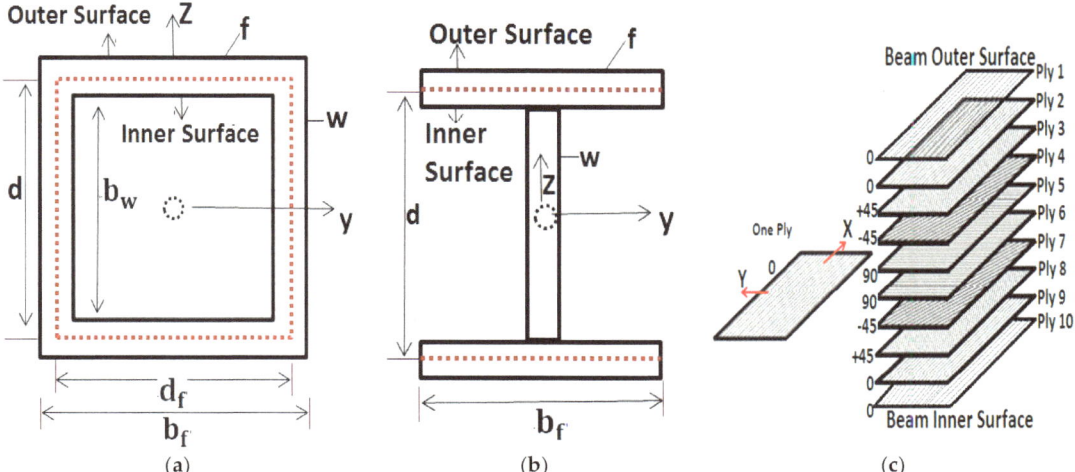

Figure 3. Beam sections: (**a**) square beam, (**b**) I-beam, and (**c**) laminate schematic of the plies for L8 = $[0_2/\pm 45/90]_s$.

Table 2 displays the E_x, G_{xy}, and EI_{yy} values for the CFRP beams with various layups, as well as for an AA beam. As shown in this table, the elastic module was (E_x = 71.0 GPa) for the AA beam and (G_{xy} = 20.7 GPa) for the shear model. The CFRP beams with the L1, L2, L4, L5, L7, and L8 layups had higher E_x values than the AA beam, and the beams with the L3 and L6 layups had lower E_x values than the AA beam. The CFRP beams with the L3 and L6 layups had a higher G_{xy} than the AA beam and the beams with the other layups had lower shear moduli than the AA beam. Overall, the results indicated that increasing the fiber direction in an angle-ply layup and adding layers in fiber directions other than zero decreases the equivalent bending stiffness. Equations (1)–(5) and the parameter values used here were adopted from references [14,18].

Table 2. Mechanical properties for CFRP and AA beams with square and I-shaped cross-sections.

Layup	E_x (GPa)	G_{xy} (GPa)	EI_{yy}^{Square} (N m^2)	EI_{yy}^{I-beam} (N m^2)
L1 = $[0,0,0,0,0]_s$	142.0	4.60	27,427.58	23,271.33
L2 = $[20,-20]_5$	90.12	17.43	17,408.40	14,770.42
L3 = $[30,-30]_5$	46.17	27.90	8919.26	7567.68
L4 = $[0_4/\pm 45/0_4]$	117.7	10.84	22,740.67	19,294.84
L5 = $[0_2/\pm 45/0]_s$	92.71	17.37	17,909.04	15,195.38

Table 2. Cont.

Layup	E_x (GPa)	G_{xy} (GPa)	EI_{yy}^{Square} (N m^2)	EI_{yy}^{I-beam} (N m^2)
L6 = [±45/0$_2$/ ± 45/0$_2$/ ± 45]	67.36	23.42	13,010.45	11,038.71
L7 = [0, 0, 0, 0, 90]$_s$	116.0	4.60	22,391.39	18,998.51
L8 = [0$_2$/ ± 45/90]$_s$	71.60	17.37	13,845.24	11,747.56
AA	71.00	20.69	13,713.13	11,635.69

5. Effect of Layups on Beam Deflection

In this section, the effect of the layups on upper link deflection was determined using FE modelling. For verification purposes, the results obtained from FEA were compared with the analytical solution results. The FE model geometry, applied loading, and boundary conditions are shown in Figure 4. The cantilever beams shown in this figure were 1.2 m long and had cross-sectional dimensions of 120 mm × 60 mm. A concentrated load of 450 N in the z-direction was applied at the free end of the cantilever. The elements adopted were Solid185 and Solid186. These elements had three degrees of freedom at each node, specifically translations in the nodal x-, y-, and z-directions, and were either linear, i.e., defined by eight nodes (Solid-185), or were a higher order (i.e., quadratic) defined by twenty nodes (Solid-186). As a default in FE software, the CFRP beams were meshed with Solid185 elements, while the AA beams were meshed with Solid186 elements. The AA and CFRP material properties, as defined in Tables A7 and A8, were assigned to these FE models.

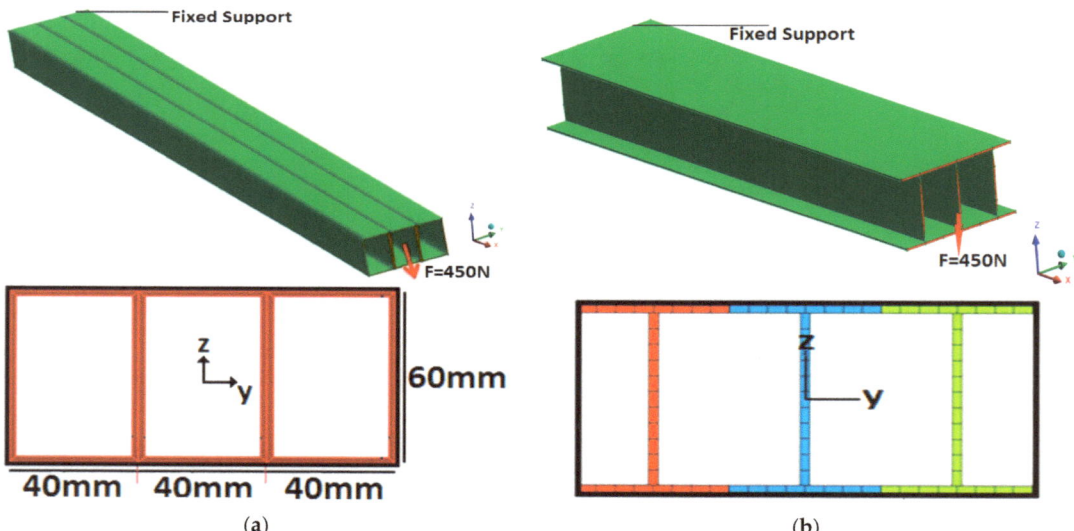

Figure 4. Finite element model geometry, applied loading, and boundary conditions: (**a**) "SSS" beam schematic, and (**b**) "III" beam schematic.

The beam's maximum deflection was calculated from Equation (6), for which the EI_{yy} values are given in Table 2.

$$\delta_b = \frac{PL^3}{3EI_{yy}} \quad (6)$$

where δ_b is the maximum bending deflection in the z-direction, P is the applied force, L is the beam length, and EI_{yy} is the equivalent bending stiffness. Table 3 shows the beam tip deflection for the different layups as calculated analytically and using FEA.

Table 3. Comparison between analytical (Equation (6)) and FEA-calculated tip deflection for the "SSS" and "III" beam.

Layup	"SSS" Beam			"III" Beam		
	δ_b^{FEA} (mm)	$\delta_b^{(6)}$ (mm)	Difference%	δ_b^{FEA} (mm)	$\delta_b^{(6)}$ (mm)	Difference%
L1	3.13	3.15	0.63	3.66	3.71	1.36
L2	4.93	4.96	0.60	5.81	5.84	0.51
L3	9.72	9.68	0.41	11.5	11.4	0.86
L4	3.91	3.79	3.06	4.59	4.47	2.61
L5	4.89	4.82	1.43	5.75	5.68	1.21
L6	6.67	6.64	0.44	7.85	7.82	0.38
L7	4.08	3.85	5.63	4.81	4.54	5.61
L8	6.31	6.26	0.79	7.40	7.37	0.40
AA	6.34	6.30	0.63	7.33	7.42	1.22

The AA "SSS" beam had a deflection (δ_b^{FEA}) of 6.34 mm and the CFRP "SSS" beams with the L1, L2, L4, L5, L7, and L8 layups had deflections lower than the AA beam. These deflections were 3.13, 4.93, 3.91, 4.89, 4.08, and 6.31 mm, respectively, as shown in Table 3. The CFRP beams with L3 and L6 layups had higher deflections than the AA beam. More specifically, the L1 layup, with ten 0° plies, as well as the L4 and L7 layups, with eight 0° plies, had the lowest deflections. The L3 and L6 layups, with ten 30° plies and six 45° plies, respectively, had the highest deflections, showing that adding layers with fibers in directions other than 0° will increase beam deflection. Similarly, the same trends were noted for the CFRP "III" beams. For the AA "III" beam, the tip deflection was 7.33 mm, versus 3.66, 5.81, 4.59, 5.75, and 4.81 mm for the L1, L2, L4, L5, and L7 layups, respectively, as shown in Table 3. Equation (7) was used to calculate the differences in Table 3:

$$\text{Difference\%} = \left| \frac{\delta_b^{FEA} - \delta_b^{(6)}}{\delta_b^{FEA}} \times 100 \right| \quad (7)$$

The values for the difference percentages provided in Table 3 ranged from 0% to 5.63%. These differences were due to the fact that the analytical solution did not take into account the effects of shear and warping deformations. As mentioned in Section 4, the square beam was 1.17 times stiffer than the I-beam, and Equation (8) was used to express the relationship between the "SSS" and "III" beams. For example, the relationship between the beams in the L1 layup was as follows:

$$\left(\frac{EI_{SSS}}{EI_{III}}\right)_{\text{Analytical}} = \left(\frac{\delta_{III}}{\delta_{SSS}}\right)_{\text{FEA}} \rightarrow \frac{82282.5}{69813.9} = \frac{3.66 \text{ mm}}{3.13 \text{ mm}} \approx 1.17 \quad (8)$$

where EI_{SSS} and δ_{SSS} are the equivalent bending stiffness and deflection values for the "SSS" beam with the L1 layup, respectively, and EI_{III} and δ_{III} are the equivalent bending stiffness and deflection values for the "III" beam with the L1 layup, respectively. In addition to the comparisons between the deflection of the AA and CFRP beams, the AA "SSS" and "III" beams had masses of 3.81 kg and 2.73 kg, whereas the CFRP "SSS" and "III" beams had masses of 2.19 kg and 1.56 kg, respectively. These results show that the "SSS" beams were 1.17 times stiffer and 1.40 times heavier than the "III" beams. The results presented in this section indicate that using CFRP beams in a robot's upper link structure could decrease the tip deflection and mass of the structure. See the discussions before Equation (7).

6. Failure Index

The aim here was to determine the first ply failure of the CFRP beams, specifically under bending. When bending is applied to a CFRP beam, all three stress components (tensile, compressive, and shear) are induced (see Appendix A.2). The allowable strength (or failure stress) is the combined effect of these stress components. Therefore, it was assumed that failure would happen in the CFRP beams when the first ply reached the allowable strength. Two failure criteria for fiber-reinforced materials, specifically Tsai–Hill and Tsai–Wu, were used to determine the failure indices [18].

In order to use a failure criterion, it is necessary to calculate the stress components in the fiber direction or a local coordinate system (1, 2), as shown in Figure 5. For this purpose, the analytical method from [14] was adopted; the details of the stress analysis are presented in Appendix A.1. Also, for verification purposes, the results obtained analytically here were compared with the FEA results. This comparison is reported in Appendix A.2.

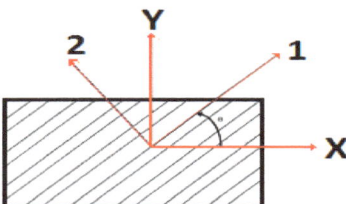

Figure 5. Local and global axes of an angle lamina.

The first failure criterion applied was Tsai–Hill, which was developed based on von Mises–Henky's distortion energy theory. The Tsai–Hill failure criterion for composites is as follows [18]:

$$\frac{\sigma_1^2}{F_1^2} + \frac{\sigma_2^2}{F_2^2} - \frac{(\sigma_1 \sigma_2)}{F_1^2} + \frac{\sigma_{12}^2}{F_6^2} = FI \qquad (9)$$

where FI is the failure index, σ_1 is the ply's longitudinal stress in the fiber direction (1 in Figure 5), σ_2 is the ply's transverse stress (2 in Figure 5), and σ_{12} is the ply's shear stress in local coordinates (1, 2) as shown in Figure 5. Here, F_1 and F_2 refer to the tensile yield strengths in directions 1 and 2, respectively, and F_6 is the shear strength. To account for the difference between the tensile and compressive strengths, the modified Tsai–Hill approach was proposed, where F_{1t}, F_{1c}, F_{2t}, and F_{2c} are the longitudinal tensile and compressive strengths, respectively:

$$F_1 = \begin{cases} F_{1t} \text{ when } \sigma_1 > 0 \\ F_{1c} \text{ when } \sigma_1 < 0 \end{cases} \qquad (10)$$

$$F_2 = \begin{cases} F_{2t} \text{ when } \sigma_2 > 0 \\ F_{2c} \text{ when } \sigma_2 < 0 \end{cases} \qquad (11)$$

An alternative approach was to use the Tsai–Wu criterion as expressed below:

$$f_{11} \sigma_1^2 + f_{22} \sigma_2^2 + 2 f_{12} (\sigma_1 \sigma_2) + f_{66} \sigma_{12}^2 + f_1 \sigma_1 + f_2 \sigma_2 = FI \qquad (12)$$

where $f_1 = \frac{1}{F_{1t}} - \frac{1}{F_{1c}}$, $f_2 = \frac{1}{F_{2t}} - \frac{1}{F_{2c}}$, $f_{11} = \frac{1}{F_{1t} F_{1c}}$, $f_{22} = \frac{1}{F_{2t} F_{2c}}$, $f_{12} = -0.5 \sqrt{f_{11} f_{22}}$, and $f_{66} = \frac{1}{F_6^2}$ are the coefficients obtained from the shearing and uniaxial strength tests [18]. Here, f_{12} is a coefficient which represents a biaxial composite beam strength in the longitudinal and transverse directions. The main difference between these criteria is that the Tsai–Wu criterion accounts for the interactions and differences between stresses, but requires biaxial testing to define f_{12}. Using the failure index defined in Equations (9) and (12), a safety ratio (SR) was defined as shown below:

$$SR = \frac{1}{FI} \qquad (13)$$

The safety ratio is a multiplier that determines when a failure occurs for a given load. For example, a safety ratio of 1.4 indicates that a failure will occur in a composite structure if the applied load, or moment, is increased by 40%.

As shown in Figure 6, the "SSS" and "III" beams were modelled using the FEA software, and a moment (My = 3500 Nm) was applied at the free end of the beams about the Y-axis to calculate the load-carrying capacity of the manipulator. The moment value was determined based on the manipulator's fully extended operational configuration: when the distance between the manipulator's end effector and its base was about 3 m in the x-direction, and the end effector's mass was about 20 kg. In order for one of the layups to meet a failure criterion, the moment value employed in the strength analysis was selected to be nearly six times this magnitude (e.g., the "III" beam with the L3 layup). The FEA results were reported for a point in the middle of the beam, at a sufficient distance away from the boundary condition effects.

Tables 4 and 5 compare the plies with the minimum safety ratios in each layup, where SR_{TW}^{ana} and SR_{TH}^{ana} refer to the analytical Tsai–Wu and Tsai–Hill safety ratios, respectively; the methods are explained in Appendix A.1. The analytical results obtained were compared with the FEA results, as presented in Tables A1–A4 in Appendix A.2. For comparison, the von Mises stress was used in the FEA software to calculate the AA beam's safety ratio. Also, since the beam was under moment loading about the Y-axis, the top and bottom flanges (layers) of the beam were under tension and compression, respectively.

Table 4. Comparison of plies with minimum safety ratios (SRs) for the "SSS" and "III" beams (flange under tension); see Figure 3c for a laminate stacking sequence sample.

Layup	Ply No. (Angle)	"SSS" (Tension)		"III" (Tension)	
		SR_{TW}^{ana}	SR_{TH}^{ana}	SR_{TW}^{ana}	SR_{TH}^{ana}
L1	Ply 1 (0°)	11.3	11.3	9.39	9.39
L2	Ply 1 (20°)	4.51	4.26	3.73	3.53
L3	Ply 1 (30°)	2.18	1.89	1.80	1.56
L4	Ply 5 (45°)	6.36	6.39	5.21	5.25
L5	Ply 3 (45°)	4.82	4.91	3.99	4.06
L6	Ply 1 (45°)	3.36	3.47	2.85	2.89
L7	Ply 5 (90°)	3.99	4.10	3.30	3.39
L8	Ply 5 (90°)	2.35	2.66	1.94	2.20
AA		1.55		1.36	

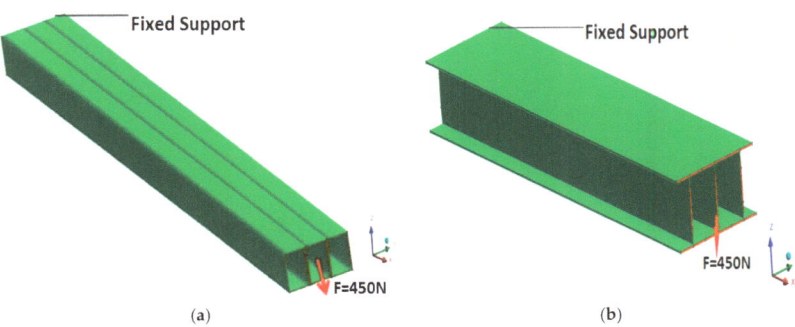

Figure 6. Proposed multi-cell beams subjected to bending loads: (**a**) "SSS" beam under the bending moment, and (**b**) "III" beam under the bending moment.

Table 5. Comparison of plies with minimum safety ratios (SRs) in "SSS" and "III" beams (flange under compression); see Figure 3c for a laminate stacking sequence sample.

Layup	Ply No. (Angle)	"SSS"		"III"	
		SR_{TW}^{ana}	SR_{TH}^{ana}	SR_{TW}^{ana}	SR_{TH}^{ana}
L1	Ply 1 (0°)	6.57	6.57	5.43	5.43
L2	Ply 1 (20°)	2.14	2.62	1.77	2.17
L3	Ply 1 (30°)	1.23	1.45	1.02	1.20
L4	Ply 1 (0°)	4.34	5.13	3.59	4.25
L5	Ply 1 (0°)	3.12	3.83	2.59	3.17
L6	Ply 3 (0°)	2.20	2.73	1.81	2.26
L7	Ply 1 (0°)	6.61	5.33	5.48	4.41
L8	Ply 1 (0°)	3.30	3.32	2.72	2.73
AA		1.55		1.36	

From Table 4, the AA "SSS" and "III" beam safety ratios were 1.55 and 1.36, respectively. For the "SSS" and "III" beams with the L1, L2, L3, and L6 layups, the first ply (or the beams' outer ply) had the lowest SR; this can be considered failure initiation under tension. In addition, for the "SSS" and "III" beams with the L4, L7, and L8 layups, the fifth ply (from the beams' outer ply) had the lowest SR. In beams with the L5 layup, the third ply (from the beams' outer ply) had the lowest SR. Furthermore, as shown in Table 4 for the "SSS" and "III" beams under tension, the safety ratios for L1 and L3 decreased as the fiber direction changed from 0° to 30°; also, the layups L1, with ten plies at 0°, and L3, with ten plies at 30°, had the highest and lowest safety ratios among all eight layups, respectively. The safety ratio dropped incrementally for the L1, L4, L5, and L6 layups, for which the number of 0° plies decreased incrementally from ten to four. When comparing L1 and L4 ($[0_4/\pm 45/0_4]$), with two 45° plies, and L7 ($[0,0,0,0,90]_s$), with two 90° plies, the safety ratio increased and subsequently decreased. Furthermore, when comparing L7 and L8 layups, both with an equal number of 90° plies, but with L8 having additional 45° plies, L8 had a decreased safety ratio. Overall, the results indicated that, for the beam under tension, increasing the fiber direction from 0° to 30° in angle-ply layups, decreasing the number of 0° plies, and adding ±45° and 90° plies decreased the load-carrying capacity (or SR) for both the "SSS" and "III" beams.

Table 5 shows a comparison of plies with the minimum safety ratio in "SSS" and "III" beams when the beam's flange (layer) is under compression. For the "SSS" and "III" beams with the L1, L2, L3, L4, L5, L7, and L8 layups, the first ply (or the beams' outer ply) had the lowest SR, while in beams with L6 layup, the third ply (from the beams' outer ply) had the lowest SR. Comparing Tsai–Wu and Tsai–Hill criteria showed that, for the L1 layup with all plies at 0°, both criteria estimated a similar safety ratio; the SR differed for cases where the stress interactions were higher because of fibers that were oriented in a direction other than zero.

7. Effect of Layup on Composite Beam's Natural Frequency

This section investigates the effects of different layups on the natural frequencies of the AA and CFRP "SSS" and "III" beams. To calculate the natural frequencies in the composite beams, the two methods reported in [14] were adopted. The first method is based on classic beam theory (CBT) and the second method is based on first shear deformation theory (FSDT). To consider the effect of shear deformation, an approximate method was used to calculate the fundamental natural frequency. Equations (14)–(17) were used to calculate the fundamental natural frequency for a cantilever beam. As shown in Figure 1, the multi-cell-section beams were symmetric about the Z-axis. The beam's fundamental natural frequency in the xz-plane, due to lateral vibration, is given by the following [14,28]:

$$w_b = \sqrt{\frac{EI_{yy}\cdot(1.875)^4}{\rho\cdot L^4}} \qquad (14)$$

$$w_s = \sqrt{\frac{S_{ZZ}\cdot(\frac{\pi}{2})^2}{\rho\cdot L^2}} \qquad (15)$$

$$\frac{1}{w_y^2} = \frac{1}{w_b^2} + \frac{1}{w_s^2} \qquad (16)$$

$$f = \frac{w_y}{2\pi} \qquad (17)$$

In Equation (14), w_b is the natural frequency of lateral vibration in the xz-plane for a cantilever beam undergoing bending deformation, EI_{yy} is the equivalent bending stiffness about the Y-axis, ρ is the mass per unit length, and L is the beam length. In Equation (15), w_s is the natural frequency of the beam due to torsional motion (considering shear deformation), where S_{ZZ} is the equivalent shear stiffness, as defined in Appendix A.3. In Equation (16), w_y is the approximate combined natural frequency of vibration in the xz-plane. This solution was obtained from Föpplms theorem [14], developed for estimating the buckling load for elastic structures. Table 6 shows the fundamental natural frequencies for AA and the eight layups discussed earlier for the "SSS" and "III" section beams. The difference was calculated using Equation (18). Here, f_b^{CBT} was obtained from Equations (14) and (17), and f_y^{FSDT} was obtained using Equations (16) and (17), respectively. Note that w_y gives the frequency in rad/s, while f is the frequency in Hz. Because the manipulator top links were made out of three beams, the EI_{yy} values in Table 6 are three times higher than the EI_{yy} values found in Table 2.

$$\text{Difference} = \left|\frac{f_b^{FEA} - f_y^{FSDT}}{f_b^{FEA}}\right| \times 100 \qquad (18)$$

Table 6. Fundamental natural frequencies of the eight different layups for "SSS" and "III" cantilever beams (see Table 2 for details of layups).

Layup	"SSS" Beam					"III" Beam				
	EI_{yy}^{SSS} (N m^2)	f_b^{CBT} (Hz)	f_y^{FSDT} (Hz)	f_b^{FEA} (Hz)	Difference (%)	EI_{yy}^{III} (N m^2)	f_b^{CBT} (Hz)	f_y^{FSDT} (Hz)	f_b^{FEA} (Hz)	Difference (%)
L1	82,282.5	82.1	78.4	78.6	0.25	69,813.9	89.9	83.4	83.7	0.36
L2	52,225.2	65.4	64.9	65.9	1.52	44,311.2	71.6	70.7	71.3	0.84
L3	26,757.6	46.8	46.7	47.9	2.51	22,703.0	51.3	51.0	51.6	1.16
L4	68,221.8	74.8	73.5	73.6	0.14	57,884.4	81.9	79.6	79.0	0.76
L5	53,727.1	66.4	65.8	66.0	0.30	45,586.1	72.6	71.6	71.7	0.14
L6	39,031.4	56.6	56.3	56.6	0.53	33,116.1	61.9	61.4	61.5	0.16
L7	67,173.9	74.2	71.4	71.5	0.14	56,995.5	81.2	76.3	76.5	0.26
L8	41,535.7	58.4	58.0	58.0	0.00	35,242.6	63.9	63.2	63.1	0.15
AA	41,139.3	44.1	44.1	44.1	0.00	34,907.1	48.3	48.1	48.1	0.00

From Equation (14), it can be seen that the angular natural frequency is proportional to the beam's equivalent bending stiffness and inversely proportional to the beam's mass, as expected. When comparing the natural frequencies of the "III" and "SSS" section beams, the "III" section beam with the lower mass had a greater natural frequency than the "SSS" section beam. Also, among L1 to L8, the L1 layup, with ten 0° plies, and the L4 and L7 layups, with eight 0° plies, had the highest fundamental natural frequencies. From Table 6, it can be observed that CFRP beams with all eight layups had higher fundamental natural frequencies than the AA beam. By comparing the CBT and FSDT results, it can be observed that the CBT method overestimated the natural frequencies and the FSDT method underestimated the natural frequencies when compared with the FEA results. The f_y^{FSDT} and f_b^{FEA} results were in a good agreement, with a maximum difference of less than 3%.

8. Selecting the Best Layup for Composite Parts

The next stage was to select the ideal layup and cross-section for the manipulator's composite parts after calculating the deflection, load capacity, and natural frequency. The "SSS" beam had a lower deflection and a higher safety ratio, whereas the "III" beam had a higher natural frequency and a lower weight. As demonstrated in [11], the "SSS" beam had a smaller angle of twist under torsional loading than the "III" beam. Therefore, both cross-sections were suitable for the manipulator's top link. In this study, the "SSS" beam was selected to model the robot's top link in order to perform a modal analysis on the 5-DOF robot. Also, the L1 ($[0,0,0,0,0]_s$), L4 $[0_4/\pm45/0_4]$, and L7 ($[0,0,0,0,90]_s$) layups were found to have the best values in terms of the minimum deflection, the plies with the highest minimum safety ratio, and the highest natural frequencies, so these three layups were selected for this analysis.

Considering the "SSS" section, the L1 layup, with ten 0° plies, had the lowest tip deflection ($\delta_b = 3.15$ mm), the highest safety ratio ($SR_{TW} = 6.57$), and the highest fundamental natural frequency (f_y^{FSDT} = 78.4 Hz). Table 7 shows that, in terms of deflection and natural frequency, L4 and L7 had similar values, but the L7 safety ratio under compression was 52% greater than the L4 safety ratio. Considering the mentioned specifications, L7, with two 90° plies and eight 0° plies, and L1, with ten 0° plies, could be ideal layups for creating the manipulator's composite parts. In the following section, the CFRP "SSS" beam with the $[0,0,0,0,90]_s$ layup was used to perform the modal analysis on the CFRP 5-DOF manipulator in order to estimate possible improvements in terms of the structural mass and the fundamental natural frequency compared with the AA 5-DOF manipulator.

Table 7. Design criteria values for selected layups with the "SSS" cross-section.

Layups	EI_{yy}^{SSS} (N·m^2)	δ_b (mm)	SR_{TW} (Compression)	w_y^{FSDT} (Hz)
L1 = $[0,0,0,0,0]_s$	82,282	3.15	6.57	78.4
L4 = $[0_4/\pm45/0_4]$	68,221	3.79	4.34	73.5
L7 = $[0,0,0,0,90]_s$	67,173	3.85	6.61	71.4

9. Modal Analysis of the Manipulator

In this section, three configurations for the 5-DOF manipulator, which are common operating conditions, were considered for the modal analysis. Figure 7 shows these three configurations. The aim here was to determine the natural frequencies and their corresponding mode shapes. From the three configurations presented, the fully extended Configuration 3 (Figure 7c) was the most important one, as it was the manipulator's working (crop monitoring) pose. Configuration 2 was the manipulator's transport mode (Figure 7b) and Configuration 1 (Figure 7a) was the manipulator's occasional use pose. In Figure 7, the green parts are AA clamps, and all members are in the xz-plane.

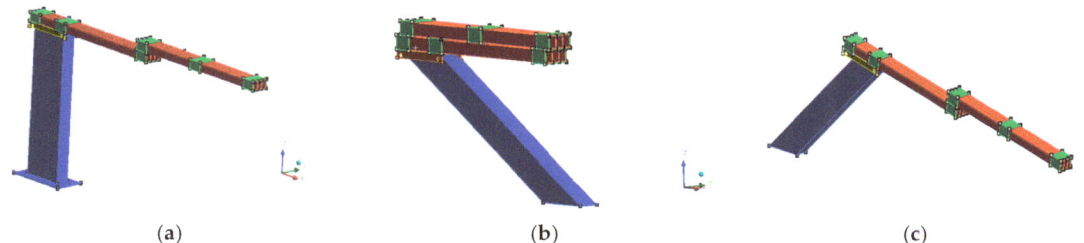

Figure 7. The 5-DOF robot manipulator's common operating configurations: (**a**) Configuration 1, (**b**) Configuration 2, and (**c**) Configuration 3.

The original manipulator [21] base was made out of steel and its top link was created from AA. The aim here was to investigate the impact of using CFRP for the top links on the manipulator's free vibration via modal analysis. For this purpose, the robot's top links were replaced with thin-walled laminated CFRP beams. Figure 8a,b depict the top link models made of AA and CFRP materials, respectively. The Solid186, CONTA174, and TARGE170 element types were used to mesh the AA model, and the Solid185, CONTA174, and TARGE170 element types were used to mesh the CFRP model. To mesh the CFRP "SSS" section top link, 104,912 elements were used, and to mesh the AA "SSS" section top link, 26,242 elements were used. The difference in the total number of elements was due to the fact that the composite beam, with a total thickness of 2 mm, was formed by stacking 10 CFRP layers, each with a thickness of 0.2 mm (Figure 8a). Each layer was meshed separately with a mesh-seed equal to 0.2 mm. The AA beam was defined with a thickness of 2 mm and a beam cross-section meshed with a mesh-seed that was equal to the cross-sectional thickness of close to 2 mm. Comparing the CFRP "SSS" section beam with the AA beam, the element ratio was 10:1 in the direction of the thickness. In addition, the joint between the beams was modeled as a fixed joint.

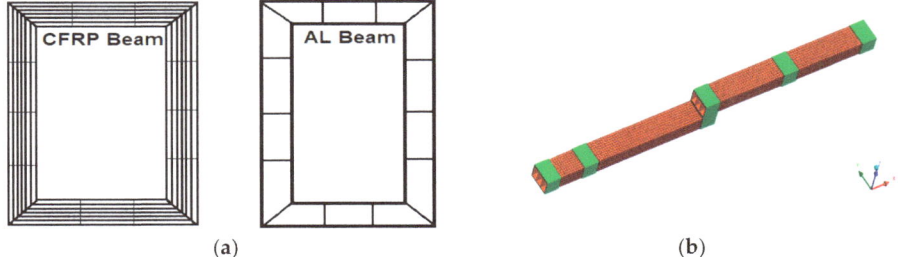

Figure 8. Mesh details for top link models created from AA and CFRP materials: (**a**) cross-sections of CFRP and AA beams, and (**b**) schematic of the 5-DOF robot manipulator's top link.

In the final step, the CFRP and AA beams were incorporated into the rest of the manipulator with a steel body. Table 8 shows the first six natural frequencies of the AA and composite CFRP manipulators. Figures 9 and 10 show the first mode shape for the three configurations. Higher mode shapes are given in Appendix C.

Table 8. Natural frequencies of the AA and CFRP manipulators under three configurations (Hz).

Mode	CFRP Robot 1st Config.	AA Robot 1st Config.	1st Config Difference (%)	CFRP Robot 2nd Config.	AA Robot 2nd Config.	2nd Config Difference (%)	CFRP Robot 3rd Config.	AA Robot 3rd Config.	3rd Config Difference (%)
1	5.33	4.46	19.5	14.6	12.4	17.7	10.3	8.63	19.3
2	17.9	16.9	5.92	20.5	19.2	6.77	18.2	17.1	6.43
3	35.2	30.3	16.1	45.0	40.6	10.8	24.8	20.7	19.8
4	55.7	48.7	14.3	64.6	64.1	0.78	63.5	61.3	3.59
5	60.5	60.1	0.67	95.1	94.1	1.06	66.1	63.3	4.42
6	95.5	95.1	0.42	97.3	94.6	2.85	94.3	93.8	0.53

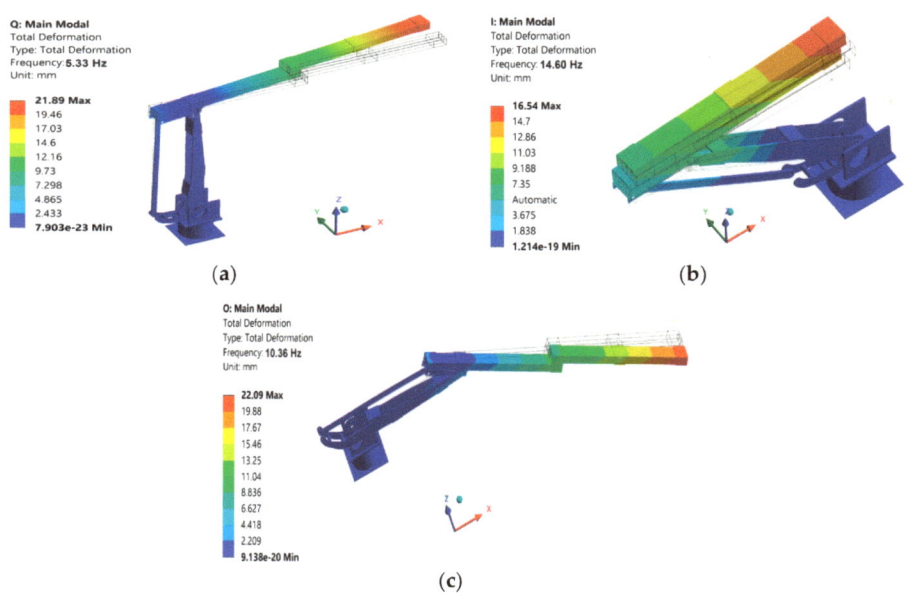

Figure 9. First mode shapes of the CFRP manipulator: (**a**) first configuration, (**b**) second configuration, and (**c**) third configuration.

The difference percentages in Table 8 were computed using Equation (19), where w_{CFRP} is the natural frequency of the CFRP manipulator and w_{AA} is the natural frequency of the AA manipulator.

$$\text{Difference} = \left| \frac{w_{AA} - w_{CFRP}}{w_{AA}} \right| \times 100 \qquad (19)$$

As mentioned earlier, the third configuration was the most important pose. Looking at the natural frequencies in Table 8 and their mode shapes, the first natural frequency of the CFRP robot was the bending vibration around the Z-axis in the xy-plane, in which the top links had the largest contribution. The second natural frequency was a mixed vibration mode formed by two vibration types: first, the rotational vibration around the Z-axis; and second, the bending around the Y-axis. The third and fourth natural frequencies were the bending vibrations around the Y-axis, with the top links having the largest displacement. Also, from the mode shapes in Appendix C, it can be seen that the robot's top links had the largest contribution among the robot components from the first to the fourth natural frequencies. Thus, the top links were the parts that were prone to vibration.

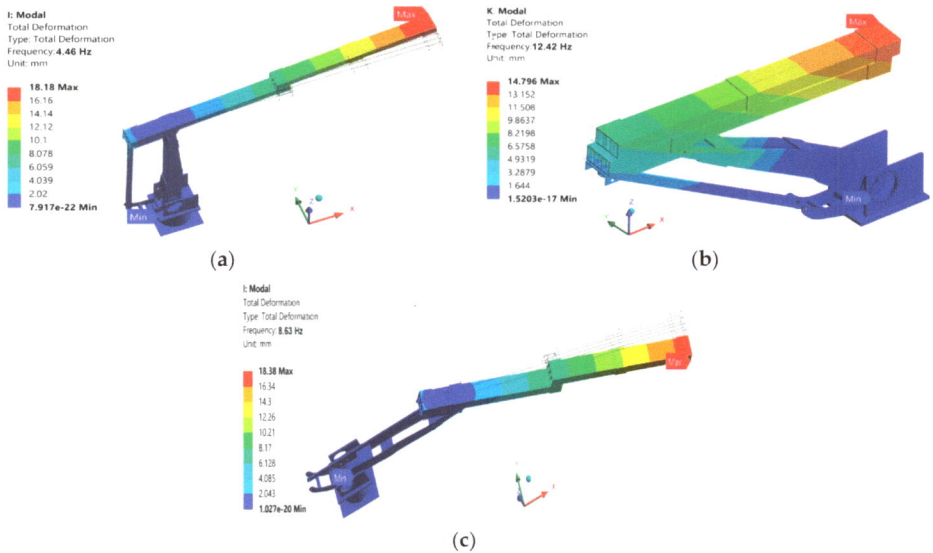

Figure 10. First mode shapes of the AA manipulator: (**a**) first configuration, (**b**) second configuration, and (**c**) third configuration.

Table 8 shows the improvement in the fundamental natural frequencies achieved by using CFRP beams in the manipulator's structure. In the third configuration, the natural frequencies from the first mode to the third mode increased by 19.3%, 6.43%, and 19.8%, respectively. Furthermore, the CFRP manipulator's mass was 51.8 kg, while the AA robot's mass was 55.2 kg, indicating a 6.15% difference in the structural mass. Overall, the natural frequencies and mode shapes depended on the mass and stiffness distributions of the structure. The structure with the higher stiffness and lower mass had the highest natural frequency. Therefore, the analysis above indicates that it is critical to decrease the top link mass as much as possible while meeting the load-carrying capacity requirements.

10. Conclusions

This study investigated the effects of thin-walled laminated CFRP beams on the static and free-vibration performance of a 5-DOF manipulator. The effects of using composite materials were considered in terms of the fundamental natural frequency, load-carrying capacity, and structural mass, and were compared with the 5-DOF aluminum alloy (AA) manipulator. It was shown how the cross-sections and layups can affect these parameters. For this purpose, AA and CFRP beams with eight layups were used to determine the structural efficiency of composite materials for a 5-DOF manipulator.

A beam safety ratio was introduced using the Tsai–Wu and Tsai–Hill failure criteria. The fundamental natural frequencies were determined using the classical beam and shear deformation theories. Three-dimensional finite element models using linear and quadratic solid elements were employed, and results were compared with the analytical results. Three working configurations of the 5-DOF manipulator were investigated using a modal analysis, and the results were reported for the CFRP and AA models. Overall, this study showed that changing the fiber direction from 90° to 0° in an angle-ply layup and adding more 0° plies while keeping the overall numbers of layers the same decreased the deflection and increased the natural frequency and load-carrying capacity. Replacing parts of an AA manipulator with CFRP can increase the fundamental natural frequencies by 19% and can decrease the structure mass by about 6.15%. See the discussions in Section 5 for details. It was shown that CFRP can be used to improve the performance of a manipulator made of

AA, without compromising its performance or structural integrity and while producing moderate gains in efficiency, i.e., increasing the natural frequency, which was desirable for this application.

Author Contributions: Conceptualization, R.F.; formal analysis, M.A.K.; funding acquisition, R.F.; investigation, M.A.K., R.F. and R.M.; methodology, M.A.K., R.F. and R.M.; software, M.A.K.; supervision, R.F. and R.M.; validation, M.A.K.; writing—review and editing, R.F. and R.M. All authors have read and agreed to the published version of the manuscript.

Funding: Financial support for this work was provided by NSERC through a grant from the Canada First Research Excellence Fund (CFREF) via the Global Institute for Food Security (GIFS), the University of Saskatchewan, Canada. Funding was also provided by an NSERC-CRD grant.

Limitation: Although the research presented here could be extended to other serial-parallel manipulators, the results should be carefully viewed when extending to other robotic arms. The research presented here is for a 5-DOF serial-parallel manipulator with its lower part (two links) made of aluminum alloy and its upper part (three links) made of CFRP. Composite materials are more difficult to fabricate and almost impossible to machine.

Informed Consent Statement: Not applicable.

Acknowledgments: This research was conducted in the robotics laboratory at the University of Saskatchewan. The role of other lab members, especially QianWei Zhang, Hedieh Badkoobehhezaveh, Joshua Cote, and Majid KhakPour, in the development and testing of the AA manipulator is gratefully acknowledged. Also, useful comments from James Johnston, Duncan Cree, and Leon Wegner in improving the presentation of this manuscript is gratefully acknowledged.

Conflicts of Interest: The authors declare no conflict of interest.

Appendix A. Detail of Stress Analysis and Verification for Laminated Composite Beam

Appendices A.1 and A.2 include the details on the stress analysis calculations and verification for Section 6. The square beam and I-beam equivalent shear stiffness equations, which were used in Section 7, are presented in Appendices A.3 and A.4, which show the calculation steps related to the composite beam's equivalent bending and shear stiffness.

Appendix A.1. Stress Analysis of Laminated Composite Beam Details

To calculate the stress and strain components for the laminated composite beams, the method in [14] was used, as described below:

Using the classical laminate theory, the laminate stiffness and compliance matrix were calculated. In Equations (A1) to (A2), A_{ij} is the laminate in-plane stiffness that relates the in-plane forces N_x, N_y, and N_{xy} to the in-plane deformations ε_x^0, ε_y^0, and γ_{xy}^0. D_{ij} is the bending stiffness that relates the moments M_x, M_y, and M_{xy} to the curvatures K_x, K_y, and K_{xy}. B_{ij} is the in-plane coupling stiffness.

$$\begin{bmatrix} N_x \\ N_y \\ N_{xy} \\ M_x \\ M_y \\ M_{xy} \end{bmatrix} = \begin{bmatrix} A_{11} & A_{12} & A_{16} & B_{11} & B_{12} & B_{16} \\ A_{12} & A_{22} & A_{26} & B_{12} & B_{22} & B_{26} \\ A_{16} & A_{26} & A_{66} & B_{16} & B_{26} & B_{66} \\ B_{11} & B_{12} & B_{16} & D_{11} & D_{12} & D_{16} \\ B_{12} & B_{22} & B_{26} & D_{12} & D_{22} & D_{26} \\ B_{16} & B_{26} & B_{66} & D_{16} & D_{26} & D_{66} \end{bmatrix} \begin{bmatrix} \varepsilon_x^0 \\ \varepsilon_y^0 \\ \gamma_{xy}^0 \\ k_x \\ k_y \\ k_{xy} \end{bmatrix} \quad (A1)$$

$$\begin{bmatrix} \varepsilon_x^0 \\ \varepsilon_y^0 \\ \gamma_{xy}^0 \\ k_x \\ k_y \\ k_{xy} \end{bmatrix} = \begin{bmatrix} a_{11} & a_{12} & a_{16} & b_{11} & b_{12} & b_{16} \\ a_{12} & a_{22} & a_{26} & b_{12} & b_{22} & b_{26} \\ a_{16} & a_{26} & a_{66} & b_{16} & b_{26} & b_{66} \\ b_{11} & b_{12} & b_{16} & d_{11} & d_{12} & d_{16} \\ b_{12} & b_{22} & b_{26} & d_{12} & d_{22} & d_{26} \\ b_{16} & b_{26} & b_{66} & d_{16} & d_{26} & d_{66} \end{bmatrix} \begin{bmatrix} N_x \\ N_y \\ N_{xy} \\ M_x \\ M_y \\ M_{xy} \end{bmatrix} \quad (A2)$$

Thin-walled laminated beams are created from a number of segments. Using the below equation, the beam's equivalent stiffness and compliance matrices were calculated. In Equations (A3) and (A4), R_k and O_k are the stiffness matrix components of each segment of a thin-walled beam; R_k is the transformation matrix, which transforms the wall segment geometry to the global coordinate system; O_k is the wall segment stiffness matrix component; and the index k refers to the wall segment number.

$$[R_k] = \begin{bmatrix} 1 & z_k & y_k & 0 \\ 0 & \cos\alpha_k & -\sin\alpha_k & 0 \\ 0 & \sin\alpha_k & \cos\alpha_k & 0 \\ 0 & 0 & 0 & 1 \end{bmatrix} \quad (A3)$$

$$[O_k] = \frac{1}{b_k}\begin{bmatrix} a_{11} & 0 & 0 & 0 \\ 0 & d_{11} & 0 & \frac{-d_{13}}{2} \\ 0 & 0 & \frac{12}{A_{11}b_k^2} & 0 \\ 0 & \frac{-d_{13}}{2} & 0 & \frac{d_{33}}{4} \end{bmatrix} \quad (A4)$$

The total stiffness matrix of the beam $[P]$ is given by Equation (A5):

$$[P] = \sum_{K=1}^{n}[R_k][R_k]^T[O_k]^{-1} \quad (A5)$$

The compliance matrix of the beam $[W]$ is given by the below equation:

$$[W] = [P]^{-1} \quad (A6)$$

By substituting the obtained compliance matrix into Equation (A7), the beam's axial and bending strain components could be calculated as shown below:

$$\begin{bmatrix} \varepsilon_x^0 \\ \varepsilon_y^0 \\ \gamma_{xy}^0 \\ k_x \\ k_y \\ k_{xy} \end{bmatrix} = W \begin{bmatrix} N_x \\ N_y \\ N_{xy} \\ M_x \\ M_y \\ M_{xy} \end{bmatrix} \quad (A7)$$

$$\begin{bmatrix} \varepsilon_x \\ \varepsilon_y \\ \gamma_{xy} \end{bmatrix} = \begin{bmatrix} \varepsilon_x^0 \\ \varepsilon_y^0 \\ \gamma_{xy}^0 \end{bmatrix} + zk\begin{bmatrix} k_x \\ k_y \\ k_{xy} \end{bmatrix} \quad (A8)$$

$$\begin{bmatrix} \sigma_x \\ \sigma_y \\ \sigma_{xy} \end{bmatrix} = \begin{bmatrix} \overline{Q}_{11} & \overline{Q}_{12} & \overline{Q}_{16} \\ \overline{Q}_{12} & \overline{Q}_{22} & \overline{Q}_{26} \\ \overline{Q}_{16} & \overline{Q}_{26} & \overline{Q}_{66} \end{bmatrix}\begin{bmatrix} \varepsilon_x \\ \varepsilon_y \\ \gamma_{xy} \end{bmatrix} \quad (A9)$$

Appendix A.2. Comparison of Analytical Safety Ratios with FEA Results

The analytically estimated safety ratios were compared to the FEA results in this appendix (related to Section 6). In Tables A1–A4, σ_1, σ_2, and σ_{12} are local stress components that were determined analytically. It should be noted that the FEA results were reported for a point in the middle of the beam, at a sufficient distance away from the effects of boundary conditions. Tables A1 and A3 show the safety ratios and stress components for a flange under tension for the "SSS" and "III" beams, while Tables A2 and A4 show the safety ratios and stress components for a flange under compression for the "SSS" and "III" beams. In the below tables, SR_{TW}^{ana} and SR_{TH}^{ana} refer to the analytical safety ratios calculated using the Tsai–Wu and Tsai–Hill methods, respectively, while SR_{TW}^{FEA} and SR_{TH}^{FEA} are the FEA safety ratios calculated using the Tsai–Wu and Tsai–Hill methods, respectively.

Table A1. Comparison of plies with minimum safety ratios (SRs) for the "SSS" beam (flange under tension; stress values are in MPa); see Figure 3c for a laminate stacking sequence sample.

Layup	Ply No. (Angle)	σ_1 (MPa)	σ_2 (MPa)	σ_{12} (MPa)	SR_{TW}^{ana}	SR_{TW}^{FEA}	SR_{TH}^{ana}	SR_{TH}^{FEA}
L1	Ply 1 (0°)	167	0	0	11.3	10.4	11.3	10.4
L2	Ply 1 (20°)	182	−14.7	−13.0	4.51	4.27	4.26	4.02
L3	Ply 1 (30°)	190	−23.1	−35.0	2.18	2.06	1.89	1.80
L4	Ply 5 (45°)	440	3.44	−10.0	6.36	5.85	6.39	5.88
L5	Ply 3 (45°)	41.8	3.43	−13.7	4.82	4.46	4.91	4.54
L6	Ply 1 (45°)	41.1	4.23	−19.8	3.36	3.15	3.47	3.22
L7	Ply 5 (90°)	−12.4	12.4	0	3.99	3.68	4.10	3.79
L8	Ply 5 (90°)	−96.2	18.5	0	2.35	2.14	2.66	2.42

Table A2. Comparison of plies with minimum safety ratios (SRs) for the "SSS" beam (flange under compression; stress values are in MPa); see Figure 3c for a laminate stacking sequence sample.

Layup	Ply No. (Angle)	σ_1 (MPa)	σ_2 (MPa)	σ_{12} (MPa)	SR_{TW}^{ana}	SR_{TW}^{FEA}	SR_{TH}^{ana}	SR_{TH}^{FEA}
L1	Ply 1 (0°)	−167	0	0	6.57	6.04	6.57	6.04
L2	Ply 1 (20°)	−181	14.6	12.9	2.14	2.01	2.62	2.44
L3	Ply 1 (30°)	−190	23.1	34.9	1.23	1.16	1.45	1.38
L4	Ply 1 (0°)	−200	3.22	−0.46	4.34	4.05	5.13	4.77
L5	Ply 1 (0°)	−254	5.83	0	3.12	2.94	3.83	3.58
L6	Ply 3 (0°)	−345	9.21	0.27	2.20	2.05	2.73	2.54
L7	Ply 1 (0°)	−206	−3.04	0	6.61	6.02	5.33	4.93
L8	Ply 1 (0°)	−331	0.04	0	3.30	3.06	3.32	3.06

Table A3. Comparison of plies with minimum safety ratios (SRs) for the "III" beam (flange under tension; stress values are in MPa); see Figure 3c for a laminate stacking sequence sample.

Layup	Ply No. (Angle)	σ_1 (MPa)	σ_2 (MPa)	σ_{12} (MPa)	SR_{TW}^{ana}	SR_{TW}^{FEA}	SR_{TH}^{ana}	SR_{TH}^{FEA}
L1	Ply 1 (0°)	202	0	0	9.39	8.89	9.39	8.89
L2	Ply 1 (20°)	219	−17.7	−15.6	3.73	3.32	3.53	3.14
L3	Ply 1 (30°)	229	−27.8	42.2	1.80	1.63	1.56	1.42
L4	Ply 5 (45°)	49.7	4.31	12.1	5.21	4.94	5.25	4.97
L5	Ply 3 (45°)	50.7	4.13	−16.6	3.99	3.75	4.06	3.83
L6	Ply 1 (45°)	49.7	5.11	23.9	2.85	2.65	2.89	2.71
L7	Ply 5 (90°)	−15	14.9	0	3.30	3.13	3.39	3.19
L8	Ply 5 (90°)	−116	22.3	0	1.94	1.84	2.20	2.07

Table A4. Comparison of plies with minimum safety ratios (SRs) for the "III" beam (flange under compression; stress values are in MPa); see Figure 3c for a laminate stacking sequence sample.

Layup	Ply No. (Angle)	σ_1 (MPa)	σ_2 (MPa)	σ_{12} (MPa)	SR_{TW}^{ana}	SR_{TW}^{FEA}	SR_{TH}^{ana}	SR_{TH}^{FEA}
L1	Ply 1 (0°)	−202	0	0	5.43	5.14	5.43	5.14
L2	Ply 1 (20°)	−219	17.7	15.7	1.77	1.55	2.17	1.88
L3	Ply 1 (30°)	−229	27.8	42.2	1.02	0.91	1.20	1.06
L4	Ply 1 (0°)	−242	3.89	−0.56	3.59	3.21	4.25	3.90
L5	Ply 1 (0°)	−307	7.04	0.02	2.59	2.31	3.17	2.88
L6	Ply 3 (0°)	−417	11.1	0.32	1.81	1.71	2.26	2.13
L7	Ply 1 (0°)	−248	−3.67	0	5.48	5.14	4.41	4.11
L8	Ply 1 (0°)	−400	0.04	0.03	2.72	2.63	2.73	2.59

Appendix A.3. Shear Compliances and Stiffness Equations for Square and I-Beams

Related to Sections 4 and 7, Equations (A10) and (A11) can be used to calculate the square beam and I-beam equivalent shear compliances ($\widehat{s_{ZZ}}$), respectively. In Equation (A12), S_{ZZ} is the beam's equivalent shear stiffness, which is equal to the inverse of $\widehat{s_{ZZ}}$.

$$\widehat{s_{ZZ}} = \frac{a_{66}}{2d} + \frac{a_{66} \cdot d_f}{6d^2 \left(1 + \frac{a_{11}d}{3a_{11}d_f}\right)^2} \tag{A10}$$

$$\widehat{s_{ZZ}} = \frac{a_{66}}{d} + \frac{a_{66} \cdot b_f}{6d^2 \left(1 + \frac{a_{11}d}{6a_{11}b_f}\right)^2} \tag{A11}$$

$$S_{ZZ} = [\widehat{s_{ZZ}}]^{-1} \tag{A12}$$

Appendix A.4. Details on Calculating the Equivalent Bending and Shear Stiffness of a Composite Beam

This section explains the steps for calculating the equivalent bending and shear stiffness of a CFRP square beam, which are related to Sections 4 and 7. The compliance matrix components for a laminate with a $[0/0/0/0/0]_s$ layup are shown in Table A5.

Using the data in Tables A5 and A6 and Equation (4), the equivalent bending stiffness (EI_{yy}) for a square beam is 27,427.58 (N·m^2).

Table A5. Laminate with $[0/0/0/0/0]_s$ layup compliance components.

a_{11} ($\frac{mm}{N}$)	a_{66} ($\frac{mm}{N}$)	d_{11} ($\frac{1}{N \cdot mm}$)
3.52×10^{-6}	1.08×10^{-4}	1.05×10^{-5}

Table A6. Square beam geometrical dimensions (mm) (Note: the cross-sectional details are shown in Figure 3).

Layup	d	b_w	b_f	d_f
$[0/0/0/0/0]_s$	58	56	40	38

Similarly, the square beam equivalent shear compliance and stiffness can be determined using Equations (A10) and (A12). The equivalent shear compliance for a square beam with a $[0/0/0/0/0]_s$ layup is 1.02693×10^{-6} (1/N), and the equivalent shear stiffness is 973,778.40 (N).

Appendix B. Material Properties

Table A7. Material properties, FAW200 prepreg moduli, and strength parameters from [17].

Property	Carbon Fiber-Reinforced Epoxy (CFRP)
Longitudinal modulus, E_1 [GPa]	142
Transverse modulus, E_2 [GPa]	9
Out-of-plane modulus, E_3 [GPa]	9
In-plane shear modulus, G_{12} [GPa]	4.6
Out-of-plane shear modulus, G_{23} [GPa]	3.08
Out-of-plane shear modulus, G_{13} [GPa]	4.6
Major in-plane Poisson's ratio v_{12}	0.32
Out-of-plane Poisson's ratio v_{23}	0.46
Out-of-plane Poisson's ratio v_{13}	0.32
Longitudinal tensile strength, F_{1t} [MPa]	1900
Transverse tensile strength, F_{2t} [MPa]	51
Out-of-plane tensile strength, F_{3t} [MPa]	51
Longitudinal compressive strength, F_{1c} [MPa]	1100
Transverse compressive strength, F_{2c} [MPa]	130
Out-of-plane compressive strength, F_{3c} [MPa]	130
In-plane shear strength, F_6 [MPa]	72
Out-of-plane shear strength, F_4 [MPa]	70
Out-of-plane shear strength, F_5 [MPa]	72

Table A8. Steel and T 6061-T6 aluminum alloy mechanical properties [18].

Property	Steel	Aluminum Alloy
E(GPa)	200	71
G (GPa)	76.92	20.69
v	0.3	0.33

Appendix C. CFRP Manipulator Mode Shapes

The natural frequencies and corresponding mode shapes for the 5-DOF manipulator with the CFRP material top links are given in Figures A1–A3 for the three configurations discussed in Section 9. The natural frequencies and mode shapes given here are the eigenvalues and eigenvectors obtained from FEA software for free vibration.

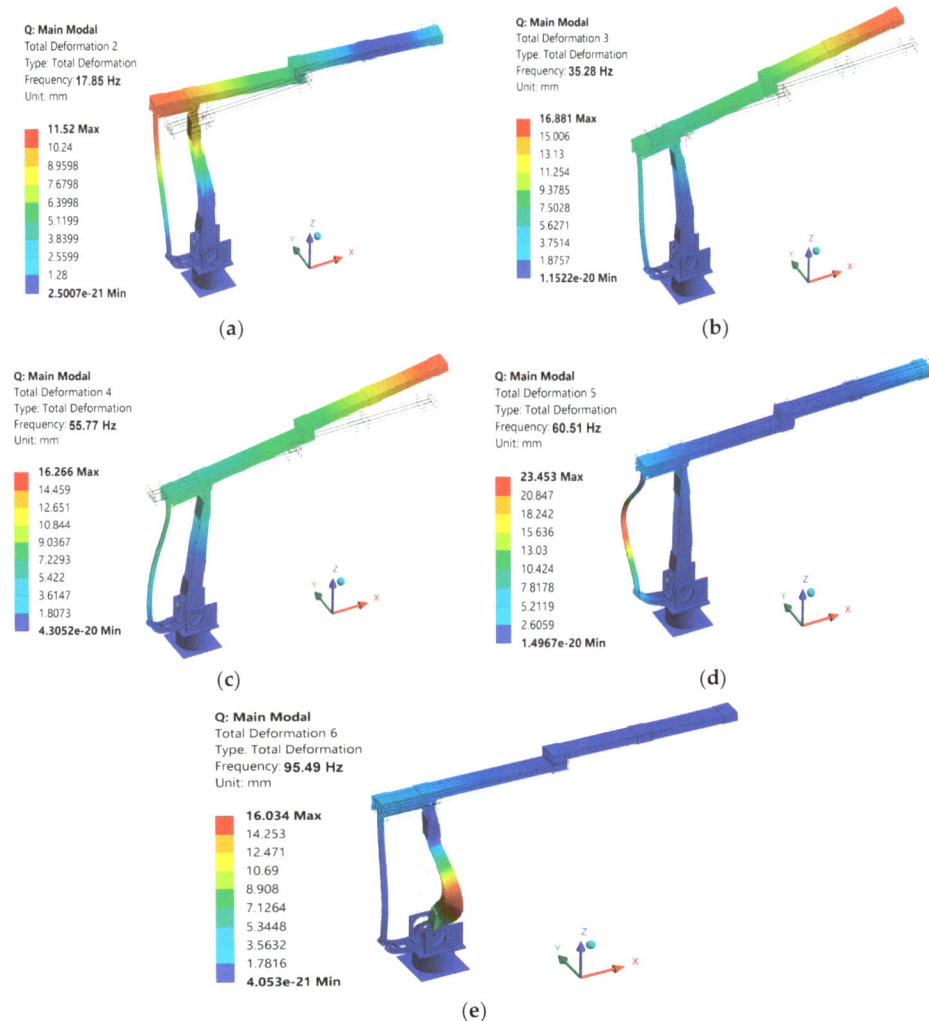

Figure A1. Natural frequencies and corresponding mode shapes for the CFRP 5-DOF manipulator in Configuration 1: (**a**) second frequency, (**b**) third frequency, (**c**) fourth frequency, (**d**) fifth frequency, and (**e**) sixth frequency.

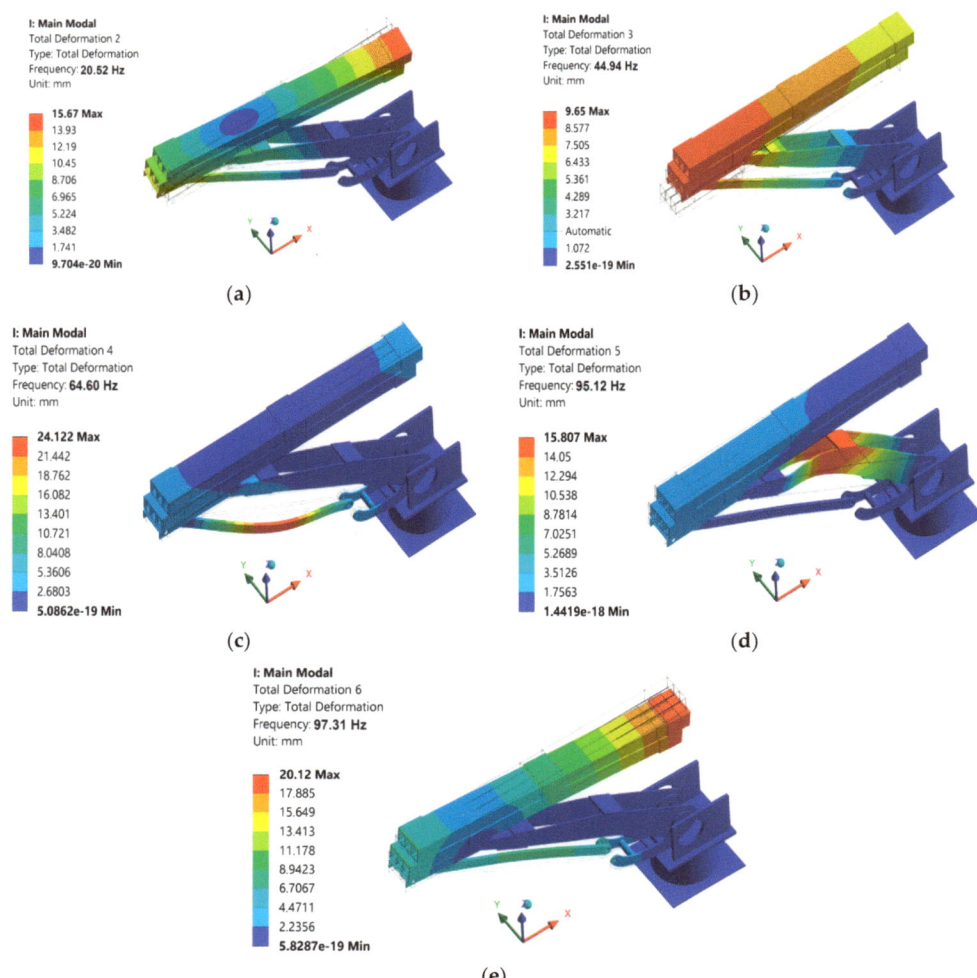

Figure A2. Natural frequencies and corresponding mode shapes for the CFRP 5-DOF manipulator in Configuration 2: (**a**) second frequency, (**b**) third frequency, (**c**) fourth frequency, (**d**) fifth frequency, and (**e**) sixth frequency.

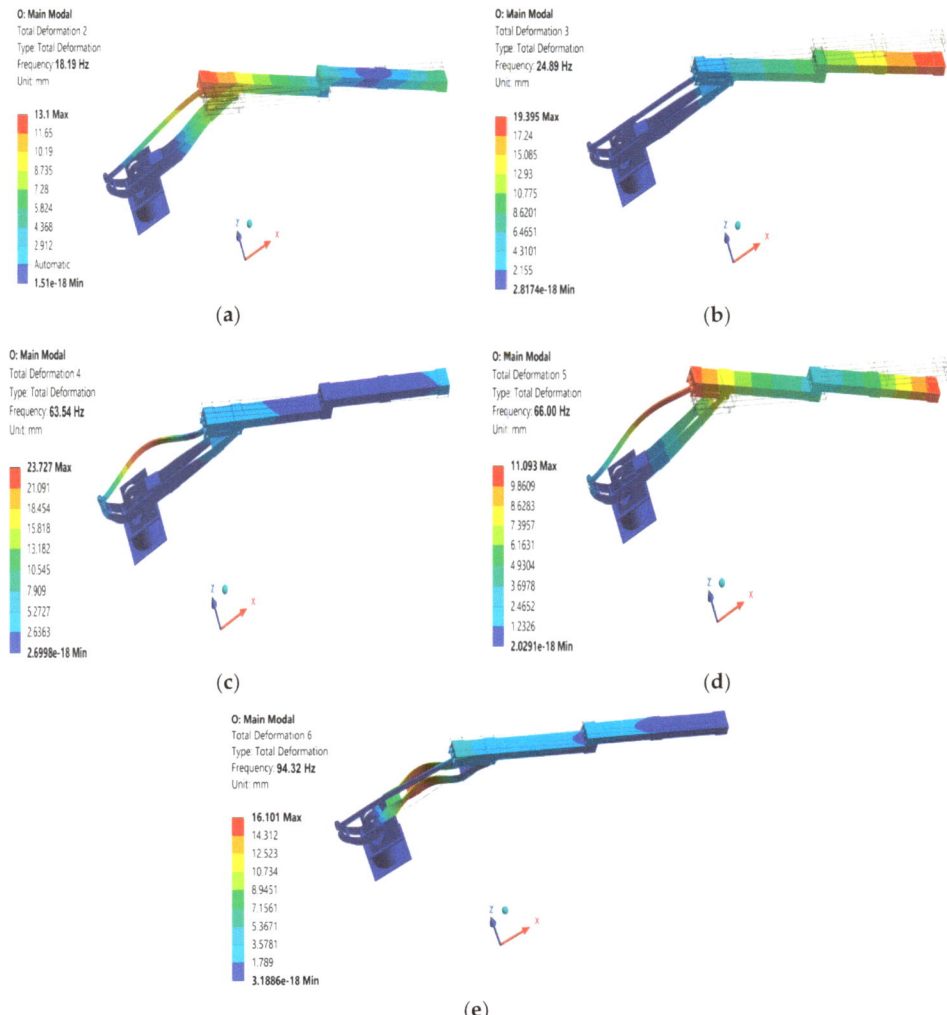

Figure A3. Natural frequencies and corresponding mode shapes for the CFRP 5-DOF manipulator in Configuration 3: (**a**) second frequency, (**b**) third frequency, (**c**) fourth frequency, (**d**) fifth frequency, and (**e**) sixth frequency.

Appendix D. Additional Details of the FEA Results

This section includes additional data of the FEA results for the "SSS" and "III" CFRP beams with the L2 and L8 layups, related to Sections 5 and 7. Figure A4 shows the static analysis deflection results, and Figure A5 shows the modal analysis results for cantilever beams. The strain results in the x-direction from the three-point bending test simulation related to Section 3 are depicted in Figure A6. Table A9 shows the mesh convergence study related to Section 3.

Table A9. Mesh convergence study related to Section 3.

Load (N)	Mesh Size (mm)	Strain in x-Direction	CPU Time (s)
435	20	0.130×10^{-4}	49.7
435	10	0.160×10^{-4}	50.8
435	5.0	0.163×10^{-4}	62.4

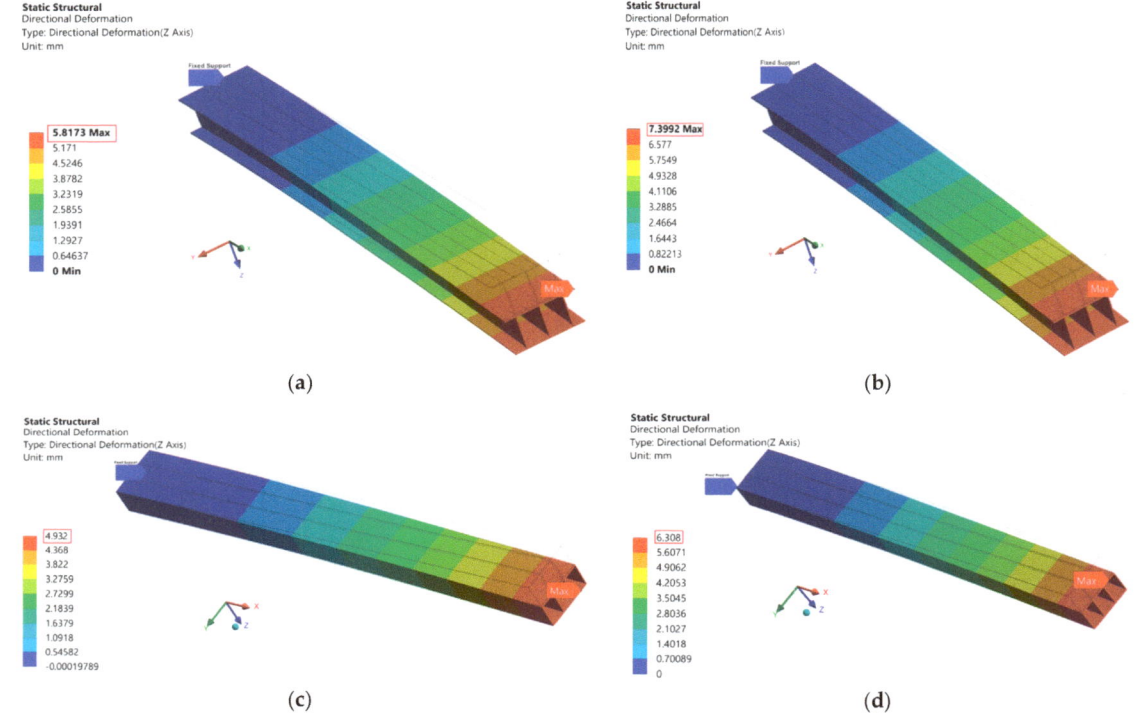

Figure A4. Static loading deflection of the CFRP beam related to Section 5 for (**a**) "III" section and "L2" layup, (**b**) "III" section and "L8" layup, (**c**) "SSS" section and "L2" layup, and (**d**) "SSS" section and "L8" layup.

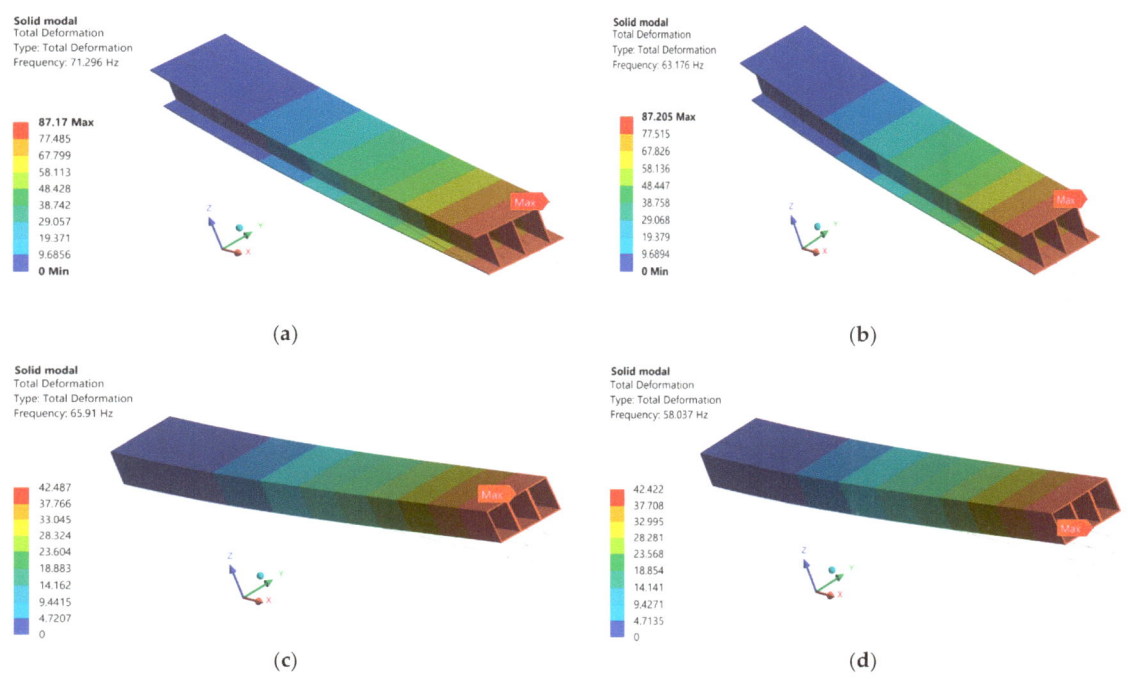

Figure A5. Natural mode shapes of the CFRP cantilever beam related to Section 7 for (**a**) "III" section and "L2" layup, (**b**) "III" section and "L8" layup, (**c**) "SSS" section and "L2" layup, and (**d**) "SSS" section and "L8" layup.

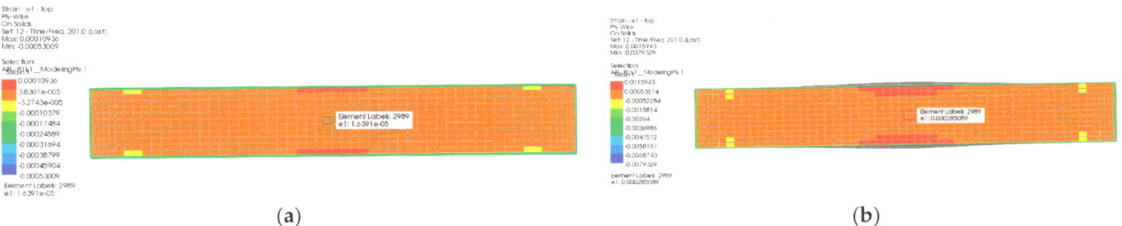

Figure A6. Three-point bending test FEA results related to Section 3: (**a**) strain in the x-direction for the simulated beam under 435N loading, and (**b**) strain in the x-direction for the simulated beam under 4024N loading.

The FEA results for "SSS" CFRP beams with the L2 and L8 layups related to Section 6 are shown in Figures A7 and A8. The normal stress components in the x- and y-directions and the shear stress in the xy-plane for the square beam with the L2 and L8 layups are shown in Figures A7a and A8a, respectively. Figures A7b and A8b show the safety ratios estimated with the FEA software using the Tsai–Wu criteria for the "SSS" beam with the L2 and L8 layups, respectively. The results are presented for a wall segmented under compression and can be compared with the results in Table A2.

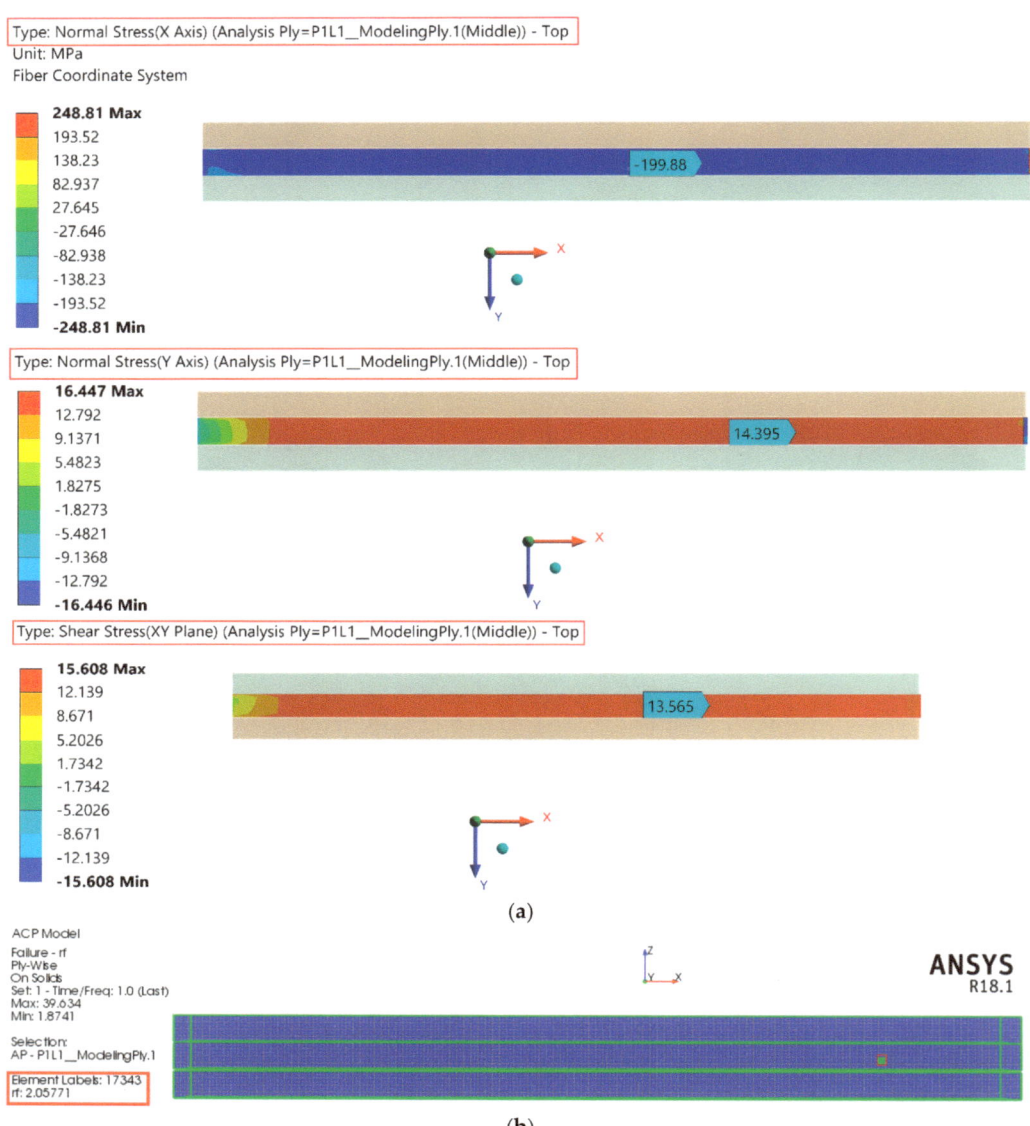

Figure A7. Stress components for a CFRP square beam with L2 layup: (**a**) normal stress components in the x- and y-directions and shear stress in the xy-plane, and (**b**) safety ratio.

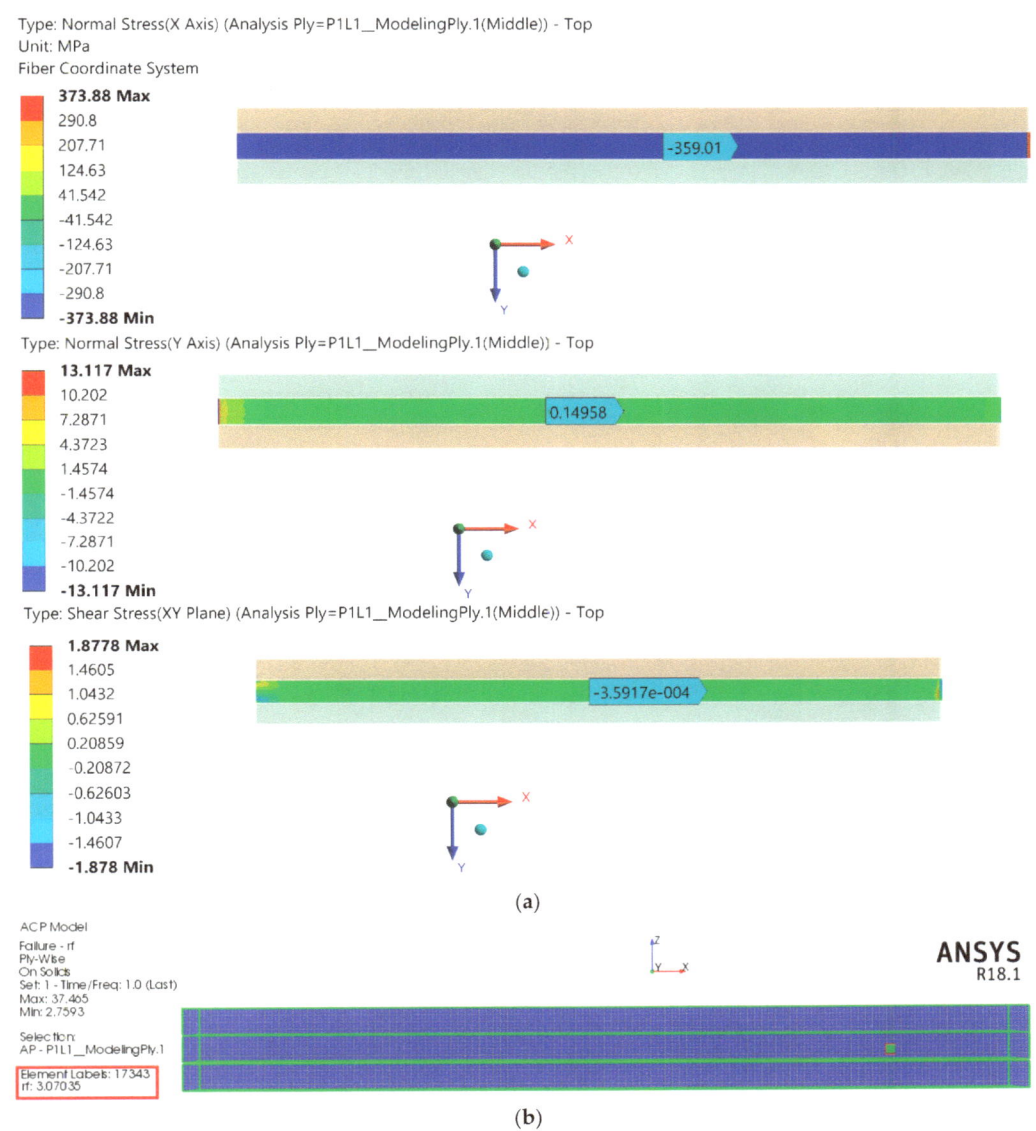

Figure A8. Stress components for a CFRP square beam with L8 layup: (**a**) normal stress components in the x- and y-directions and shear stress in the xy-plane, and (**b**) safety ratio.

References

1. Zenkov, E. Investigation of the stress–strain state of racks of light steel thin-walled structures by the method of digital image correlation. *Mater. Today Proc.* **2021**, *38*, 1375–1378. [CrossRef]
2. Hou, Y.; Li, Z.; Ni, S.; Gong, J. Structural responses of a modular thin-walled steel trestle structure. *J. Constr. Steel Res.* **2019**, *158*, 502–521. [CrossRef]
3. Kasiviswanathan, M.; Upadhyay, A. Global buckling behavior of blade stiffened compression flange of FRP box-beams. *Structures* **2021**, *32*, 1081–1091. [CrossRef]
4. Ascione, L.; Berardi, V.P.; Giordano, A.; Spadea, S. Local buckling behavior of FRP thin-walled beams: A mechanical model. *Compos. Struct.* **2013**, *98*, 111–120. [CrossRef]

5. An, H.; Singh, J.; Pasini, D. Structural efficiency metrics for integrated selection of layup, material, and cross-section shape in laminated composite structures. *Compos. Struct.* **2017**, *170*, 53–68. [CrossRef]
6. Keller, A.; Geissberger, R.; Studer, J.; Leone, F.; Stefaniak, D.; Pascoe, J.; Dransfeld, C.; Masania, K. Experimental and numerical investigation of ply size effects of steel foil reinforced composites. *Mater. Des.* **2021**, *198*, 109302. [CrossRef]
7. Huang, Z.; Li, Y.; Zhang, X.; Chen, W.; Fang, D. A comparative study on the energy absorption mechanism of aluminum/CFRP hybrid beams under quasi-static and dynamic bending. *Thin-Walled Struct.* **2021**, *163*, 107772. [CrossRef]
8. Qin, H.; Guo, Y.; Liu, Z.; Liu, Y.; Zhong, H. Shape optimization of automotive body frame using an improved genetic algorithm optimizer. *Adv. Eng. Softw.* **2018**, *121*, 235–249. [CrossRef]
9. Cui, X.; Zhang, H.; Wang, S.; Zhang, L.; Ko, J. Design of lightweight multi-material automotive bodies using new material performance indices of thin-walled beams for the material selection with crashworthiness consideration. *Mater. Des.* **2011**, *32*, 815–821. [CrossRef]
10. Bao, Y.; Wang, B.; He, Z.; Kang, R.; Guo, J. Recent progress in flexible supporting technology for aerospace thin-walled parts: A review. *Chin. J. Aeronaut.* **2022**, *35*, 10–26. [CrossRef]
11. Moazed, R.; Khozeimeh, M.A.; Fotouhi, R. Simplified Approach for Parameter Selection and Analysis of Carbon and Glass Fiber Reinforced Composite Beams. *J. Compos. Sci.* **2021**, *5*, 220. [CrossRef]
12. Lin, Y.; Bai, H.; Lin, J.; Wang, M.; Lu, H.; Min, J. A lightweight method of thin-walled beams based on cross-sectional characteristic. *Procedia Manuf.* **2018**, *15*, 852–860. [CrossRef]
13. Chen, Z.; Li, G.-Q.; Bradford, M.A.; Wang, Y.-B.; Zhang, C.; Yang, G. Local buckling and hysteretic behavior of thin-walled Q690 high-strength steel H-section beam-columns. *Eng. Struct.* **2022**, *252*, 113729. [CrossRef]
14. Kollár, L.P.; Springer, G. *Mechanics of Composite Structures*; Cambridge University Press: Cambridge, UK, 2003.
15. Xiao, Y.; Wen, X.; Liang, D. Failure modes and energy absorption mechanism of CFRP Thin-walled square beams filled with aluminum honeycomb under dynamic impact. *Compos. Struct.* **2021**, *271*, 114159. [CrossRef]
16. Samal, P.K.; Pruthvi, I.; Suresh, B. Effect of fiber orientation on vibration response of glass epoxy composite beam. *Mater. Today Proc.* **2021**, *43*, 1519–1525. [CrossRef]
17. Ding, G.; Zhang, Y.; Zhu, Y. Experimental and numerical investigation of the flexural behavior of CFRP box girders. *Adv. Compos. Lett.* **2019**, *28*, 2633366X19891171. [CrossRef]
18. Daniel, I.M.; Ishai, O. *Engineering Mechanics of Composite Materials*; Oxford University Press: New York, NY, USA, 1994.
19. Gliszczyński, A.; Kubiak, T. Load-carrying capacity of thin-walled composite beams subjected to pure bending. *Thin-Walled Struct.* **2017**, *115*, 76–85. [CrossRef]
20. Debski, H.; Kubiak, T.; Teter, A. Experimental investigation of channel-section composite profiles' behavior with various sequences of plies subjected to static compression. *Thin-Walled Struct.* **2013**, *71*, 147–154. [CrossRef]
21. Zhang, Q.; Fotouhi, R.; Cote, J.; Pour, M.K. Lightweight Long-Reach 5-DOF Robot Arm for Farm Application. In Proceedings of the ASME 2019 International Design Engineering Technical Conferences and Computers and Information in Engineering Conference, Anaheim, CA, USA, 18–21 August 2019. [CrossRef]
22. Wyatt, H.; Wu, A.; Thomas, R.; Yang, Y. Life Cycle Analysis of Double-Arm Type Robotic Tools for LCD Panel Handling. *Machines* **2017**, *5*, 8. [CrossRef]
23. Hagenah, H.; Böhm, W.; Breitsprecher, T.; Merklein, M.; Wartzack, S. Construction and manufacture of a lightweight robot arm, 8th CIRP Conference on Intelligent Computation in Manufacturing Engineering. *Procedia CIRP* **2013**, *12*, 211–216. [CrossRef]
24. Yin, H.; Liu, J.; Yang, F. Hybrid Structure Design of Lightweight Robotic Arms Based on Carbon Fiber Reinforced Plastic and Aluminum Alloy. *IEEE Access* **2019**, *7*, 64932–64945. [CrossRef]
25. Lee, C.S.; Lee, D.G.; Oh, J.H.; Kim, H.S. Composite wrist blocks for double arm type robots for handling large LCD glass panels. *Compos. Struct.* **2002**, *57*, 345–355. [CrossRef]
26. Zeng, W.; Yan, J.; Hong, Y.; Cheng, S.S. Numerical analysis of large deflection of the cantilever beam subjected to a force pointing at a fixed point. *Appl. Math. Model.* **2021**, *92*, 719–730. [CrossRef]
27. Yang, W.; Hu, Y.; Zhang, J.; Ding, G.; Song, C. Analytical model for flexural damping responses of CFRP cantilever beams in the low-frequency vibration. *J. Low Freq. Noise Vib. Act. Control* **2018**, *37*, 669–681. [CrossRef]
28. Kollár, L.P. Flexural–torsional vibration of open section composite beams with shear deformation. *Int. J. Solids Struct.* **2001**, *38*, 7543–7558. [CrossRef]
29. *Ansys®Academic Research Mechanical, Release 18.1, Help System, Composite Materials Analysis User Guide*; ANSYS, Inc.: Canonsburg, PA, USA, 2017.

Article

The Machinability Characteristics of Multidirectional CFRP Composites Using High-Performance Wire EDM Electrodes

Ramy Abdallah [1,2], Richard Hood [1] and Sein Leung Soo [1,*]

[1] Machining Research Group, Department of Mechanical Engineering, School of Engineering, University of Birmingham, Edgbaston, Birmingham B15 2TT, UK; ramy_abdulnabi@h-eng.helwan.edu.eg (R.A.); r.hood@bham.ac.uk (R.H.)

[2] Production Engineering Department, Faculty of Engineering, Helwan University, Helwan 11795, Egypt

* Correspondence: s.l.soo@bham.ac.uk; Tel.: +44-121-4144196

Abstract: Due to the abrasive nature of the material, the conventional machining of CFRP composites is typically characterised by high mechanical forces and poor tool life, which can have a detrimental effect on workpiece surface quality, mechanical properties, dimensional accuracy, and, ultimately, functional performance. The present paper details an experimental investigation to assess the feasibility of wire electrical discharge machining (WEDM) as an alternative for cutting multidirectional CFRP composite laminates using high-performance wire electrodes. A full factorial experimental array comprising a total of 8 tests was employed to evaluate the effect of varying ignition current (3 and 5 A), pulse-off time (8 and 10 µs), and wire type (Topas Plus D and Compeed) on material removal rate (MRR), kerf width, workpiece surface roughness, and surface damage. The Compeed wire achieved a lower MRR of up to ~40% compared with the Topas wire when operating at comparable cutting parameters, despite having a higher electrical conductivity. Statistical investigation involving analysis of variance (ANOVA) showed that the pulse-off time was the only significant factor impacting the material removal rate, with a percentage contribution ratio of 67.76%. In terms of cut accuracy and surface quality, machining with the Compeed wire resulted in marginally wider kerfs (~8%) and a higher workpiece surface roughness (~11%) compared to the Topas wire, with maximum recorded values of 374.38 µm and 27.53 µm Sa, respectively. Micrographs from scanning electron microscopy revealed the presence of considerable fibre fragments, voids, and adhered re-solidified matrix material on the machined surfaces, which was likely due to the thermal nature of the WEDM process. The research demonstrated the viability of WEDM for cutting relatively thick (9 mm) multidirectional CFRP laminates without the need for employing conductive assistive electrodes. The advanced coated wire electrodes used in combination with higher ignition current and lower pulse-off time levels resulted in an increased MRR of up to ~15 mm^3/min.

Keywords: electrical discharge machining; carbon fibre; composites; material removal rate; kerf width; surface roughness

Citation: Abdallah, R.; Hood, R.; Soo, S.L. The Machinability Characteristics of Multidirectional CFRP Composites Using High-Performance Wire EDM Electrodes. *J. Compos. Sci.* **2022**, *6*, 159. https://doi.org/10.3390/jcs6060159

Academic Editor: Jiadeng Zhu

Received: 15 March 2022
Accepted: 24 May 2022
Published: 27 May 2022

Publisher's Note: MDPI stays neutral with regard to jurisdictional claims in published maps and institutional affiliations.

Copyright: © 2022 by the authors. Licensee MDPI, Basel, Switzerland. This article is an open access article distributed under the terms and conditions of the Creative Commons Attribution (CC BY) license (https://creativecommons.org/licenses/by/4.0/).

1. Introduction

Carbon-fibre-reinforced plastic (CFRP) composites have become widely used in a variety of applications, particularly within the aerospace and automotive industries due to favourable properties including low density (1.5 to 2 gm/cm^3), high strength-to-weight ratio (~785 kN·m/kg), as well as strong resistance to fatigue and corrosion [1–3]. This has enabled some of the latest commercial aircraft such as the Airbus A350XWB and Boeing 787 to fly with considerably reduced fuel consumption and lower gas emissions, as up to 50% of their airframes are composed of CFRP materials instead of heavier metallic alloys [4,5]. Despite efforts in developing net shape manufacturing technologies for CFRP components, machining operations remain essential to meet the required dimensional accuracy, geometrical tolerances, and for part assembly [6,7]. Commonly employed machining operations include routing, slot milling, and drilling, with the latter utilised for producing bolt or rivet holes, which can number up to 400,000 for private jets and in-excess of 1 million in larger transport/cargo planes [8]. However, in addition to the anisotropic and inhomogeneous properties of CFRP, the highly abrasive nature of carbon fibres results in the rapid wear of tools in conventional machining processes, leading to severe workpiece defects such as delamination, fibre pull-out, matrix degradation, burr formation, splintering, and micro cracking [9,10], thus compromising the strength as well as integrity of the workpiece. While recent research investigating the machining of CFRP using nontraditional cutting processes encompassing ultrasonic vibration-assisted drilling (UVAD), abrasive water jet machining (AWJM), and laser beam machining (LBM) have reported some benefits over conventional processes such as lower cutting forces, longer tool life, and reduced surface damage, there were also several drawbacks observed including abrasive particle embedment in the matrix, the presence of craters/ridges/valleys, and the formation of heat-affected zones (HAZ) [11–13].

Another nonconventional process that has seen increasing research interest as a potential alternative for cutting CFRP composites is electrical discharge machining (EDM). When EDM drilling of CFRP using copper and graphite electrodes, Sheikh-Ahmed and Shinde [8] investigated the impact of varying current and pulse duration on the resulting tool wear, material removal rate (MRR), workpiece delamination, and hole tapering/size deviation. In terms of MRR, the graphite electrode was found to outperform the copper tool by up to 38% when operating at a current of 2 A and pulse-on time of 190 µs, albeit at the cost of a 500% increase in electrode wear rates in the former. When machining with a low current of 0.4 A and pulse duration ranging between 20 and 105 µs, a reduced delamination factor of 1.15 from a maximum of ~1.3 was recorded for both electrodes. However, the highest deviation from the nominal hole size (~4.5%) together with a maximum taper of 0.048 mm was observed when using the graphite electrode at elevated current (2 A) and pulse duration (190 µs) levels. In related work, epoxy decomposition, fibre breakage, and sublimation together with Joule heating generated by short-circuiting of fibres were identified as the main material removal mechanisms [14]. Delamination was most apparent at the hole entry, which was substantially influenced by changes in the discharge current, while pulse-on time had the strongest impact on HAZ and hole tapering.

According to a study conducted by Yue et al. [15], material removal in the EDM of CFRP was caused not only by plasma and Joule heating, but also through mechanical and chemical mechanisms where evaporated gases from epoxy and carbon fibre sublimation (due to the localised high temperatures in the discharge zone) form a high-speed gaseous jet. The energy from impact of the gas stream caused fibres in the workpiece to rupture, particularly when machining perpendicular to the fibre orientation. In addition, the oxygen released following dissociation of the deionised water dielectric during sparking further contributed to increased thermal energy and hence faster material erosion. Kumaran et al. [16] investigated the impact of incorporating carbon black (1 and 2 vol%) and graphite (5 and 10 vol%) fillers into the CFRP matrix on EDM performance in terms of MRR, electrode wear, and hole quality. Drilling was carried out using a 1 mm diameter brass electrode at negative polarity with variable parameters of pulse-off time (p = 20, 30, and 42 µs) and maximum current (I_{max} = 46, 91, and 153 A). When cutting with pulse-off times above 20 µs, workpieces containing a filler typically exhibited a higher MRR compared to the conventional reference CFRP (no filler added), regardless of current levels, implying that the flushing efficiency was most likely improved. Furthermore, electrode wear and thermal damage around the hole exit were reduced when machining workpieces with the filler material, probably as a result of increased matrix thermal conductivity leading to a greater diffusion of heat over the workpiece, thus preventing/minimising matrix degradation. Kumar et al. [17] studied the viability of EDM in the micro-drilling of CFRP using a 110 µm diameter electrode for a hole depth of 1.2 mm (aspect ratio of 10.9). The principal variable parameters were voltage, capacitance, and electrode/tool rotational speed. In terms of maximising MRR and minimising electrode wear rate, both voltage and capacitance were found to be statistically significant factors, while tool speed had no significant impact on either response. Similarly, Makudapathy and Sundaram [18] reported a maximum mean hole aspect ratio of over 11 during the micro-EDM drilling of CFRP (2.5 mm thick) utilising a tungsten carbide electrode (Ø 120 µm) at a feed rate and voltage of 6 µm/s and 70 V, respectively. An increasing voltage and lowering feed rate resulted in larger overcuts, while a positive electrode polarity was found to reduce tool wear.

Lau and Lee [19] reported some of the earliest research on the wire electrical discharge machining (WEDM) of CFRP, which was evaluated alongside laser cutting. Even though laser machining exhibited a substantially higher MRR (95 mm^2/min) compared to WEDM (12 mm^2/min), the latter revealed a better edge quality and reduced HAZ/surface damage. The influence of fibre orientation with respect to cut direction was assessed by Yue et al. [15] using uncoated brass in the WEDM of 2 mm thick unidirectional (UD) CFRP laminates. Cutting speed was found to be ~18% higher when machining parallel to the fibre orientation due to the higher electrical and thermal conductivity of the fibres along the axial direction. More recently, the influence of open volage, current, and pulse-on/off time when WEDM thicker UD-CFRP laminates (8.4 mm) was investigated when cutting parallel to the fibre direction using zinc-coated brass wire [20]. Analysis of variance (ANOVA) of the results revealed that ignition current had a statistically significant effect on MRR and kerf width, with the pulse-off time also having a substantial impact on MRR. A further study by Abdallah et al. [21] to assess WEDM performance when cutting perpendicular to the fibre orientation found that MRR was ~14% lower than when machining parallel to the fibre direction, which was similar to that previously reported by Yue at al. [15]. However, workpiece surface roughness was considerably better when cutting perpendicular to the fibres (6.84 µm Sa) instead of the parallel direction (9.48 µm Sa), although surface defects including uncut/fractured fibre fragments, matrix/fibre loss, and re-solidified resin were prevalent on all of the machined surfaces.

In order to enhance workpiece electrical conductivity and aid spark initiation, several researchers have evaluated the use of 'assisting electrodes' where the CFRP workpiece is stacked or clamped together with metallic plates. Apart from increased cutting efficiency/productivity [22], the technique was also found to prevent/avoid frequent wire breakage and improve cut profile accuracy [4,23]. Based on the comparatively limited published research on the WEDM of CFRP to date, the work presented here aimed to investigate the effect of key process parameters (ignition current and pulse-off time) on the machinability of 9 mm thick multidirectional CFRP composite laminates when utilising two different high-performance wire electrodes. Process performance and capability were assessed with respect to an in-depth analysis of resulting kerf widths, surface roughness, and condition together with material removal rates.

2. Materials and Methods

The workpiece materials used in the experiments were 9 mm thick, square (122 mm × 122 mm) multidirectional CFRP laminate plates comprising 36 UD pre-pregs/plies, each with a thickness of 0.25 mm. The plies were composed of high-tensile-strength (HTS) pre-impregnated carbon fibres (6–8 µm diameter) aligned within a toughened epoxy resin matrix. The commercial matrix/fibre designation of the pre-pregs was ACG MTM44-1/HTS-268-12K, which had a fibre volume fraction (V_f) of 56.5%. As shown in Figure 1, the laminates were initially manually laid up according to an orientation sequence of $[45°/0°/135°/90°/45°/0°]_{3S}$. The workpiece samples were then cured for 30 min at a steady temperature of 80 °C, followed by a gradual increase in temperature to 135 °C at a rate of 1 °C per minute. This temperature was subsequently kept constant for 4 h under a vacuum pressure of 0.9 bar [24]. The resulting key physical/mechanical properties of the cured CFRP laminate plates are detailed in Table 1 [24,25].

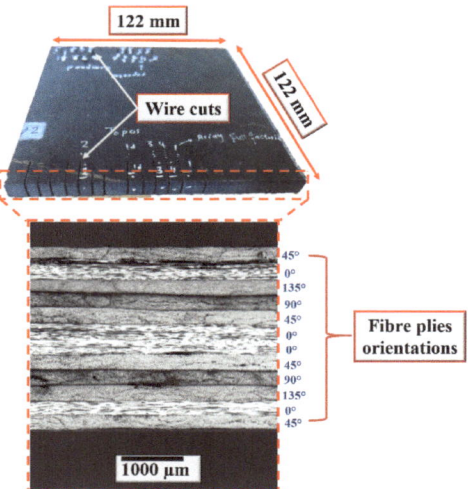

Figure 1. Multidirectional CFRP workpiece with micrograph of cross-section showing lay-up orientation of the plies [24,25].

Table 1. Key physical/mechanical properties of the post cure CFRP laminates [24,25].

Property	Details
Density	1.6 g/cm^3
Hardness	60–65 Barcol
Ultimate tensile strength	2000 MPa
Interlaminate shear strength	14 MPa
Modulus of elasticity	150 GPa
Thermal conductivity	1 W/mK $\perp$ and 70 W/mK//to fibre
Workpiece electrical resistivity (X, Y, Z)	0.0833 $\Omega\cdot$cm, 0.092 $\Omega\cdot$cm, 1980.1 $\Omega\cdot$cm
Coefficient of thermal expansion (CTE)	Up to 10 μm/mK

As low cutting speeds and frequent wire rupture have previously been reported as the primary shortcomings in the WEDM of CFRP composites, two high-performance electrodes involving zinc-coated brass (Topas Plus D) and copper/brass-coated steel (Compeed) wires were selected for evaluation in the present work. The Topas Plus D wire was composed of a brass core (20%Zn80%Cu) with a dual layer of zinc-rich diffused β and γ-phase coatings, giving a tensile strength of 800 to 900 MPa and electrical conductivity of 16.24×10^6 S/m. This has been found to provide a 30–35% increase in cutting speed compared to conventional uncoated brass wires [26,27]. In contrast, the Compeed wire has a steel core with a twin layer of copper and diffused β-phase brass coating, resulting in an electrical conductivity of 29×10^6 S/m and room-temperature tensile strength of 800 MPa. Despite comparable tensile strengths, the Compeed wire has greater resistance to breakage than the Topas Plus D due to the higher fracture toughness of the steel core. The diameter of both wires was 0.25 mm, with their respective compositions shown in Figure 2.

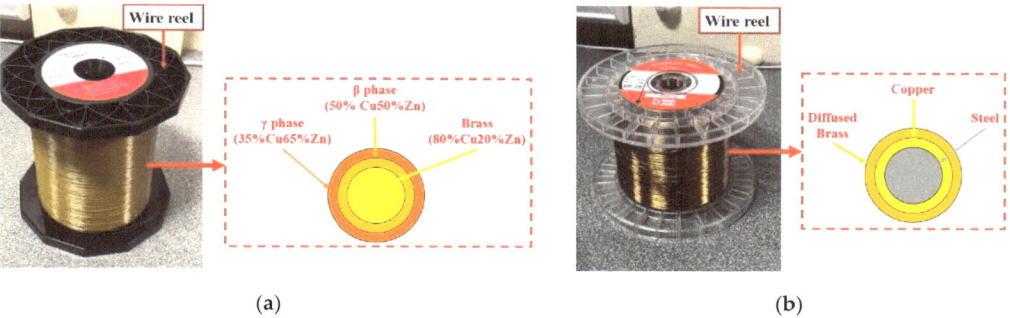

Figure 2. Composition of wire electrodes: (**a**) Topas Plus D and (**b**) Compeed.

All of the machining trials were undertaken on an AgieCharmilles Robofil FI240cc wire EDM machine, as shown in Figure 3. The CFRP workpiece was immersed in deionised water dielectric with an electrical conductivity of ~5 µS/cm during machining. An Alicona G5 InfiniteFocus microscope was used to measure the machined kerf width as well as to scan and generate 3D topographical plots of the cut surfaces to determine areal roughness (Sa) at magnifications of 5× and 20×, respectively. A toolmakers microscope connected to a digital camera was employed to capture optical micrographs of machined kerfs on the top and bottom faces of the workpiece, while a JEOL JCM-6000 Plus scanning electron microscope (SEM) was utilised to obtain high-resolution micrographs of the machined surfaces.

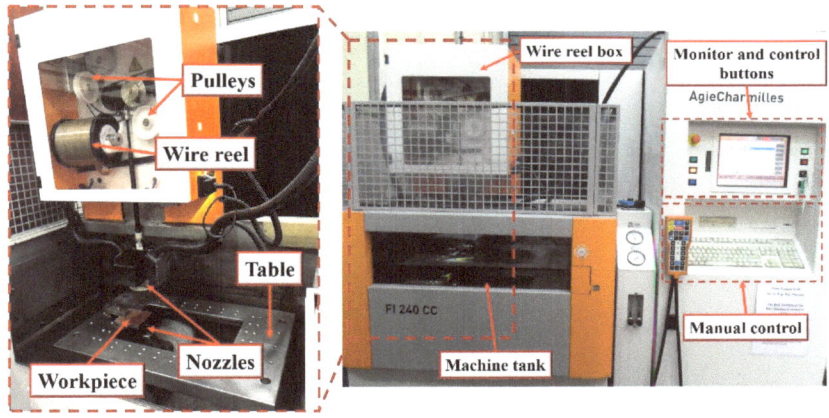

Figure 3. AgieCharmilles Robofil FI240cc wire EDM machine.

The experiments carried out were based on a full factorial design involving three variable factors of ignition current, pulse-off time, and wire electrode type, with each at 2 levels. Table 2 details both the variable and fixed parameters together with the corresponding levels selected for the trials, which were based on results by Abdallah et al. [20,21] following the WEDM of UD-CFRP laminates. Process performance was evaluated in terms of material removal rate, average kerf width (Wa), and workpiece surface roughness following a cut length of 12 mm in each test, with statistical analysis of the results using Minitab software employed to determine the significance (at the 5% level) of variable parameters. The full factorial test array is outlined in Table 3.

Table 2. Variable and constant parameters.

	Parameters	Symbol (Unit)	Level 1	Level 2
Variable factors	Ignition current	IAL (A)	3	5
	Pulse-off time	B (μs)	8	10
	Wire type	W	Topas Plus D (T)	Compeed (C)
Constant factors	Open gap voltage	Vo (V)	100	-
	Pulse-on time	A (μs)	0.6	-
	Servo voltage	Aj (V)	15	-
	Wire tension	WB (N)	13	-
	Wire speed	WS (m/min)	10	-
	Flushing pressure	INJ (bar)	16	-
	Frequency	FF (%)	10	-

Table 3. Full factorial experimental array.

Test No.	Ignition Current, IAL (A)	Pulse-off Time, B (μs)	Wire Type, W
1	3	8	T
2	5	8	T
3	3	10	T
4	5	10	T
5	3	8	C
6	5	8	C
7	3	10	C
8	5	10	C

The machining time for each test (over the 12 mm cut length) was recorded using a stopwatch. Following the conclusion of each trial, an air blower and hair drier were utilised to dry and evaporate the absorbed water in the CFRP plates. A digital scale (operating range of 0.5 to 3500 g) was used to weigh the mass of the workpiece samples before and after each test (precision of 0.01 g). The MRR was subsequently determined using Equation (1) [28,29]:

$$\text{MRR} = \frac{m_b - m_a}{\rho \times t_m} \quad (1)$$

where m_b and m_a are the masses of workpiece material before and after machining, respectively (g), ρ is the density of the workpiece (g/mm^3), and t_m is the machining time (min). The delamination factor (Fd) as a result of the damage on the top and bottom faces of the workpiece was calculated using Equation (2) [30]:

$$\text{Fd} = \frac{W_{max}}{W_a} \quad (2)$$

where Wmax is the maximum width of delamination of the machined kerf and Wa is the average kerf width.

The average kerf width was calculated based on a total of 18 measurements taken at equal intervals along the machined length on the top and bottom of the workpiece (9 readings per surface). Following completion of kerf width measurements, the CFRP workpieces were cross-sectioned using a diamond disc cutter for surface roughness (Sa) measurements. This was evaluated as an average of three readings, each over a scan area of 13.2 mm^2 on the machined surface and a cut-off length of 887 μm.

3. Results and Discussion

3.1. Material Removal Rate and Workpiece Surface Damage

Figure 4 shows the effect of varying ignition current and pulse-off time on MRR when cutting multidirectional CFRP laminates using the Topas and Compeed wire electrodes. When operating at equivalent ignition current and pulse-off time levels, the MRR achieved by the Topas wire was between 11% and 40% higher than that of the Compeed electrode, with maximum values recorded of 14.82 mm^3/min (Test 2) and 13.31 mm^3/min (Test 6), respectively. These cutting rates were considerably faster compared to previously reported results when machining unidirectional CFRP composites [21], which was primarily attributed to the lower electrical resistivity of the multidirectional lay-up configuration of the workpiece in the present work. As expected, the main effects plot for the mean of MRR shown in Figure 5 indicates that the cutting rate increased at higher ignition current and lower pulse-off time levels. Conversely, the MRR was generally lower when machining with the Compeed wire, despite having a higher electrical conductivity. The elevated conductivity of the Compeed wire likely resulted in sparks with greater intensity, thereby causing increased workpiece erosion and debris formation consisting of the decomposed nonconductive epoxy resin within the machining gap, which also accumulated/fused over the fibre edges on the cut surface [31]. Furthermore, the rate of fibre detachment was possibly higher due to a greater spark collision with fibres at the different ply orientations (45°, 90°, and 135°) [29]. This build-up of debris comprising the melted epoxy resin and broken fibres in the cutting zone consequently led to poor discharge efficiency and hence the lower MRR.

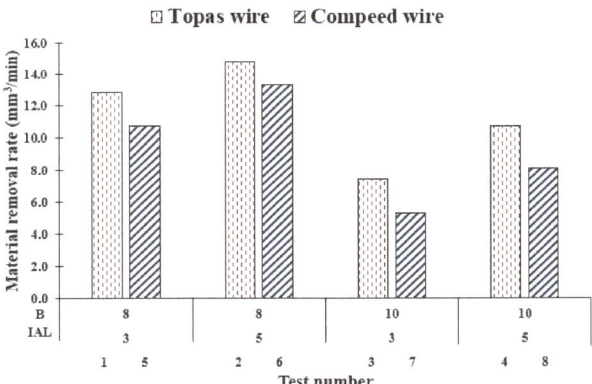

Figure 4. Material removal rate in each test for the Topas and Compeed wire electrodes.

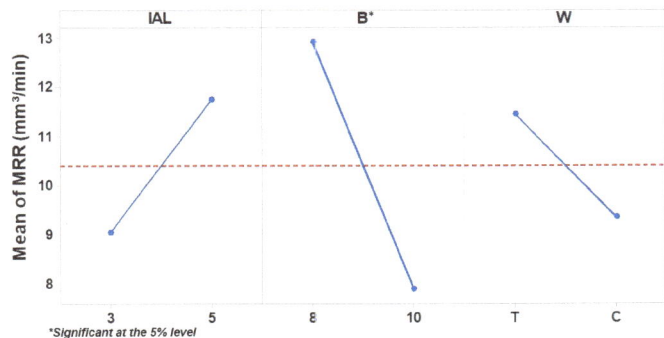

Figure 5. Main effects plot for MRR.

The ANOVA considering the effect of the main factors as well as interactions between the parameters on MRR is detailed in Table 4. According to the results, the pulse-off time was the only significant factor at the 5% level, with a corresponding percentage contribution ratio (PCR) of 67.76%. Similarly, none of the interactions were shown to have any significant influence. The derived regression model exhibited strong agreement with the experimental data based on a coefficient of determination (R^2) of 99.79%. Additionally, the similarly high adjusted coefficient of determination (Adj R^2) of 98.55% indicated that the model likely included all the relevant terms for good correlation between the responses and machining variables.

Table 4. ANOVA for material removal rate including interactions.

Source	DF	Seq SS	Adj SS	Adj MS	F-Value	*p*-Value	PCR
Model	6	74.0556	74.0556	12.3426	80.07	0.085	99.79%
Linear	3	73.597	73.597	24.5323	159.15	0.058	99.17%
IAL	1	14.5094	14.5094	14.5094	94.13	0.065	19.55%
B	1	50.281	50.281	50.281	326.19	0.035 *	67.76%
W	1	8.8065	8.8065	8.8065	57.13	0.084	11.87%
2-Way Interactions	3	0.4587	0.4587	0.1529	0.99	0.611	0.62%
IAL*B	1	0.2877	0.2877	0.2877	1.87	0.402	0.39%
IAL*W	1	0.001	0.001	0.001	0.01	0.949	0.00%
B*W	1	0.17	0.17	0.17	1.1	0.484	0.23%
Error	1	0.1541	0.1541	0.1541			0.21%
Total	7	74.2098					100.00%
Model equation		MRR = 34.40 − 0.36 IAL − 3.266 B + 0.22 W + 0.190 IAL*B + 0.011 IAL*W − 0.146 B*W					
Model summary							
S	R^2		Adj R^2		PRESS		Pred R^2
0.392615	99.79%		98.55%		9.86536		86.71%

* Significant at the 5% level.

Figure 6 shows optical micrographs of the kerf generated on the top and bottom surfaces of the workpiece machined using the Topas wire in Test 2, which recorded the highest MRR. The top surface exhibited severe delamination along the kerf edges and at the end of the cut, see Figure 6a. The kerf edges were also irregular/uneven with evidence of adhered debris/contaminants on the surface, with similar damage also found on the bottom surface, although to a lesser degree, as shown in Figure 6b.

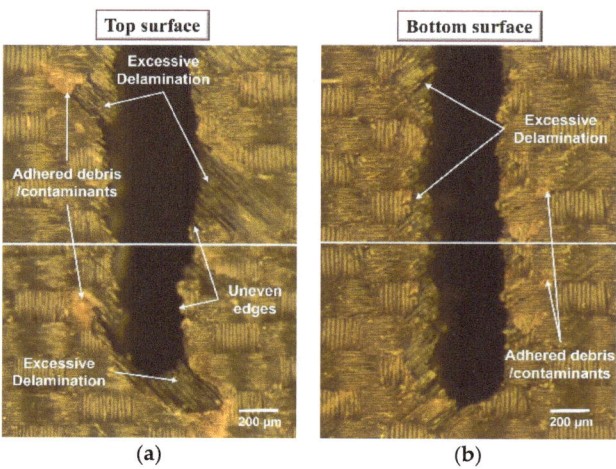

Figure 6. Optical micrographs of the kerf machined using Topas wire in Test 2 at the (**a**) top and (**b**) bottom surface.

The slots machined using the Compeed wire at equivalent process parameters (Test 6) also exhibited considerable delamination along the length of kerf edges together with irregular/uneven edges and the presence of frayed fibres and bronze-coloured debris attached on both the top and bottom surfaces of the workpiece, as detailed in Figure 7. The delamination typically occurred in epoxy-rich areas due to the pressure of gases generated from resin decomposition, as described by Sheikh-Ahmad [14]. The gases together with wire tension, as well as electrostatic and dielectric flushing forces, led to wire vibration [32], which probably contributed to the irregular/uneven cut edges. The damage on the machined kerfs was generally more severe in tests utilising the Compeed wire, which was probably due to the increased discharge intensity as a result of its higher electrical conductivity. This corresponded to the higher delamination factor (Fd) levels obtained of up to 2.94 for slots machined with the Compeed wire compared to a maximum value of 2.5 in tests involving the Topas wire. In addition, the fractured/frayed fibre damage observed was also partly attributed to the impact of high-speed gaseous jets generated during vaporisation and sublimation of epoxy/carbon fibres, which have also been outlined by other researchers [14,15]. While the delamination factor levels were up to ~49% higher compared to previously reported results when WEDM of UD-CFRP [21], the degree of workpiece damage was found to be lower when using WEDM as opposed to laser machining, according to Lau and Lee [19].

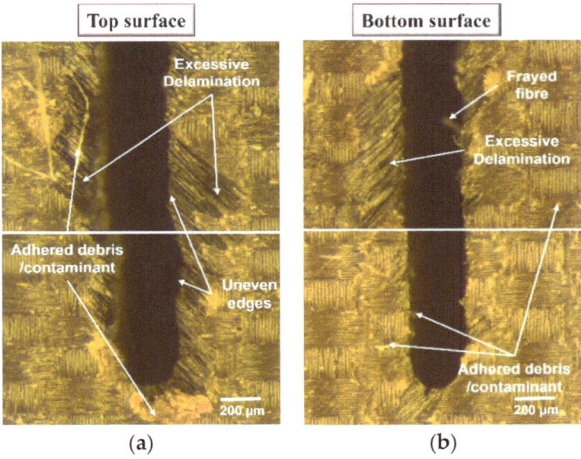

Figure 7. Optical micrographs of the kerf machined using Compeed wire in Test 6 at the (**a**) top and (**b**) bottom surface.

Figure 8a shows a SEM micrograph of the surface machined in Test 3, which recorded the lowest MRR (7.42 mm^3/min) when using the Topas wire. In addition to significant areas with voids due to fibre/matrix loss, the surface was covered with broken fibre fragments at different ply orientations together with accumulated re-solidified resin at the fibre ends. Conversely, loose long fibres were predominant on the workpiece machined in Test 2 (highest MRR), together with large cavities and re-solidified resin dispersed over the cut surface. In addition, resin particles were seen on the ends of some fibres at the 45°/135° ply orientation, as shown in Figure 8b.

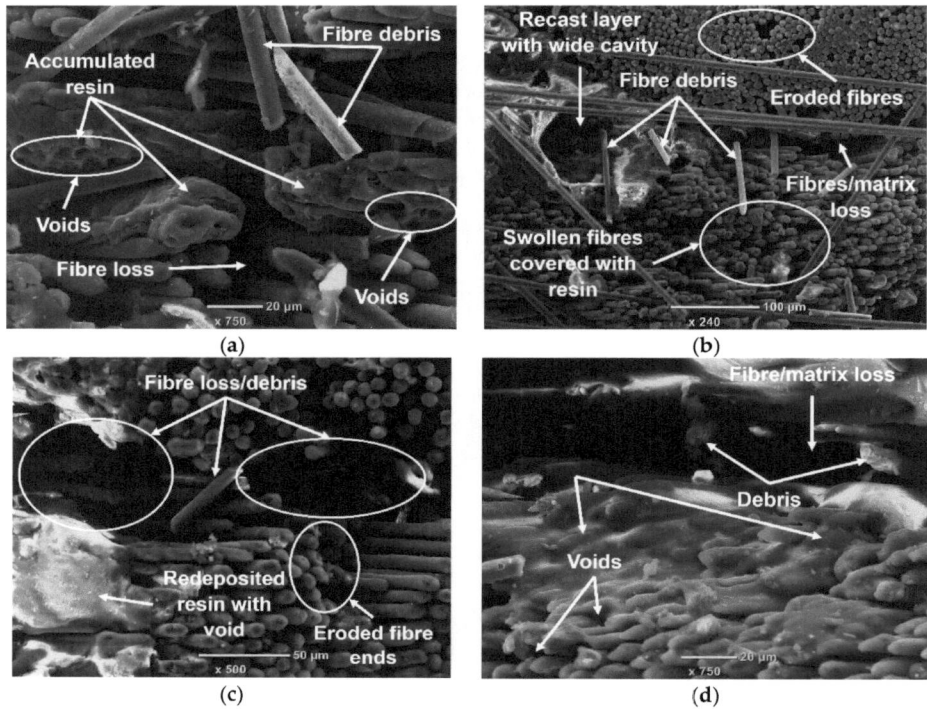

Figure 8. Sample SEM micrographs of surfaces machined using the Topas wire in (**a**) Test 3 and (**b**) Test 2 and Compeed wire in (**c**) Test 7 and (**d**) Test 6.

The damage on the surface machined with the Compeed wire having the lowest MRR (5.29 mm^3/min) in Test 7 included substantial fibre loss particularly at the ply interface regions and fibre fragments together with bulk re-solidified resin covering some of the eroded fibre sections, see SEM micrograph in Figure 8c. The machined surface condition/quality further deteriorated particularly in Test 6 at the highest MRR, which was characterised by the presence of deep cavities as well as extensive areas of accumulated re-solidified debris/matrix material, as shown in Figure 8d. In contrast, Ablyaz et al. [33] reported minimal surface damage/defects following the WEDM of 2 mm thick polymer composite plates sandwiched between two layers of conductive titanium plates (each 1 mm thick). Despite the improved surface quality, the use of a sandwich/stack configuration may not always be practical/possible in a production environment, depending on the component geometry.

3.2. Kerf Width

Figure 9 highlights the average kerf widths obtained in each of the tests for both wire electrodes. Marginally larger kerfs ranging from ~334 to 374 µm were produced when cutting with the Compeed wire as opposed to slot widths measuring ~333 to 347 µm from tests utilising the Topas electrode. The wider kerfs were typically associated with operating at low pulse-off times and high ignition currents, which most likely resulted in greater discharge energies. This trend was exemplified by the main effects plot for average kerf width shown in Figure 10. While increasing the pulse-off time generally improves flushing efficiency for removing debris from the discharge gap, it also reduces the active spark-workpiece interaction time, thereby resulting in a narrower kerf. Previous studies involving the WEDM of UD-CFRP at comparable cutting conditions reported smaller average kerf widths, which did not exceed 300 µm [20,21].

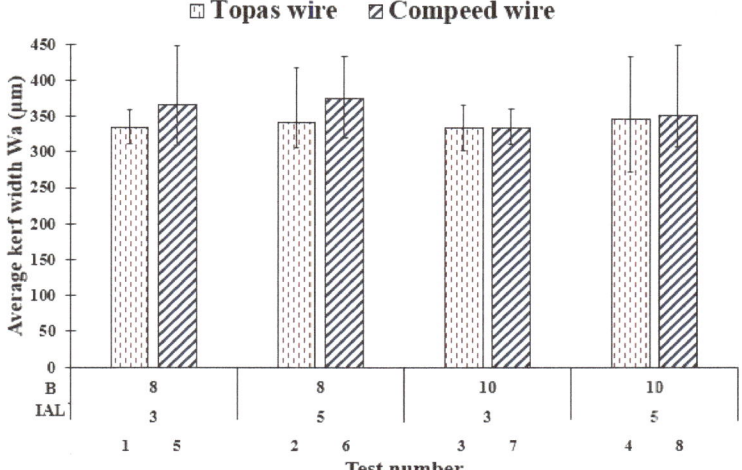

Figure 9. Average kerf width in each test for the Topas and Compeed wire electrodes.

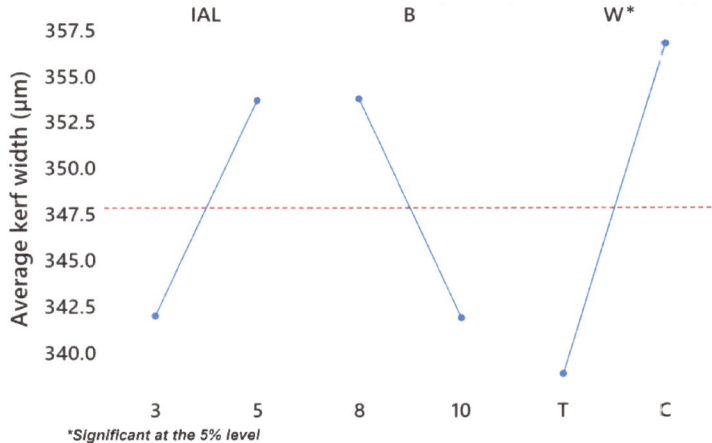

Figure 10. Main effects plot for average kerf width.

According to the ANOVA for average kerf width outlined in Table 5, wire type and its interaction with pulse-off time were found to be significant at the 5% level, with a PCR of 37.69% and 27.39%, respectively. The corresponding regression model derived revealed a strong correlation with the experimental data based on the high R^2 and Adj R^2 values of 99.88% and 99.14%, respectively.

Table 5. ANOVA for average kerf width including interactions.

Source	DF	Seq SS	Adj SS	Adj MS	F-Value	p-Value	PCR
Model	6	1701.86	1701.86	283.643	135.34	0.066	99.88%
Linear	3	1195.78	1195.78	398.592	190.19	0.053	70.18%
IAL	1	272.42	272.42	272.422	129.99	0.056	15.99%
B	1	281.11	281.11	281.111	134.13	0.055	16.50%
W	1	642.24	642.24	642.244	306.45	0.036 *	37.69%
2-Way Interactions	3	506.08	506.08	168.693	80.49	0.082	29.70%
IAL*B	1	34.33	34.33	34.332	16.38	0.154	2.01%
IAL*W	1	5.08	5.08	5.084	2.43	0.363	0.30%
B*W	1	466.66	466.66	466.664	222.67	0.043 *	27.39%
Error	1	2.1	2.1	2.096			0.12%
Total	7	1703.95					100.00%
Model equation		\multicolumn{7}{l}{Wa = 452.4 − 12.81 IAL − 14.21 B + 74.51 W + 2.072 IAL*B + 0.797 IAL*W − 7.638 B*W}

Model summary				
S	R^2	Adj R^2	PRESS	Pred R^2
1.44767	99.88%	99.14%	134.129	92.13%

* Significant at the 5% level.

3.3. Workpiece Surface Roughness

The measured surface roughness for each trial detailed in Figure 11 exhibited similar trends to the average kerf width results, where the Compeed wire produced rougher surfaces compared to the Topas wire, with the exception of Test 7, which was possibly due to variation in spark stability. The higher surface roughness obtained when machining with the Compeed wire was attributed to the increased energy/heat generated in the machining gap as a consequence of its higher electrical conductivity, which produced deeper craters and considerable adhered debris on the surface; see sample 3D topographical maps of workpieces machined with the Topas (Test 2) and Compeed (Test 6) wires in Figure 12. Evidence of voids/craters was present on both surfaces due to vaporisation of the matrix phase, while more significant levels of frayed/loose fibres were visible on the surfaces machined with the Compeed wire; see Figure 12b.

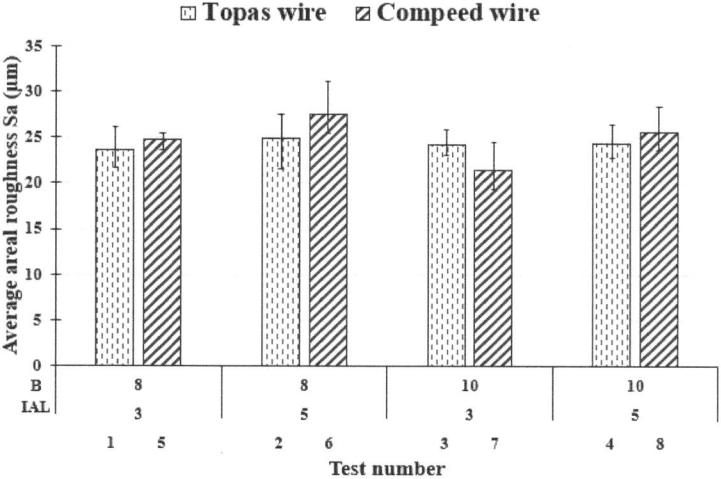

Figure 11. Average areal surface roughness in each test for the Topas and Compeed wire electrodes.

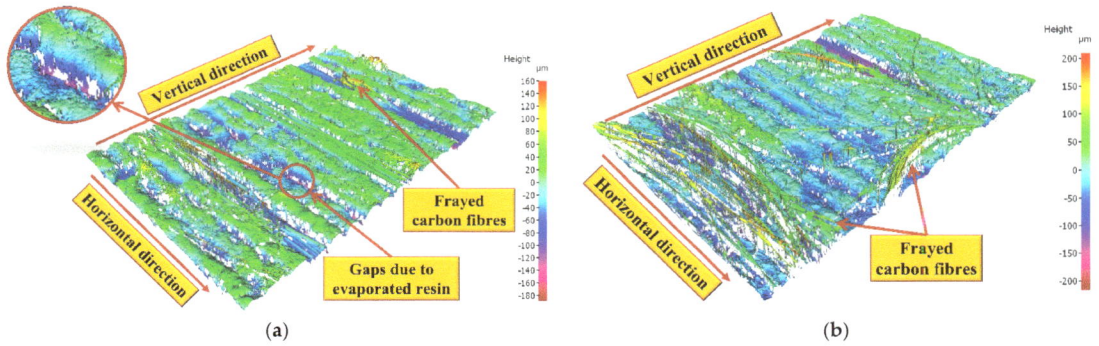

Figure 12. Surface topographical map of workpiece machined with (**a**) Topas (Test 2) and (**b**) Compeed (Test 6) wires.

The main effects plot in Figure 13 indicates that utilising low ignition current and high pulse-off time levels with the Topas wire resulted in lower surface roughness. However, the associated ANOVA in Table 6 showed that none of the linear factors and corresponding interactions were statistically significant with respect to surface roughness.

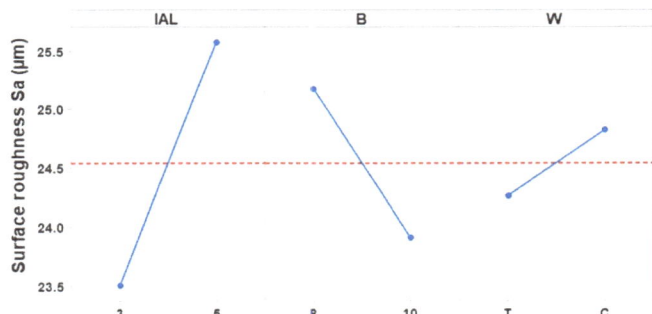

Figure 13. Main effects plot for average surface roughness.

Table 6. ANOVA for average surface roughness.

Source	DF	Seq SS	Adj SS	Adj MS	F-Value	*p*-Value	PCR
Model	6	19.9873	19.9873	3.33121	4.27	0.355	96.24%
Linear	3	12.4798	12.4798	4.15992	5.33	0.306	60.09%
IAL	1	8.6433	8.6433	8.6433	11.07	0.186	41.62%
B	1	3.2123	3.2123	3.21231	4.12	0.292	15.47%
W	1	0.6242	0.6242	0.62416	0.8	0.536	3.01%
2-Way Interactions	3	7.5075	7.5075	2.5025	3.21	0.385	36.15%
IAL*B	1	0.0039	0.0039	0.00389	0	0.955	0.02%
IAL*W	1	4.0162	4.0162	4.0162	5.14	0.264	19.34%
B*W	1	3.4874	3.4874	3.4874	4.47	0.281	16.79%
Error	1	0.7806	0.7806	0.7806			3.76%
Total	7	20.7679					100.00%
Model equation		Sa = 26.9 + 0.84 IAL − 0.72 B + 3.39 W + 0.022 IAL*B + 0.709 IAL*W − 0.660 B*W					
Model summary							
S		R^2	Adj R^2	PRESS		Pred R^2	
0.883518		96.24%	73.69%	49.9587		0.00%	

4. Conclusions

The impact of two key operating parameters: ignition current and pulse-off time, in addition to two different wire electrodes when WEDM multidirectional CFRP laminate composites was assessed with regard to material removal rate, average kerf width, and workpiece surface roughness. The following were the main outcomes and observations.

- Material removal rates ranged between 7.42 and 14.82 mm^3/min when utilising the Topas wire, while a marginally lower MRR of 5.29 to 13.31 mm^3/min was recorded in tests involving the Compeed wire. The lower cutting rates in the latter was possibly due to increased debris generation in the spark gap resulting in reduced machining efficiency. Pulse-off time was found to be the only statistically significant factor influencing MRR, with a PCR of 67.76%. The regression model considering both linear factors as well as two-way interactions between the factors exhibited a strong correlation to the experimental results with a R^2 of 99.79%.
- Considerable levels of delamination were observed on the top and bottom surfaces of the machined workpieces in all tests, although delamination factor calculations indicated somewhat higher values of up to 2.94 for workpieces machined using the Compeed wire, whilst a corresponding maximum Fd of 2.5 was obtained when employing the Topas wire.
- Analysis of the cut surfaces revealed the presence of significant fibre debris, voids, and aggregated re-solidified resin material, particularly for workpieces machined with the Topas wire. In similar tests involving the Compeed wire, additional defects were prevalent in the form of large cavities in the vicinity of ply interfaces due to resin evaporation and fibre loss.
- Marginally higher kerf widths of up to ~8% were observed when machining with the Compeed wire. The corresponding ANOVA highlighted that wire type as well as its interaction with pulse-off time were significant at the 5% level, with PCRs of 37.69% and 27.39%, respectively. The derived regression model for kerf width showed a high R^2 of 99.88%, which suggests a strong fit with the experimental data.
- The resulting average workpiece surface roughness was relatively high, irrespective of cutting conditions and wire type, with maximum values of 24.86 µm and 27.53 µm Sa for the Topas and Compeed wire, respectively. None of the variable factors or interactions, however, were found to have a significant influence on surface roughness, despite ignition current having a considerable PCR of 41.62%.

Author Contributions: Conceptualization, S.L.S. and R.H.; methodology, R.A. and S.L.S.; validation, R.A.; formal analysis, R.A.; investigation, R.A.; resources, S.L.S. and R.H.; data curation, R.A.; writing—original draft preparation, R.A.; writing—review and editing, S.L.S. and R.H.; visualization, R.A. and S.L.S.; supervision, S.L.S. and R.H.; project administration, S.L.S.; funding acquisition, S.L.S. and R.H. All authors have read and agreed to the published version of the manuscript.

Funding: This research received no external funding.

Institutional Review Board Statement: Not applicable.

Informed Consent Statement: Not applicable.

Data Availability Statement: Not applicable.

Acknowledgments: The authors would like to express their gratitude to the Egyptian Cultural and Educational Bureau for awarding a research studentship to R.A. to undertake the work.

Conflicts of Interest: The authors declare no conflict of interest.

Nomenclature

Acronym	Description
3D	Three dimensional
A	Pulse-on time
ACG	Advanced Composite Group
Adj R^2	Adjusted coefficient of determination
Aj	Servo voltage
ANOVA	Analysis of variance
AWJM	Abrasive water jet machining
B	Pulse-off time
C	Compeed
CFRP	Carbon-fibre-reinforced plastic
Cu	Copper
EDM	Electrical discharge machining
F_d	Delamination factor
HAZ	Heat-affected zone
HTS	High tensile strength
IAL	Ignition current
INJ	Injection pressure
LBM	Laser beam machining
m_a	Mass after machining
m_b	Mass before machining
MRR	Material removal rate
PCR	Percentage contribution ratio
R^2	Coefficient of determination
Sa	Arithmetic 3D areal roughness
SEM	Scanning electron microscope
tm	Machining time
UD-CFRP	Unidirectional carbon-fibre-reinforced plastic
UVAD	Ultrasonic-vibration-assisted drilling
Vf	Fibre volume fraction
Vo	Open voltage
W	Wire type
Wa	Average kerf width
WB	Wire tension
WEDM	Wire electrical discharge machining
Wmax	Maximum damage width
Ws	Wire speed
Zn	Zinc
ρ	Density

References

1. Sekaran, S.C.; Yap, H.J.; Liew, K.E.; Kamaruzzaman, H.; Tan, C.H.; Rajab, R.S. Haptic-based virtual reality system to enhance actual aerospace composite panel drilling training. In *Structural Health Monitoring of Biocomposites, Fibre-Reinforced Composites and Hybrid Composites*; Jawaid, M., Thariq, M., Saba, N., Eds.; Elsevier: Duxford, UK, 2019; pp. 113–128.
2. Mohee, F.M.; Al-Mayah, A.; Plumtree, A. Anchors for CFRP plates: State-of-the-art review and future potential. *Compos. Part B Eng.* **2016**, *90*, 432–442. [CrossRef]
3. Wang, H.; Zhang, X.; Duan, Y. Investigating the Effect of Low-Temperature Drilling Process on the Mechanical Behavior of CFRP. *Polymers* **2022**, *14*, 1034. [CrossRef]
4. Dutta, H.; Debnath, K.; Sarma, D.K. Investigation on cutting of thin carbon fiber-reinforced polymer composite plate using sandwich electrode-assisted wire electrical-discharge machining. *Proc. Inst. Mech. Eng. Part E J. Process. Mech. Eng.* **2021**, *235*, 1628–1638. [CrossRef]
5. El-Hofy, M.H.; Soo, S.L.; Aspinwall, D.K.; Sim, W.M.; Pearson, D.; M'Saoubi, R.; Harden, P. Tool temperature in slotting of CFRP composites. *Procedia Manuf.* **2017**, *10*, 371–381. [CrossRef]
6. Kim, D.; Beal, A.; Kwon, P. Effect of tool wear on hole quality in drilling of carbon fiber reinforced plastic–titanium alloy stacks using tungsten carbide and polycrystalline diamond tools. *J. Manuf. Sci. Eng.* **2016**, *138*, 031006. [CrossRef]

7. Kumar, D.; Singh, K.K. An approach towards damage free machining of CFRP and GFRP composite material: A review. *Adv. Compos. Mater.* **2015**, *24*, 49–63. [CrossRef]
8. Sheikh-Ahmad, J.Y.; Shinde, S.R. Machinability of carbon/epoxy composites by electrical discharge machining. *Int. J. Mach. Mach. Mater.* **2016**, *18*, 3–17.
9. Teicher, U.; Müller, S.; Münzner, J.; Nestler, A. Micro-EDM of carbon fibre-reinforced plastics. *Procedia CIRP* **2013**, *6*, 320–325. [CrossRef]
10. Chen, L.; Li, M.; Yang, X. The feasibility of fast slotting thick CFRP laminate using fiber laser-CNC milling cooperative machining technique. *Opt. Laser Technol.* **2022**, *149*, 107794. [CrossRef]
11. Hejjaji, A.; Zitoune, R.; Toubal, L.; Crouzeix, L.; Collombet, F. Influence of controlled depth abrasive water jet milling on the fatigue behavior of carbon/epoxy composites. *Compos. Part A Appl. Sci. Manuf.* **2019**, *121*, 397–410. [CrossRef]
12. Cao, S.; Li, H.N.; Huang, W.; Zhou, Q.; Lei, T.; Wu, C. A delamination prediction model in ultrasonic vibration assisted drilling of CFRP composites. *J. Mater. Process. Technol.* **2021**, *302*, 117480. [CrossRef]
13. Leone, C.; Mingione, E.; Genna, S. Laser cutting of CFRP by Quasi-Continuous Wave (QCW) fibre laser: Effect of process parameters and analysis of the HAZ index. *Compos. Part B Eng.* **2021**, *224*, 109146. [CrossRef]
14. Sheikh-Ahmad, J.Y. Hole quality and damage in drilling carbon/epoxy composites by electrical discharge machining. *Mater. Manuf. Process.* **2016**, *31*, 941–950. [CrossRef]
15. Yue, X.; Yang, X.; Tian, J.; He, Z.; Fan, Y. Thermal, mechanical and chemical material removal mechanism of carbon fiber reinforced polymers in electrical discharge machining. *Int. J. Mach. Tools Manuf.* **2018**, *133*, 4–17. [CrossRef]
16. Kumaran, V.U.; Kliuev, M.; Billeter, R.; Wegener, K. Influence of carbon-based fillers on EDM machinability of CFRP. *Procedia CIRP* **2020**, *95*, 437–442. [CrossRef]
17. Kumar, R.; Agrawal, P.K.; Singh, I. Fabrication of micro holes in CFRP laminates using EDM. *J. Manuf. Process.* **2018**, *31*, 859–866. [CrossRef]
18. Makudapathy, C.; Sundaram, M. High aspect ratio machining of carbon fiber reinforced plastics by electrical discharge machining process. *J. Micro Nano-Manuf.* **2020**, *8*, 041005. [CrossRef]
19. Lau, W.S.; Lee, W.B. A comparison between EDM wire-cut and laser cutting of carbon fibre composite materials. *Mater. Manuf. Process.* **1991**, *6*, 331–342. [CrossRef]
20. Abdallah, R.; Soo, S.L.; Hood, R. A feasibility study on wire electrical discharge machining of carbon fibre reinforced plastic composites. *Procedia CIRP* **2018**, *77*, 195–198. [CrossRef]
21. Abdallah, R.; Soo, S.L.; Hood, R. The influence of cut direction and process parameters in wire electrical discharge machining of carbon fibre–reinforced plastic composites. *Int. J. Adv. Manuf. Technol.* **2021**, *113*, 1699–1716. [CrossRef]
22. Dutta, H.; Debnath, K.; Sarma, D.K. A study of wire electrical discharge machining of carbon fibre reinforced plastic. In *Advances in Unconventional Machining and Composites*; Shunmugam, M.S., Kanthababu, M., Eds.; Springer: Singapore, 2020; pp. 451–460.
23. Wu, C.; Gao, S.; Zhao, Y.J.; Qi, H.; Liu, X.; Liu, G.; Guo, J.; Li, H.N. Preheating assisted wire EDM of semi-conductive CFRPs: Principle and anisotropy. *J. Mater. Process. Technol.* **2021**, *288*, 1169159. [CrossRef]
24. Kuo, C.L. Drilling of Ti/CFRP/Al Multilayer Stack Materials. Ph.D. Thesis, University of Birmingham, Birmingham, UK, 2014.
25. Shyha, I.S.E.M. Drilling of Carbon Fibre Reinforced Plastic Composites. Ph.D. Thesis, University of Birmingham, Birmingham, UK, 2010.
26. GF Machining Solutions. Available online: https://www.gfms.com/content/dam/gfms/pdf/lifecycle-services/operate/en/CS-certified-wires_en.pdf (accessed on 2 February 2022).
27. Bedra Intelligent Wires. Available online: http://www.bedra.hk/UploadImage/edit/files/topasplusflyer_GB.pdf (accessed on 2 February 2022).
28. Ji, R.; Liu, Y.; Diao, R.; Xu, C.; Li, X.; Cai, B.; Zhang, Y. Influence of electrical resistivity and machining parameters on electrical discharge machining performance of engineering ceramics. *PLoS ONE* **2014**, *9*, e110775. [CrossRef] [PubMed]
29. Habib, S.; Okada, A.; Ichii, S. Effect of cutting direction on machining of carbon fibre reinforced plastic by electrical discharge machining process. *Int. J. Mach. Mach. Mater.* **2013**, *13*, 414–427. [CrossRef]
30. Davim, J.P.; Reis, P. Damage and dimensional precision on milling carbon fiber-reinforced plastics using design experiments. *J. Mater. Process. Technol.* **2005**, *160*, 160–167. [CrossRef]
31. Lau, W.; Wang, M.; Lee, M. Electrical discharge machining of carbon fibre composite materials. *Int. J. Mach. Tools Manuf.* **1990**, *30*, 297–308. [CrossRef]
32. Habib, S.; Okada, A. Experimental investigation on wire vibration during fine wire electrical discharge machining process. *Int. J. Adv. Manuf. Technol.* **2016**, *84*, 2265–2276. [CrossRef]
33. Ablyaz, T.R.; Shlykov, E.S.; Muratov, K.R.; Sidhu, S.S. Analysis of wire-cut electro discharge machining of polymer composite materials. *Micromachines* **2021**, *12*, 571. [CrossRef] [PubMed]

Article

Experimental Analysis of Residual Stresses in CFRPs through Hole-Drilling Method: The Role of Stacking Sequence, Thickness, and Defects

Tao Wu [1,*], Roland Kruse [2], Steffen Tinkloh [3], Thomas Tröster [3], Wolfgang Zinn [1], Christian Lauhoff [1] and Thomas Niendorf [1]

1. Institute of Materials Engineering, University of Kassel, Mönchebergstr. 3, 34125 Kassel, Germany; zinn@uni-kassel.de (W.Z.); lauhoff@uni-kassel.de (C.L.); niendorf@uni-kassel.de (T.N.)
2. Institute of Applied Mechanics, Technical University of Braunschweig, Pockelsstraße 3, 38106 Brauschweig, Germany; r.kruse@tu-braunschweig.de
3. Institute for Lightweight Design with Hybrid Systems, University of Paderborn, Mersinweg 7, 33100 Paderborn, Germany; steffen.tinkloh@uni-paderborn.de (S.T.); thomas.troester@upb.de (T.T.)
* Correspondence: wutao202@hotmail.com

Citation: Wu, T.; Kruse, R.; Tinkloh, S.; Tröster, T.; Zinn, W.; Lauhoff, C.; Niendorf, T. Experimental Analysis of Residual Stresses in CFRPs through Hole-Drilling Method: The Role of Stacking Sequence, Thickness, and Defects. *J. Compos. Sci.* **2022**, *6*, 138. https://doi.org/10.3390/jcs6050138

Academic Editor: Jiadeng Zhu

Received: 19 March 2022
Accepted: 5 May 2022
Published: 9 May 2022

Publisher's Note: MDPI stays neutral with regard to jurisdictional claims in published maps and institutional affiliations.

Copyright: © 2022 by the authors. Licensee MDPI, Basel, Switzerland. This article is an open access article distributed under the terms and conditions of the Creative Commons Attribution (CC BY) license (https:// creativecommons.org/licenses/by/ 4.0/).

Abstract: Carbon fiber reinforced plastics (CFRPs) gained high interest in industrial applications because of their excellent strength and low specific weight. The stacking sequence of the unidirectional plies forming a CFRP laminate, and their thicknesses, primarily determine the mechanical performance. However, during manufacturing, defects, e.g., pores and residual stresses, are induced, both affecting the mechanical properties. The objective of the present work is to accurately measure residual stresses in CFRPs as well as to investigate the effects of stacking sequence, overall laminate thickness, and the presence of pores on the residual stress state. Residual stresses were measured through the incremental hole-drilling method (HDM). Adequate procedures have been applied to evaluate the residual stresses for orthotropic materials, including calculating the calibration coefficients through finite element analysis (FEA) based on stacking sequence, laminate thickness and mechanical properties. Using optical microscopy (OM) and computed tomography (CT), profound insights into the cross-sectional and three-dimensional microstructure, e.g., location and shape of process-induced pores, were obtained. This microstructural information allowed for a comprehensive understanding of the experimentally determined strain and stress results, particularly at the transition zone between the individual plies. The effect of pores on residual stresses was investigated by considering pores to calculate the calibration coefficients at a depth of 0.06 mm to 0.12 mm in the model and utilizing these results for residual stress evaluation. A maximum difference of 46% in stress between defect-free and porous material sample conditions was observed at a hole depth of 0.65 mm. The significance of employing correctly calculated coefficients for the residual stress evaluation is highlighted by mechanical validation tests.

Keywords: residual stresses; incremental hole-drilling method; CFRP; stacking sequence; laminate thickness; defect population

1. Introduction

Carbon fiber reinforced plastics (CFRPs), i.e., carbon fiber used as reinforcement elements in a polymer matrix, have found extensive use in the aviation and automotive industries because of their outstanding properties, such as high strength and stiffness, low density, high fatigue resistance as well as low thermal expansion coefficient [1]. However, the laminate characteristics, i.e., the thickness of each ply, fiber orientation, as well as the number of layers, have to be considered carefully for acquiring these excellent mechanical properties in a CFRP laminate. By choosing adequate laminate characteristics, the CFRP's material properties can be directly tailored. In the past years, a great amount of work has been conducted to improve the mechanical performance of CFRP components, e.g., fatigue resistance [2] and energy absorption capacity [3].

In general, residual stresses are formed in the course of the manufacturing of CFRPs for various reasons. In particular, the differences in the coefficient of thermal expansion (CTE) between the carbon fibers and the resin system, the differences in the mechanical properties arising from variation in the fiber orientation, and the cure shrinkages of the resin are of utmost importance [4,5]. The process-induced residual stresses, in turn, can cause fiber waviness, transverse cracking, delamination, and geometric distortion. Moreover, process-induced pores have to be considered in the CFRPs, as they can have a significant influence on the mechanical performance of a wide range of composite materials [6]. The formation of pores in the manufacturing of CFRPs is a complex phenomenon, and some of the involved elementary processes are still not well understood. The underlying mechanisms vary from manufacturing process to manufacturing process because of the differences in thermodynamic and rheological phenomena [7]. Taking the *pre-preg process technology* as an example, the formation of pores is mainly induced by an insufficient curing process, air entrapment either during impregnation or during laying up, and moisture dissolved in the resin [7]. For further details on pore formation mechanisms in different CFRPs manufacturing processes and the effect on the mechanical performance, the reader is referred to [7].

The determination of residual stresses in CFRPs using robust experimental approaches paves the way not only for optimizing the manufacturing parameters in order to reduce residual stresses but also for a damage-tolerant design by exploiting residual stresses instead of increasing safety margins and overall dimensions (resulting in cost and weight increase), respectively. In the past years, several methods have been developed for measuring residual stresses in CFRP components [8]. One commonly used method is to estimate the residual stresses from the experimentally determined curvature of a structure through a simplified model. However, this method can only be used for unsymmetric CFRP composites. Moreover, this approach cannot provide residual stress values in each individual ply [9]. The layer-removal method is another common technique for analyzing residual stresses, involving the measurement of the curvature following the progressive removal of thin layers from the sample surface [10]. Using this method, evaluation of ply level residual stresses is feasible. However, the complete and precise removal of thin material layers is challenging. Furthermore, this method is limited to flat coupons.

Alternatively, the hole-drilling method (HDM) is a well-known approach for measuring the residual stresses in composites, being capable of providing in-depth residual stress profiles [11]. The HMD has been standardized in ASTM E837-13a [12], specifying measurement procedure and range, minimum requirements of instrumentation, algorithms, and coefficients for the computation of uniform and nonuniform stress distributions, as well as error sources. As described in ASTM E837-13a, the HDM was originally developed for residual stress measurement in isotropic materials, with the assumption that the relaxed strain response has a simple trigonometric form. However, this assumption does not hold true for orthotropic materials. For accurate analysis of the residual stress state in orthotropic materials, such as CFRPs, the standard model [12] has been adapted, as can be seen in [1,11,13].

With respect to approaches for determining the calibration coefficients, the integral method provides more accurate in-depth residual stress results compared with the differential method [14]. The calibration coefficients need to be calculated through finite element analysis (FEA) based on the material type and thickness of the sample. The influence of material properties and thickness of metallic samples on the calculated calibration coefficients and the resulting residual stress values have been reported in recent years [14–16]. For instance, in [14], the influence of Poisson's ratio on the calibration coefficients for the HDM was studied. Depending on the hole depth, it was found that the maximum difference in calibration coefficients for Poisson's ratios of 0.2 and 0.4, compared with an assumed value of 0.3, can be as high as 17% for individual calibration coefficients. However, an almost ideally linear behavior of the calibration coefficients can be observed within Poisson's ratio range of 0.2 to 0.45. Magnier et al. [15] calculated the calibration coefficients of metallic sheets featuring different thickness values. Afterward, the coefficients obtained were used for the evaluation of the residual stresses using the HDM. If the real thickness of the samples was not accounted for, pronounced errors were found in that study. In particular, this holds true for thicknesses below 1 mm, while only minor errors were noticed for thicknesses above 1.6 mm. In the ASTM E837 [17], residual stress measurement by the HDM assumes that the drilled hole has a flat bottom geometry. However, the commonly employed cutting tools for the HDM are end mills, which usually have chamfers or fillets on the edges. Therefore, chamfers or fillets are directly transferred to the blind hole bottom geometry. Blödorn et al. [18] calculated calibration coefficients through FEA, considered the real hole geometry, and performed residual stress evaluation for A36 steel, AISI304L stainless steel, and AA6061 aluminum alloy. Considerable differences in residual stresses were found, in particular in the two first hole depth increments with a 50% deviation. Recently, Kümmel et al. [19] adopted the HDM to measure the residual stresses in ultrafine-grained laminated metal composites, where steel layers were positioned at the top and bottom surface of the sample, and an aluminum layer was positioned in the middle. When the aluminum core was not considered for the calculation of the calibration coefficients, assuming that composite is a homogeneous material, the evaluated residual stresses were overestimated by about 50% compared with the model of the actual structure. A very limited number of studies are available in the literature detailing this kind of analysis for composite materials. Sicot et al. [20] measured residual stresses by HDM in $[0_2/90_2]_S$ and $[0_8]$ CFRP laminates fabricated by different cure cycles. The calibration coefficients were calculated by considering the real ply stacking sequence. It was found that for unidirectional CFRP laminate, the residual stress level remains low for different cooling conditions, and the influence of cooling conditions on residual stress values is small. With respect to cross ply laminates, it was observed that the cooling conditions have an important effect on the residual stress level, and the stress increases considerably with the cooling rate. The measured residual stresses were compared to those predicted by a thermoviscoelastic approach to the classical theory of laminate. Both results were in good agreement. In other studies [13,21], calibration coefficients for glass/carbon fiber reinforced plastics (GFRP/CFRPs) were calculated based on their real geometry, thickness, and stacking sequence. The relation between the coefficients and the residual stresses was studied. Only when the actual geometry was taken into account the correct calibration coefficients were obtained, as demonstrated by mechanical bending tests in combination with FEA.

In high-tech applications requiring high strength and stiffness and, concomitantly, low weight, thin-walled CFRP laminates with complex stacking sequence are designed and used. Commonly, the manufacturing of CFRP components and structures leads to pores and defects. The objective of this work is to provide a comprehensive study on the residual stress measurement results using HDM, considering the roles of pores, stacking sequence, and thickness. Focus is given to the determination of calibration coefficients, taking into account the aforementioned factors, as well as their influence on the subsequent residual stress estimation. Furthermore, the two- and three-dimensional microstructures of CFRP laminates are characterized, which allows for a thorough understanding of the experimentally determined strain and stress results, in particular at the transition zone between the individual plies. The reliability of the residual stress measurement is validated by mechanical bending tests in combination with FEA. Potentially, this method can be used in the majority of cases concerning composite laminate materials.

2. Materials and Methods
2.1. Theory of the Hole-Drilling Method

The HDM is a well-known approach to measure residual stresses by successively drilling a hole in the surface of a sample. Employing HDM, the residual stresses cannot be directly measured but are derived from strains, which are relaxed during the drilling process and concomitantly measured as a function of depth. A small hole is drilled at the geometrical center of a strain gauge used for strain measurement. Eventually, the relationship between residual stress and relaxed strain can be established via calibration coefficients, which need to be calculated through FEA. Figure 1 illustrates the workflow of measuring the residual stresses using HDM.

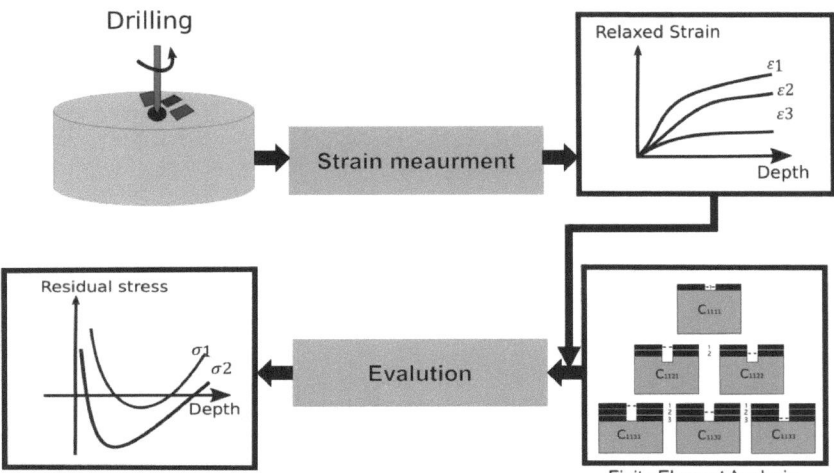

Figure 1. Framework of measuring the residual stresses by the hole-drilling method.

For establishing the stress–strain relationship in isotropic materials, the relaxed strain response is assumed to be of simple trigonometric form. However, this assumption is not valid in anisotropic materials [22]. For analyzing and measuring the residual stress in CFRPs featuring highly anisotropic characteristics and assuming that the material behaves elastically the stress–strain relationship can be expressed as follows:

$$\begin{pmatrix} \varepsilon_1 \\ \varepsilon_3 \\ \varepsilon_2 \end{pmatrix} = \begin{bmatrix} C_{11} & C_{12} & C_{13} \\ C_{21} & C_{22} & C_{23} \\ C_{31} & C_{32} & C_{33} \end{bmatrix} \cdot \begin{pmatrix} \sigma_x \\ \sigma_y \\ \tau_{xy} \end{pmatrix} = [C]\cdot(\sigma) \quad (1)$$

In Equation (1), the directions of strains ε_1, ε_2, and ε_3 and stresses σ_x, σ_y, and τ_{xy} are defined as depicted in Figure 2a. Furthermore, matrix C (constituted of the coefficients C_{kl}) represents the relationship between the residual stresses and the relaxed strains and depends on the material properties and thickness of the sample, the depth of the hole, as well as the geometry of the strain gauge. More details on calculating the calibration coefficients C_{kl} by separately imposing boundary conditions can be found in [11].

The general stress–strain relation in Equation (1) can only be applied for stresses being uniformly distributed over the thickness. In order to account for in-depth nonuniform residual stresses, the incremental HDM can be employed. This method is based on the measurement of the surface deformation upon a sequence of drilling steps, i.e., a small hole at the surface of a stressed material is incrementally drilled. For this purpose, Equation (1) can be adapted by an incremented integral formulation, where i indicates the actual drilling step and j denotes all steps up to the current one:

$$(\varepsilon)_i = \begin{pmatrix} \varepsilon_1 \\ \varepsilon_3 \\ \varepsilon_2 \end{pmatrix}_i = \sum \begin{bmatrix} C_{11} & C_{12} & C_{13} \\ C_{21} & C_{22} & C_{23} \\ C_{31} & C_{32} & C_{33} \end{bmatrix}_{ij} \cdot \begin{pmatrix} \sigma_x \\ \sigma_y \\ \tau_{xy} \end{pmatrix}_j = \sum [C]_{ij}\cdot(\sigma)_j \quad 1 \leq j \leq i \quad (2)$$

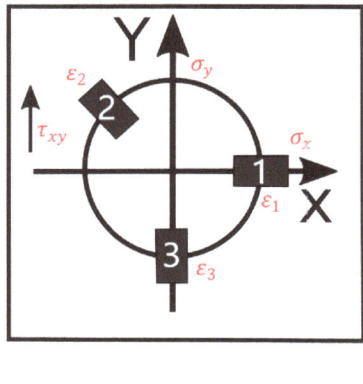 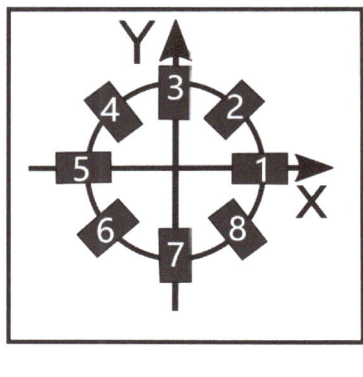

(a) (b)

Figure 2. Schematic of (**a**) a typical three-element strain gauge with corresponding coordinate system and (**b**) a special strain gauge rosette with eight grids used for strain measurement in the present work.

For the incremental HDM, the coefficients C_{klij} in Equation (2) are determined not only by the residual stress in the last drill increment i but also by the residual stresses relaxed in all previous increments j. For clarity, Figure 3 schematically depicts the procedure to constitute the calibration coefficients matrix $[C_{11}]_{ij}$. In the first drilled layer, C_{1111} indicates the influence of the stress σ_1 (being present in the 1st increment) on the measured strain ε_1. When the second layer is drilled, C_{1121} and C_{1122} define the influence of the stress σ_1 imposed in both the 1st and the 2nd increment on the strain ε_1. The same procedure can be used to calculate C_{1131}, C_{1132}, and C_{1133}.

Due to the presence of defects in CFRP laminates, unexpected strain measurements might be observed during the drilling process [13]. In this work, a strain gauge rosette with 8 gauges is used to provide redundancy and, thus, improve the reliability of the strain analysis [13]. Figure 2b illustrates the setup of 8-rosette strain gauge already used in previous works [1,13,23]. Please note that only strain information in three directions is required for the residual stress evaluation using Equation (2). Eight different combinations of strain gauge directions are employed in the present work: (1,2,3), (2,3,4), (3,4,5), (4,5,6), (5,6,7), (6,7,8), (7,8,1), and (8,1,2).

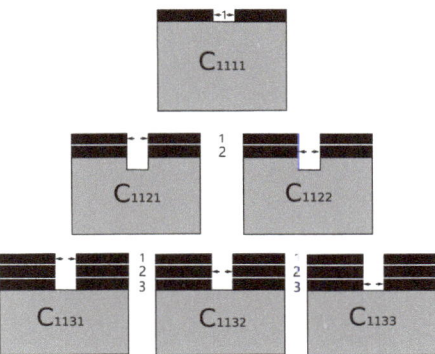

Figure 3. Schematic detailing the stress states and hole depths considered to obtain the calibration coefficients for the three initial drilling increments.

In order to determine the compliance matrices $[C]_{ij}$, the corresponding coefficients need to be calculated using FEA. In the present work, the commercial software ABAQUS was used. A cylinder with 500 mm in diameter (thickness according to the actual thickness of the sample) was discretized with eight-node solid elements of type C3D8R. Figure 4 shows the finite element model, including local mesh refinement in the direct vicinity of the drilled hole. The mesh refinement was performed manually by directly specifying the number of elements based on the partitioning function in ABAQUS. The number of elements was iteratively increased until changes in the calculated matrix coefficients became negligible. The mesh in the model remained the same during the entire simulation process. With respect to the prescribed mechanical boundary conditions, the circular boundaries are fixed in all directions, while the part of the model below the hole is allowed to deform freely.

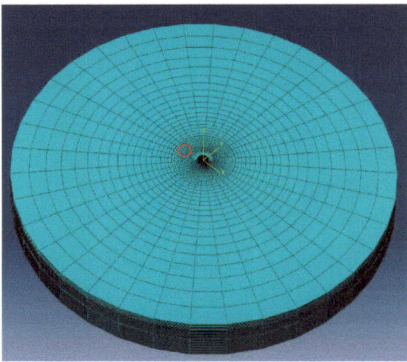

Figure 4. Finite element analysis model in ABAQUS software for calculating the calibration coefficients in orthotropic materials, displaying local mesh refinement in the direct vicinity of the drilled hole (recompiled from [23]).

In the present work, a hole drilled in 25 increments was considered: 10 steps of 20 μm and then 15 steps of 40 μm thickness. The calculation time for these 25 increments was about 7 days using a dual-processor Intel Xeon ×5660. The step size of 20 μm was not chosen for the whole simulation in order to avoid high computational cost. In the experiment, however, a step size of 20 μm was employed in the entire drilling procedure, ensuring a smooth strain profile as a function of the drilling depth, in particular across the interface between two adjacent layers. The experimentally determined strains were approximated and smoothed with a polynomial of 6th order for residual stress evaluation. The strains at a given depth, for which the calibration coefficients were calculated, were chosen for evaluating the residual stresses.

Properties required as input include the stacking sequence and the actual material properties of the unidirectional plies. Note that each unidirectional ply is considered to be homogeneous and orthotropic, and plies differ only by orientation. The (residual) stresses are prescribed as the boundary conditions on the hole walls for every actual total hole depth. The relaxed strains follow from the displacement data. The drilling process is simulated by removing material (elements) with the "Remove" function, using a Python script written by the authors. This procedure is repeated until all calibration coefficients are determined. More information on the calculation of the coefficients using FEA can be found in [11].

2.2. Sample Manufacturing

One objective of the present work is to compare the residual stresses of samples featuring different stacking sequences and thicknesses. One CFRP sample [0/45/90/135/180/−135/−90/−45/0] (9 plies) with dimensions of $300 \times 30 \times 2$ mm^3 was manufactured by the *pre-preg-process* technology employing a pressure of 0.5 MPa at a temperature of 200 °C. For further details on the manufacturing process, the reader is referred to [23]. After manufacturing, the sample was cooled down under a laboratory atmosphere with a cooling rate of approximately 100 °C/min. The carbon fiber pre-pregs used were of type C U255-0/NF-E322/37% (with an epoxy resin content of 37%). In order to investigate the influence of the stacking sequence on the residual stress state, a unidirectional CFRP laminate (9 plies) was manufactured with the same process parameters. Furthermore, CFRP laminates with different thicknesses of 1 mm (4 plies), 2 mm (8 plies), and 3 mm (12 plies) were fabricated with stacking sequence $[0/90]_s$. Again, the manufacturing and cooling parameters are the same as detailed before. The mechanical properties of the unidirectional plies used for fabrication of the different laminates are E_x = 126.15 (GPa), E_y = 7.97 (GPa), E_z = 7.97 (GPa), ν_{xy} = 0.37, ν_{xz} = 0.37, ν_{yz} = 0.37, G_{xy} = 7.11 (GPa), G_{xz} = 7.11 (GPa), and G_{yz} = 2.9 (GPa), experimentally determined according to DIN EN ISO 527 [13]. Figure 5 shows the dimensions and stacking sequence of the multidirectional CFRP sample. As this sample is symmetrically constructed and, thus, characterized by a self-balanced residual stress state, pronounced deformation (bending) was not observed after the sample was removed from the die. Schematics for the other samples are not shown to avoid redundancy.

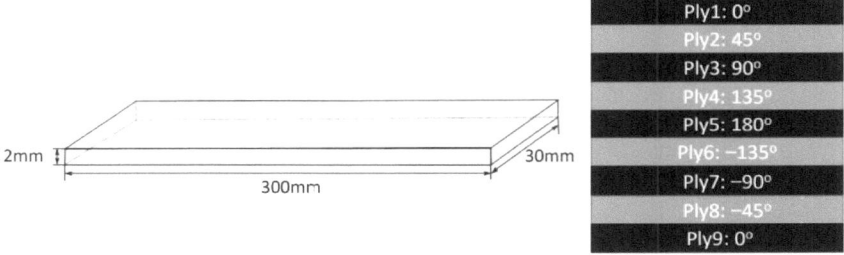

Figure 5. Dimensions and stacking sequence of the multidirectional CFRP [0/45/90/135/180/ −135/−90/−45/0] laminate.

2.3. Hole-Drilling Process

For the incremental HDM, a small hole at the geometrical center of a strain gauge rosette featuring eight grids (Hottinger Baldwin Messtechnik, Figure 2b) was incrementally drilled. The strain gauges were attached to the laminate surface and connected to a bridge amplifier, using a supply voltage of 1 V. A carbide tungsten tool (ref H2.010, Komet) with a nominal diameter of 1 mm was used for drilling. An air turbine was used with a drilling speed of about 300,000 rpm (at 3 bar) and an orbital movement. Using this setup, the formation of new stresses in the material can be effectively avoided since chips produced by the side surface of the cutting tool have a shorter and less constrained exit path [13]. Because of the orbital movement, the final hole diameter is around 2 mm, which is in accordance with the inner diameter of the strain gauge rosette. Small drilling steps of 20 μm were chosen in order to have a smooth strain profile across the piles. In order to ensure the accuracy of the strain measurement used for residual stress evaluation, it should be emphasized that some further aspects have to be carefully considered: (1) precise alignment of the strain gauge alongside the fiber direction (errors induced by misalignment of strain gauges have been discussed in [13,23]), (2) exact position of the drill to be placed at the geometric center of the strain gauge rosette [12], (3) adoption of the correct zero-depth setting (an overestimation of residual stresses was reported for an incorrectly chosen zero-depth setting [13]), and (4) the measured strains are approximated and smoothed with a polynomial of 6th order for residual stress evaluation.

2.4. Microstructure Characterization

The two-dimensional cross-sectional microstructure was characterized using optical microscopy (OM). Following manufacturing, samples were cut and embedded in an epoxy resin. The embedded samples were ground down using 400, 600, 800, and 1500 grit size silicon carbide abrasive paper and finally polished with 1 μm diamond paste. Defect analysis within the sample volume was carried out using an RX Solutions EasyTom 160-150 computed tomography (CT) system. For the investigations, samples with dimensions of 2.8 × 2.4 × 2 mm^3 were cut from the center of the original laminates. CT scans were conducted at 70 kV and 100 µA. These settings enabled a spatial resolution of smaller than 4 µm. A total of 1440 projections were collected at a pixel size of 4.5 µm. The volumetric (absorption contrast) images were reconstructed using the Feldkamp, Davis, and Kres (FDK) algorithm [24], cropped to the sample size, and normalized in the greyscale range. For visualization of the pores, threshold-based segmentation was employed since the contrast between ambient air and the CRFP samples was high. However, it should be noted that pores being in size close to the voxel size cannot be detected reliably because of the partial volume effect and noise. Thus, only pores with at least 9 µm in diameter (two voxels) were retained.

3. Results and Discussion

3.1. Microstructure Characterization

In Figure 6, a cross-sectional optical micrograph of the multidirectional CFRP sample (c.f. Figure 5) is shown, providing information on the thicknesses of the individual ply. Beside the stacking sequence, these values are used as input parameters to the FEA model for calculating the calibration coefficients (cf. Section 2.1). It can be directly seen that the thickness of each ply is not exactly the same. The cross-sectional microstructures of the other samples are not shown for brevity.

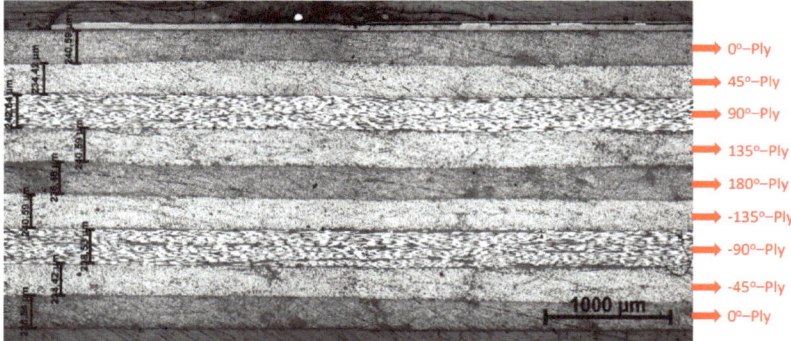

Figure 6. Cross-sectional microstructure characterization of the multidirectional CFRP sample.

Figure 7 shows an example of the cross-sectional microstructure of a drilled hole in the multidirectional CFRP sample, detailing the quality and the depth of the drilled hole. The results are representative for the other samples (not shown here) as well. The drilled hole quality is of utmost importance for the accuracy of the residual stress evaluation. According to ASTM–E837 [12], a reliable measurement (in the case of nonuniform residual stresses) can be made up to a maximum depth of 0.4 times the diameter of the hole. Since the hole diameter is 2 mm in the present work (cf. Section 2.3), a depth limit of around 800 µm is set. The depth of the hole shown in Figure 7, measuring 0.8 mm, is evidence of the good controllability and high precision of the drilling device used. Moreover, the hole is orthogonal to the bottom face of the sample, and substantial damage is not observed. The aforementioned characteristics imply that the drilled hole is of good quality. However, the used cutting tool is an end cutter with a chamfer on the edges (not shown), and related effects are potentially enlarged by the air turbine vibration. Consequently, chamfers or fillets are directly transferred to the blind hole bottom geometry, as can be seen at a depth of 700 to 800 µm, shown in Figure 7. Thus, the results in this depth range are excluded from the analysis.

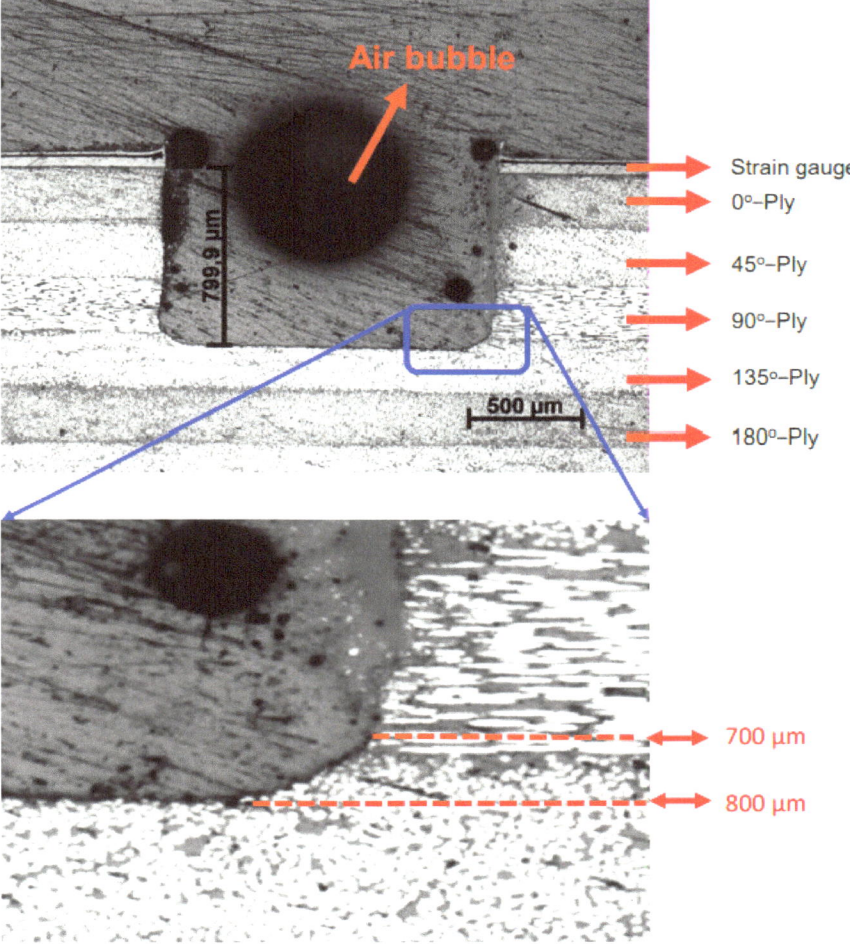

Figure 7. Cross-sectional optical micrographs of a drilled hole in the multidirectional CFRP sample, embedded in resin.

In this section, the process-induced defect population in the uni- and multidirectional CFRP samples is characterized, and the effect of stacking sequence on the porosity and pore morphology is shown. Please be reminded that both samples consisted of nine plies (same thickness) and were fabricated with the same process parameters (cf. Section 2.2). The *pre-preg* technology used for fabrication is known for a variety of phenomena finally contributing to pore formation [25]. Thus, the porosity and pore morphology are dependent on material properties, reinforcement structure, stacking sequence, thickness, and processing parameters (temperature, pressure, and curing time). Figure 8 depicts two- and three-dimensional CT images. On the one hand, it is well-known that two-dimensional images are section-biased, i.e., dependent on the cutting direction. However, cross-sectional views can provide clear information on the volume fraction and shape of the pores. With respect to the pore distribution, three-dimensional images provide additional information. Therefore, both two- and three-dimensional data are shown.

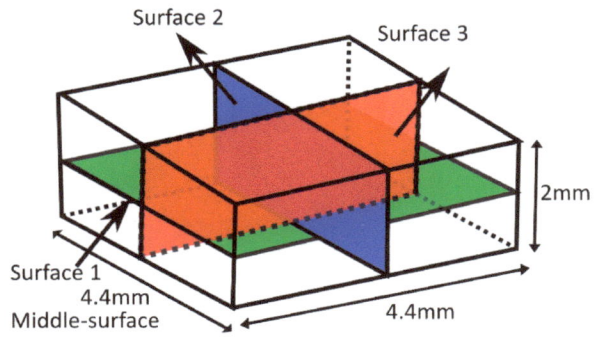

Figure 8. *Cont.*

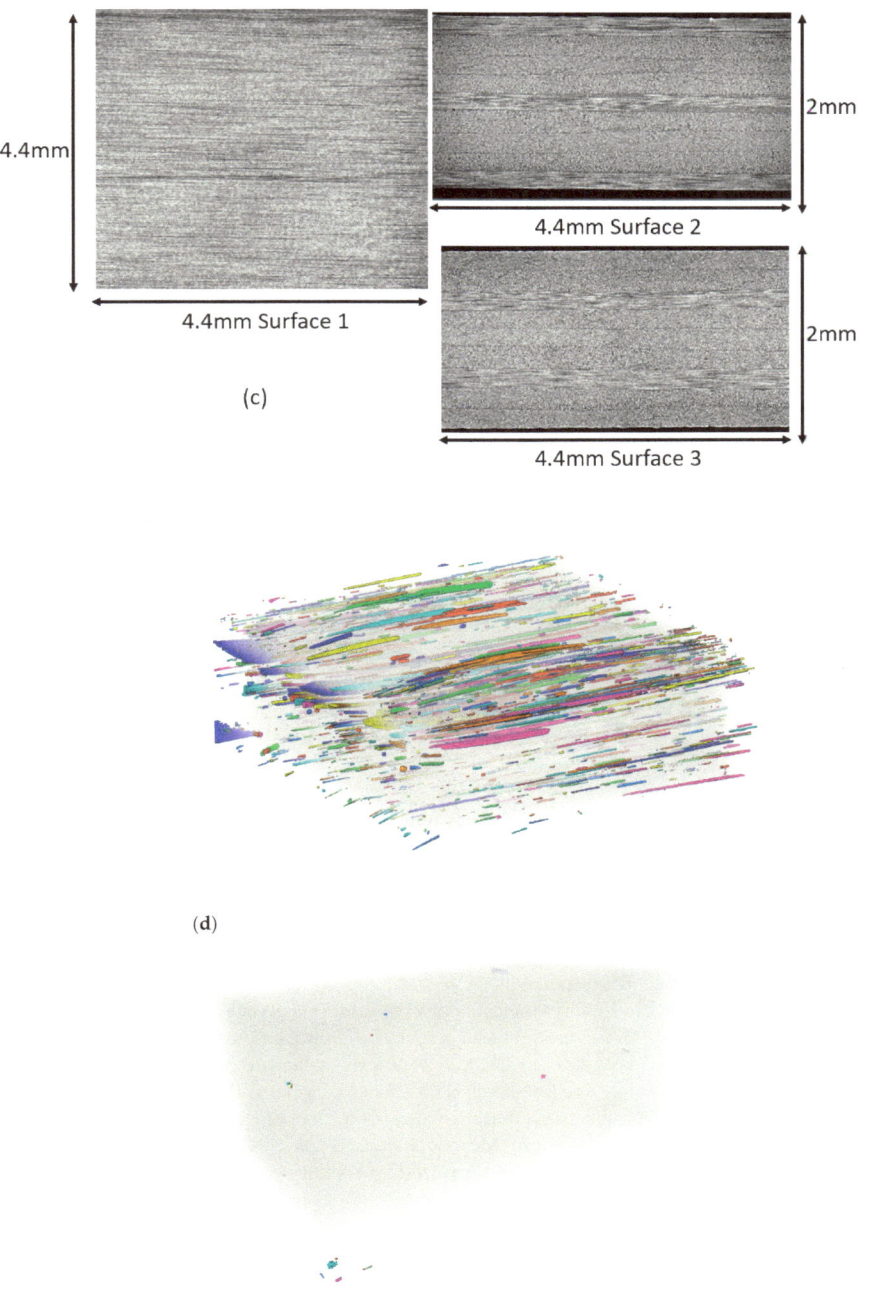

Figure 8. CT analysis of the CFRP samples: (**a**) Sketch of the cut planes for two-dimensional microstructure characterization. Two-dimensional microstructure of unidirectional (**b**) and multi-directional (**c**) CFRP. Three-dimensional microstructure of unidirectional (**d**) and multidirectional (**e**) CFRP with pores shown in color.

Figure 8a sketches the cut planes for two-dimensional microstructure characterization. Figure 8b,c show cross sections along the three orthogonal directions for the uni- and multidirectional CFRP, respectively. The section orientations are the same for both samples, and black areas represent pores. In Figure 8b, it can be seen that many pores are present in the middle area. This observation can be explained by an inhomogeneous distribution of the temperature and curing degree in the middle of the sample, as well as by the fact that the trapped pores in the middle of the sample have more difficulties in migrating to the edge during curing. In contrast to the unidirectional CFRP, only a few pores are found in the multidirectional sample. As reported in previous work [7], more pores could be induced in multidirectional CFRP than in unidirectional CFRP owing to the following reasons: (1) because of the complex stacking sequence, the air entrapment in multidirectional CFRP is more pronounced, and (2) a nonuniform distribution of the pressure and temperature leads to a nonuniform curing degree in multidirectional CFRP. However, the reverse case is observed in the present work, which can be explained by the employed process parameters not being optimized for both types of laminates. However, the present work focuses on the residual stress analysis, while process parameter optimization was out of scope.

A three-dimensional image of the unidirectional CFRP sample is shown in Figure 8d, revealing interlaminar and intralaminar pores. Most of the pores are elongated in the adjacent fiber direction and are quite long (needle-like voids). In line with the two-dimensional images in Figure 8c, the three-dimensional image of the multidirectional CFRP (Figure 8d) shows hardly any pores at all, keeping in mind that pores of dimensions similar to the voxel size are not displayed. Based on the assessment of CT data, the porosity of the uni- and multidirectional CFRP is 1% and 0.05%, respectively.

3.2. Calibration Coefficients: Relation between Residual Stress and Surface Strain

Details about the calculation of the calibration coefficients have been introduced in Section 2.1. In the following, the influence of the stacking sequence and the presence of pores on the calculated calibration coefficients is presented.

3.2.1. Effect of Stacking Sequence

Although strain gauges with eight grids have been employed in the present work in order to provide redundant strain information during drilling, only three grids are required for residual stress evaluation. To assess the effect of stacking sequence on the calibration coefficients, uni- and multidirectional CFRP samples with the same theoretical thickness (nine plies) were investigated. For clarity, only C_{11ij} of Equation (2) will be in focus here. Furthermore, only grids No. 1 to 3 are considered, featuring angles with respect to the fiber direction of $0°$, $45°$, and $90°$, respectively.

Table 1 shows the calculated calibration coefficients C_{11ij} for the uni- (a) and multidirectional (b) CFRP, as well as the relative difference between the two samples (c) up to a maximum hole depth of 400 μm. At this depth, two plies are drilled, whereas, in the multidirectional CFRP sample, the first ply is at $0°$ fiber direction and the second ply at $45°$ fiber direction. From Table 1, it can be seen that the values of the upper triangular matrix are zero (cf. Figure 3 for a schematic explanation). Moreover, in each row, the absolute values decrease from left to right. This observation can be rationalized by a smaller deformation induced on the sample surface by force imposed on the bottom layer relative to an upper layer. Consequently, it can be concluded that there is a limit to the maximum depth, as also reported in [12]. Furthermore, the absolute values increase from top to bottom. This tendency is based on the fact that the compliance of the material is increased when more layers are drilled. Generally speaking, a difference in the calibration coefficients between uni- and multidirectional CFRP can be observed. Since even in the first ply, featuring the same fiber orientation in both samples, differences are present, a significant influence of the stacking sequence is demonstrated. For quantitative analysis, the percentage deviations are shown in Table 1c, where the values of the unidirectional CFRP are taken as reference. In the first ply, the highest difference of approximately 45% is at the first removed increment, as the surface strain response is very sensitive.

Table 1. Calculated coefficients C_{11ij} of the unidirectional (**a**) and multidirectional (**b**) CFRP sample and (**c**) the relative difference between both samples in percentage.

Depth (μm)	20	40	60	80	100	120	140	160	180	200	240	280	320	360	400
20	−2.42	0	0	0	0	0	0	0	0	0	0	0	0	0	0
40	−2.74	−2.58	0	0	0	0	0	0	0	0	0	0	0	0	0
60	−3.02	−2.90	−2.69	0	0	0	0	0	0	0	0	0	0	0	0
80	−3.29	−3.18	−3.02	−2.76	0	0	0	0	0	0	0	0	0	0	0
100	−3.54	−3.43	−3.29	−3.10	−2.80	0	0	0	0	0	0	0	0	0	0
120	−3.78	−3.67	−3.54	−3.37	−3.15	−2.81	0	0	0	0	0	0	0	0	0
140	−4.01	−3.90	−3.77	−3.61	−3.42	−3.17	−2.81	0	0	0	0	0	0	0	0
160	−4.23	−4.11	−3.98	−3.84	−3.66	−3.44	−3.17	−2.78	0	0	0	0	0	0	0
180	−4.43	−4.31	−4.18	−4.04	−3.87	−3.67	−3.43	−3.14	−2.74	0	0	0	0	0	0
200	−4.62	−4.50	−4.38	−4.23	−4.07	−3.88	−3.65	−3.40	−3.10	−2.68	0	0	0	0	0
240	−4.97	−4.85	−4.73	−4.59	−4.42	−4.24	−4.04	−3.82	−3.57	−3.29	−5.51	0	0	0	0
280	−5.27	−5.16	−5.03	−4.89	−4.73	−4.55	−4.36	−4.15	−3.92	−3.68	−6.54	−5.16	0	0	0
320	−5.55	−5.42	−5.29	−5.15	−4.99	−4.83	−4.64	−4.43	−4.22	−3.98	−7.23	−6.14	−4.78	0	0
360	−5.77	−5.65	−5.53	−5.38	−5.23	−5.05	−4.87	−4.67	−4.46	−4.24	−7.77	−6.78	−5.70	−4.38	0
400	−5.97	−5.85	−5.72	−5.58	−5.43	−5.26	−5.07	−4.87	−4.67	−4.45	−8.23	−7.28	−6.29	−5.24	−3.98

(**a**)

Depth (μm)	20	40	60	80	100	120	140	160	180	200	240	280	320	360	400
20	−1.33	0	0	0	0	0	0	0	0	0	0	0	0	0	0
40	−1.59	−1.47	0	0	0	0	0	0	0	0	0	0	0	0	0
60	−1.83	−1.74	−1.57	0	0	0	0	0	0	0	0	0	0	0	0
80	−2.06	−1.97	−1.84	−1.63	0	0	0	0	0	0	0	0	0	0	0
100	−2.28	−2.2	−2.08	−1.92	−1.69	0	0	0	0	0	0	0	0	0	0
120	−2.5	−2.41	−2.3	−2.16	−1.98	−1.72	0	0	0	0	0	0	0	0	0
140	−2.72	−2.63	−2.52	−2.38	−2.23	−2.03	−1.76	0	0	0	0	0	0	0	0
160	−2.93	−2.83	−2.73	−2.6	−2.45	−2.28	−2.08	−1.8	0	0	0	0	0	0	0
180	−3.14	−3.05	−2.94	−2.82	−2.68	−2.52	−2.34	−2.14	−1.87	0	0	0	0	0	0
200	−3.38	−3.3	−3.2	−3.08	−2.95	−2.8	−2.64	−2.48	−2.29	−2.05	0	0	0	0	0
240	−3.53	−3.44	−3.34	−3.22	−3.08	−2.94	−2.78	−2.62	−2.46	−2.27	−3.69	0	0	0	0
280	−3.62	−3.54	−3.43	−3.31	−3.18	−3.03	−2.88	−2.73	−2.57	−2.39	−4.15	−3.16	0	0	0
320	−3.7	−3.61	−3.51	−3.39	−3.26	−3.11	−2.96	−2.81	−2.65	−2.49	−4.4	−3.62	−2.77	0	0
360	−3.77	−3.68	−3.57	−3.45	−3.32	−3.17	−3.02	−2.87	−2.72	−2.56	−4.57	−3.87	−3.19	−2.45	0
400	−3.82	−3.73	−3.62	−3.5	−3.37	−3.23	−3.08	−2.93	−2.78	−2.62	−4.72	−4.06	−3.45	−2.87	−2.24

(**b**)

Depth (μm)	20	40	60	80	100	120	140	160	180	200	240	280	320	360	400
20	0.450	0	0	0	0	0	0	0	0	0	0	0	0	0	0
40	0.419	0.430	0	0	0	0	0	0	0	0	0	0	0	0	0
60	0.394	0.4	0.416	0	0	0	0	0	0	0	0	0	0	0	0
80	0.373	0.380	0.390	0.409	0	0	0	0	0	0	0	0	0	0	0
100	0.355	0.358	0.367	0.380	0.396	0	0	0	0	0	0	0	0	0	0
120	0.338	0.343	0.350	0.359	0.371	0.387	0	0	0	0	0	0	0	0	0
140	0.321	0.325	0.331	0.340	0.347	0.359	0.373	0	0	0	0	0	0	0	0
160	0.307	0.311	0.314	0.322	0.330	0.337	0.343	0.352	0	0	0	0	0	0	0
180	0.291	0.292	0.296	0.301	0.307	0.313	0.317	0.318	0.317	0	0	0	0	0	0
200	0.268	0.266	0.269	0.271	0.275	0.278	0.276	0.270	0.261	0.235	0	0	0	0	0
240	0.289	0.290	0.293	0.298	0.303	0.306	0.311	0.314	0.310	0.310	0.330	0	0	0	0
280	0.313	0.313	0.318	0.323	0.327	0.334	0.339	0.342	0.344	0.350	0.365	0.387	0	0	0
320	0.333	0.333	0.336	0.341	0.346	0.356	0.362	0.365	0.372	0.374	0.391	0.410	0.420	0	0
360	0.346	0.348	0.354	0.358	0.365	0.372	0.379	0.385	0.390	0.396	0.411	0.429	0.440	0.440	0
400	0.360	0.362	0.367	0.372	0.379	0.385	0.392	0.398	0.404	0.411	0.426	0.442	0.442	0.442	0.437

(**c**)

3.2.2. Effect of Pores

The underlying mechanisms for the formation of pores in CFRP have been detailed in Section 3.1. As will be shown in the present section, the process-induced pores significantly affect the residual stress measurement results obtained by HDM if they appear in high density and large size in the direct vicinity of the drilled hole. In order to estimate this effect, the FEA has been modified as follows: the elements to be drilled at a depth between 60 and 120 µm were set as pores, assigning Young's modulus of 1 MPa and a Poisson's ratio of 0. Afterwards, the FEA was conducted with boundary conditions as previously described. Since the multidirectional CFRP laminate is characterized by a very low degree of porosity (Figure 8a,b), only coefficients for the unidirectional laminate have been calculated.

Table 2 shows the calculated coefficients C_{11ij} for the unidirectional CFRP being characterized by pronounced porosity at a depth between 60 and 120 µm. It is obvious that the coefficients in the items of the matrix defined as pores (marked in red) have constant values. In the incremental HDM, the coefficients are not only a function of the current drilling step but also of all previous increments. Therefore, the coefficients of a porous sample are not the same as in the case without pores. The influence of pores on the estimated residual stresses will be shown in a subsequent section.

Table 2. Calculated coefficients C_{11ij} of unidirectional CFRP under consideration of pores.

Depth (µm)	20	40	60	80	100	120	140	160	180	200	240	280	320	360	400
20	−1.51	0	0	0	0	0	0	0	0	0	0	0	0	0	0
40	−1.76	−1.64	0	0	0	0	0	0	0	0	0	0	0	0	0
60	−1.98	−1.88	−1.75	0	0	0	0	0	0	0	0	0	0	0	0
80	−1.98	−1.88	−1.75	−1.59	0	0	0	0	0	0	0	0	0	0	0
100	−1.98	−1.88	−1.75	−1.59	−1.35	0	0	0	0	0	0	0	0	0	0
120	−1.98	−1.88	−1.75	−1.59	−1.35	−1.32	0	0	0	0	0	0	0	0	0
140	−2.30	−2.20	−2.08	−1.93	−1.76	−1.55	−1.27	0	0	0	0	0	0	0	0
160	−2.44	−2.34	−2.22	−2.08	−1.91	−1.72	−1.50	−1.22	0	0	0	0	0	0	0
180	−2.57	−2.47	−2.34	−2.20	−2.04	−1.86	−1.66	−1.44	−1.16	0	0	0	0	0	0
200	−2.69	−2.58	−2.46	−2.32	−2.16	−1.99	−1.80	−1.59	−1.37	−1.09	0	0	0	0	0
240	−2.89	−2.78	−2.66	−2.53	−2.37	−2.20	−2.02	−1.83	−1.63	−1.43	−2.18	0	0	0	0
280	−3.06	−2.96	−2.83	−2.69	−2.53	−2.37	−2.19	−2.01	−1.82	−1.64	−2.72	−1.90	0	0	0
320	−3.21	−3.09	−2.96	−2.82	−2.67	−2.50	−2.33	−2.15	−1.97	−1.79	−3.06	−2.37	−1.63	0	0
360	−3.32	−3.20	−3.08	−2.94	−2.78	−2.61	−2.44	−2.26	−2.08	−1.91	−3.31	−2.66	−2.04	−1.39	0
400	−3.41	−3.29	−3.17	−3.03	−2.87	−2.70	−2.53	−2.35	−2.18	−2.00	−3.51	−2.88	−2.30	−1.75	−1.18

3.2.3. Strain Results by HDM

Figure 9a and b show the strain-depth results for the uni- and multidirectional CFRP samples, respectively. The absolute strain values increase with increasing drilling depth because of the relief of the residual stresses. Here, strain gauges 1 and 5 of the strain rosette employed (cf. Figure 2b) were aligned parallel with the fiber direction (of the top ply), whereas strain gauges 3 and 7 were aligned orthogonal to the fiber. Theoretically, strain values obtained by strain gauges 1 and 5 should be the same. However, a difference is present, which can be explained by local defects or heterogeneity. This phenomenon shows the significance of using the special strain gauge rosette to obtain sufficient strain information in all directions. Figure 9b depicts the strain results of the multidirectional CFRP sample up to a depth of 700 µm, crossing the first three plies with fiber directions of 0°, 45°, and 90°, respectively. For clarity, the positions of the transition zones between the neighboring plies are highlighted in the figure. At the first ply with the fiber direction of 0°, the strain values obtained by strain gauges 1 and 5, i.e., in fiber direction, increase with increasing depth, while strains decrease in other directions. This strain–depth behavior is different from that in the unidirectional CFRP (Figure 9a) and is supposed to result from the more complex stress state induced by the complex stacking sequence. In the second ply with a fiber direction of 45°, it is seen that the strains of strain gauges 1 and 5 start to decrease. Moreover, a change from compressive to tensile strain components can be deduced from strain gauges 3 and 7. Consequently, the strong influence of a change in the fiber direction, i.e., the stacking sequence, on the strain–depth response and, thus, the internal stress state of CFRP laminates is seen.

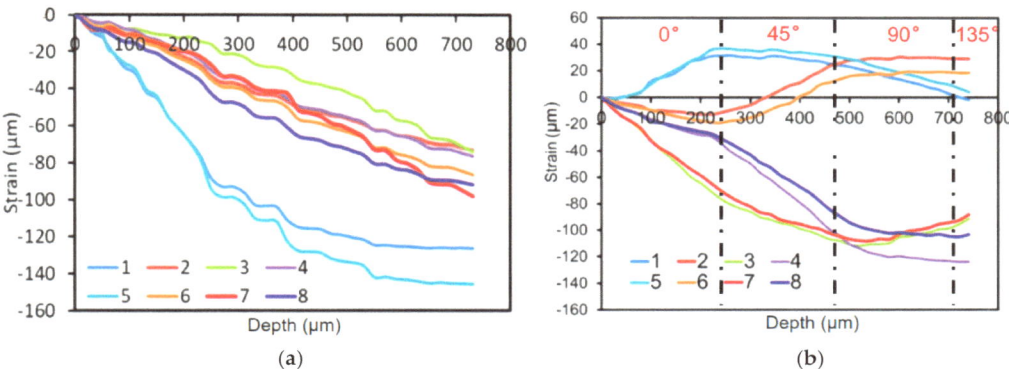

Figure 9. Measured surface strains in 8 directions as a function of drilling depth for (**a**) the unidirectional and (**b**) multidirectional (**b**) CFRP.

3.2.4. Effect of the Stacking Sequence on the Residual Stresses

The residual stresses σ_x and σ_y of the uni- and multidirectional CFRP samples are shown in Figure 10. Eight different combinations of strain gauges (c.f. Section 2.1) were used for assessment. The residual stresses were evaluated using the experimentally determined strains shown in Figure 9, and the calibration coefficients were calculated based on ply elastic parameters, stacking sequence, and thickness of the samples. As can be seen from the in-depth residual stress σ_x of the unidirectional CFRP (Figure 10a), at the surface of the sample, a tensile residual stress of around 13 MPa is present. All strain gauge combinations show the same results, except the combinations 2 and 6 (cf. Figure 2b), as those do not contain the strain information in the fiber direction. With increasing depth, the residual stress component continually decreases to about 0 MPa. When the depth reaches 0.7 mm, the stress value starts to diverge strongly as the surface strain response becomes insensitive. In direct comparison to σ_x, σ_y features a tensile value of around 4 MPa at the surface (Figure 10b), which is smaller than σ_x. With increasing depth, however, σ_y steadily increases to around 8 MPa. Note that the positions of the interfaces between plies in different samples are slightly different. After the residual stress measurement, the position of the interfaces between plies was carefully characterized.

Figure 10c shows the depth profile of the residual stress component σ_x in the multidirectional CFRP. At the surface, σ_x is characterized by tensile stress with a value of around 10 MPa. With increasing depth, then, the stress component switches several times from tension to compression and vice versa. This is mainly induced by the variation of the fiber orientation. Reaching the third ply with a fiber direction of 90°, tensile stress is seen, and this value increases with depth. In comparison, Figure 10d shows the residual stress component σ_y in the multidirectional CFRP sample. Close to the surface, the residual stress features a tensile value of around 25 MPa, decreasing then with depth. At the transition zone between the first and second ply, σ_y starts to increase. Close to the transition zone between the second and third ply, σ_y changes from tensile to compressive. A smooth change across the interface is obtained, which can be related to the following reasons: (1) small drilling increments, (2) the measured strains are smoothed with a polynomial of sixth-order, and the approximated strains are used for residual stress evaluation, (3) the step size in the calculation of the coefficients is consistent with the one used in the measurement, (4) an advanced formalism to evaluate the in-depth residual stresses for orthotropic materials, and (5) high-quality manufacturing of the sample with only a few defects near the interfaces.

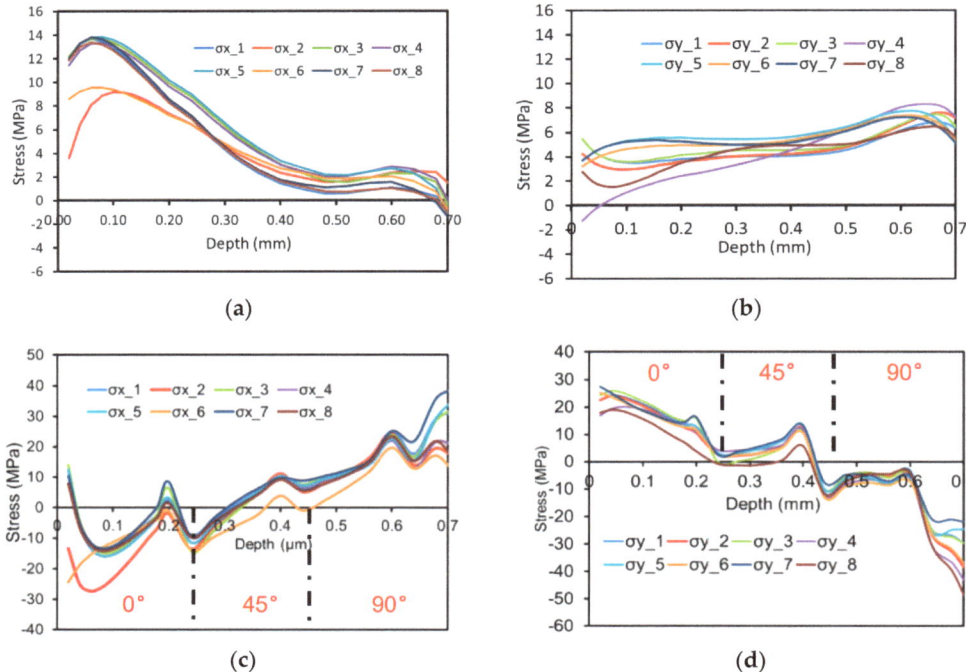

Figure 10. Depth-dependent residual stresses σ_x (**a,c**) and σ_y (**b,d**) of uni- (**a,b**) and multidirectional (**c,d**) CFRP, respectively, for 8 different combinations of strain gage directions.

Comparing the residual stress profiles between the uni- (a,b) and multidirectional (c,d) CFRPs in Figure 10, it is found that the course of σ_x of the two samples is quite different. Furthermore, absolute values of σ_y in the multidirectional CFRP sample are higher than in the unidirectional sample. These differences can be explained by the following reasons: (1) the complex stacking sequence in multidirectional CFRP increases the mechanical strength in multiple directions, (2) the complex stacking sequence leads to the increase in internal self-constraint force and, thus, promotes the evolution of higher residual stresses, and (3) the nonuniform distribution of the curing degree results in a more complex stress profile. In particular, the transition zones seem to have a significant influence only on the stress profile in the multidirectional laminate.

3.2.5. Effect of the Laminate Thickness on the Residual Stresses

As will be discussed in the following, the thickness of a CFRP laminate has a significant influence on the formation of residual stresses because of differences in temperature gradient and curing degree. In the present work, CFRP laminate samples with a stacking sequence of $[0/90]_s$ and with thicknesses of 1 mm, 2 mm, and 3 mm were investigated. More details of the samples can be found in Section 2.2. Note that all calibration coefficients were calculated based on the actual thickness and stacking sequence. Figure 11 shows the stress components σ_x and σ_y experimentally determined for the CFRP samples investigated. Regarding the samples with thicknesses of 2 mm and 3 mm (Figure 11c,d and Figure 11a,b, respectively), the stresses are characterized by similar profiles. At both sample surfaces, a tensile stress with a similar absolute value for σ_x can be seen. In contrast, for σ_y it is seen that the stress in the sample with a thickness of 3 mm is larger than in the sample of 2 mm. Possible explanations are (a) complex temperature gradients and degree of cure, which prevail in the thickness direction of the laminate during the curing process; these local differences induce a spatially resolved material response and a viscoelastic stress development, and (b) the effect of chemical shrinkage on residual stress development is increased with an increase in laminate thickness [5]. In the sample with 1 mm thickness, in turn, compressive residual stresses σ_x with small values (−5 MPa) are found, which can be explained by the sheer force between die and sample and the release of the residual stresses after the sample is removed from the die. Both factors are more apparent in the sample with 1 mm thickness because of the lower stiffness compared with other samples.

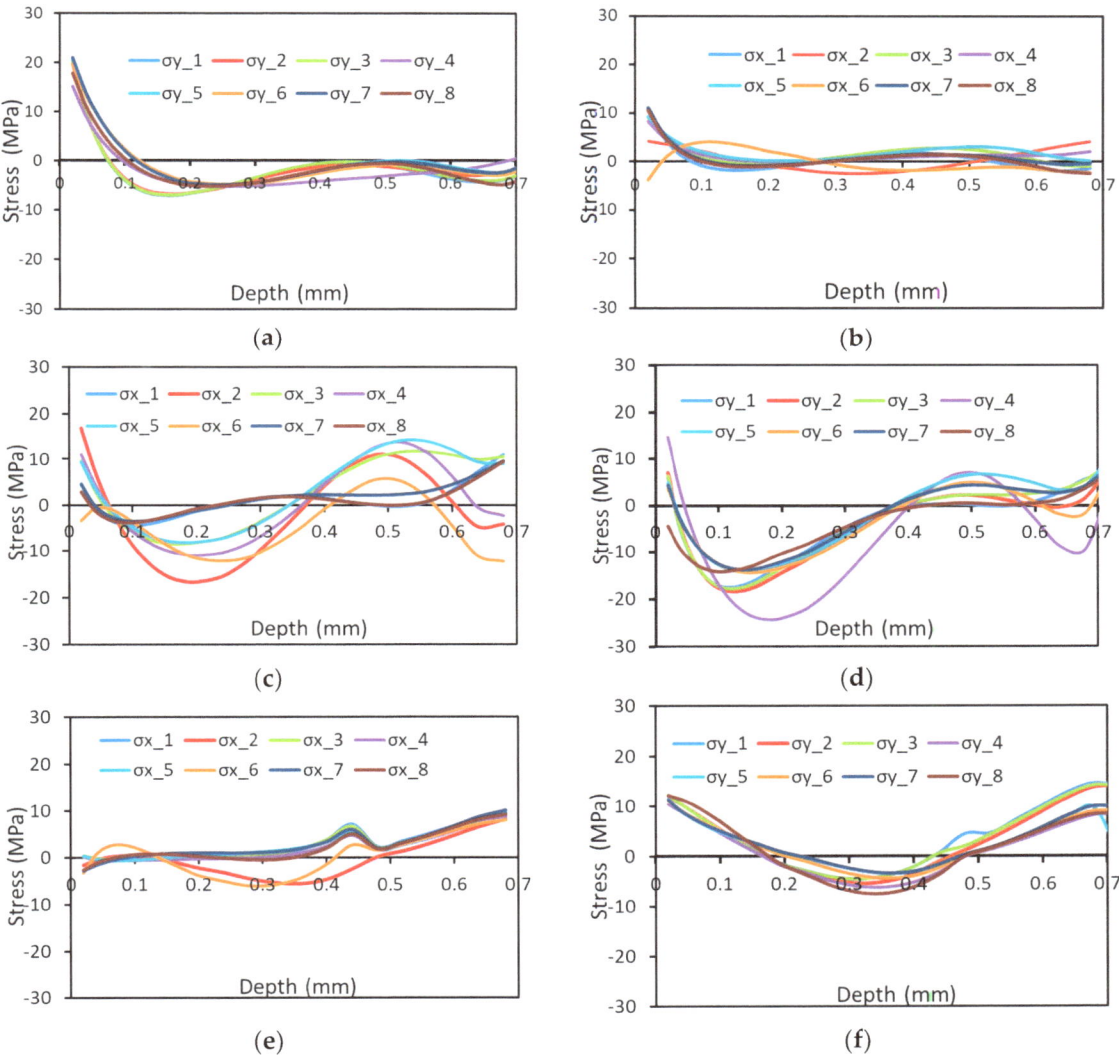

Figure 11. Depth-dependent residual stresses σ_x (**a,c,e**) and σ_y (**b,d,f**) of CFRP with thickness of 3 mm (**a,b**), 2 mm (**c,d**), and 1 mm (**e,f**), respectively.

3.2.6. Effect of Porosity on the Residual Stresses

As already detailed before, pores can be induced in the manufacturing process. These can affect the residual stress measurement using HDM. To consider the pores in the FEA geometry for the calculation of the calibration coefficients, ideally, the defect population within a sample is to be measured before drilling, using any nondestructive technology, e.g., CT. In practice, however, this procedure is not viable. Because of the need for a high-resolution CT system to resolve pores in CFRP, the sample in question suffers strict limitations regarding its dimensions. Furthermore, strain gauges cannot be attached to the surface of a CT sample for the same reason. Therefore, following the analysis shown in Section 3.2.2 detailing the influence of artificially defined pores on the calibration coefficients, those results are used in the present work to evaluate the effect of pores on the resulting residual stresses as well. We choose the unidirectional CFRP sample for analysis. Figure 12 shows the residual stress profile of σ_x in the sample using calibration coefficients of defect-free and porous material for evaluation. The results are derived from strain gauges 1, 2, and 3. As mentioned in Section 3.2.2, pores at a depth of 0.06 mm to 0.12 mm were accounted for in the model. As a consequence of these differences, a maximum difference of 46% in stress between the two sample conditions can be observed at a hole depth of 0.65 mm. Eventually, based on the theory used in incremental HDM, any defects present in the upper layers also affect the results in the deeper parts of the sample. Note that the labels in Figure 12, "With pore" and "Without pore", refer to the consideration of pores in the calculation of calibration coefficients.

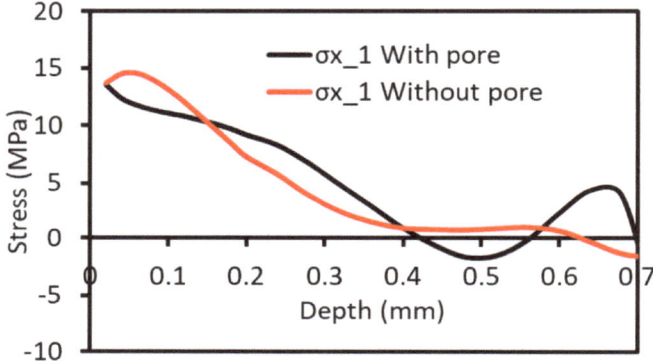

Figure 12. Residual stress profiles for the material being free of pores and featuring porosity.

3.3. Validation of Residual Stress Measurement

The objective of this section is the validation of the residual stress measurements (in multidirectional CFRP) using bending tests employing a defined loading condition. The stacking sequence and dimension of the sample can be found in Section 2.2. This procedure was already used successfully for thin metal sheets and polycarbonate material [1,23]. As shown in Figure 13, for validation, a residual stress measurement is initially conducted at a distance of 40 mm from the free edge of the CFRP beam. One side of the sample is clamped, while the other side is loaded (cantilever beam). With the given load, the bending stresses within the whole sample can be calculated via FEA. In the present study, a maximum force of 814 N is applied, leading to the stress of about 100 MPa in the longitudinal direction on the surface at the drilling point. Under constant load, another hole is drilled next to the first one (being drilled without external load) at a distance of 5 mm. Without loading, the residual stresses of two adjacent points are supposed to be very similar. Therefore, the stress induced by loading is obtained as the difference between the measured residual stress values (without external load) and the total stress values (with external load).

Figure 13. Experimental setup used for validation of the residual stress profiles.

In Figure 14, the measured stress values in the longitudinal direction are directly compared to the values from the FEA, where the dashed line corresponds to the FEA results. In general, a good agreement between experimental and numerical results can be seen. With respect to the first ply featuring a fiber orientation of 0°, i.e., being parallel to the longitudinal direction, the discrepancy between simulation and experimental results at a depth of 0.1 mm is around 5.2%. Close to the first interface between the first two plies, an apparent discrepancy can be seen. At the second ply with a fiber orientation of 45° with respect to the longitudinal direction, the stress level is reduced as the mechanical strength in the longitudinal direction is decreased. While a good agreement can still be observed within the second ply, again, a clear difference between numerical and experimental results is present at the second interface. At the third ply with fiber orientation of 90°, i.e., perpendicular to the longitudinal direction, the lowest stresses, which are in good agreement with the simulation, can be observed as the mechanical strength perpendicular to the fiber direction is low. In the following, however, with reaching a depth of 0.55 mm, a clear discrepancy between the numerical and experimental results appears. This tendency results from the increasing insensitivity of the strain measurement on the sample surface with increasing hole depth.

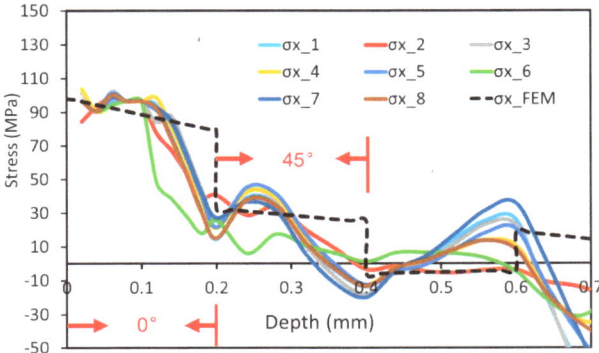

Figure 14. Validation of the residual stress measurements in the multidirectional CFRP sample based on the assessment of stresses introduced by superimposed load and a direct comparison to FEA. Solid lines show experimentally determined bending stress, while the dashed line illustrates the FEA results. See text for details.

It can be concluded that, in general, a good agreement between numerical and experimental can be achieved, implying that the measurement results through HDM are reliable. The apparent discrepancy at the interface between plies can be attributed to the following reasons: (1) the assumption that the laminate is perfectly bonded in the calculation of the calibration coefficients, (2) the local defect population and heterogeneity of the CFRP samples affect the experimental results; however, they are not considered here (see discussion before), and (3) the experimentally determined strains are approximated with a polynomial of sixth order for residual stress evaluation. Thus, it is capable of providing a smooth residual stress profile but losing some information at the interface.

4. Conclusions

In the present work, residual stress measurements of carbon fiber reinforced plastics (CFRPs) are conducted and analyzed. The residual stresses were measured through the incremental hole-drilling method (HDM), adopting a formalism for nonuniform in-depth stress analysis in orthotropic materials. Special strain gauges with eight grids were employed for recording strains released by drilling in multiple directions to improve the reliability of the analysis. The calibration coefficients (compliance matrix) were calculated by a finite element analysis (FEA) based on single-ply material properties, stacking sequence, and thickness of the sample. In addition, the two- and three-dimensional microstructures of uni- and multidirectional CFRP were characterized.

A comprehensive analysis of the effects of stacking sequence, thickness, and the presence of pores on the residual stresses is presented. The following conclusions can be drawn from the results shown:

- The two- and three-dimensional microstructures of unidirectional and multidirectional CFRP samples were characterized by computed tomography (CT). Pores were found in the samples, indicating the significance of taking into account these pores in residual stress analysis. Analysis of the effect of pores was implemented in the calibration procedure. Pores were artificially defined for the calculation of the calibration coefficients in a depth of 0.06 mm to 0.12 mm. Those results were used to evaluate the effect of pores on the resulting residual stresses. A maximum difference of 46% in stress between defect-free and porous material sample conditions can be observed at a hole depth of 0.65 mm;
- Based on FEA, the effect of the stacking sequence and the presence of pores on the calibration coefficients were studied. The stacking sequence and overall dimensions of the CFRP samples have a significant influence on the residual stress state;
- For validating the reliability of the measured residual stress through incremental HDM, a bending test applying a defined load was carried out. The residual stress measurements were compared with the stress values calculated by FEA (beam theory). A good agreement could be found in individual plies. The present apparent discrepancy at the interface between plies is due to the following reasons: (i) the laminate is assumed to be perfectly bonded, and (ii) the experimentally determined strains are approximated with a polynomial of sixth order for residual stress evaluation, losing some information at the interface.

Author Contributions: Conceptualization, T.N. and T.T.; methodology, T.W.; software, T.W.; validation, T.W., W.Z. and S.T.; formal analysis, T.W.; investigation, T.W. and R.K.; writing—original draft preparation, T.W.; writing—review and editing, T.W., C.L. and T.N.; supervision, T.N. and T.T.; project administration, T.W., W.Z. and S.T.; funding acquisition, T.N. and T.T. All authors have read and agreed to the published version of the manuscript.

Funding: This research was funded by the Deutsche Forschungsgemeinschaft (DFG) with project number 399304816.

Institutional Review Board Statement: Not applicable.

Informed Consent Statement: Not applicable.

Data Availability Statement: Data are available upon reasonable requests.

Conflicts of Interest: The authors declare no conflict of interest.

References

1. Alam, P.; Mamalis, D.; Robert, C.; Floreani, C.; Brádaigh, C. The fatigue of carbon fibre reinforced plastics-A review. *Compos. B. Eng.* **2019**, *166*, 555–579. [CrossRef]
2. Cutolo, A.; Carotenuto, A.R.; Palumbo, S.; Esposito, L.; Minutolo, V.; Fraldi, M.; Ruocco, E. Stacking sequences in composite laminates through design optimization. *Meccanica* **2021**, *56*, 1555–1574. [CrossRef]
3. Wang, H.; Duan, Y.; Abulizi, D.; Zhang, X. Design optimization of CFRP stacking sequence using a multi-island genetic algorithms under low-velocity impact loads. *J. Wuhan Univ. Technol. Sci. Ed.* **2017**, *32*, 720–725. [CrossRef]
4. Parlevliet, P.P.; Bersee, H.E.N.; Beukers, A. Residual stresses in thermoplastic composites—A study of the literature. Part III: Effects of thermal residual stresses. *Compos. Part A Appl. Sci. Manuf.* **2007**, *38*, 1581–1596.
5. Parlevliet, P.P.; Bersee, H.E.N.; Beukers, A. Residual stresses in thermoplastic composites—A study of the literature—Part I: Formation of residual stresses. *Compos. Part A Appl. Sci. Manuf.* **2006**, *37*, 1847–1857. [CrossRef]
6. Li, Y.; Li, Q.; Ma, H. The voids formation mechanisms and their effects on the mechanical properties of flax fiber reinforced epoxy composites. *Compos. Part A Appl. Sci. Manuf.* **2015**, *72*, 40–48. [CrossRef]
7. Mehdikhani, M.; Gorbatikh, L.; Verpoest, I.; Lomov, S.V. Voids in fiber-reinforced polymer composites: A review on their formation, characteristics, and effects on mechanical performance. *J. Compos. Mater.* **2019**, *53*, 1579–1669. [CrossRef]
8. Seers, B.; Tomlinson, R.; Fairclough, P. Residual stress in fiber reinforced thermosetting composites: A review of measurement techniques. *Polym. Compos.* **2021**, *42*, 1631–1647. [CrossRef]
9. Cowley, K.D.; Beaumont, P.W. The measurement and prediction of residual stresses in carbon-fibre/polymer composites. *Compos. Sci. Technol.* **1997**, *57*, 1445–1455. [CrossRef]
10. Dreier, S.; Benkena, B. Determination of Residual Stresses in Plate Material by Layer Removal with Machine-integrated Measurement. *Procedia CIRP* **2014**, *24*, 103–107. [CrossRef]
11. Wu, T.; Tinkloh, S.R.; Tröster, T.; Zinn, W.; Niendorf, T. Residual stress measurements in GFRP/steel hybrid components. In Proceedings of the 4th International Conference Hybrid 2020: Materials and Structures, Web-Conference, 28–29 April 2020; pp. 1–6.
12. Sibisi, P.N.; Popoola, A.P.I.; Arthur, N.K.K.; Pityana, S.L. Review on direct metal laser deposition manufacturing technology for the Ti-6Al-4V alloy. *Int. J. Adv. Manuf. Technol.* **2020**, *107*, 1163–1178. [CrossRef]
13. Wu, T.; Tinkloh, S.; Tröster, T.; Zinn, W.; Niendorf, T. Measurement and Analysis of Residual Stresses and Warpage in Fiber Reinforced Plastic and Hybrid Components. *Metals* **2021**, *11*, 335. [CrossRef]
14. Nau, A.; von Mirbach, D.; Scholtes, B. Improved Calibration Coefficients for the Hole-DrillingMethod Considering the Influenceof the Poisson Ratio. *Exp. Mech.* **2013**, *53*, 1371–1381. [CrossRef]
15. Magnier, A.; Zinn, W.; Niendorf, T.; Scholtes, B. Residual Stress Analysis on Thin Metal Sheets Using the Incremental Hole Drilling Method–Fundamentals and Validation. *Exp. Tech.* **2019**, *43*, 65–79. [CrossRef]
16. Nau, A.; Scholtes, B. Evaluation of the High-Speed Drilling Technique for the Incremental Hole-DrillingMethod. *Exp. Mech.* **2013**, *53*, 531–542. [CrossRef]
17. ASTM E837-13; Standard Test Method for Determining Residual Stresses by the Hole-Drilling Strain-Gage Method. ASTM: West Conshohocken, PA, USA, 2013.
18. Blödorn, R.; Bonomo, L.A.; Viotti, M.; Schroeter, R.B.; Albertazzi, A. Calibration Coefficients Determination Through Fem Simulations for the Hole-Drilling Method Considering the Real Hole Geometry. *Exp. Tech.* **2017**, *41*, 37–44. [CrossRef]
19. Kümmel, F.; Magnier, A.; Wu, T.; Niendorf, T.; Höppel, H.W. Residual stresses in ultrafine-grained laminated metal composites analyzed by X-ray diffraction and the hole drilling method. *Adv. Eng. Mater.* **2022**, in press. [CrossRef]
20. Sicot, O.; Gong, X.L.; Cherouat, A.; Lu, J. Determination of Residual Stress in Composite Laminates Using the Incremental Hole-drilling Method. *J. Compos. Mater.* **2003**, *37*, 831–844. [CrossRef]
21. Magnier, A.; Wu, T.; Tinkloh, S.; Tröster, T.; Scholtes, B.; Niendorf, T. On the reliability of residual stress measurements in unidirectional carbon fibre reinforced epoxy composites. *Polym. Test.* **2021**, *97*, 107146. [CrossRef]
22. Schajer, G.S.; Yang, L. Residual-stress measurement in orthotropic materials using the hole-drilling method. *Exp. Mech.* **1994**, *34*, 324–333. [CrossRef]
23. Wu, T.; Tinkloh, S.; Tröster, T.; Zinn, W.; Niendorf, T. Determination and Validation of Residual Stresses in CFRP/Metal Hybrid Components Using theIncremental Hole Drilling Method. *J. Compos. Sci.* **2020**, *4*, 143. [CrossRef]
24. Feldkamp, L.A.; Davis, L.C.; Krss, J.W. Practical Cone-Beam Algorithm. *J. Opt. Soc. Am. A* **1984**, *1*, 612–619. [CrossRef]
25. Howe, C.A.; Paton, R.J.; Goodwin, A.A. A Comparison between Voids in RTM and Prepreg Carbon/Epoxy Laminates. In Proceedings of the Eleventh International Conferenceon Composite Materials (ICCM11), Gold Coast, Australia, 14–18 July 1997.

Article

Tailoring the Local Design of Deep Water Composite Risers to Minimise Structural Weight

Chiemela Victor Amaechi [1,2,*], Nathaniel Gillet [3,*], Idris Ahmed Ja'e [4,5] and Chunguang Wang [6]

1. Department of Engineering, Lancaster University, Bailrigg, Lancaster LA1 4YR, UK
2. Standards Organisation of Nigeria (SON), 52 Lome Crescent, Wuse Zone 7, Abuja 900287, Nigeria
3. Department of Production Engineering, Trident Energy, Wilton Road, London SW1V 1JZ, UK
4. Department of Civil Engineering, Universiti Teknologi PETRONAS, Seri Iskander 32610, Malaysia; idris_18001528@utp.edu.my
5. Department of Civil Engineering, Ahmadu Bello University, Zaria 810107, Nigeria
6. School of Civil and Architectural Engineering, Shandong University of Technology, Zibo 255000, China; cgwang@sdut.edu.cn
* Correspondence: c.amaechi@lancaster.ac.uk (C.V.A.); gillettnathaniel@gmail.com (N.G.)

Citation: Amaechi, C.V.; Gillet, N.; Ja'e, I.A.; Wang, C. Tailoring the Local Design of Deep Water Composite Risers to Minimise Structural Weight. *J. Compos. Sci.* **2022**, *6*, 103. https://doi.org/10.3390/jcs6040103

Academic Editor: Jiadeng Zhu

Received: 17 December 2021
Accepted: 22 March 2022
Published: 26 March 2022

Publisher's Note: MDPI stays neutral with regard to jurisdictional claims in published maps and institutional affiliations.

Copyright: © 2022 by the authors. Licensee MDPI, Basel, Switzerland. This article is an open access article distributed under the terms and conditions of the Creative Commons Attribution (CC BY) license (https://creativecommons.org/licenses/by/4.0/).

Abstract: Following the rising technological advancements on composite marine structures, there is a corresponding surge in the demand for its deployment as ocean engineering applications. The push for exploration activities in deep waters necessitates the need for composite marine structures to reduce structural payload and lessen weights/loads on platform decks. This gain is achieved by its high strength–stiffness modulus and light-in-weight attributes, enabling easier marine/offshore operations. Thus, the development of composite marine risers considers critical composite characteristics to optimize marine risers' design. Hence, an in-depth study on composite production risers (CPR) is quite pertinent in applying composite materials to deep water applications. Two riser sections of 3 m and 5 m were investigated under a 2030 m water depth environment to minimise structural weight. ANSYS Composites ACP was utilized for the CPR's finite element model (FEM) under different load conditions. The choice of the material, the fibre orientation, and the lay-up configurations utilised in the modelling technique have been reported. In addition, the behaviour of the composite risers' layers under four loadings has been investigated under marine conditions. Recommendations were made for the composite tubular structure. Results on stresses and weight savings were obtained from different composite riser configurations. The recommended composite riser design that showed the best performance is AS4/PEEK utilising PEEK liner, however more work is suggested using global design loadings on the CPR.

Keywords: composite riser; tailored local design; finite element model (FEM); marine pipeline risers; composite marine structures; numerical modelling; advanced composite material; stress

1. Introduction

Presently, the global demand for oil/gas products has induced the incremental need for rapid technological advances in novel materials [1–6]. Such an advancement in composite materials has been induced by the increased application of marine composites [7,8] and the shift in the fluid-transfer operations from shallow water to deep water [9–13]. This trend is particularly evident in the developments achieved on deep water composite risers [14–18]. However, the deep water operations require more riser lengths, resulting in a considerable increment in weight. The number of the risers needed depends on the type of offshore structure, the type of risers required, the riser configuration considered, and the water depth [19–22]. Improving the riser technology is a challenge, as such composite materials are recommended as an excellent choice. The composites proffer superior positives, which can be utilised. These benefits include weight gains, high strength attributes, high fatigue resistance, low bending stiffness, high corrosion resistance, and light weight [23–30]. Due to the exciting features of composite materials, there is growth in the application and recent research on offshore composites, marine composites, and composite risers [31–40]. However, recent studies on these structures reflect the necessity to design composite risers by considering the loadings including environmental loads [41–48]. A typical composite riser pipe technology called M-pipe developed by Magma is shown in Figure 1.

Figure 1. Composite riser pipe section with Magma m-pipe end fittings, showing (**a**) end-fitting with fittings, pipe, seal and smaller flange (**b**) end-fitting coupled with pipe and bigger flange with unique bolt ends (courtesy: Technip FMC's Magma Global).

Composite marine risers have been considered for deployment, utilisation in deep waters, and as composite production risers [49–58]. These studies reflect different novel approaches on composite riser design and analysis. These studies also underpin the stress deformations alongside the buckling attributes of the structure. Research on composite risers stems from previous studies on shells, composite tubulars, and cylindrical structures [34,59–65] as seen in current related standards [66–71]. The first successful deployment of composite risers was a joint for composite risers installed on the Heidrun offshore platform [2,23,72–74]. Subsequently, the design evolution of composite risers has been achieved progressively for over 30 years [2,14–18,23]. The solutions on weight reduction and efficiency improvement were conducted using fibre reinforcements at an angle of $+/-55°$ by Doris Engineering [15]. The authors presented their design that assumed the fibres in each layer were load-bearing, according to netting theory [25,74–78]. However, there were no stresses in the transverse direction, and that resulted in an efficient angle of $+/-54.7°$. This study helped in the application of the filament winding technique on composite risers [60,61], design of prototypes [17,38,79,80], experimental tests [81–86], and the global–local design [42–46,87–91] and safety-reliability assessments [92–95]. Similarly, the developments led to advances in end-fitting designs, the metal–composite interface (MCI), debonding, delamination, and composite riser joint design. In a nutshell, more studies focused on the strength performance, buckling, material choice, and fibre orientation and optimization of composite risers [96–105]. The technology for composite risers is not fully enabled for full deployment. It is currently used as a hybrid riser system, since it is a supporting or enabling technology which implies that it requires more investigation for its qualification for deep water applications [106]. Jha et al. [107] presented various optimised hybrid composite pipe solutions by appreciating the variety of composite pipe design concepts and addressing key concerns bearing on both the manufacturing and customer acceptability point of view. Tan et al. [44] investigated 1500 m composite risers by utilising coupling analysis in both global and local analysis, whereby the authors found that titanium liner had lesser root mean square (RMS) strains than aluminium liner. However, the technology challenges for composite risers are based on the weight of risers limiting their lengths and motion of offshore operation facilities when applied further into deep waters [35–37,47,89]. An illustration of the composite marine risers with loads acting on the structure is represented in Figure 2.

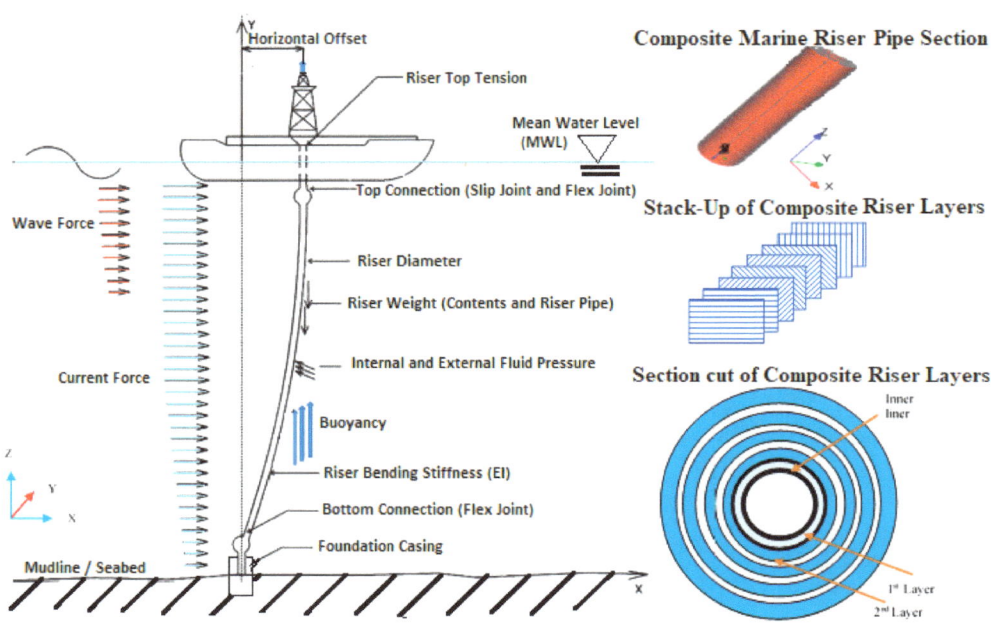

Figure 2. Typical composite riser showing composite stack-up, cross-section, and some loads (not drawn to scale).

Another engineering challenge is the qualification of lengthy CPRs, as more novel optimized designs are required. Some discussions on optimization were presented in different literature [97–104]. Harte et al. [101,102] carried out some optimisation techniques on composite pipelines using a safety factor of 4.5. The authors observed that to lessen the composite pipeline's weight, optimisation had to be conducted around the composite pipeline joint (CPJ). They also found a significant reduction in the peak stresses and the weight of the pipeline. Composite riser designs evolved into optimized designs as presented with a prototype using bonded and unbonded armours [108]. Amaechi et al. [87,88] introduced a novel numerical approach for the 18-layered composite riser structure using the ANSYS Composites ACP module and presented some safety factors on the structure under six unique loading cases and concluded that the finite element model was recommended for the validated design. Wang et al. [45–48,89] investigated global and local design of composite risers using ANSYS APDL and presented stress profiles for some configurations used to investigate the reduction in the structural weight under critical load cases. In addition, this technique considered the manual tailoring effects by using multiple variables for less, which was later optimised using surrogate-assisted evolutionary algorithm (SAEA) optimisation [104], which led to a weight reduction of 25% from the designs. Singh and Ahmad [91] presented a numerical design of carbon epoxy composite production risers carried out using ABAQUS AQUA in random water waves and investigated the global and local design using steel liner by considering the limit state failure criteria and validated the study with a composite riser model by Kim [108]. Other numerical methods deployed include homogenization of the multilayered composite offshore production risers [109,110]. In another study on the failure analysis of composite cylindrical structure by Bhavya et al. [111], the failure criteria applied were the Tsai-Wui and the maximum normal stress theories. They studied the influence of diameter to thickness ratio under pressure loads on four-layered and six-layered cylindrical structures using the finite element model.

The tailored local design on composite risers for deep water environments to minimise the structural weight is presented herein. In this paper, Section 1 presents the introduction,

Section 2 presents the analytical model, Section 3 presents the numerical model, Section 4 presents the results and discussion, while Section 5 presents the concluding remarks. The study will aid the development of the global design of composite risers under different configurations for riser installation and deployment.

2. Analytical Model

2.1. Stress and Deformation

Consider a thick laminate of the composite riser formed by more plies for the composite laminas. In theory, the laminas are made up of two components, called M and N, where M is the representative volume of the total composite riser body and N is the number of fibre orientations. The material is considered in both the cylindrical coordinate system and cartesian coordinate system. The z-axis lies perpendicular to the plane of the laminae as given in Figure 3. With regards the effective macroscale of the composite material, the stress and strain definitions for the material are respectively given by Equations (1) and (2) [91]:

$$\overline{\sigma}_{ij} = \frac{1}{M} \int_M \sigma_{ij} dM \qquad (1)$$

$$\overline{\varepsilon}_{ij} = \frac{1}{M} \int_M \varepsilon_{ij} dM \qquad (2)$$

Let us consider the point where both the stress and strain values on each of the composite riser lamina are constant. At this point, integrate the set of equations in Equations (1) and (2) to obtain Equations (3) and (4) [91]:

$$\overline{\sigma}_{ij} = \sum_{k=1}^{N} M_k \sigma_{ij}^{(k)} \qquad (3)$$

$$\overline{\varepsilon}_{ij} = \sum_{k=1}^{N} M_k \varepsilon_{ij}^{(k)} \qquad (4)$$

In this research, the 18 layers of the composite riser under internal pressure load are represented by N = 18. The material model of the cylindrical composite riser is depicted by a cylindrical coordinate system, where r represents the radial, θ is the hoop, and z is the axial coordinate. The strains and stresses in cylindrical riser pipe do not depend on θ when it is asymmetrically loaded.

Both the radial displacement and axial displacements are independent on the radial coordinates (r) and axial coordinates (z), respectively. Thus, the displacement field for the composite riser can be enunciated in Equation (5), where u_r, $u_θ$, and u_z are the displacements for the radial, hoop and axial directions, respectively.

$$u_r = u_r(r), \quad u_θ = u_θ(r, z), \quad u_z = u_z(z) \qquad (5)$$

where the reference frames are denoted by MPCS (material principal coordinate system) and CPCS (cylindrical coordinate system), as in Figure 3.

The characteristic effects of both twist (shear twist) and Poisson ratio are considered for each lamina of the anisotropic material. For the N-layered cylindrical composite riser, 2 N + 2 unknown constants are to be obtained for each of the layers. This is determined based on stress–strain relationship, stress–displacement relationship, and the boundary conditions for the system.

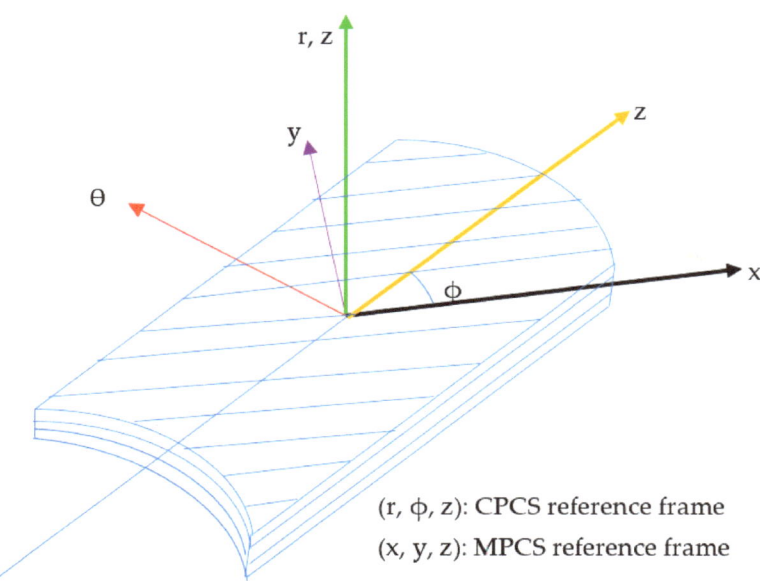

Figure 3. Representation of cylindrical coordinate and material principal coordinate systems.

2.2. Elastic Solution of Stresses

The classical laminate theory has been utilised in the design approach for the CPR. This is used to evaluate the stack-up and laminate properties by considering the laminate forces. The composite riser is considered as a shell model with nodal deformations. Adequate bonding of the layers is necessary to ensure less stress on the laminates. The material properties for the composite riser are provided for each layer with different thickness, as described in Section 3. The mechanical behaviour of the composite riser is dependent on the material attributes, the orientation angle of the lay-ups, and the laminate thicknesses. Although risers are slender structures, we design the composite riser as a shell with characteristic values. These are given as the in-plane laminate stiffness and flexural laminate stiffness, respectively. Under operational conditions, we can formulate the boundary conditions for the composite riser.

Consider the three-dimensional material properties for multilayered composite materials, having engineering constants where the matrix elements for the material modulus are given by C_{ij} (i, j = x, y, z) and G_{ii} (i = x, y, z). Assume transverse isotropy for the unknown equivalent characteristics along the y-z axis of the unidirectional composite material(s). The denotation for the engineering constant is E_i (i = x, y, z), the material's principal axis along the fibre direction is x, and the material's principal axis along the transverse direction is y; thus the illustration in Figure 3.

2.3. Constitutive Equation

For the local design of the composite risers, consider a composite hose tube acted upon by internal pressure; thus, it will be subject to the orientation of the composite laminate and the winding angle ∅. However, when there is axisymmetric loading acting on the composite tube, such as internal pressure loads, the relationship for the stresses and strains becomes dependent on θ, such that Equation (6) holds, where u, σ, and ε represent the displacement vector, stress tensor, and strain tensor, respectively.

$$\frac{\partial u}{\partial \theta} = 0, \ \frac{\partial \sigma}{\partial \theta} = 0, \ \frac{\partial \varepsilon}{\partial \theta} = 0 \tag{6}$$

There will be a resultant force that acts laterally on the material of the composite hose tube or composite riser pipe when it is subjected to constant internal pressure. That resultant force is proportional to the constant tension force which produces a constant axial strain. Considering the stress analysis of composite tubes and composite risers, it is determined based on the thickness of the wall. For the present study, the stress analysis of thick-walled composite is applied. Details on the numerical modelling approach are discussed in Section 3. The results of this analytical model were computed and compared with the numerical model as presented in Section 4.1. See analytical formulations in [112].

3. Numerical Model
3.1. The Model Design

The model designed for the CPR in this research is the local design investigated utilising the ANSYS workbench. It was used to design the riser by considering it as a multi-layered structure. A comparative study between the tailored model and the conventional model of the composite riser was conducted in the local design. Optimisation and global design were also carried out but not detailed in the present study. The particulars for the CPR are detailed in Table 1. This design was carried out for a CPR operating in a water depth of 2030 m. The effective weight of the riser was taken into account in the riser's tension computation depending on the wall thickness(es) utilised. The extreme cases for the burst analysis are not included, and the tension of the riser was designed differently. In the design of the composite riser, three approaches were considered: numerical design, conventional design, and analytical design. The constitutive model for the composite riser was derived using the analytical design. The traditional design of composites is based on orthogonal design, in which laminate reinforcements are exclusively positioned along the hoop and axial fibre directions. In this approach, the composite risers' plies are oriented similarly, wherein both the axial and hoop layers are oriented at $0°$ and $90°$, respectively.

The composite riser design was analysed by applying maximal stress distribution as the failure criteria. This was on each composite lamina (layer), as the design targets the implementation in deep waters. The effects of the fibre orientation and material combinations were investigated in in-plane shear, transverse, and fibre directions. The distribution was formulated under four different load cases. In order to analyse the stresses in composite riser body of the composite materials, we used the finite element model (FEM) software, utilising ANSYS ACP (versions R18.2 and R1 2021) [113,114]. Two riser sections of 3 m and 5 m were applied in the local design, and the consequence of the burst case was validated numerically. The riser design was considered for application in deep waters of about 2030 m depth. The water depth is vital for global design, which is not conducted herein; however, further considerations should include the global design of the composite riser.

Table 1. Parameters for the composite riser model.

Particulars	Value	Unit
Material	Composites	-
Number of Layers	17, 18, and 21	-
Failure Criteria	Max. Stress	-
Innermost Layer	Liner	-
Outer Diameter	0.305	m
Length of Riser Section	3.000	m
Ocean Depth	2030.000	m
Surface Area	7.661	m^2

3.2. Material Properties

The material properties considered in this local design are used to model the composite riser. The setup for the material model includes the materials for the matrix, the fibre reinforcements, and the liners considered in the design, as presented in Tables 2 and 3. The scientific basis for selecting parameters in Tables 2 and 3 is based on specifications recommended in the ABS industry standard [70]. The design considerations for the conditions for the composite production riser (CPR) are pressure resistance, fluid tightness, and fluid capillarity. As such, the choice of design materials for the CPR, particularly the liner, includes the prevention of fluid leakage. On that note, the key considerations are not only the choice of the fibre reinforcements and the matrix but also the materials selection. Hence, the materials selected must be able to withstand different mechanical loadings and different environmental conditions, as specified in DNV and ABS standards [66–70]. Considering a composite marine riser that is fully operational, high pressure conditions must be considered. The elastic constants considered in the material properties were utilised based on the theory applied by Sun and Li [115]. Based on the classical laminate theory, the laminates for the composite riser were efficiently analysed. The walls of this composite marine riser are considered as thick laminates. Thus, higher-order mathematical theories (HOMT) with three-dimensional (3D) material characteristics are required. The homogenous solid was used for a characteristic lengthwise span of the CPR. Due to deformation in the global composite riser, there is comparative deformation when compared to composite materials of shorter lengths. Considering the classical laminate theory (CLT), constants on the in-plane effective properties of the composite body can be generated. Thus, there must be three-dimensional effective properties. The material properties of the composite riser depend on time, static loads (tension and pressure), and environmental conditions (such as temperature, chemicals, or water). These properties were obtained via several validated scientific studies, as in referenced publications and technical reports [116–119]. These selected composite materials were also validated for the composite riser model, using verified modelling methods [3–5,87–91].

Table 2. Material characteristics for unidirectional FRP lamina.

Parameter/ Description	Fibre Volume Fraction	Density (kg/m^3)	E_1 (GPa)	$E_2 = E_3$ (GPa)	$G_{12} = G_{13}$ (GPa)	G_{23} (GPa)	σ_1^T (GPa)	σ_1^c (GPa)	σ_2^T (GPa)	σ_2^c (GPa)	τ_{12} (GPa)	$v_{12} = v_{13}$	v_{23}
(APC2) IM7/PEEK	0.55	1320.0	172.00	8.30	5.50	2.80	2900	1300	48.3	152.0	68.0	0.27	0.48
(V2021) Carbon fibre/Epoxy	0.55	1580.0	10.32	10.32	7.97	3.70	4900	1470	69.0	146.0	98.0	0.27	0.50
(APC2) P75/PEEK	0.55	1773.0	280.00	6.70	3.43	1.87	668	364	24.8	136.0	68.0	0.30	0.69
(T700) Carbon fibre/Epoxy	0.58	1580.0	230.00	20.90	27.60	2.70	4900	1470	69.0	146.0	98.0	0.20	0.27
(APC2) AS4/PEEK	0.58	1561.0	131.00	8.70	5.00	2.78	1648	864	62.4	156.8	125.6	0.28	0.48
(S-2) Glass fibre/Epoxy	0.55	2464.0	87.93	16.00	9.00	2.81	4890	1586	55.0	148.0	70.0	0.26	0.28
(938) P75/Epoxy	0.55	1776.0	310.00	6.60	4.10	2.12	720	328	22.4	55.2	176.0	0.29	0.70
(938) AS4/Epoxy	0.60	1530.0	135.40	9.37	4.96	3.20	1732	1256	49.4	167.2	71.2	0.32	0.46

FRP—fibre-reinforced plastics; S-2—AGY glass fibre; T700—Toray carbon fibre; PEEK—polyether-ether-ketone. 1st subscript—fibre path; 2nd subscript—transverse path; 3rd subscript—in-plane shear path. Composite ply calculations were on stress components with orientations in these 3 directions called stress directions or stress paths. v_1, v_2, v_3—Poisson's ratios; G_{12}, G_{13}, G_{23}—shear moduli; E_2, E_2, E_3—elastic moduli. HS—high-strength; superscript C—compression; superscript T—tension.

Table 3. Material characteristics for liners.

Parameter/Description	Poisson's Ratio, υ	Elongation at Break (%)	Ultimate Stress (MPa)	Yield Stress (MPa)	Elastic Modulus (MPa)	Density (kg/m³)
(Victrex) PEEK	0.400	45.00	125.0	110.0	4.0	1300.0
HDPE	0.460	10.00	43.0	1350.0	565.0	995.0
(Nylon PA) PA12	0.400	10.00	54.0	1500.0	540.0	1010.0
PVDF	0.400	10.00	54.0	1540.0	550.0	1780.0
(X80) Steel	0.300	5.90	950.0	880.0	207.0	7850.0
(1953T1) Aluminium alloy	0.300	7.50	540.0	480.0	71.0	2780.0
(Ti6Al4V) Titanium alloy	0.342	14.00	950.0	880.0	113.8	4430.0

PEEK—polyether-ether-ketone; HDPE—high-density polyethelene; PVDF—polyvinylidene fluoride; PA12—polyamide 12.

3.3. Stack-Up Sequence

Modelling of the composite riser layers was conducted by considering the arrangement of the plies. The layers were arranged by considering the stack-up sequence in Table 4. This included the arrangement of the matrix and the fibre reinforcements. Sketch of stack-up sequence of laminate ply showing different layer orientations is shown in Figure 4. Table 4 only shows the matrix mix using titanium liner, but other liners were also used as seen in Section 4. The rules of matrix used in this study are not discussed herein. However, it is noteworthy that the rules of matrix should be considered in selecting the matrix mix, liners and the stacking sequence (see related literature [87,88,112]).

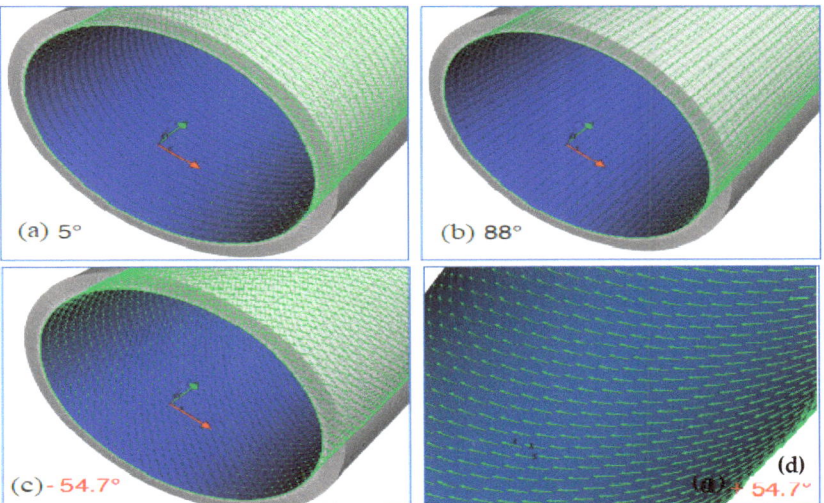

Figure 4. Fibre reinforcements for (**a**) axial ply, (**b**) hoop ply, and (**c**,**d**) off-axis plies.

Table 4. Selected design, stack-up sequences, material configurations, and CPR's composite lamina orientations.

Liner Material	Liner Thickness (mm)	Fibre	Matrix	Lay-Up	Lamina Thickness (mm)		
					0°	±53.5°	90°
Titanium	9	AS4	PEEK	[90$_3$,(±53.5)$_5$,0$_5$]	1.84	1.48	0.6
Titanium	9	IM7	PEEK	[90$_3$,(±53.5)$_5$,0$_5$]	1.84	1.48	0.6
Titanium	9	P75	PEEK	[90$_3$,(±53.5)$_5$,0$_5$]	1.84	1.48	0.6
Titanium	9	AS4	Epoxy	[90$_3$,(±53.5)$_5$,0$_5$]	1.84	1.48	0.6
Titanium	9	IM7	Epoxy	[90$_3$,(±53.5)$_5$,0$_5$]	1.84	1.48	0.6
Titanium	9	P75	Epoxy	[90$_3$,(±53.5)$_5$,0$_5$]	1.84	1.48	0.6

3.4. Design Load Cases

Five (5) different loading cases have been investigated on the local design in Table 5. The loadings on this CPR designed in this research were conducted using the load cases stipulated in industry recommendations on designing CPRs [66–70]. Figure 2 depicts the loads acting on a typical CPR. The fixed ends of the composite riser section include a top boundary and bottom boundary as in Figure 5. The boundary condition for the burst load was developed using one fixed end while the other was free with an end effect. Both the pressure and tension loads considered on the CPR are tabulated in Table 5. Both the composite body and the liner contribute to the effective weight of the CPR. According to the ABS Standard [70], effective tensions are established from dual trials. The first is the riser model's effective weight when the annulus is filled with mud. A safety factor of 1.5 was used, as recommended by the standards. The second factor is the effective weight of the riser model with an oil-filled annulus. A safety factor (S.F) of 2.25 is indicated in recommendations from the ABS standard for the tension study, but S.F. of 2 was applied.

Table 5. The design loadings for modelling the composite riser.

Design Loads	Parameters	Detailed Information of Load
Design Load 01	Tension Load	Using 2.25 as factor of load and the max. tensions
Design Load 02	Internal Pressure (Burst) + effect of load at ends	Using the int. pres. at 155.25 MPa is utilised
Design Load 03	External Pressure (Collapse)	Using the ext. pres. at 60.00 MPa is utilised
Design Load 04	Combined—Tension cum Internal Pressure	Using the int. pres. at 155.25 MPa for the tensions
Design Load 05	Combined—Tension cum External Pressure	Using 2.25 as factor of load for 19.50 MPa ext. pres.

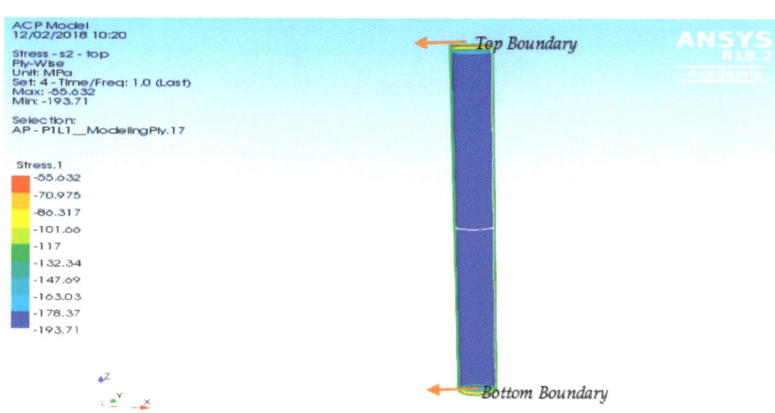

Figure 5. Stress result for ply 17 in ANSYS ACP.

3.5. Design Method

In the numerical approach, the composite riser's reinforcements were developed in three directions: axial, angled, and hoop. This provided different orientations for this composite riser as designed. Furthermore, the plies' stacking lay-up pattern and angles of the fibres for the composite riser's structure were meticulously planned and specially arranged. This is depicted in Table 4. The liner properties are given in Tables 2–4, and different orientations were used. The riser design is a multilayered tubular structure having eighteen (18) layers. Figure 6 presents the simplified methodology for the design approach used. This procedure was followed in order to obtain the optimum model for the project, as researched. This method is iterative with an advantage of presenting the strength performance of more layers. After inputting the basic design parameters, a finite element (FE) investigation was undertaken. With these variable values, the boundary conditions were set as fixed at one end and open at the other end. The burst load case was first conducted to calculate the composite riser's thickness. At this point, a decision was taken, leading to a corresponding update to be carried out on the CPR model. Hoop-, angled- (or off-axis), and axial-reinforced bracings were considered in the tailored local design. In addition, the hoop- and axial-fibre-reinforced bracings were considered in the conventional orthogonal method. In the design, the burst case dictates the performance results of the composite riser. It is the fundamentally crucial loading for the design and also firstly investigated among other loadings. When the burst case results show good performance, the other load cases in the model will also have good performance. Thus, the burst load case is considered the most sensitive of all the load cases.

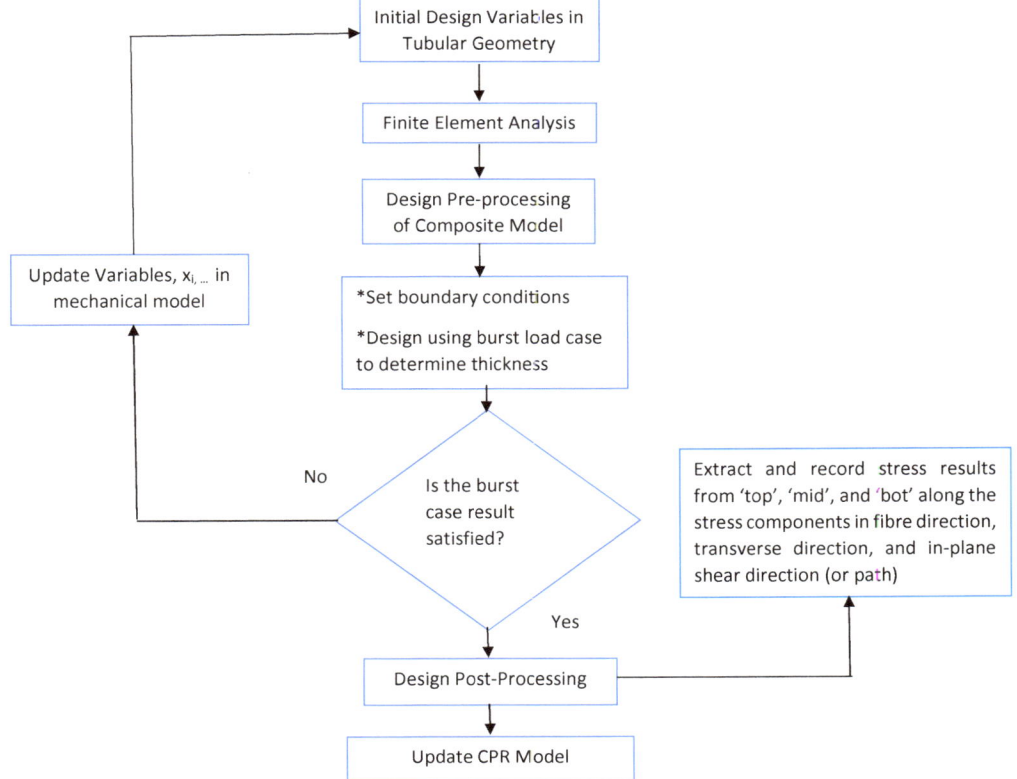

Figure 6. Simplified methodology for the design approach.

3.6. Stress Plot Criteria

The stresses for each layer were obtained and plotted in series of the composite riser. The stress distributions were obtained from the top, middle, and bottom locations of the composite riser layers as depicted in Figure 7. The results for each of the load cases are presented in Section 4. The results are from the laminae directions, as illustrated in Figure 8. The typical stress result is depicted in Figure 5. Maximum stress criterion was used to determine the Factor of Safety (FOS) for failure in the first ply. Using ANSYS ACP module, the design criterion was chosen with caution, by incorporating numerous failure criteria such as maximum stress, LaRC, and Puck rather than employing a single criterion. For each load instance, these criteria were utilised to consider all the out-of-plane and in-plane shear stress components in each composite lamina. This method was then carried out again by guesswork to optimise the design. Thus, the trials involved the use of different layer thicknesses and configurations until a minimal safety factor was obtained as 1.0.

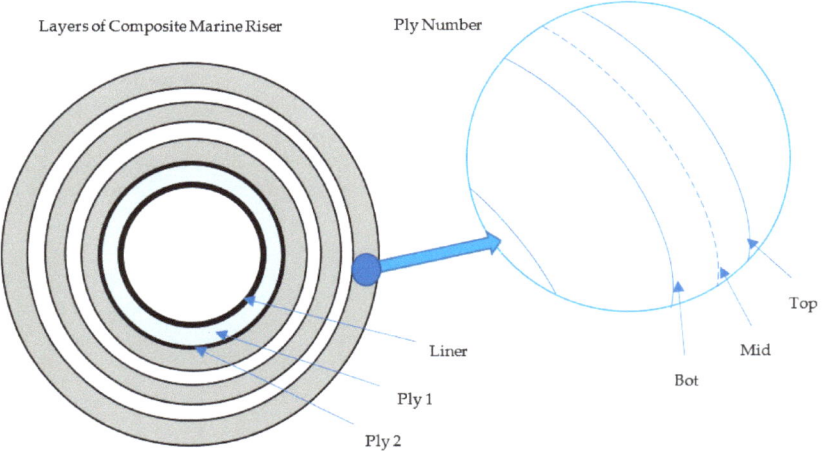

Figure 7. Layers of composite riser depicting locations of top, mid, and bottom for stress plots.

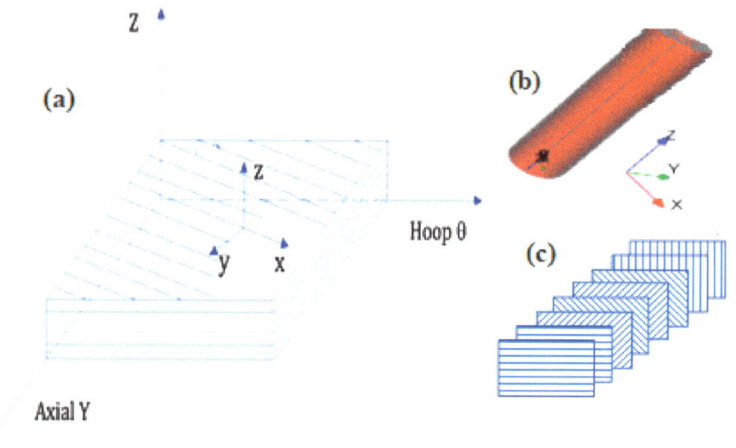

Figure 8. Illustrations of composite laminae showing (**a**) axial cum hoop directions, (**b**) the composite tubular and (**c**) its lay-up.

3.7. Finite Element Modelling

Table 1 lists the characteristics of the composite riser. The composite riser's finite element modelling (FEM) was carried out numerically in ANSYS Composites ACP. For the FEA, 3D layered structural solid elements called Solid 186 elements were used. This class of element can allow quadratic displacements as well as 3-degrees-of-freedom translation motion around the vertices. These linking vertices (or nodes) are 20 in number. Solid-186-layered components were used to simulate CPR laminates. Therefore, Solid-186-homogenous element was utilised to simulate solitary elements such as radial axis liners. The circumference of the sketched CPR section in Figures 4–8 were designed in ANSYS Design Modeler. Figures 4 and 5 show that the CPR model has two end-edges located at the top and two edges located at the bottom created using two semicircles at each end. Next, two side edges were designed to define the circumferential divisions on the outer wall of the composite riser. The FEM was performed utilising a ratio of axial to circumferential divisions as 50:65 per semicircle. 6000 nodes and 6500 elements were applied in the first finite element model, whereas in the second model, there were 8632 nodes and 8684 elements in the finite element model. In ANSYS ACP, the composite riser was modelled as a shell structure to apply the composite materials. The material layup for the composite riser was designed for different configurations with each having 18 layers. Various liners were considered on the designs studied. Figure 4 represents the resulting plot depicting plies in ANSYS ACP for the composite riser's FEM. The load cases were used to obtain stress values for every composite layer across various thicknesses as part of the local design technique. Specific initial values for the composite riser layers were estimated and used in the numerical method by first guessing the behaviour. The next step was ensuring the carryout of the analysis for the different load cases. Figure 8 represents the fibre direction and orientation for the composite riser at 90°. The green arrows represent the fibre direction while the purple arrows represent the orientation. The values of the layers along the axial, angled, and hoop paths had increased and decreased magnitudes based on the values, orientation, and condition used in the design.

3.8. Mesh and Convergence

The mesh convergence study on the 3D CPR model was implemented using two methods as presented in this section. The mesh model for the CPR is shown in Figure 4. A convergence investigation was conducted using the mesh of the CPR. The goal of the investigation includes mesh convergence. Establishing the strength-wise behaviour of the composite riser requires the convergence study. This was calculated using the highest value of the aggregate deformation. Figure 9a,b depicts the convergence study using different mesh numbers of divisions for obtaining the mesh sizes. The convergence results show that convergence occurred at maximum total deformation of 2.4599. The divisions were in two directions: the axial divisions and circumferential divisions. In Figure 9, the convergence occurred for the mesh case for 50 axial divisions and 65 circumferential divisions. This method enables the fibres to be formed around the circumferential axis of the composite riser. It can be observed also from the results that when less nodes and elements are used, the stress values obtained from the mesh elements are less, as in the case of 30 axial divisions and 55 circumferential divisions. This is observed to be consistent in the hoop, angled, and axial layers. When the element and nodes are smallest, more time is consumed in processing the finite element analysis. However, with the convergence study, the best mesh size has been determined.

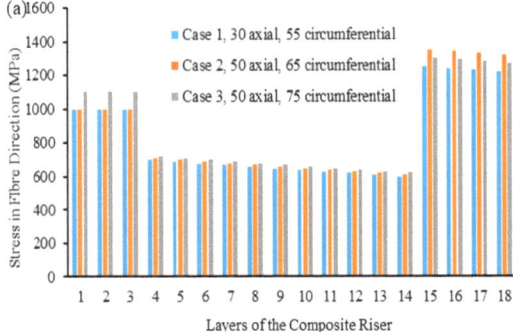 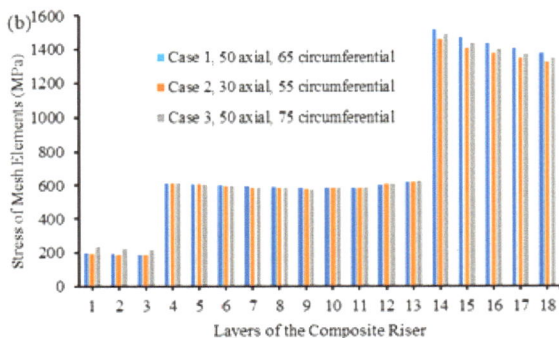

Figure 9. Convergence study using different mesh numbers of divisions for two different composite riser models, showing (**a**) Model 1 having Case 1 (30 axial, 55 circumferential), Case 2 (50 axial, 65 circumferential), and Case 3 (50 axial, 75 circumferential), and (**b**) Model 2 having Case 1 (50 axial, 65 circumferential), Case 2 (30 axial, 55 circumferential), and Case 3 (50 axial, 75 circumferential).

3.9. Validation

According to DNV [66] and ABS [70], the factor of safety (F.S) approach could be utilised to evaluate the stress distribution of composite riser layers. The results acquired for the CPR's local design from the current research were compared to the research findings from Wang et al. [45]. The ratioed proportion of permissible (or allowable) strength to the actual strength is known as the factor of safety (F.S). The output of the numerical analysis on the composite riser model under the burst load case are presented in Figure 10a. The results show good agreement and, as also observed, this model had 18 layers represented as layers 1–18, while the CPR model by Wang et al. [45] had 17 layers, represented as layers 2–18. In addition, layers 1–4 represent the axial layers for both models, but of material properties. Thus, the factor of safety values have some variance. This is due to the difference in the material's modulus used, the material's thickness utilised, and the length of the composite tube. Thus, the behaviour of the reinforcements considering the factor of safety shows that the method applied in this design is in good agreement with that of Wang's model. The safety factor method and stress component magnitudes were used in validating this study and they were shown to be in good agreement across the off-axis lamina and the hoop lamina. In addition, layers 1–4 represent the axial layers for both models, with variance in the factor of safety. However, despite the similarity in the models, there was some variance observed across the axial lamina. This is due to the difference in the modulus of the material used, the material's thickness used, and the composite tube's length. In this current research modelling, 5 m composite riser was used, while Wang et al. [45] employed a 3 m composite riser. Thus, the behaviour of the reinforcements considering the factor of safety is also a function of the length of the composite riser. The study shows that the method applied in this design is in good agreement with Wang's CPR model [45]. The second method considered was the element sizes in the investigation as depicted in Figure 10b. The ANSYS model was thus specified utilising the thickness for the PEEK liner as 6 mm. It was configured as $[0_3,(\pm 52)_{10},90_4]$ laminates having 17 lamina layers for the hoop, off-axis, and axial layers in the ratio 90°:52°:0°. The fibre-reinforced bracings had respective thicknesses of 1.6 mm, 1.30 mm, and 1.40 mm. The same laminate materials were employed in this scenario, but the ANSYS model's liner assumed PEEK polymer characteristics. Figure 10b contrasts the findings of both analyses in the burst instance, illustrating the validity of the FE model applied in this research. The hoop, off-axis, and axial laminae all had analogous tensile stresses along the fibre direction, with an average difference of 2.096% across all layers. In Figure 10b, the maximum tensile stress distributions for the burst instance are compared during the CPR's local analysis, and the element size of 30 mm is used versus a 5 mm element size. There were 158,316 nodes and 158,000 elements generated. The stress values at the top and bottom of each ply were obtained for both mesh sizes. From the results of the stresses obtained, no considerable variation was seen. For the 5 mm mesh size, the stress in the hoop plies (14–18) was almost 20.0 MPa less. Since improving the mesh very slightly improves the validity of the findings, this verifies that the 30 mm element size findings were adequately convergent.

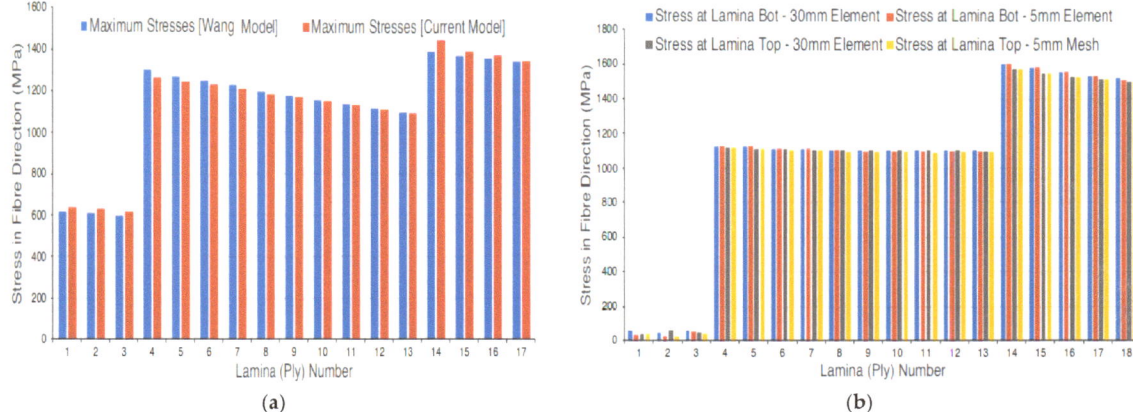

Figure 10. Validation of the model using (**a**) current ANSYS ACP model compared against Wang's ANSYS APDL model and (**b**) burst instance stress profiles of the original mesh (30 mm) and the improved mesh compared to convergence study (5 mm).

4. Results

The results are presented using bar charts depicting the maximal stresses across each lamina (ply). Starting with the deepest inner lamina encircling the liner and culminating at the uttermost outer lamina of the laminate, the layers are numbered along the horizontal plane. The maximal interlaminar stresses are investigated using pressures located along the outermost (top), innermost (bottom), and centre (midpoint). This was recorded across the whole CPR plies; they are known respectively as top, bot, and mid. The maximal stresses can be calculated using the highest stress in the middle of each lamina.

4.1. Studies on the Design Models

4.1.1. Result of Conventional Design

The analytical method was computed by utilising the exact elastic solution that considers 3D anisotropic elasticity by Xia M. et al. [61] and the analytical CPR model by Wang C. [112]. It was studied via a comparative investigation by utilising the FEM computed via ANSYS ACP and ANSYS Structural modules. The geometry and orientation of the composite layers under conventional design were configured as $[90/(0/90)_4]$, as tabulated in Table 6. The stresses were obtained based on the simulation using the 3 m geometry of the composite riser. From the finite element analysis (FEA), stresses were obtained along the various layers, as presented in Figure 11a,b. In Figure 11c,d, the safety factors for the same configuration are taken from different rosette orientation to show the difference along each component from FEA. It can be observed that this FEM presents relatively more stress values in comparison to the analytical method. The burst case was conducted by considering an internal pressure of 155.25 MPa for this conventional design. From this investigation, the PEEK liner was recorded to have a stress value of 98.90 MPa from the analytical method and 99.70 MPa from the finite element method. Thus, both methods presented a variance of 0.79%. However, the deviations along the fibre path and the transverse path were respectively >5% and >2% on these stress distributions across the 21 layers of the CPR model.

Table 6. Geometry and orientation of the [90/(0/90)$_4$] composite plies for conventional design.

Ply/Layer	Name of Layer	Inclination Angle/Orientation (°)	Thickness (mm)
00	The Liner		2.00
1, 3, 5, 7, 9, 11, 13, 15, 17, 19, 21	Hoop Layers	90	1.62
2, 4, 6, 8, 10, 12, 14, 16, 18, 20	Axial Layers	0	1.58

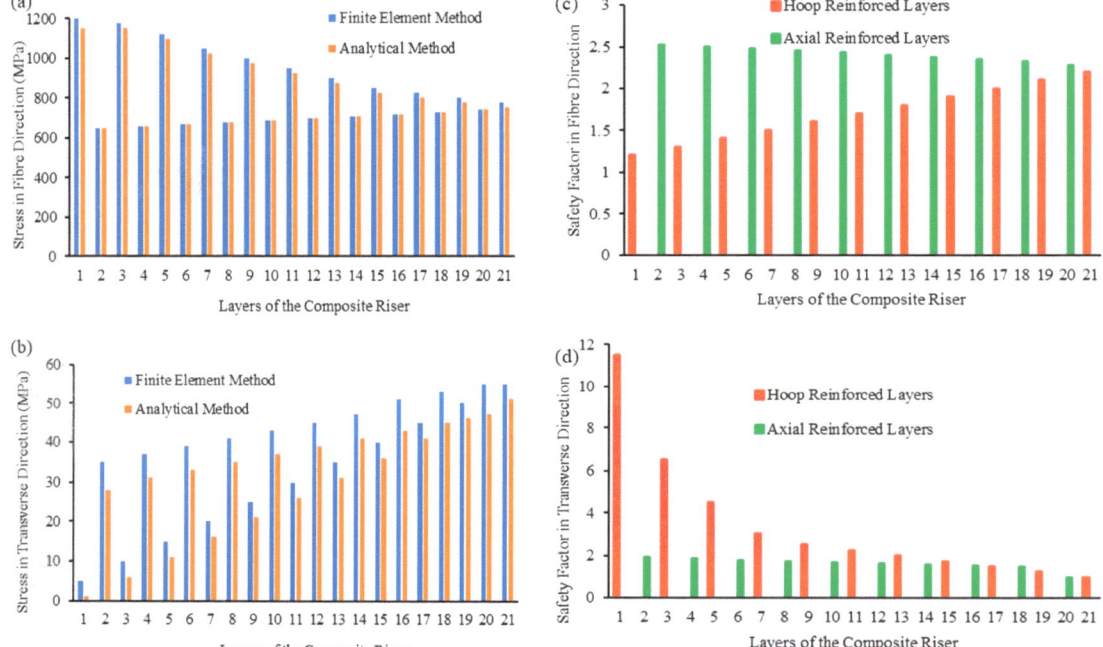

Figure 11. Result of composite riser model's conventional design for AS4/PEEK utilising PEEK liner under burst case showing the axial and hoop reinforcements along (**a**,**c**) fibre and (**b**,**d**) transverse directions for finite element method vs. analytical method.

4.1.2. Result of Tailored Local Design

A comparison between the analytical method (AM) using the analytical CPR model by Wang C. [112] versus the finite element method (FEM) in this study was performed. It was investigated by utilising the tailored local design based on the [0_4,(± 53.5)$_5$,90_4] configuration as presented in Figure 12a–c. In Figure 12d–f, the safety factors for the same configuration with 17 laminas were taken from different rosette orientations, to show the difference along with each component from the FEA. It can be observed that the FEM shows relatively more stress values in comparison to the analytical method. For the local design, the stack-up sequence and fibre thicknesses in Table 7 to obtain the results are presented in Figure 12. The burst case was conducted by considering an internal pressure of 155.25 MPa for this conventional design. From this investigation, the PEEK liner was recorded with a stress value of 115.80 MPa from the analytical method and 115.70 MPa from the FEM. Thus, both methods presented a variance of 0.79%. However, the deviations along the fibre path, the transverse path, and the in-plane shear path were respectively >1.5%, >7.5%, and >0.5% along the stress distributions across the 21 laminas of the CPR model.

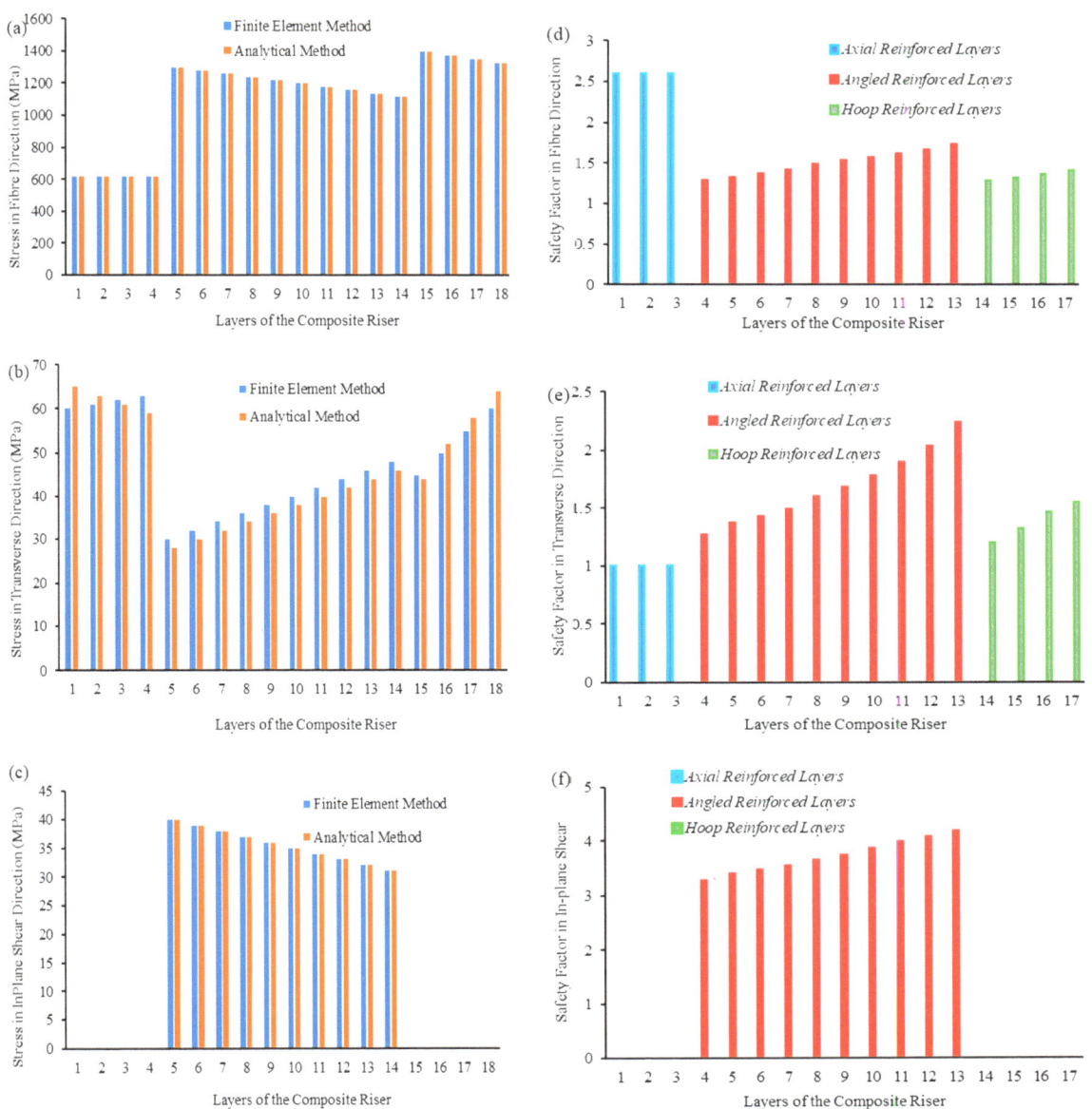

Figure 12. Result of composite riser model's local design for AS4/PEEK utilising PEEK liner under burst case showing the axial, angled, and hoop reinforcements along (**a**,**d**) fibre, (**b**,**e**) transverse, and (**c**,**f**) in-plane shear directions for finite element method vs. analytical method.

Table 7. Lay-up/ply inclination angle for local design for the $[0_4,(\pm53.5)_5,90_4]$ and $[90_3,(\pm53.5)_5,0_5]$ configurations.

Design Ply/Layer	Configuration Name of Layer	$[0_4, (\pm53.5)_5, 90_4]$ Inclination Angle/Orientation (°)	Thickness (mm)	$[90_3,(\pm53.5)_5,0_5]$ Inclination Angle/Orientation (°)	Thickness (mm)
00	The Liner	0.0	2.00	0.0	9.00
01	Axial Layers	0.0	1.58	0.0	1.84
02	Axial Layers	0.0	1.58	0.0	1.84
03	Axial Layers	0.0	1.58	0.0	1.84
04	Axial Layers	0.0	1.58	0.0	1.84
05	Angled Layers (or Off-axis Layers)	53.5	1.88	53.5	1.48
06	Angled Layers (or Off-axis Layers)	−53.5	1.88	−53.5	1.48
07	Angled Layers (or Off-axis Layers)	53.5	1.88	53.5	1.48
08	Angled Layers (or Off-axis Layers)	−53.5	1.88	−53.5	1.48
09	Angled Layers (or Off-axis Layers)	53.5	1.88	53.5	1.48
10	Angled Layers (or Off-axis Layers)	−53.5	1.88	−53.5	1.48
11	Angled Layers (or Off-axis Layers)	53.5	1.88	53.5	1.48
12	Angled Layers (or Off-axis Layers)	−53.5	1.88	−53.5	1.48
13	Angled Layers (or Off-axis Layers)	53.5	1.88	53.5	1.48
14	Angled Layers (or Off-axis Layers)	−53.5	1.88	−53.5	1.48
15	Hoop Layers	90.0	1.62	90.0	1.60
16	Hoop Layers	90.0	1.62	90.0	1.60
17	Hoop Layers	90.0	1.62	90.0	1.60
18	Hoop Layers	90.0	1.62	90.0	1.60

4.2. Result of Load Cases

4.2.1. Result of Burst Case

This subsection presents the findings from the preliminary CPR design, which was the development's first model. Figure 13a,b shows a preliminary design with 18 laminae (plies) and a comparatively thick 2 mm titanium liner but greater layer thicknesses. Figure 13a shows the maximal stressed profiles along the fibre path. This was conducted using the same titanium liner on the CPR structure subjected to burst loads. Tensile stresses were found in the laminae as a result of the internal pressure causing significant tension in the riser. The highest maximal stress along the fibre path was 1480.8 MPa, recorded from hoop's ply 14 across the bottom section, whereas the least maximal stress was 1125.9 MPa, recorded from hoop's ply 3 across the top portion. However, across the axial plies, there were high stresses of 2597 MPa which have more weight. Thus, there is a need for weight reduction, or to optimise the model by minimising the weight as presented in Section 4.3. Along the fibre path, the maximal tensile permissible stress was 1648 MPa for AS4/PEEK unidirectional composites. This maximum stress value was much below that, as along axial's layers 1–4, there were relatively minor tensile stresses. Figure 13b depicts the maximal stressed profiles along the transverse path. This was also conducted using the same burst load on the laminae. Along the transverse path, the minimal stress recorded was 74.89 MPa, recorded from axial's plies 1–4 located across the top portion, whereas the maximal stress recorded across the hoop's plies 15–18 were 193.84 MPa, located across the bottom portion. From Figure 13a,b, the axial layers had the maximum stress at 1017.7 MPa along the fibre path and 517.7 MPa along the transverse path. 60 MPa was the value of the external pressure introduced to the collapse instance. The results show similar behaviour for the fibre and transverse paths (or directions). However, the stress values in the fibre direction were higher than the stress values along the transverse direction. It was also observed that the axial layers had the highest stress values, but the stresses decreased when it reached the angled layers. This is due to debonding between the different materials and the orientation angle. As shown in Table 8, the liner's corresponding stress was 550.0 MPa, which is about 62% of the yield strength. This titanium liner yielded a maximal deformation of 3.37 mm. Since this liner's thickness is fairly considerable, it was lowered in subsequent updates to minimise weight. From this investigation, there is the need to minimise the structural weight of the CPR based on the preliminary designs while maintaining the integrity of the CPR structure and without violating acceptable limits of stress.

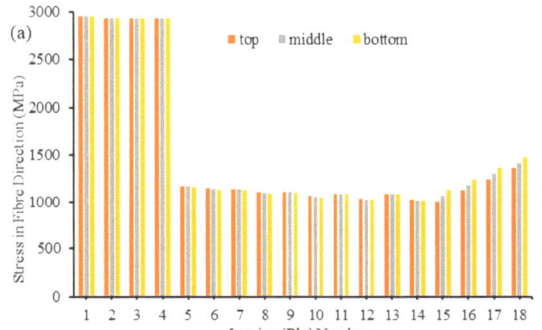
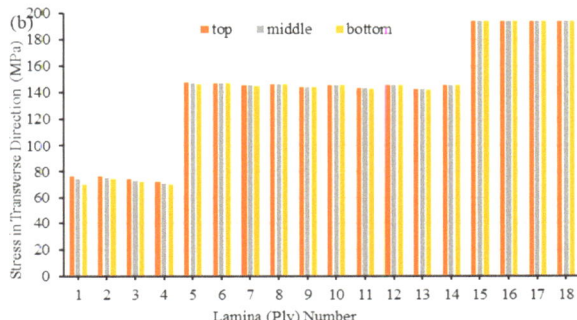

Figure 13. Stress profiles for the top, middle, and bottom lamina along the (**a**) fibre and (**b**) transverse directions under burst case.

Table 8. Riser deformation and liner stress during burst event's preliminary design.

Particulars	Value	Unit
Maximum Riser Deformation	3.37	mm
Ultimate Stress of Titanium Alloy	950.00	MPa
Yield Stress of Titanium Alloy	880.00	MPa
Titanium Liner's Equivalent Stress	550.00	MPa

4.2.2. Result of Collapse Case

From Figure 14, the titanium alloy liner is observed to be important in the reinforcement of the composite body. The axial layers in the fibre direction had the highest stress, up to 2930 MPa, while the titanium liner's stress was 1057.7 MPa. The value of the maximum stress decreased for the angled layers to 1174.8 MPa. The stresses spiked up to 1362.5 MPa for the hoop layers. As seen in Figure 14, the burst case for the transverse direction had a different pattern. The highest stress occurred at the titanium liner, up to 239.5 MPa. This shows that there was a lot of pressure applied, up to 155.25 MPa, and the other layers could not withstand much of the pressure load, so the liner had to take in a lot of the pressure. The stress value in the axial layers (layers 2–4) went low to 76.71 MPa. This shows that at that stage, the liner took in much of the pressure. However, the angle's layers (layers 5–14) had stresses of up to 145.85 MPa, while the hoop's layers had pressure effects of up to 193.8 MPa. Figure 6 represents the stress postprocessing of ply 17. Since the fibre-reinforced bracings perpendicularly align across that of the direction of the load, the highest stresses occur across the axial plies. The laminate had insignificant tensile stresses profiled across the fibre path. Additionally, all the maximal stresses that were highest were recorded across compressed portions of the laminates, thus yielding compressive laminae. The highest stresses via the transverse path of each ply, on the other hand, were tensile, while the compressive stresses via the transverse path were of low capacity. The stresses in the transverse direction were tensile because the laminate was perpendicular to the imparted compressive force in this direction, resulting in tension. This resulted in low compressive stress along the fibre path and higher tensile stress via the transverse path. The highest stresses recorded were via the transverse path of the plies during the collapse instance. Across the transverse path, it was recorded that the axial's ply 1 has the highest maximal stress of 54.0 MPa. On the other hand, the hoop's ply 14 has the least maximal stress of 28.0 MPa. The laminate's maximal tensile long-term strength was 62.4 MPa, correspondingly about 85.0% that of the maximal stress. As seen in Figure 14, the collapse case satisfied the design requirement, but there were some inconsistencies between the top, middle, and bottom layers, which was due to some design nonlinearities and environmental factors. The comparable stress in the titanium alloy liner in the collapse instance was 525.0 MPa, about 60% of the yield strength, as shown in Table 9.

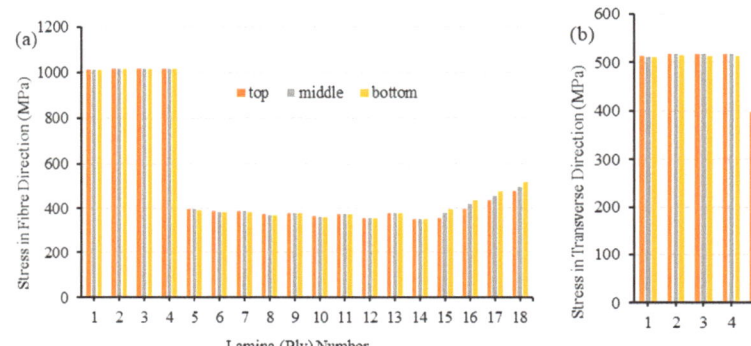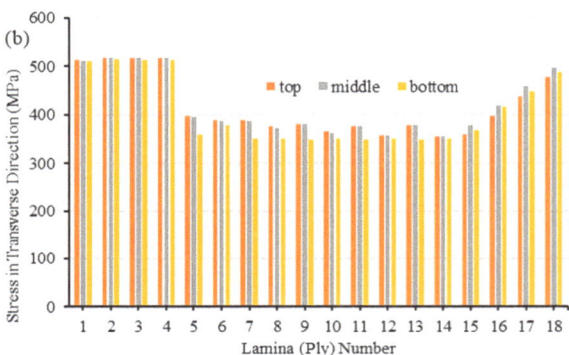

Figure 14. Stress profiles for the top, middle, and bottom lamina along the (**a**) fibre and (**b**) transverse directions under collapse case.

Table 9. Riser deformation and liner stress during collapse event's preliminary design.

Particulars	Value	Unit
Riser's Maximum Deformation	1.34	mm
Ultimate Stress of Titanium Alloy	950.00	MPa
Yield Stress of Titanium Alloy	880.00	MPa
Titanium Liner's Equivalent Stress	525.00	MPa

4.2.3. Result of Tension Case

The calculations for the tension cases were achieved by using the effective weight of the riser when carrying fluid and the riser's net weight. The composite riser's effective weight can be referred to as a function of the wall thickness of the composite riser being analysed. It was utilised to compute the tension of the riser. The maximum stress profiles measured in the laminae's fibre path under tension are shown in Figure 15a,b. At 4584 kN, the riser was fully tensioned and exposed. The laminae's tensile stress profiles were examined for the composite's long-term tensile stress. The least peak stress along the fibre path was 59.4 MPa across the hoop plies 1–4 and 55.1 MPa across the angled plies, followed by 254.5 MPa across the axial plies 1–4. The maximum long-term allowable tensile stress in the fibre direction of the AS4/PEEK unidirectional composite is around 8% greater than this maximum stress figure of 1648.0 MPa. It is worth noting that the hoop layers, plies 14–18, exhibited compressive stresses. The majority of the load was carried in pure tension by the fibres that were orientated in the force direction; as shown by the profiles in Figure 15, the axial laminae had the highest tensile stresses. Since the hoop fibres were perpendicularly aligned to the load, compressive stresses were found along the fibre direction. The average maximum compressive stress across all hoop plies was 48.2 MPa, which is considerably less than the AS4/PEEK composite's maximal long-term compressive strength of 864.0 MPa. For the tension scenario, Figure 15 illustrates the potential stresses present in the transverse direction of the laminae. The least maximum stress in the transverse direction was 1 MPa in axial ply 1 and the largest maximum stress was 12 MPa in hoop ply 14. This maximum stress value is 80% less than the unidirectional AS4/PEEK laminate's maximum long-term allowable tensile strength of 62 MPa. The highest tensile stress in plies 4–17 is greater than the maximum stresses in the mid and top sections of each lamina, which could be an exception. In comparison to the collapse instance, there were no considerable compressive stresses across the transverse path. From the results presented in Figure 15a,b, the axial lamina along the fibre path for the tension instance had the maximum stresses of 252.6 MPa. Based on the angled lamina and hoop lamina, the stresses had a maximum value of 61.74 MPa, and they were fairly uniform. However, different behaviour is observed in the transverse direction for the tension instance, as recorded in Figure 15. The highest stress values recorded across the axial, angled, and hoop lamina were 3.74 MPa, 11.36 MPa, and 15.15 MPa, respectively. The comparable stress in the titanium alloy liner in the pure tension scenario was 162 MPa, or about 20% of the yield strength, as shown in Table 10. This demonstrates titanium alloy's exceptional tensile behaviour in pure tension.

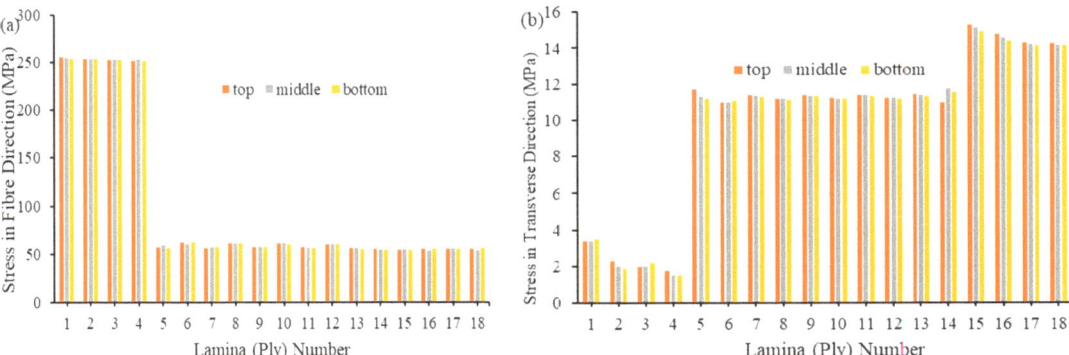

Figure 15. Stress profiles for the top, middle, and bottom lamina along the (**a**) fibre and (**b**) transverse directions under tension case.

Table 10. Riser deformation and liner stress during tension event's preliminary design.

Particulars	Value	Unit
Riser's Maximum Deformation	7.78	mm
Ultimate Stress of Titanium Alloy	950.00	MPa
Yield Stress of Titanium Alloy	880.00	MPa
Titanium Liner's Equivalent Stress	162.00	MPa

4.2.4. Result of Tension cum External Pressure Case

The highest stresses present along the fibre direction of the laminae for the tension cum external pressure condition are shown in Figure 16a,b. The riser was subjected to a maximum tensile load of 2037.0 kN and an external pressure of 19.5 MPa. The composite's long-term strength values were then compared to the maximum stress in each ply. Tensile stresses existed in axial laminae 1–4, but compressive stresses existed in off-axis laminae

4–13 and hoop laminae 14–18. The rationale here is due to two factors: when the CPR is tensioned, the axial fibres are tensile and the hoop fibres are in compression, as previously stated. Furthermore, an external pressure load compresses both the hoop and off-axis laminae. Due to the 88° hoop fibre reinforcements, these fibres contribute to the CPR's collapse and burst capacity, with the hoop laminae offering the best collapse resistance. The stress distribution illustrated in Figure 16 clearly demonstrates this behaviour; the axial and hoop laminae had the highest tensile and compressive stresses, in respective order. The hoop laminae provided the best collapse resistance due to the 88° hoop fibre reinforcing. Off-axis ply 7 had the lowest maximum fibre direction compressive stress of 72 MPa, followed by hoop ply 14 with 278 MPa. The highest stress value was around 30% higher than the maximum long-term compressive stress value that can be tolerated at 864.0 MPa. The highest tensile stress along the fibre direction was 300.0 MPa on average, which is about 20% of the maximum long-term tensile strength of 1648.0 MPa. For the tension condition, Figure 16a,b depicts the maximum stresses in the plies' transverse direction. There were no substantial tensile stress distributions throughout the laminate, and the majority of the stresses across the transverse path were compressive. The least maximum stress in the transverse path was 5.90 MPa in angled ply 5 down to 1.50 MPa at angled ply 14, while the largest maximum stress was 12.77 MPa in hoop ply 17. This maximum stress is around 75% less than the maximum long-term permitted compressive stress of 156.80 MPa via the transverse path of unidirectional AS4/PEEK laminate. From the results presented in Figure 16a,b, the axial layers across the fibre path for the tension instance had the maximum stresses of 374.1 MPa. For the angled layer, the stresses had a maximum value of 87.2 MPa, and the minimum value was fairly uniform. However, a different behaviour was observed in the transverse direction for the tension instance. The highest stress values for the axial, angled, and hoop lamina were 3.74 MPa, 11.36 MPa, and 15.15 MPa, respectively. The comparable stress in the titanium alloy liner in the tension and external pressure situation was 253 MPa, which is about 30% of the yield strength, as shown in Table 11.

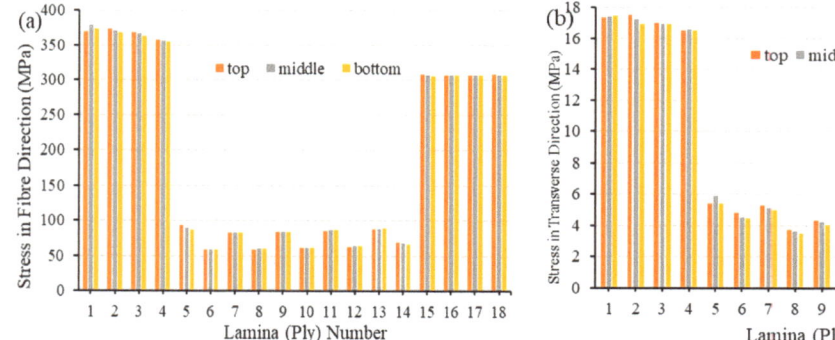
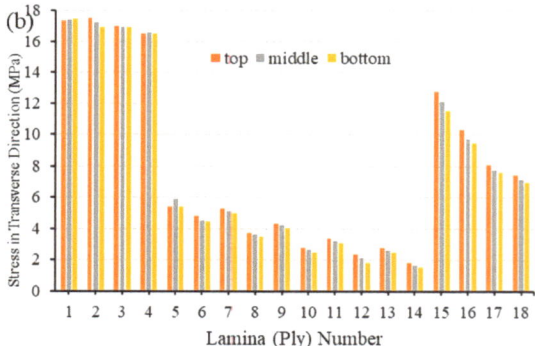

Figure 16. Stress profiles for the top, middle, and bottom lamina along the (**a**) fibre and (**b**) transverse directions under tension and external pressure case.

Table 11. Riser deformation and liner stress during tension cum external pressure event's preliminary design.

Particulars	Value	Unit
Riser's Maximum Deformation	11.80	mm
Ultimate Stress of Titanium Alloy	950.00	MPa
Yield Stress of Titanium Alloy	880.00	MPa
Titanium Liner's Equivalent Stress	253.00	MPa

4.3. Result of Minimum Weight Study

4.3.1. Effect of Weight per Unit Area

Figure 17 represents the distribution of the thickness of layers and the weight per unit area for 12 different composite riser configurations. This study was investigated using the same liner material—titanium liner—on AS4/PEEK material. From this study, it can be observed that different configurations have different weight per unit areas and thicknesses, which also has an implication on the structural strength of the composite riser. As observed in Figure 17 and Table 12, the best configuration based on this analysis was DesignCase7, as it had the least weight per unit area and would be the lightest with a weight/area value of 6.60×10^{-8} kg/mm^2. However, more investigation is suggested on the detailed weight savings of the riser models from the local design against the conventional design.

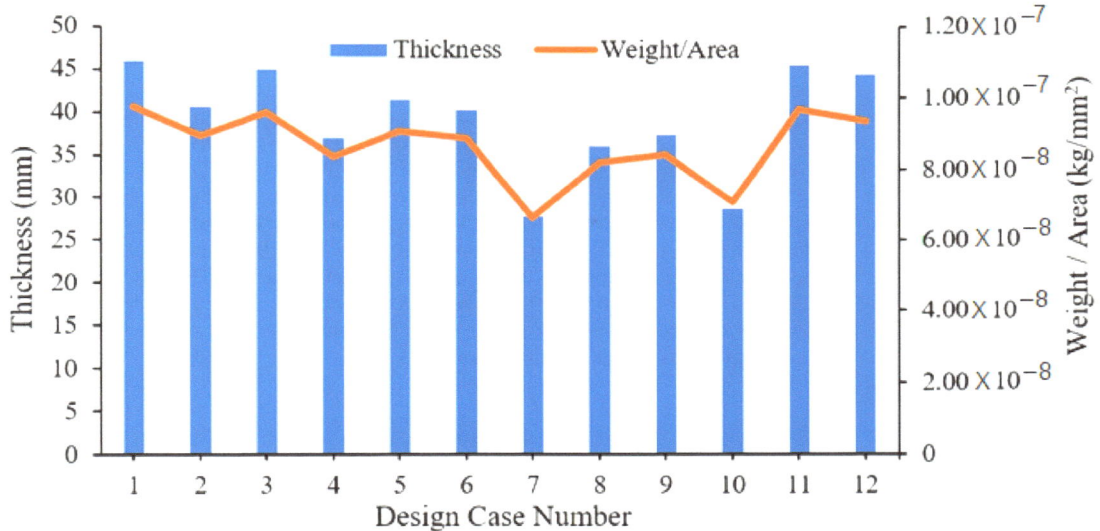

Figure 17. Effect of weight per unit area for 12 different composite riser configurations.

Table 12. Considerations for effect of weight per unit area for 12 design cases.

Design Case	Liner Material	Fibre	Matrix	Lay-Up	Lamina Thickness (mm)		
					0°	±53.5°	90°
Case 1	Titanium	AS4	PEEK	[0$_5$,(±53.5)$_5$,90$_5$]	2.49	1.7	0.6
Case 2	Titanium	AS4	PEEK	[90$_3$,(±53.5)$_5$,0$_5$]	2.49	1.7	1.84
Case 3	Titanium	AS4	PEEK	[90$_3$,(±53.5)$_5$,0$_5$]	2.49	1.7	3.25
Case 4	Titanium	AS4	PEEK	[0$_5$,(±53.5)$_5$,90$_5$]	2.49	1.7	1.7
Case 5	Titanium	AS4	PEEK	[90$_5$,(±53.5)$_5$,0$_5$]	2.49	1.7	1.84
Case 6	Titanium	AS4	PEEK	[(±53.5)$_5$,0$_5$,90$_5$]	1.15	1.7	1.84
Case 7	Titanium	AS4	PEEK	[0$_6$,(±53.5)$_5$,90$_4$]	2.49	1.7	0.6
Case 8	Titanium	AS4	PEEK	[90$_5$,(±53.5)$_5$,0$_5$]	2.49	1.7	0.6
Case 9	Titanium	AS4	PEEK	[90$_5$,(±53.5)$_5$,0$_5$]	1.15	1.7	1.84
Case 10	Titanium	AS4	PEEK	[(±53.5)$_5$,90$_5$,0$_5$]	1.15	1.7	1.84
Case 11	Titanium	AS4	PEEK	[(±53.5)$_5$,90$_5$,0$_5$]	1.15	1.7	3.25
Case 12	Titanium	AS4	PEEK	[0$_6$,(±53.5)$_5$,90$_4$]	2.49	1.7	1.7

4.3.2. Effect of Matrix Cracking

The failure order of the liner, fibre, or matrix is numerically predictable; however, it is dependent on the composite material, composition of the lay-up, and matrix cracking [87,88]. Thus, there is a need to investigate further on the failure pattern by utilising matrix cracking. Matrix cracking could be taken as a time-dependent phenomenon that can ensue prior to failure or post failure of the liner. In most cases, it begins to propagate from the hoop lamina when the resultant forces perpendicularly act along the fibre path. The axial lamina (or layers) support the displaced load as a result of this damage. According to the research, matrix cracking can cause a 10% drop in laminate stiffness [29,32,108]. As a result of the unpredictability as it is uncertain, the local analysis used lower long-term strength and stiffness parameters to account for matrix cracking with fatigue, wear, and service failure over time. The influence of matrix cracking on the CPR model was comparatively studied. This considered the structural weight using PEEK liners on AS4/PEEK and AS4/Epoxy designed with matrix cracking and without matrix cracking permissions, as presented in Figure 18. Matrix cracking study is important to avoid any failure due to liner leakages, according to specifications in design standards. As such, matrix cracking can be considered as the failure mode that is most critical for composite riser designs. As such, the design considered in this research does not permit matrix cracking. The comparison between the conventional design versus the tailored local design shows that both designs give some weight gains when there are permissions for matrix cracking. It can also be observed that when there was no matrix cracking, lower weight profiles were generated. In the design application of AS4/PEEK configured utilising PEEK liner, the ratio of percentage weight savings for the conventional to the tailored local design was 68.5%:83.5% without matrix cracking. However, permitting matrix cracking offered some weight savings of about 5.88%:8.33% for the tailored designs for AS4/PEEK and AS4/Epoxy respectively. Thus, the study satisfied this design requirement.

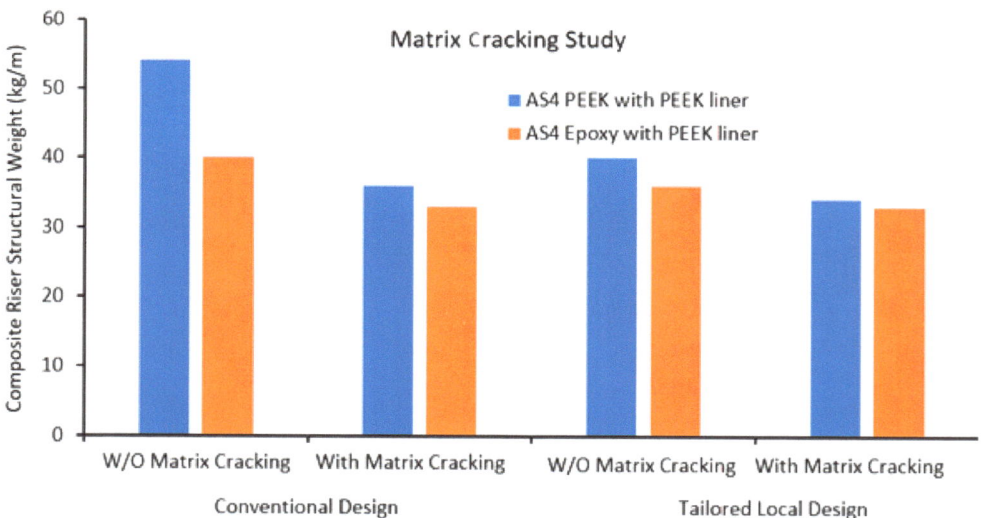

Figure 18. Comparative study on structural weight using PEEK liners on AS4/PEEK and AS4/Epoxy designed with matrix cracking and without matrix cracking permissions.

4.3.3. Result of Structural Weight

The investigation on the structural weight of the layers for the composite riser was conducted using P75/PEEK and AS4/PEEK for six different liners as presented in Figure 19. It can be observed that the optimization of the structural weight with the tailored local design proved to be effective as it presented lower structural weight for the two fibre reinforcements investigated. It can be also observed that the P75/PEEK had higher structural weight than the AS4/PEEK because it had higher strength modulus. Based on the liners, it was observed that the combination with thermoplastic liners such as PA12 liners was comparatively less than the combination with the metallic liners. It was recorded that the AS4/PEEK with PEEK liner had a normalised weight saving ratio of the conventional and tailored designs of 0.30:0.24. In addition, the structural weight savings for the composite riser configured using the metallic liners—aluminium, titanium, and steel—are respectively 25%:23%:23% for the tailored design versus the conventional design. On the other hand, the structural weight savings for the thermoplastic liners—PA12, PEEK, and PVDF—are respectively 23%:20%:20% for the tailored design versus the conventional design.

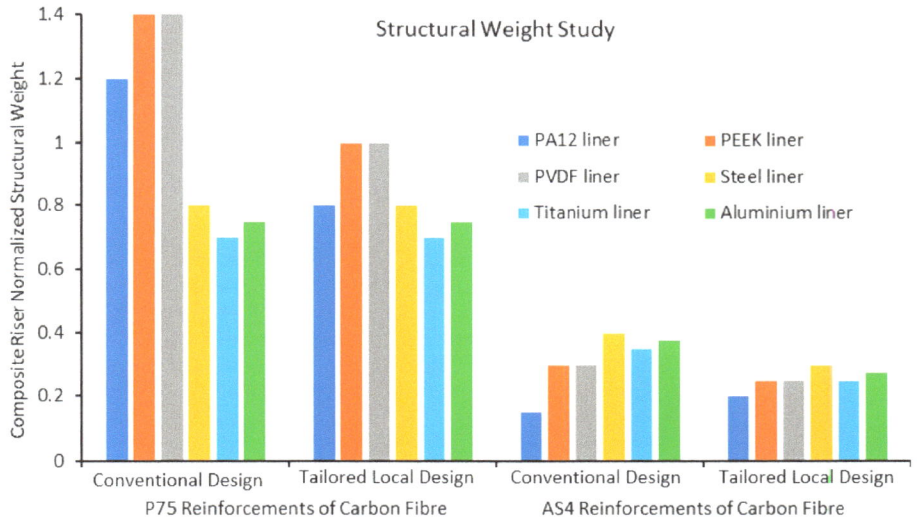

Figure 19. Comparative study at optimized angled plies for normalized structural weight.

4.3.4. Result of Aggregate Thickness

The investigation on the aggregate thickness of the layers of the composite riser was conducted using P75/PEEK and AS4/PEEK for six different liners as presented in Figure 20. It can be observed that the optimization of the aggregate thickness with the tailored local design proved to be effective as it presented lower normalized values for the two fibre reinforcements investigated. It can be also observed that the P75/PEEK had higher magnitudes than the AS4/PEEK because it had higher strength modulus. In addition, composite risers had higher thickness than the steel model, as increasing the thickness reflected to be relative to the type of liner combination and the reinforcement matrix. The worst thickness was reflected by the P75/PEEK model, as the thickness is over four times (4×) higher than that of AS4/PEEK model, as well as that of steel riser model. Thus, it will be the least desirable material combination to be considered. Based on the liners, it was recorded that the AS4/PEEK with PEEK liner had the best performance in normalised aggregate thickness and it is the recommended configuration.

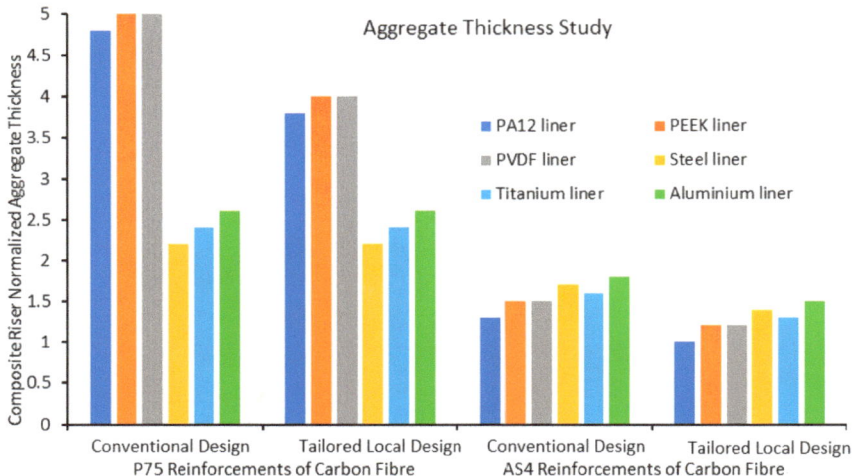

Figure 20. Comparative study at optimized angled plies for normalized aggregate thickness.

4.3.5. Result of Stack-Up Sequence

The influence of the stack-up sequence on the composite riser was conducted as presented in Figure 21. It was conducted comparing composite riser models at optimized angled plies considering the normalized structural weights and the aggregate thickness. Four locations for the angled plies were considered in the investigation: "inner&outer", "outer", "middle", and "inner" locations. The combination of "inner&outer" location is the axial fibre in the inner layer with the hoop fibre in the outer layer. In this study, the most efficient angle was obtained as ±53.5° as in the $[(0)_4,(\pm 53.5)_5,(90)_4]$ model configuration and was considered in the design. From this investigation, it can be observed that the combination using "inner&outer" location had the least normalized values for both the aggregate thickness and the structural weight.

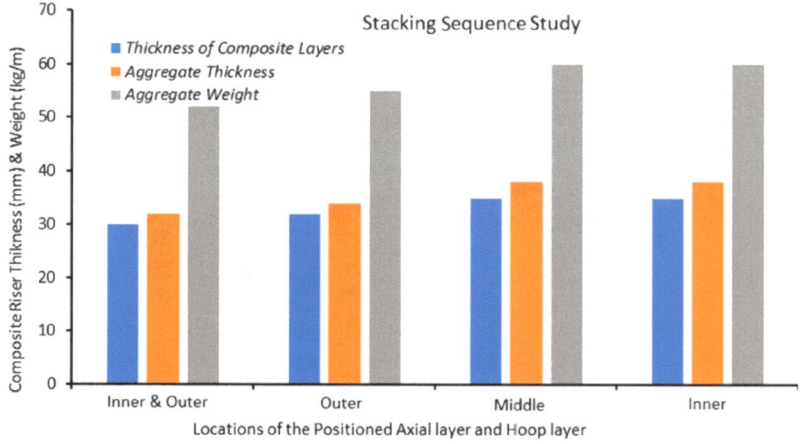

Figure 21. Comparative study at optimized angled plies for normalized structural weights and thickness.

4.3.6. Result of CPR Weight Reduction

An investigation of the CPR weight reduction in steel riser against composite riser was conducted as presented herein. The minimum structural weights of the steel riser joints used in TTR production riser strings are listed in Table 13. The weights of the initial and final CPR designs are compared to the weight limit of a TTR of an X80 steel riser in Table 14. This global design for the Truss SPAR used was detailed in earlier studies [25,30]. These results are close to the 50%–70% savings from weight reduction reported in the literature for composite risers. In challenging conditions such as the Gulf of Mexico (GOM), the North Sea (NS), or Offshore West Africa (OWA), which are the loading conditions for which this CPR is intended, thicker steel risers are normally employed. As a result, the indicated weight savings are the absolute least that this CPR design can provide. Lastly, a comparative study of the structural weight of three design configurations presented in Figure 22 shows that the tailored local design also proffered more weight savings for composite risers. Between the three configurations, the AS4/Epoxy with titanium liner had the highest weight, followed by the AS4/Epoxy configured utilising aluminium liner, and the least was the AS4/PEEK configured utilising PEEK liner. Thus, the least configuration is the best design as shown.

Table 13. Comparison of deep water TTR steel riser under 2000 m water depth against its structural weight.

Depth (m)	Thickness (mm)	Weight (kg/m)
0–1000	155	23
1000–1600	162	24
1600–2000	170	25
~2000 (approx. range)	160	

Table 14. Comparison of weights between a typical TTR steel riser and composite riser.

Design	Configuration	OD, Outer Diameter	OD Increase	Riser Weight	Weight Saving
Steel TTR	$[90/(0/90)_4]$	273–275	-	160	-
Preliminary	$[5_4,(\pm 54.7)_{10},88_4]$	314.6	13%	73	54%
Second	$[0_4,(\pm 53.5)_5,90_4]$	315.2	13.5%	70	57%
Final	$[3_3,(\pm 62)_{10},88_5]$	316.6	14%	54	66%

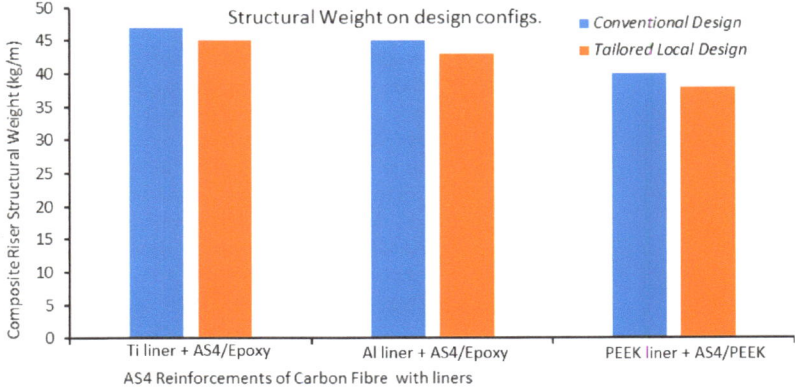

Figure 22. Comparative study of different designs using structural weights on design configurations.

5. Concluding Remarks

The 18-layered composite production riser (CPR) was designed specifically for deep water conditions. The tailored design of the composite riser was conducted for deep water environments. This research presents different CPR configurations designed to minimise the structural weight. To achieve this, the material attributes listed in Section 3.2 were utilised. The CPR is comprised of a liner and a multilayered body. Three lamina designs were numerically studied using ANSYS Composites ACP R1 2021, consisting of 17, 18, and 21 layers. In this research, a variety of liners, including titanium and aluminium alloys, were investigated. The stresses on the CPR lamina in both the fibre and transverse directions were calculated utilising four distinct loadings. The stress profiles were found to be affected differently depending on the design orientation. The results for all stress profiles in the fibre and transverse orientations for the CPR model were also provided. The results were validated and show good agreement. In the current model, two models of 3 m and 5 m composite riser were considered.

The model highlights include the following: Firstly, composite materials applied in the local analysis of composite marine risers are useful in prototype production. Secondly, the local design of composite risers using different materials minimises structural weight was presented. Thirdly, there is novelty in the analysis of composite riser assessment to minimise the weight, useful in the global design. Fourthly, the comparative study of composite riser designs is presented with detailed mechanical characterization for deep water application. Lastly, the study uses indicators such as safety factors to investigate the stress magnitudes and make recommendations for standards development on composite risers.

From this model, it is observed that there is an influence of metal cracking and metal–composite interface on the lamina, between the hoop plies, off-axis plies, and axial plies. Moreover, the results show that the stresses along the fibre directions are higher than the stresses in the hoop layers. Overall, the technique for this design revealed the composite risers' stress profiles for various load scenarios, guiding offshore designers on composite risers. The angled layers were more affected by tension and external pressure loading, as shown in Section 4.2. The loads operating on the riser lamina (or plies) led to the behaviours along the direction of the applied forces. In addition, the design revealed that throughout the burst instance, the liner withstood considerable pressure. This means that the inner liners do not need to be reinforced any further. For the first load instance (burst), the results showed high stress magnitudes and safety factor values along the fibre direction acting upon the axial plies. The implication is that the CPR will be extremely strong and able to tolerate harsh weather conditions. It is noteworthy that the stress profiles of the inner axial lamina, alternated angled lamina, and outer hoop lamina show substantial variation in all of the results.

However, more research into the global design of the CPR under various ocean conditions is recommended. This is desirable to explain the possible cause leading to particular observations. Finally, the analysis indicates that adapting the design of the CPR to appropriate fibre reinforcement angles and liner combinations can save enough weight. This study shows sufficient weight savings on the CPR design by tailoring the designs to achieve suitable fibre reinforcements, fibre orientation angles, unique lay-up sequences and liner combinations. Further study is recommended on cost with factors for the cost model for CPRs, as the cost model and manufacturing may be considered in the future. In addition, data-driven optimization methods for composite shells should be supplemented with proper-orthogonal-decomposition-based buckling analysis and optimization of hybrid fiber composite risers [120–122]. These could include a multi-fidelity competitive sampling method for surrogate-based stacking sequence optimization of composite shells with multiple cut-outs, as well as other techniques. Lastly, further work on the composite riser should include robust global design [42–46], fatigue analysis of the composite riser [93,123–126], and the structural integrity/reliability of the composite riser [94,127].

Supplementary Materials: The following supporting information can be downloaded at: https://www.mdpi.com/article/10.3390/jcs6040103/s1, There are supplementary data available for the model used and summary: Amaechi, Chiemela Victor; Wang, Chunguang (2022), "Composite CLT theory MATLAB codes", Mendeley Data, V1, doi:10.17632/pdjytm9gzx.1. Also, Amaechi Chiemela Victor (2022), "Dataset on Composite Risers for Deep Water Application", Mendeley Data, V1, doi:10.17632/4rvvwkjs27.1.

Author Contributions: Conceptualization, C.V.A. and N.G.; methodology, C.V.A., N.G., I.A.J. and C.W.; software, C.V.A., N.G., I.A.J. and C.W.; validation, C.V.A., N.G. and C.W.; formal analysis, C.V.A., N.G. and C.W.; investigation, C.V.A., N.G. and C.W.; resources, C.V.A., N.G. and C.W.; data curation, C.V.A. and N.G.; writing—original draft preparation, C.V.A. and N.G.; writing—review and editing, C.V.A., N.G., I.A.J. and C.W.; visualization, C.V.A., N.G., I.A.J. and C.W.; supervision, C.V.A.; project administration, C.V.A. and N.G.; funding acquisition, C.V.A. All authors have read and agreed to the published version of the manuscript.

Funding: The Department of Engineering, Lancaster University, UK and Engineering and Physical Sciences Research Council (EPSRC)'s Doctoral Training Centre (DTC), UK are highly appreciated. In addition, the funding of the Overseas Postgraduate Scholarship by Niger Delta Development Commission (NDDC), Nigeria is also appreciated. The authors also recognizes the support of the Standards Organisation of Nigeria (SON), F.C.T Abuja, Nigeria. The funding support on the following projects and for supporting this study are also acknowledged as follows: (a) Shandong Provisional Natural Science Foundation (ZR2020ME269), (b) Key Research and Development Program of Shandong Province (2019GHY112076), and (c) Shandong Provincial Key Laboratory of Ocean Engineering with grant No. KLOE202005. Lastly, the funding support for the APC charges to Author 1, C.V.A, from MDPI's JCS is well appreciated.

Institutional Review Board Statement: Not applicable.

Informed Consent Statement: Not applicable.

Data Availability Statement: The data for this study are not shared because they are still part of a present study on this research. However, see supplementary materials for some data on the study.

Acknowledgments: The authors acknowledge the technical support with computational resources from Lancaster University Engineering Department. We also appreciate the approvals and necessary technical permissions obtained during the COVID-19 pandemic. The feedback and supervision by Jianqiao Ye of Lancaster University on this research is highly acknowledged. This paper is a product of a BEng dissertation and a research project. The feedback of Charles Odijie of MSCM Limited, UK in an earlier version was noted. In addition, the authors acknowledge the project contributions of Armin Kanani in the development of the ANSYS composite riser model. Lastly, the ANSYS support team is appreciated for technical support in the global marine riser design.

Conflicts of Interest: The authors declare no conflict of interest. The funders had no role in the design of the study; in the collection, analyses, or interpretation of data; in the writing of the manuscript; or in the decision to publish the results.

References

1. Zhang, H.; Tong, L.; Addo, M.A. Mechanical Analysis of Flexible Riser with Carbon Fiber Composite Tension Armor. *J. Compos. Sci.* **2020**, *5*, 3. [CrossRef]
2. Ochoa, O.; Salama, M. Offshore composites: Transition barriers to an enabling technology. *Compos. Sci. Technol.* **2005**, *65*, 2588–2596. [CrossRef]
3. Savari, A. Failure analysis of composite repaired pipes subjected to internal pressure. *J. Reinf. Plast. Compos.* **2022**, 1–20, ahead-of-print. [CrossRef]
4. Amaechi, C.V.; Chesterton, C.; Butler, H.O.; Gu, Z.; Odijie, A.C.; Wang, F.; Hou, X.; Ye, J. Finite Element Modelling on the Mechanical Behaviour of Marine Bonded Composite Hose (MBCH) under Burst and Collapse. *J. Mar. Sci. Eng.* **2022**, *10*, 151. [CrossRef]
5. Amaechi, C.V.; Chesterton, C.; Butler, H.O.; Gu, Z.; Odijie, A.C. Numerical Modelling on the Local Design of a Marine Bonded Composite Hose (MBCH) and Its Helix Reinforcement. *J. Compos. Sci.* **2022**, *6*, 79. [CrossRef]
6. Odijie, A.C.; Wang, F.; Ye, J. A review of floating semisubmersible hull systems: Column stabilized unit. *Ocean Eng.* **2017**, *144*, 191–202. [CrossRef]
7. Johnson, D.; Salama, M.; Long, J.; Wang, S. Composite Production Riser—Manufacturing Development and Qualification Testing. In Proceedings of the Offshore Technology Conference, Houston, TX, USA, 4–7 May 1998; pp. 113–123. [CrossRef]

8. Johnson, D.B.; Baldwin, D.D.; Long, J.R. Mechanical Performance of Composite Production Risers. In Proceedings of the Offshore Technology Conference, Houston, TX, USA, 3–6 May 1999; pp. 1–10. [CrossRef]
9. Amaechi, C.V. Novel Design, Hydrodynamics and Mechanics of Marine Hoses in Oil/Gas Applications. Ph.D. Thesis, Lancaster University, Engineering Department, Lancaster, UK, 2021.
10. Amaechi, C.V.; Chesterton, C.; Butler, H.O.; Wang, F.; Ye, J. Review on the design and mechanics of bonded marine hoses for Catenary Anchor Leg Mooring (CALM) buoys. *Ocean Eng.* **2021**, *242*, 110062. [CrossRef]
11. Amaechi, C.V.; Wang, F.; Ja'E, I.A.; Aboshio, A.; Odijie, A.C.; Ye, J. A literature review on the technologies of bonded hoses for marine applications. *Ships Offshore Struct.* **2022**, 1–32, ahead-of-print. [CrossRef]
12. Amaechi, C.V.; Chesterton, C.; Butler, H.O.; Wang, F.; Ye, J. An Overview on Bonded Marine Hoses for sustainable fluid transfer and (un)loading operations via Floating Offshore Structures (FOS). *J. Mar. Sci. Eng.* **2021**, *9*, 1236. [CrossRef]
13. Amaechi, C.V.; Wang, F.; Ye, J. Mathematical Modelling of Bonded Marine Hoses for Single Point Mooring (SPM) Systems, with Catenary Anchor Leg Mooring (CALM) Buoy Application—A Review. *J. Mar. Sci. Eng.* **2021**, *9*, 1179. [CrossRef]
14. Pham, D.-C.; Sridhar, N.; Qian, X.; Sobey, A.; Achintha, M.; Shenoi, A. A review on design, manufacture and mechanics of composite risers. *Ocean Eng.* **2016**, *112*, 82–96. [CrossRef]
15. Picard, D.; Hudson, W.; Bouquier, L.; Dupupet, G.; Zivanovic, I. Composite Carbon Thermoplastic Tubes for Deepwater Applications. In Proceedings of the Offshore Technology Conference, Houston, TX, USA, 30 April–3 May 2007; pp. 1–9. [CrossRef]
16. Rasheed, H.; Tassoulas, J. Strength Evaluation of Composite Risers. In Proceedings of the Offshore Technology Conference, Houston, TX, USA, 1–4 May 1995; pp. 215–222. [CrossRef]
17. Carpenter, C. Qualification of Composite Pipe. *J. Pet. Technol.* **2016**, *68*, 56–58. [CrossRef]
18. Wilkins, J. Qualification of Composite Pipe. In Proceedings of the Offshore Technology Conference, Houston, TX, USA, 2–5 May 2016; pp. 1–15, Paper Number: OTC-27179-MS. [CrossRef]
19. Bai, Y.; Bai, Q. *Subsea Pipelines and Risers*, 1st ed.; Elsevier Science Ltd.: Kidlington, UK, 2005; pp. 3–19. [CrossRef]
20. Dareing, D.W. *Oilwell Drilling Engineering*; ASME: New York, NY, USA, 2019. [CrossRef]
21. Dareing, D.W. *Mechanics of Drillstrings and Marine Risers*, 1st ed.; ASME Press: New York, NY, USA, 2012. [CrossRef]
22. Sparks, C.P. *Fundamentals of Marine Riser—Basic Principles and Simplified Analyses*, 2nd ed.; PennWell: Oklahoma, TN, USA, 2018.
23. Amaechi, C.V.; Chesterton, C.; Butler, H.O.; Gillet, N.; Wang, C.; Ja'E, I.A.; Reda, A.; Odijie, A.C. Review of Composite Marine Risers for Deep-Water Applications: Design, Development and Mechanics. *J. Compos. Sci.* **2022**, *6*, 96. [CrossRef]
24. Amaechi, C.V. A review of state-of-the-art and meta-science analysis on composite risers for deep seas. *Ocean Eng.* **2022**. under review.
25. Gillett, N. Design and Development of a Novel Deepwater Composite Riser. BEng Dissertation, Engineering Department, Lancaster University, Lancaster, UK, 2018.
26. Chesterton, C. A Global and Local Analysis of Offshore Composite Material Reeling Pipeline Hose, with FPSO Mounted Reel Drum. BEng Dissertation, Engineering Department, Lancaster University, Lancaster, UK, 2020.
27. Amaechi, C.V.; Ye, J. A numerical modeling approach to composite risers for deep waters. In Proceedings of the International Conference on Composite Structures (ICCS20), ICCS20 20th International Conference on Composite Structures, Paris, France, 4–7 September 2017; Published in Structural and Computational Mechanics Book Series. Ferreira, A.J.M., Larbi, W., Deu, J.-F., Tornabene, F., Fantuzzi, N., Eds.; Societa Editrice Esculapio: Bologna, Italy, 2017; pp. 262–263.
28. Ye, J.; Cai, H.; Liu, L.; Zhai, Z.; Amaechi, C.V.; Wang, Y.; Wan, L.; Yang, D.; Chen, X.; Ye, J. Microscale intrinsic properties of hybrid unidirectional/woven composite laminates: Part I experimental tests. *Compos. Struct.* **2021**, *262*, 113369. [CrossRef]
29. Ward, E.G.; Ochoa, O.; Kim, W.; Gilbert, R.M.; Jain, A.; Miller, C.; Denison, E. *A Comparative Risk Analysis of Composite and Steel Production Risers*; MMS Project 490, Minerals Management Service (MMS); Texas A&M University: College Station, TX, USA, 2007. Available online: https://www.bsee.gov/sites/bsee.gov/files/tap-technical-assessment-program/490ab.pdf (accessed on 4 March 2022).
30. Amaechi, C.V.; Gillett, N.; Odijie, A.C.; Wang, F.; Hou, X.; Ye, J. Local and Global Design of Composite Risers on Truss SPAR Platform in Deep waters, Paper 20005. In Proceedings of the 5th International Conference on Mechanics of Composites, Lisbon, Portugal, 1–4 July 2019; pp. 1–3. Available online: https://eprints.lancs.ac.uk/id/eprint/136431 (accessed on 4 March 2022).
31. Sparks, C.; Odru, P.; Metivaud, G.; Le Floc'h, C. Composite Riser Tubes: Defect Tolerance Assessment and Nondestructive Testing. Paper presented at the Offshore Technology Conference, Houston, TX, USA, 4 May 1992; pp. 191–198.
32. Ochoa, O.O. *Composite Riser Experience and Design Guidance*; Prepared for MMS as a Guideline for Composite Offshore Engagements. Final Project Report. MMS Project Number 490, Offshore Technology Research Center; Texas A&M University: College Station, TX, USA, 2006; pp. 1–42. Available online: https://www.bsee.gov/sites/bsee.gov/files/tap-technical-assessment-program//490aa.pdf (accessed on 18 March 2022).
33. Pham, D.C.; Narayanaswamy, S.; Qian, X.; Zhang, W.; Sobey, A.; Achintha, M.; Shenoi, R.A. Composite riser design and development—A review. In *Analysis and Design of Marine Structures V*; CRC Press: Boca Raton, FL, USA, 2015; pp. 651–660. Available online: https://eprints.soton.ac.uk/372798/1/DC%2520Pham%2520et%2520al%2520%2520Composite%2520riser%2520design%2520and%2520development%2520%2520A%2520review_2.pdf (accessed on 18 March 2022). [CrossRef]
34. Bakaiyan, H.; Hosseini, H.; Ameri, E. Analysis of multi-layered filament-wound composite pipes under combined internal pressure and thermomechanical loading with thermal variations. *Compos. Struct.* **2009**, *88*, 532–541. [CrossRef]

35. Baldwin, D.; Newhouse, N.; Lo, K.; Burden, R. Composite Production Riser Design. In Proceedings of the Offshore Technology Conference, Houston, TX, USA, 5–8 May 1997; pp. 1–8. [CrossRef]
36. Baldwin, D.D.; Johnson, D.B. Rigid Composite Risers: Design for Purpose Using Performance-Based Requirements. In Proceedings of the Offshore Technology Conference, Houston, TX, USA, 6–9 May 2002; pp. 1–10. [CrossRef]
37. Baldwin, D.; Lo, K.; Long, J. Design Verification of a Composite Production Riser. In Proceedings of the Offshore Technology Conference, Houston, TX, USA, 4–7 May 1998; pp. 103–112. [CrossRef]
38. Chen, Y.; Seemann, R.; Krause, D.; Tay, T.-E.; Tan, V. Prototyping and testing of composite riser joints for deepwater application. *J. Reinf. Plast. Compos.* **2016**, *35*, 95–110. [CrossRef]
39. Salama, M.M.; Murali, J.; Baldwin, D.D.; Jahnsen, O.; Meland, T. Design Consideration for Composite Drilling Riser. In Proceedings of the Offshore Technology Conference, Houston, TX, USA, 3–6 May 1999; pp. 1–11. [CrossRef]
40. Salama, M. Lightweight Materials For Deepwater Offshore Structures. In Proceedings of the Offshore Technology Conference, Houston, TX, USA, 5–8 May 1986; pp. 297–304. [CrossRef]
41. Salama, M.M.; Spencer, B.E. Metal Lined Composite Risers in Offshore Applications. U.S. Patent 20040086341A1, 6 May 2004. Available online: https://patentimages.storage.googleapis.com/ce/83/5d/acfffc9dadc312/US20040086341A1.pdf (accessed on 4 March 2022).
42. Chen, Y.; Bin Tan, L.; Jaiman, R.K.; Sun, X.; Tay, T.E.; Tan, V.B.C. Global-Local Analysis of a Full-Scale Composite Riser During Vortex-Induced Vibration. In Proceedings of the ASME 2013 32nd International Conference on Ocean, Offshore and Arctic Engineering. Volume 7: CFD and VIV, Nantes, France, 9–14 June 2013. [CrossRef]
43. Akula, V.M.K. Global-Local Analysis of a Composite Riser. In Proceedings of the ASME 2014 Pressure Vessels and Piping Conference. Volume 3: Design and Analysis, Anaheim, CA, USA, 20–24 July 2014; pp. 1–9. [CrossRef]
44. Tan, L.; Chen, Y.; Jaiman, R.K.; Sun, X.; Tan, V.; Tay, T. Coupled fluid–Structure simulations for evaluating a performance of full-scale deepwater composite riser. *Ocean Eng.* **2015**, *94*, 19–35. [CrossRef]
45. Wang, C.; Shankar, K.; Morozov, E.V. Global design and analysis of deep sea FRP composite risers under combined environmental loads. *Adv. Compos. Mater.* **2017**, *26*, 79–98. [CrossRef]
46. Wang, C.; Shankar, K.; Morozov, E.V. Design of deep sea composite risers under combined environmental loads. In Proceedings of the 9th Composites Australia & CRC-ACS conference Diversity in Composites Conference, Leura, Australia, 15–16 March 2012; pp. 1–14.
47. Wang, C.; Shankar, K.; Morozov, E.V. Local design of composite risers under burst, tension and collapse cases. In Proceedings of the 18th International Conference on Composite Materials (ICCM18), Jeju, Korea, 22 August 2011; pp. 1–6. Available online: http://www.iccm-central.org/Proceedings/ICCM18proceedings/data/2.%20Oral%20Presentation/Aug22%28Monday%29/M04%20Applications%20of%20Composites/M4-1-IF0161.pdf (accessed on 4 March 2022).
48. Wang, C.G.; Shankar, K.; Morozov, E.V. Tailoring of Composite Reinforcements for Weight Reduction of Offshore Production Risers. *Appl. Mech. Mater.* **2011**, *66*, 1416–1421. [CrossRef]
49. Calash & MagmaGlobal. *Commercial Review of 8 Riser SLOR System: Magma M-Pipe Versus Steel Pipe*; CALASH Report; Prepared for MagmaGlobal; Calash & MagmaGlobal: Portsmouth, UK, 2015; pp. 1–16.
50. MagmaGlobal. Ocyan-Magma CompRisers. 2016. Available online: https://www.magmaglobal.com/risers/ocyan-compriser/ (accessed on 23 May 2021).
51. MagmaGlobal. Magma M-Pipe® End Fittings, Monitoring and Bend Test. 2015. Available online: https://www.youtube.com/watch?v=kNoM32UZcgc (accessed on 22 May 2021).
52. Salama, M.M.; Johnson, D.B.; Long, J.R. Composite Production Riser—Testing and Qualification. *SPE Prod. Facil.* **1998**, *13*, 170–177. [CrossRef]
53. Beyle, A.I.; Gustafson, C.G.; Kulakov, V.L.; Tarnopol'Skii, Y.M. Composite risers for deep-water offshore technology: Problems and prospects. 1. Metal-composite riser. *Polym. Mech.* **1997**, *33*, 403–414. [CrossRef]
54. Smits, A.; Neto, T.B.; de Boer, H. Thermoplastic Composite Riser Development for Ultradeep Water. In Proceedings of the Offshore Technology Conference, Houston, TX, USA, 30 April–3 May 2018; pp. 1–9. [CrossRef]
55. Van, O.M.; Gioccobi, S.; de Boer, H. Evaluation of the first deployment of a composite downline in deepwater Brazil. In Proceedings of the IBP1852_14, Rio Oil & Gas Conference 2014, Rio De Janeiro, Brazil, 15–18 September 2014.
56. Van Onna, M.; O'Brien, P. A New Thermoplastic Composite Riser for Deepwater Application. In *Proceedings of the Subsea UK Conference*; Subsea UK News: Aberdeen, UK, 2011; pp. 1–23. Available online: https://www.subseauk.com/documents/martinvanonnasubsea2011presentation.pdf (accessed on 4 March 2022).
57. Van, O.M.; Lyon, J. Installation of World's 1st Subsesa Thermoplastic Composite Pipe Jumper on Alder 2011. 2017. Available online: https://www.subseauk.com/documents/presentations/martin%20van%20onna%20-%20fields%20of%20the%20future%20-%20airborne.pdf (accessed on 4 March 2022).
58. Steuten, B.; van, O.M. Reduce Project and Life Cycle Cost with TCP Flowline. In Proceedings of the Offshore Technology Conference Asia, Kuala Lumpur, Malaysia, 22–25 March 2016; pp. 1–10. [CrossRef]
59. Ye, J.; Soldatos, K.P. Three-dimensional buckling analysis of laminated composite hollow cylinders and cylindrical panels. *Int. J. Solids Struct.* **1995**, *32*, 1949–1962. [CrossRef]
60. Bhudolia, S.; Fischer, S.; He, P.; Yue, C.Y.; Joshi, S.C.; Yang, J. Design, Manufacturing and Testing of Filament Wound Composite Risers for Marine and Offshore Applications. *Mater. Sci. Forum* **2015**, *813*, 337–343. [CrossRef]

61. Xia, M.; Takayanagi, H.; Kemmochi, K. Analysis of multi-layered filament-wound composite pipes under internal pressure. *Compos. Struct.* **2001**, *53*, 483–491. [CrossRef]
62. Jones, R.M. *Mechanics of Composite Materials*, 2nd ed.; Taylor & Francis: Philadelphia, PA, USA, 1999. [CrossRef]
63. Kaw, A.K. *Mechanics of Composite Materials*, 2nd ed.; CRC Press Imprint; Taylor & Francis: Boca Raton, FL, USA, 2005. [CrossRef]
64. Ye, J. *Laminated Composite Plates and Shells: 3D Modelling*; Springer: London, UK, 2003.
65. Ye, J. *Structural and Stress Analysis: Theories, Tutorials and Examples Second*; CRC Press: New York, NY, USA, 2016.
66. DNV. *Recommended Practice: Composite Risers DNV-RP-F202 October*; Det Norske Veritas: Oslo, Norway, 2010. Available online: https://rules.dnv.com/docs/pdf/dnvpm/codes/docs/2010-10/RP-F202.pdf (accessed on 4 March 2022).
67. DNV. *DNV-OS-C501: Composite Components*; Det Norske Veritas: Oslo, Norway, 2013. Available online: https://rules.dnv.com/docs/pdf/dnvpm/codes/docs/2013-11/OS-C501.pdf (accessed on 4 March 2022).
68. DNVGL. *Recommended Practice: Thermoplastic Composite Pipes—DNVGL-RP-F119 December*; Det Norske Veritas & Germanischer Lloyd: Oslo, Norway, 2015. Available online: https://www.dnvgl.com/oilgas/download/dnvgl-st-f119-thermoplastic-composite-pipes.html (accessed on 4 March 2022).
69. DNV. *Dynamic Risers: Offshore Standard DNV-OS-F201, October*; Det Norske Veritas: Oslo, Norway, 2010. Available online: https://rules.dnv.com/docs/pdf/dnvpm/codes/docs/2010-10/Os-F201.pdf (accessed on 4 March 2022).
70. ABS. *Guide for Building and Classing Subsea Riser Systems*, 3rd ed.; American Bureau of Shipping: New York, NY, USA, 2017. Available online: https://ww2.eagle.org/content/dam/eagle/rules-and-guides/current/offshore/123_guide_building_and_classing_subsea_riser_systems_2017/Riser_Guide_e-Mar18.pdf (accessed on 16 May 2021).
71. Schuett, C.; Paternoster, A. Full Generic Qualification of Nylon 12 Carbon Fiber Composite for Dynamic Thermoplastic Composite Pipe and Hybrid Flexible Pipe Applications. In Proceedings of the Offshore Technology Conference, Houston, TX, USA, 16–19 August 2021. [CrossRef]
72. Salama, M.M.; Stjern, G.; Storhaug, T.; Spencer, B.; Echtermeyer, A. The First Offshore Field Installation for a Composite Riser Joint. In Proceedings of the Offshore Technology Conference, Houston, TX, USA, 6–9 May 2002. [CrossRef]
73. Bybee, K. The First Offshore Installation of a Composite Riser Joint. *J. Pet. Technol.* **2003**, *55*, 72–74. [CrossRef]
74. Echtermeyer, A.; Steuten, B. Thermoplastic Composite Riser Guidance Note. In Proceedings of the Offshore Technology Conference, Houston, TX, USA, 6–9 May 2013; pp. 1–10. [CrossRef]
75. Hossain, R.; Carey, J.; Mertiny, P. Framework for a Combined Netting Analysis and Tsai-Wu-Based Design Approach for Braided and Filament-Wound Composites. *J. Press. Vessel Technol.* **2013**, *135*, 031204. [CrossRef]
76. Evans, J.T.; Gibson, A.G. Composite angle ply laminates and netting analysis. *Proc. R. Soc. A Math. Phys. Eng. Sci.* **2002**, *458*, 3079–3088. [CrossRef]
77. Tew, B.W. Preliminary Design of Tubular Composite Structures Using Netting Theory and Composite Degradation Factors. *J. Press. Vessel Technol.* **1995**, *117*, 390–394. [CrossRef]
78. DOD. *Military Handbook, MIL-HDBK-17-3F: Composite Materials Handbook*; Polymer Matrix Composites Materials Usage, Design and Analysis; U.S. Department of Defense (DOD): Arlington, VA, USA, 17 June 2002; Volume 3 of 5, pp. 43–53. Available online: https://www.library.ucdavis.edu/wp-content/uploads/2017/03/HDBK17-3F.pdf (accessed on 4 March 2022).
79. Andersen, W.; Anderson, J.; Landriault, L. Full-Scale Testing of Prototype Composite Drilling Riser Joints-Interim Report. In Proceedings of the Offshore Technology Conference, Houston, TX, USA, 4–7 May 1998; pp. 147–154. [CrossRef]
80. Roberts, D.; Hatton, S.A. Development and Qualification of End Fittings for Composite Riser Pipe. In Proceedings of the Offshore Technology Conference, Houston, TX, USA, 6–9 May 2013. [CrossRef]
81. Cederberg, C.A.; Baldwin, D.D.; Bhalla, K.; Tognarelli, M.A. Composite-Reinforced Steel Drilling Riser for Ultra-Deepwater High Pressure Wells. In Proceedings of the Offshore Technology Conference, Houston, TX, USA, 6–9 May 2013. [CrossRef]
82. Pham, D.C.; Su, Z.; Narayanaswamy, S.; Qian, X.; Huang, Z.; Sobey, A.; Shenoi, A. Experimental and numerical studies of large-scaled filament wound T700/X4201 composite risers under bending. In Proceedings of the ECCM17—17th European Conference on Composite Materials, Munich, Germany, 26–30 June 2016. Available online: https://www.researchgate.net/publication/307631336_Experimental_and_numerical_studies_of_large-scaled_filament_wound_T700X4201_composite_risers_under_bending (accessed on 15 February 2022).
83. Huang, Z.; Zhang, W.; Qian, X.; Su, Z.; Pham, D.-C.; Sridhar, N. Fatigue behaviour and life prediction of filament wound CFRP pipes based on coupon tests. *Mar. Struct.* **2020**, *72*, 102756. [CrossRef]
84. Huang, Z.; Qian, X.; Su, Z.; Pham, D.C.; Sridhar, N. Experimental investigation and damage simulation of large-scaled filament wound composite pipes. *Compos. Part B Eng.* **2020**, *184*, 107639. [CrossRef]
85. Sparks, C.; Odru, P.; Bono, H.; Metivaud, G. Mechanical Testing Of High-Performance Composite Tubes For TLP Production Risers. In Proceedings of the Offshore Technology Conference, Houston, TX, USA, 2–5 May 1988; pp. 467–472. [CrossRef]
86. Tamarelle, P.; Sparks, C. High-Performance Composite Tubes for Offshore Applications. In Proceedings of the Offshore Technology Conference, Houston, TX, USA, 27–30 April 1987; pp. 255–260. [CrossRef]
87. Amaechi, C.V.; Gillett, N.; Odijie, A.C.; Hou, X.; Ye, J. Composite risers for deep waters using a numerical modelling approach. *Compos. Struct.* **2019**, *210*, 486–499. [CrossRef]
88. Amaechi, C.V. Local tailored design of deep water composite risers subjected to burst, collapse and tension loads. *Ocean Eng.* **2022**, in press. [CrossRef]

89. Wang, C.; Shankar, K.; Morozov, E.V. Design of composite risers for minimum weight. Publisher: World Academy of Science, Engineering and Technology. *Int. J. Mech. Mechatron. Eng.* **2012**, *6*, 2627–2636. Available online: https://publications.waset.org/4236/pdf (accessed on 17 March 2022).
90. Wang, C.; Sun, M.; Shankar, K.; Xing, S.; Zhang, L. CFD Simulation of Vortex Induced Vibration for FRP Composite Riser with Different Modeling Methods. *Appl. Sci.* **2018**, *8*, 684. [CrossRef]
91. Singh, M.; Ahmad, S. Local Stress Analysis of Composite Production Riser Under Random Sea. In Proceedings of the ASME 2014 33rd International Conference on Ocean, Offshore and Arctic Engineering. Volume 4B: Structures, Safety and Reliability, San Francisco, CA, USA, 8–13 June 2014. [CrossRef]
92. Singh, M.; Ahmad, S. Bursting Capacity and Debonding of Ultra Deep Composite Production Riser: A Safety Assessment. In Proceedings of the ASME 2014 33rd International Conference on Ocean, Offshore and Arctic Engineering. Volume 6A: Pipeline and Riser Technology, San Francisco, CA, USA, 8–13 June 2014. [CrossRef]
93. Singh, M.; Ahmad, S. Probabilistic Analysis and Risk Assessment of Deep Water Composite Production Riser Against Fatigue Limit State. In Proceedings of the ASME 2015 34th International Conference on Ocean, Offshore and Arctic Engineering. Volume 3: Structures, Safety and Reliability, St. John's, NL, Canada, 31 May–5 June 2015. [CrossRef]
94. Ragheb, H.; Goodridge, M.; Pham, D.; Sobey, A. Extreme response based reliability analysis of composite risers for applications in deepwater. *Mar. Struct.* **2021**, *78*, 103015. [CrossRef]
95. Ragheb, H.; Sobey, A. Effects of extensible modelling on composite riser mechanical responses. *Ocean Eng.* **2021**, *220*, 108426. [CrossRef]
96. Amaechi, C.V.; Wang, F.; Hou, X.; Ye, J. Strength of submarine hoses in Chinese-lantern configuration from hydrodynamic loads on CALM buoy. *Ocean Eng.* **2019**, *171*, 429–442. [CrossRef]
97. Sonmez, F.O.; 2017. Optimum Design of Composite Structures: A Literature Survey. *J. Reinf. Plast. Compos.* **2017**, *36*, 3–39. [CrossRef]
98. Da Silva, R.F.; Teófilo, F.A.F.; Parente, E., Jr.; de Melo, A.M.C.; de Holanda, Á.S. Optimization of composite catenary risers. *Marine Struct.* **2013**, *33*, 1–20. [CrossRef]
99. Ghiasi, H.; Fayazbakhsh, K.; Pasini, D.; Lessard, L. Optimum stacking sequence design of composite materials Part II: Variable stiffness design. *Compos. Struct.* **2010**, *93*, 1–13. [CrossRef]
100. Ghiasi, H.; Pasini, D.; Lessard, L. Optimum stacking sequence design of composite materials Part I: Constant stiffness design. *Compos. Struct.* **2009**, *90*, 1–11. [CrossRef]
101. Harte, A.; McNamara, J.; Roddy, I. Evaluation of optimisation techniques in the design of composite pipelines. *J. Mater. Process. Technol.* **2001**, *118*, 478–484. [CrossRef]
102. Harte, A.; McNamara, J.; Roddy, I. Application of optimisation methods to the design of high performance composite pipelines. *J. Mater. Process. Technol.* **2003**, *142*, 58–64. [CrossRef]
103. Teófilo, A.F.F.; Parente, E., Jr.; de Melo, A.M.C.; de Holanda, Á.S.; da Silva, R.F. Premilinary Design of Composite Catenary Risers Using Optimization Techniques. Mecánica Computacional, XXIX. 2010, pp. 7927–7948. Available online: https://repositorio.ufc.br/bitstream/riufc/5474/1/2010_eve_amcmelo.pdf (accessed on 18 March 2022).
104. Wang, C.; Shankar, K.; A Ashraf, M.; Morozov, E.V.; Ray, T. Surrogate-assisted optimisation design of composite riser. *Proc. Inst. Mech. Eng. Part L J. Mater. Des. Appl.* **2016**, *230*, 18–34. [CrossRef]
105. Wang, C.; Shankar, K.; Morozov, E.V. Tailored design of top-tensioned composite risers for deep-water applications using three different approaches. *Adv. Mech. Eng.* **2017**, *9*, 1–18. [CrossRef]
106. Hatton, S. Carbon fibre—A riser system enabler. *Offshore Eng.* **2012**, *37*, 42–43. Available online: http://www.oedigital.com/engineering/item/696-carbon-fibre---a-riser-system-enabler (accessed on 4 March 2022).
107. Jha, V.; Finch, D.; Dodds, N.; Latto, J. Optimized Hybrid Composite Flexible Pipe for Ultra-Deepwater Applications. In Proceedings of the ASME 2015 34th International Conference on Ocean, Offshore and Arctic Engineering. Volume 5A: Pipeline and Riser Technology, St. John's, NL, Canada, 31 May–5 June 2015. [CrossRef]
108. Kim, W.K. Composite Production Riser Assessment. Ph.D. Thesis, Department of Mechanical Engineering, Texas A & M University, College Station, TX, USA, 2007. Available online: https://core.ac.uk/download/pdf/4272879.pdf (accessed on 18 March 2022).
109. Sun, X.S.; Chen, Y.; Tan, V.B.C.; Jaiman, R.K.; Tay, T.-E. Homogenization and Stress Analysis of Multilayered Composite Offshore Production Risers. *J. Appl. Mech.* **2013**, *81*, 031003. [CrossRef]
110. Sun, X.S.; Tan, V.B.C.; Chen, Y.; Tan, L.B.; Jaiman, R.K.; Tay, T.E. Stress analysis of multi-layered hollow anisotropic composite cylindrical structures using the homogenization method. *Acta Mech.* **2013**, *225*, 1649–1672. [CrossRef]
111. Bhavya, S.; Kumar, P.R.; Kalam, A. Failure analysis of a composite cylinder. *IOSR J. Mech. Civil Eng.* **2012**, *3*, 1–7. Available online: https://www.iosrjournals.org/iosr-jmce/papers/vol3-issue3/A0330107.pdf (accessed on 4 March 2022). [CrossRef]
112. Wang, C. Tailored Design of Composite Risers for Deep Water Applications. Ph.D. Thesis, School of Engineering and Information Technology, The University of New South Wales, Canberra, Australia, 2013. Available online: http://unsworks.unsw.edu.au/fapi/datastream/unsworks:11345/SOURCE01?view=true (accessed on 15 February 2022).
113. ANSYS. *ANSYS Composite PrepPost User's Guide Release 18.2*; ANSYS Inc.: Canonsburg, PA, USA, 2017.
114. ANSYS. *ANSYS Meshing User's Guide, Release 18.2*; ANSYS Inc.: Canonsburg, PA, USA, 2017.
115. Sun, C.T.; Li, S. Three-dimensional effective elastic constant for thick laminates. *J. Compos. Mater.* **1988**, *22*, 629–639. [CrossRef]

116. Toray. T700S Data Sheet, Santa Ana, CA, USA, 2008. Available online: https://www.toraycma.com/file_viewer.php?id=4459%0A (accessed on 12 April 2018).
117. Hartman, D.; Greenwood, M.E.; Miller, D.M. *High Strength Glass Fibers 2006 Repri*; AGY: Aiken, SC, USA, 1996. Available online: https://www.agy.com/wp-content/uploads/2014/03/High_Strength_Glass_Fibers-Technical.pdf (accessed on 4 March 2022).
118. MatWeb.AS4 PEEK Plus Carbon Fiber Reinforced Unidirectional—MatWeb Material Property Data. MatWeb Material Property Data. 2018. Available online: http://www.matweb.com/search/datasheet.aspx?matguid=1e8a25336d7645d8a24cdbd10ed2dd29&ckck=1 (accessed on 12 April 2018).
119. MatWeb. Solvay 934 Epoxy-S2 Glass Fiber Reinforced Unidirectional—MatWeb Material Property Data. MatWeb Material Property Data. 2021. Available online: http://www.matweb.com/search/datasheettext.aspx?matguid=0e86c9201a2d45f7bb61f85b28a3e681 (accessed on 17 March 2018).
120. Da Silva, R.F.; Da Rocha, I.B.C.M.; Parente, E., Jr.; De Melo, A.M.C. Optimum Design of Composite Risers Using A Genetic Algorithm. In Proceedings of the 10th World Congress on Computational Mechanics, Sao Paulo, Brazil, 8–13 July 2012. Available online: http://pdf.blucher.com.br.s3-sa-east-1.amazonaws.com/mechanicalengineeringproceedings/10wccm/18916.pdf (accessed on 4 March 2022).
121. Meniconi, L.; Reid, S.; Soden, P. Preliminary design of composite riser stress joints. *Compos. Part A Appl. Sci. Manuf.* **2001**, *32*, 597–605. [CrossRef]
122. Ja'E, I.A.; Ali, M.O.A.; Yenduri, A.; Nizamani, Z.; Nakayama, A. Optimisation of mooring line parameters for offshore floating structures: A review paper. *Ocean Eng.* **2022**, *247*, 110644. [CrossRef]
123. Singh, M.; Ahmad, S. Fatigue Life Calculation of Deep Water Composite Production Risers by Rain Flow Cycle Counting Method. In Proceedings of the ASME 2015 34th International Conference on Ocean, Offshore and Arctic Engineering. Volume 5B: Pipeline and Riser Technology, St. John's, NL, Canada, 31 May–5 June 2015. [CrossRef]
124. Singh, M.; Ahmad, S.; Jain, A.K. S-N Curve Model for Assessing Cumulative Fatigue Damage of Deep-Water Composite Riser. In Proceedings of the ASME 2020 39th International Conference on Ocean, Offshore and Arctic Engineering. Volume 1: Offshore Technology, Online, 3–7 August 2020. [CrossRef]
125. Khan, R.A.; Ahmad, S. Nonlinear Dynamic and Bilinear Fatigue Performance of Composite Marine Risers in Deep Offshore Fields. In Proceedings of the ASME 2020 39th International Conference on Ocean, Offshore and Arctic Engineering. Volume 2A: Structures, Safety, and Reliability, Online, 3–7 August 2020. [CrossRef]
126. Khan, R.A.; Ahmad, S. Bilinear Fatigue Performance of Composite Marine Risers in Deep Offshore Fields Due to Vortex Induced Vibrations. In Proceedings of the ASME 2021 40th International Conference on Ocean, Offshore and Arctic Engineering. Volume 2: Structures, Safety, and Reliability, Online, 21–30 June 2021. [CrossRef]
127. Hastie, J.C.; Guz, I.A.; Kashtalyan, M. Structural integrity of deepwater composite pipes under combined thermal and mechanical loading. *Procedia Struct. Integr.* **2020**, *28*, 850–863. [CrossRef]

Article

Influence of Line Processing Parameters on Properties of Carbon Fibre Epoxy Towpreg

Murat Çelik, Thomas Noble, Frank Jorge, Rongqing Jian, Conchúr M. Ó Brádaigh and Colin Robert *

School of Engineering, Institute for Materials and Processes, The University of Edinburgh, Edinburgh EH9 3FB, UK; m.celik@ed.ac.uk (M.Ç.); tnoble2@exseed.ed.ac.uk (T.N.); f.a.d.jorge@sms.ed.ac.uk (F.J.); r.jian@sms.ed.ac.uk (R.J.); c.obradaigh@ed.ac.uk (C.M.Ó.B.)
* Correspondence: colin.robert@ed.ac.uk

Abstract: This paper explores the performance of low-cost unidirectional carbon fibre towpregs with respect to line production speed and fibre volume fraction. Using an automated production line, towpregs were produced at different production speeds, resulting in modified fibre volume fractions. The towpregs were used to manufacture unidirectional composite plates, which were then tested to evaluate mechanical performance. The fibre straightness and interfacial void ratio of the composite plates were determined by statistical analysis of the samples' optical micrographs. The results demonstrate that adjusting the line production speed enables targeted fibre volume fractions (FVF) to be reached, resulting in the composites having different mechanical performances (2039 MPa and 2186.7 MPa tensile strength, 1.26 and 1.21 GPa flexural strength for 59.8% and 64.4% FVF, respectively). It was shown that at lower production speeds and FVF, composites exhibit good consolidation and low porosity, which is highlighted by the better interlaminar shear strength performances (8.95% increase), indicating the limitations of manufacturing very high FVF composites. Furthermore, it was concluded that fibre straightness plays a key role in mechanical performance, as samples with a lesser degree of fibre straightness showed a divergence from theoretical tensile properties.

Keywords: fibre volume fraction; towpreg; advanced composite manufacturing; powder epoxy

Citation: Çelik, M.; Noble, T.; Jorge, F.; Jian, R.; Ó Brádaigh, C.M.; Robert, C. Influence of Line Processing Parameters on Properties of Carbon Fibre Epoxy Towpreg. *J. Compos. Sci.* **2022**, *6*, 75. https://doi.org/10.3390/jcs6030075

Academic Editor: Jiadeng Zhu

Received: 31 January 2022
Accepted: 28 February 2022
Published: 2 March 2022

Publisher's Note: MDPI stays neutral with regard to jurisdictional claims in published maps and institutional affiliations.

Copyright: © 2022 by the authors. Licensee MDPI, Basel, Switzerland. This article is an open access article distributed under the terms and conditions of the Creative Commons Attribution (CC BY) license (https://creativecommons.org/licenses/by/4.0/).

1. Introduction

Out-of-autoclave (OOA) towpreg epoxy composites are ideal candidates to provide adequate strength and stiffness for large composite structures, such as tidal/wind turbine blades and marine or automotive applications, while being cost-effective due to the scalability and compatibility with inexpensive mould heating systems [1]. In recent years, the need for low-cost, high-performance OOA prepreg composites has led to an interest in powder-epoxy systems, which are solid at room temperature, melt between 40 and 60 °C, and cure at 180 °C. It has been demonstrated that powder–epoxy systems can be used to produce vacuum-bag-only (VBO) prepregs, and that they exhibit comparable mechanical performance to conventional epoxy composites [2,3]. Furthermore, low viscosity and low curing exotherms of powder–epoxies make them attractive for thick section composites [2–4]. Owing to their low viscosity, a complete epoxy wet-out within the mould is possible with powder–epoxy systems under vacuum only, which results in superior strength and fracture toughness due to the strong interfacial bonding between the matrix and the fibres [5]. Furthermore, the powder can be stored at room temperature thanks to its thermal stability, and little to no volatile organic compounds (VOC) are released during the production [6].

In this context, a novel powder–epoxy-based pilot towpregging line has been developed [7] to manufacture unidirectional carbon fibre towpregs (or tapes) that are compatible with advanced composite manufacturing methods such as Automated Fibre Placement

(AFP), pultrusion, or filament winding. The main goal of this system is to allow for low-cost, high-quality, and high-speed manufacturing of unidirectional carbon fibre-reinforced polymer (UD-CFRP) towpregs. Automation of the reported system [7] is in progress, which entails better process control and higher production speeds, therefore lowering the cost of the overall composite production. On one hand, higher manufacturing speeds increase the fibre volume fraction (FVF) of the composite and alter the mechanical properties remarkably. Fibres display better mechanical properties than resin; thus, a higher FVF allows for higher strength and stiffness due to a higher fibre-to-resin ratio. On the other hand, too high FVF results in a lack of wetting for individual fibres, macroscopically leading to early interfacial failure due to poor fibre–matrix interface bonding and energy transfer ability [8]. As such, it was deduced that the optimal FVF for UD-CFRP plates in terms of the ultimate tensile strength (UTS) is 56–60%, as UTS decreases for higher FVFs due to the lack of completely wet-out fibres [9]. Since there is no additional compaction pressure as in the autoclave process, reaching the desired FVF values with OOA systems is not an easy task. Courter et al. [10] demonstrated that significantly higher FVF values could be obtained by autoclave curing when compared to oven curing for the same type of prepreg materials. Still, up to 60% FVF can be achieved with VBO prepregs [11]; however, the out-life of the prepregs plays a key role in the composite production phase, and a few weeks of out-life substantially decreases the processibility of the prepregs [12] due to partial curing increasing the minimal pre-gel viscosity, therefore reducing the wetting ability.

The powder–epoxy system used in this study is solid and stable at ambient temperature and will not start to cure before 145 °C, due to its heat-activated curing mechanism [7]. Therefore, it has excellent storage performance at room temperature; moreover, a consistent FVF can be obtained by carefully adjusting the towpreg line process parameters such as speed, tension, or temperature. Most of the standard VBO prepreg resins are highly reactive and cannot maintain low viscosities for a long time [11], whereas powder–epoxy viscosity remains very low at lower temperatures [4], and complete wet-out of the fibre bed can be achieved with vacuum only—without the need for additional compaction pressure. Controlling the FVF while maintaining a high production speed in the towpregging line would allow the manufacture of high-performance composites, such as wind or tidal turbines, at a low cost. In this study, a consistent towpreg FVF was maintained for a certain production speed in a powder–epoxy based towpregging line. Two production speeds, 3 and 5 m/min, were used to produce towpreg to investigate the FVF/production speed relationship and its influence on the overall mechanical performances, while all other processing parameters, such as tension and temperature, were kept constant and monitored.

2. Experimental Section

2.1. Materials

Powder epoxy (PE6405, 1220 kg/m^3) supplied by FreiLacke and designed by Swiss CMT AG was used to manufacture the towpreg and then the CFRP plates. One of the main advantages of the powder–epoxy is that it starts to melt at 40–60 °C, reaches the minimum viscosity at 120 °C, but does not cure until 145 °C [4]. This feature provides versatility in the production phase by allowing the potential of separating the impregnation and curing of the composite parts. In addition, it is possible to co-cure different composite parts using powder–epoxy [13] instead of using adhesives [7]. Finally, as the powder epoxy is solid and stable at room temperature, storage costs are substantially lower than standard prepreg systems.

For the carbon fibre tows, commercially available Toray T700S-24K-50C (1% sizing agent) was used, as it showed the best performances comparatively to other fibre systems [14].

2.2. Towpreg Production

The powder–epoxy towpregging pilot line was originally developed by Robert et al. [7]. Since then, a second more instrumented tapeline has been developed in the framework of a

CIMComp Hub fellowship project to automate the production process. New sensors were added, including OS-PC30-2M-1V infrared temperature sensors (OMEGA Engineering), an SFD 500T tension sensor (Hans Schmidt & Co GmbH, Waldkraiburg, Germany), and a P500 rotary sensor (Positek, Cheltenham, UK). Control hardware included a B35-HM magnetic particle brake (Placid Industries Inc., Elmira, NY, USA) and EM-175 DC motor controllers. Automated temperature and tension control was achieved using software-side PIDs on LabVIEW (National Instruments Corp., Austin, TX, USA). The towpreg production in the line can be summarised in the following steps, as shown in Figure 1:

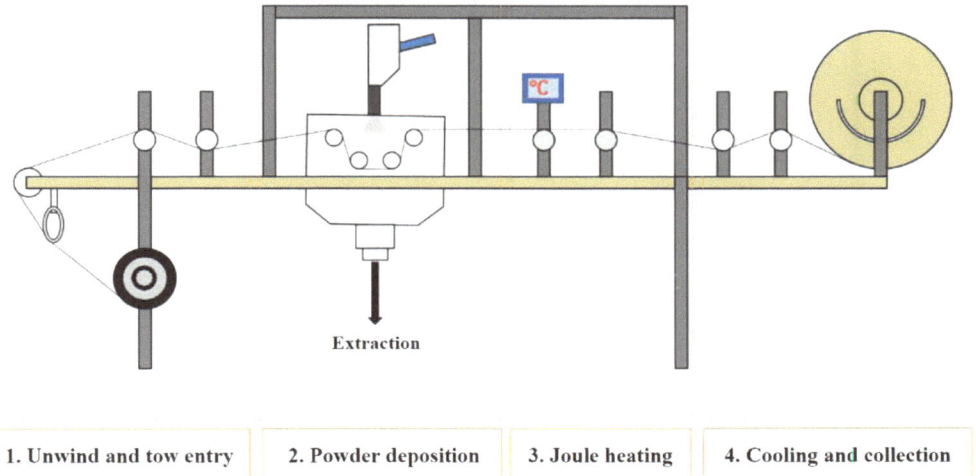

Figure 1. Diagram of the tapeline system.

- Carbon fibre tows are unwound from the reel, and the tension is maintained by a magnetic brake that automatically adjusts the tension based on the data from the tension sensor.
- Powder–epoxy is electrostatically charged and sprayed on the carbon fibre tow as it moves.
- An electrical current is supplied via a power controller between two conductive metal rollers to initiate powder melt by the Joule effect [15]. The temperature can be adjusted according to data from an infrared sensor.
- Cooled towpreg is collected on a drum after it passes through a series of rollers.

Automation of the pilot towpregging line is necessary for increasing the productivity and accuracy of the production while lowering costs. Additionally, the towpreg quality can be maintained for high production volumes as the entire system is being monitored with the data obtained from the sensors. In order to control and monitor the system, a human–machine interface (HMI) has been created using LabVIEW software, which allows the user to observe and record key process parameters, such as tension, temperature, or speed. All the parameters can be controlled either manually or by PID controllers via the HMI.

Reliability and consistency were the goals for towpreg production, which are easier to achieve at slower production speeds. Therefore, the towpreg was produced at two different speeds, 3 and 5 m/min, while keeping all other parameters constant (temperature, tension, etc.). The produced towpreg was cut into equal-sized so-called "strips" (55 cm length), and the FVF of the strip samples (FVF_{strip}) was calculated for each production run using Equation (1) [16]:

$$FVF = \frac{\rho_m}{\rho_m + \rho_f \left(\frac{1}{FWF} - 1\right)} \quad (1)$$

where ρ_m is the matrix density, ρ_f is the fibre density, and FWF is the fibre weight fraction, which are determined by a precision scale.

2.3. Composite Plate Production

For further mechanical characterisation tests, unidirectional carbon fibre composite (UD-CFRP) plates were manufactured from the produced powder epoxy towpreg strips. Composite plate production consisted of the following stages, as shown in Figure 2:

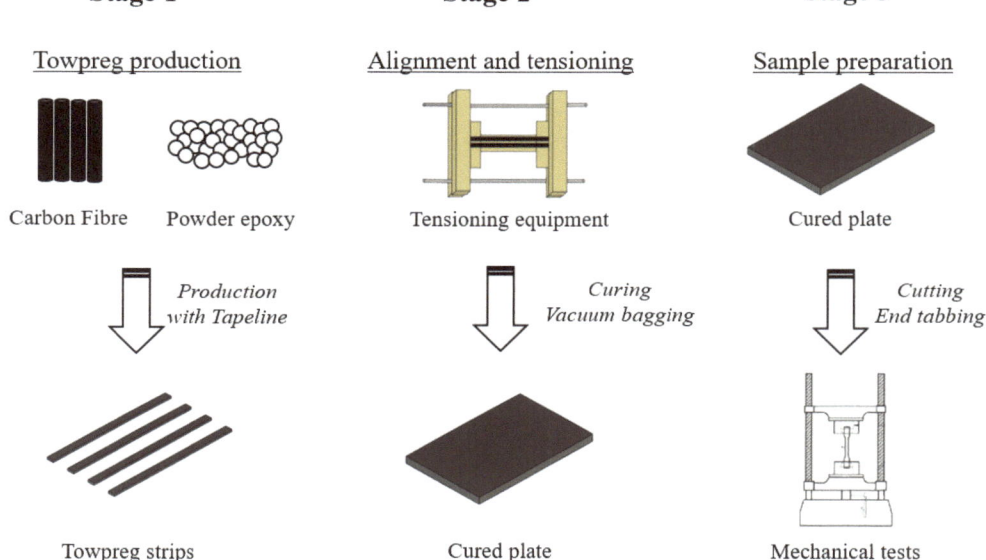

Figure 2. Different stages of the UD composite plate production.

Stage 1: Towpreg was produced with the tapeline system at different production speeds; then, it was cut into strips.

Stage 2: Towpreg strips were aligned by custom tensioning equipment [14]. Then, the tensioned preform was vacuum-bagged and cured to manufacture unidirectional composite plates, with 5 layers of towpreg strips being used to produce 1 mm thick composite plates. The thermal cycle for curing consisted of a drying stage dwell at 35 °C for 5 h, which was followed by an isothermal dwell at 120 °C for 2 h for the sintering and melting stage and another isothermal dwell at 180 °C for 2 h to complete curing. One hour ramping time was used between all dwells.

Stage 3: To prepare samples for mechanical characterisation, UD-CFRP composite plates were cut using a wet saw to ensure smooth edges. Then, mechanical tests were carried out for performance characterisation.

After the production of cured plates and extraction of samples, UD-CFRP samples were weighed again, and the FVF of the composite plates (FVF_{plate}) was calculated using Equation (1). Peel ply was used as a separation layer between the preform and tensioning equipment (Figure 3a). The excess resin was absorbed by the peel ply during the curing stage (Figure 3c), which resulted in an FVF increase compared to the initial FVF value (FVF_{strip}). Table 1 summarises the FVF of the towpreg samples and composite plates at different process parameters including speed, tension, temperature, and electrostatic spray gun settings. As can be seen from the table, both the FVF_{strip} and FVF_{plate} increase with increasing production speed, since the duration that the carbon fibre tows spend within the electrostatic spraying chamber becomes shorter for higher speeds, resulting in less powder deposited in the towpreg. For the electrostatic powder deposition, the electrostatic gun capacity was set to 40% for all experiment sets.

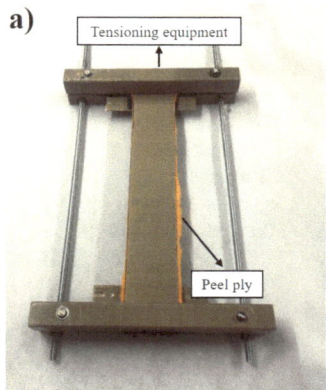 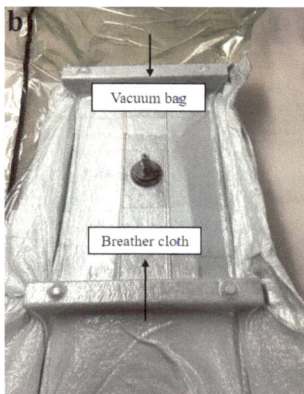 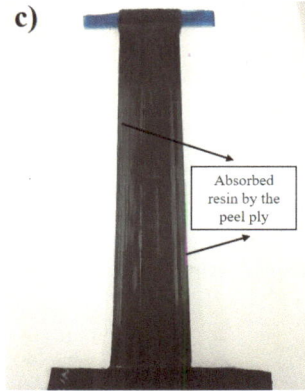

Figure 3. (**a**) Towpregs in the tensioning equipment (enclosed within the mould and not visible) and peel ply (orange). (**b**) Vacuum bagging of the tensioning equipment. (**c**) Cured composite plate and visible spots from absorbed resin by the peel ply (sample set 2 case).

Table 1. FVF of the towpreg strips and composite plates at different processing parameters.

	Line Speed (m/min)	Line Tension (N)	Heating Temperature (°C)	Gun Flow Air (%)	Total Flow (Air + Powder) (%)	Strip FVF	Plate FVF
Sample set 1	3	20	120	99	40	53.1%	56.5%
Sample set 2	5	20	120	99	40	62.5%	64.2%

2.4. Tensile Tests

Tensile tests were carried out on 190 × 15 mm (length and thickness, respectively) samples according to BS EN ISO 527-5 standard. Glass–epoxy laminates were bonded to the samples for end tabbing. For each set, 5 samples were tested in the fibre (0°) direction using an MTS Criterion model 45 (C45.305) electromechanical universal test system at a constant crosshead speed of 2 mm/min. A digital image correlation (DIC) module (Imetrum Advantage video extensometer) was used for the 2D strain measurement of the samples. The samples were collected after the test to analyse the fracture surfaces via Scanning Electron Microscopy (SEM).

2.5. Flexural Tests

Flexural performance of the 90 mm × 15 mm samples was characterised by four-point bending tests according to BS EN ISO 14125. Twelve samples (6 for each speed) were tested with Instron Model 3369 Dual column Tabletop Test system. The test fixture had an 81 mm outer span and a 27 mm inner span. The DIC module (Imetrum Advantage video extensometer) captured the deflection, and the crosshead speed was 2 mm/min.

2.6. Interlaminar Shear Strength

The interlaminar shear strength (ILSS) of the UD-CFRPs plates was evaluated according to BS EN ISO 14130:1998 on twelve 20 mm × 13 mm sized samples. The same Instron test machine was used for ILSS tests with a crosshead speed of 1 mm/min. The load cell capacity for the system was 50 kN.

2.7. Scanning Electron Microscopy (SEM)

The surface topography of the samples and fracture cross-sections were analysed by a Hitachi TM4000 Plus Scanning Electron Microscope (SEM) in Back Scattered Electrons (BSE) mode with an acceleration voltage of 15 kV under vacuum conditions.

2.8. Optical Microscopy

The fibre distribution, FVF, and porosity of the UD-CFRP samples were analysed by optical microscopy. Before the analysis, the samples were potted using liquid epoxy resin and non-stick moulds; then, they were polished by a surface grinder. The polishing procedure included sanding with varying grit sizes (P800, P1200, and P1600) followed by polishing by diamond-based dispersion (5, 3, and 1 micron). Then, the samples were analysed by a Zeiss Axioskop 2 MAT model microscope that is connected to a computer via an AxioCam MRc 5 camera. Fifty images were taken for each sample set, aiming to eliminate errors from a low sample size. The images were post-processed with ImageJ software. Firstly, images were converted to black and white to create a better contrast, as shown in Figure 4; then, the threshold for the pixels was set accordingly to capture fibres, matrixes, and voids in the sample. Then, the area of each component was calculated by ImageJ to validate the FVF values calculated with Equation (1). The void fraction and fibre circularity was also calculated with ImageJ, using the micrographs.

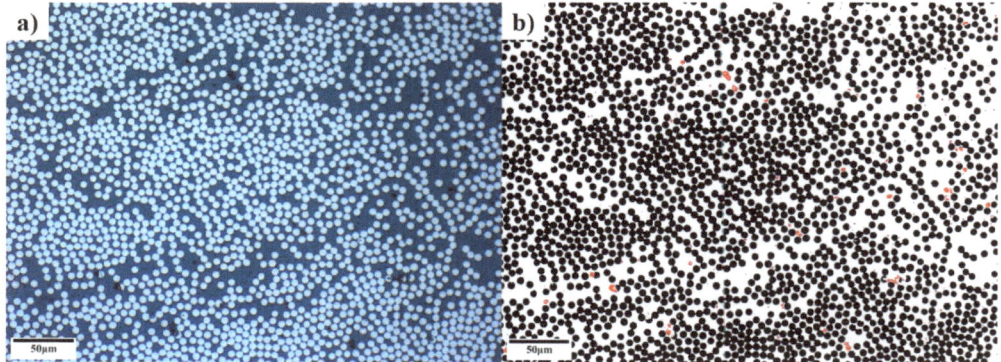

Figure 4. (**a**) Optical microscopy of the UD-CFRP sample cross-section for 59.8% FVF. (**b**) Resin (white), fibres (black), and voids (red) identified by ImageJ software filters.

2.9. Interfacial Void Content Analysis

Voids are one of the main manufacturing defects in composites, influencing the mechanical properties considerably by acting as failure initiation points [17]. The tensile, compressive, flexural, and interlaminar shear strength of the composites are adversely affected by the voids [18]; moreover, fatigue life deteriorates seriously [19]. The spatial distribution of the voids is a key factor for the performance of the composite. In the presence of voids at the fibre–matrix interface, adhesion between the fibre and matrix is weakened, and load transfer is impaired [20]. As it is dominated by the properties of matrix and matrix–fibre interface [19], interlaminar shear strength is susceptible to the interfacial voids in particular. For the samples that were produced at different production speeds, interfacial void content was determined using ImageJ and a custom LabVIEW algorithm. Using the samples analysed by the Zeiss Axioskop 2 MAT model microscope, ImageJ was used to find geometric information on the fibres, matrix, and voids. Data for all samples was exported to a csv file. This included information on the locations of the fibres and voids, the maximum diameter of each fibre (using a maximum calliper rule), and the area of the fibres and voids.

An algorithm was applied to the data from each sample image, which isolated interfacial voids from bulk voids. This was achieved using the following methodology.

- The coordinates of a fibre were compared to the locations of the voids for a sample using Equation (2).

$$\sqrt{\left(x_f - x_v\right)^2 + \left(y_f - y_v\right)^2} = d \qquad (2)$$

where:
- x_f The x-coordinate of the fibre centre point
- x_v The x-coordinate of the void centre point
- y_f The y-coordinate of the fibre centre point
- y_v The y-coordinate of the void centre point
- d Distance between the centre points of the fibre and void

- Voids that were outside the fibre radius r_f and within a set distance of the outer edge of the fibre r_{f+a} were recorded using comparator Equation (3), along with their respective area (Figure 5). These were classified as interfacial voids.

$$if\ d > r_f\ and\ d < r_{f+a} \qquad (3)$$

where:
- r_f The radius of the fibre using a maximum calliper
- r_{f+a} The radius of the fibre plus a set distance (0.25, 0.5, and 1 μm)

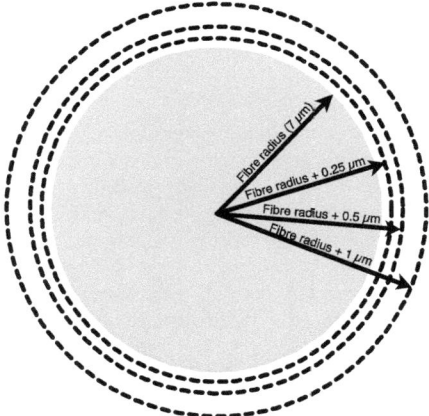

Figure 5. Boundary regions used to search for interfacial voids.

- This process was repeated for every fibre in a sample picture, and the individual areas were summed to provide a total area of interfacial voids (Equation (4)).

$$\sum A_{n\ interfacial\ voids} = A_1 + A_2 + \ldots + A_n \qquad (4)$$

- Then, the interfacial void area was compared with the total void area found previously using optical microscopy to find the percentage of interfacial voids compared to the total void fraction.

$$\frac{A_{interfacial\ voids}}{A_{voids}} \times 100 = \%\ of\ interfacial\ voids\ to\ bulk\ voids \qquad (5)$$

- This process was repeated for the other sample pictures, and an average was taken of the percentage of interfacial voids at the two towpreg line speeds, 3 and 5 m/min. Figure 6 illustrates the locations of interfacial and bulk voids that were captured by the algorithm (circles sizes are not representing the void areas).

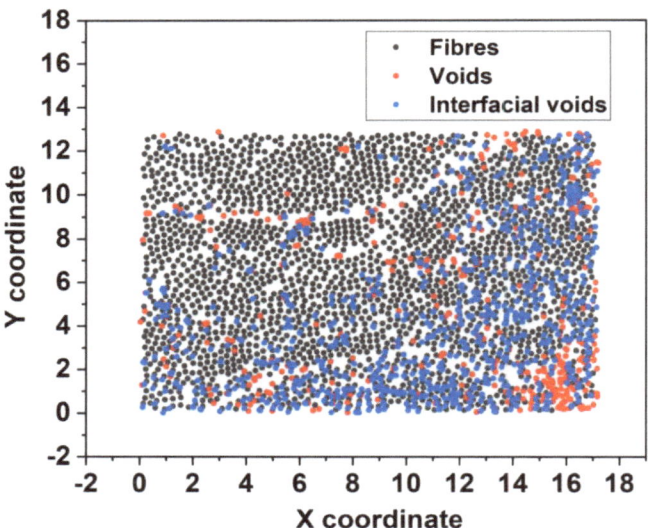

Figure 6. Bulk and interfacial voids detected by the algorithm.

3. Results and Discussion

3.1. Fibre Volume Fractions: Strip, Plate, and Bulk

Two different production line speeds (3 and 5 m/min) were used in this study. As expected, FVF_{strip} for the 3 m/min samples was lower, 53.1%, since the amount of powder supplied decreases with increasing production speed, as the towpreg is coated with epoxy powder for less time. An FVF_{strip} value of 62.5% was obtained with the 5 m/min samples, which indicates that the resin may not have completely saturated the fibres. After the curing stage, fibre volume fractions increased to 56.5% and 64.2% for 3 m/min and 5 m/min samples, respectively, which were named FVF_{plate}. This increment in FVF, as mentioned above, was caused by the fact that the peel ply absorbed some fraction of the resin from the tensioned preform during curing. Then, the calculated values of FVF_{plate} from Equation (1) were compared with the post-processing of optical microscopy images using ImageJ software. The FVF values obtained by the optical microscopy images are presented in Table 2. FVF values calculated by the optical microscopy for the 5 m/min samples matched very well with FVF_{plate}, which were estimated from Equation (1), whereas a deviation was observed for 3 m/min samples. Due to the higher resin content of the 3 m/min samples, resin-rich zones were found around sample boundaries (edges). During the optical microscopy analysis, boundaries of the samples were not considered; thus, an increment of the FVF was observed for 3 m/min samples, since the resin-rich zones were excluded. Having less resin content, 5 m/min samples did not exhibit such resin-rich zones; hence, the FVF value calculated by the optical microscopy did not differ significantly from the previous FVF values. Then, the FVF values obtained by the optical microscopy were regarded as the best representation of the bulk composite fibre volume fraction and named as $FVF_{Optical}$.

Table 2. FVF values of the composite plates, estimated by optical microscopy.

Line Speed (m/min)	FVFstrip (%)	FVFplate (%)	FVFOptical (%)	Porosity (%)
3	53.1%	56.5%	59.8 ± 4.04	0.62 ± 0.09
5	62.5%	64.2%	64.4 ± 2.20	0.82 ± 0.19

As mentioned before, voids in a composite material are detrimental to its performance; ILSS has been shown to decrease by 6% per 1% of void (up to 4% void fraction) [21]. Optical microscopy was used to investigate the porosity of the UD-CFRP composite laminate samples; in addition, the FVF and fibre distribution were also analysed. Post-processing was conducted using ImageJ software. When compared with other conventional and commercial methods [19], the void fraction of the samples is relatively low (<1%), and it can be inferred that consolidation is excellent in the manufactured plates.

3.2. Interfacial Voids

Results for the percentage of interfacial voids as a fraction of the total voids are shown below in Figure 7. Regardless of the search criterion (distance outside fibre radius), generally, a higher interfacial void content was found for samples produced at the higher speed of 5 m/min. The ratios of interfacial voids were 3.21% and 6.07% at 0.25 µm, 15.9% and 22.3% at 0.5 µm, 53.01% and 56.29% at 1 µm distance outside fibre radius at 3 m/min and 5 m/min production speed, respectively. The higher interfacial void ratio at the higher production speed (i.e., higher FVF) was attributed to the lesser resin content within the towpreg compared to 3 m/min production speed samples.

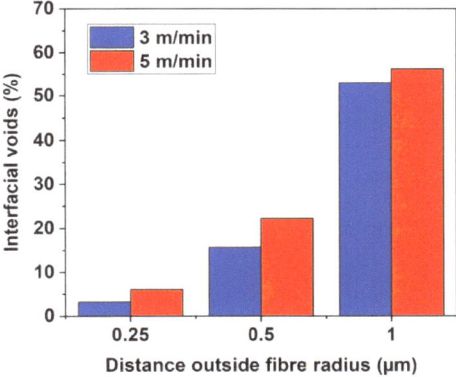

Figure 7. Percentage of interfacial voids as a fraction of the total voids.

3.3. Fibre Straightness

Mamalis et al. [14] found that the fibre straightness of unidirectional carbon fibre tows had a significant impact on the mechanical performance of the composite. Hence, ImageJ software was used to analyse the degree of fibre straightness in the samples. The minimum ($D_{i,min}$) and maximum diameter ($D_{i,max}$) of each fibre in the optical micrographs were measured, which were used to calculate the ratio $D_{i,total} = D_{i,min}/D_{i,max}$. Then, the following equation [14] can be used to obtain the direction parameter f:

$$f = \frac{\sum_{i=1}^{n} D_{i,total}}{n} \qquad (6)$$

where n is the number of carbon fibre filaments investigated for each micrograph. The direction parameter f indicates how far a fibre deviates from a perfect circle and can be used as a measure of fibre straightness. A total of 75 micrographs were analysed, each containing

≈2000 visible fibres. The number of fibres analysed (n) was greater than 150,000, therefore providing strong analytical results. The estimated direction parameter f for each FVF is shown in Figure 8. For both sample sets, a similar value of f (≈0.91) was obtained, which was influenced by the tensioning conditions during curing. For similar powder–epoxy towpreg and the same fibre type (T700S-24K-50C), an f value of 0.95 was reported for a tensioned system [14]. A lower degree of fibre straightness points to a slight misalignment of fibres caused by the tensioning apparatus, resulting in lower mechanical performances comparatively [14].

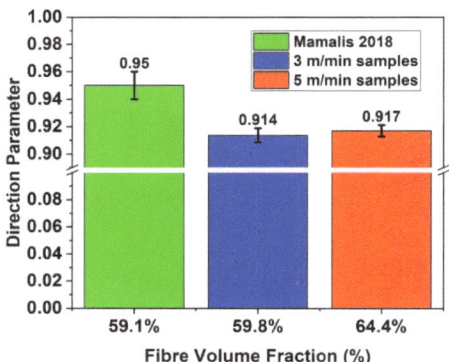

Figure 8. Direction parameter for different FVF samples compared to that of Mamalis et al. [14].

3.4. Tensile Performance

Different production speeds of the towpregging line resulted in different mechanical and fractographic behaviour. The average tensile strength of 59.8% and 64.4% FVF samples were 2039 MPa and 2187 MPa, whereas the average modulus values were 116 GPa and 138 GPa, respectively. As illustrated in Figure 9a, both strength and stiffness increased by 7.24% and 9.28%, respectively, for the higher FVF material, or in other words, with the production speed. Moreover, 64.4% FVF samples exhibited slightly larger standard deviation in modulus (±16.3 GPa compared to ±9.79 GPa), which is believed to be a result of lower homogeneity across the samples due to less consistent consolidation.

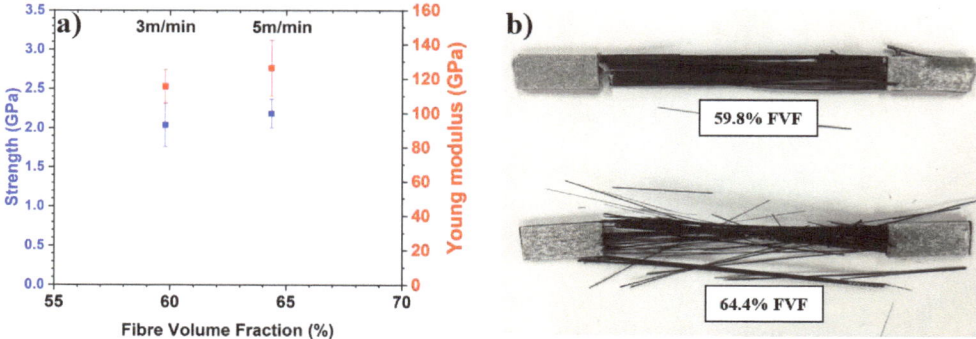

Figure 9. (**a**) Tensile test results of the UD-CFRP plates. (**b**) Different splitting modes of samples.

The Rule of Mixtures (ROM) is a simple method to estimate the mechanical properties of the composites, which can be defined for the longitudinal direction as [16]:

$$\sigma_c = \sigma_f FVF + \sigma_m(1 - FVF) \quad (7)$$

$$E_c = E_f FVF + E_m(1 - FVF) \quad (8)$$

where σ_c is the tensile strength of the composite, σ_f is the tensile strength of fibre, σ_m is the tensile strength of matrix, E_c is the modulus of the composite, E_f is the modulus of the fibre, E_m is the modulus, and FVF is the fibre volume fraction. Obtained FVF values from the experiments were used in Equations (7) and (8) to estimate the tensile strength and modulus of the composite plates. The discrepancy between the theoretical and experimental results stems from the fact that the ROM equation assumes perfect interface bonding of the fibre and matrix, which is not true in reality. As can be seen from the Table 3, the tensile test results deviated from the theoretical value by 30.73%, whereas the deviation for 64.4% FVF samples was 30.96%. As previously stated, fibre straightness has been found to have a considerable impact on tensile performance, with tensile strength dropping from 2650 to 1980 MPa (a 25% reduction) when the direction parameter f is reduced from 0.95 to 0.86 [14]. Given the calculated value of the samples' direction parameter (0.91), a divergence in tensile performance from optimal ROM values is to be expected. Lower values for tensile strength and stiffness can be attributed to the lower degree of fibre straightness of the samples.

Table 3. Comparison of measured tensile properties with theoretical values.

Method	Tensile Strength at 59.8% FVF (MPa)	Modulus at 59.8% FVF (GPa)	Tensile Strength at 64.4% FVF (MPa)	Modulus at 64.4% FVF (GPa)	Fibre Direction Parameter f
Measured	2039.04	115.63	2186.74	126.36	0.91
ROM	2913.67	138.75	3132.95	149.19	1

The adhesion between the fibres and the matrix depends on chemical bonding that can be improved by the addition of sizing agents to increase the bonding strength of intrinsically inert [22] and hydrophobic [23] carbon fibres, mechanical interlocking, or a combination of different factors [24]. Figure 9b shows standard fractured samples in tension for both 59.8% and 64.4% FVF. Higher FVFs mean lower resin content, and the interfaces between the fibre and the matrix may not be sufficient to provide an effective stress transfer. It was observed that samples with 64.4% FVF splintered in smaller bits than its lower FVF counterpart. The behaviour is consistent with a progressive fracture mode, which is caused by the inconsistent load distribution leading to premature failure of the weakest portions of the cross-section before spreading to the entire area. In contrast, samples with the lower FVF (59.8%) fractured in splitting mode and retain a better macroscopic consistency, which is compatible with an explosive failure profile, indicating a better interfacial bonding of the fibre and resin matrix.

3.5. Flexural Performance

Flexural properties are of great importance for many composite applications. For instance, tidal turbine blades endure extreme flapwise bending moments (up to six times larger than edgewise moments) because of the thrust loadings by the water [25]. For the same output, tidal turbine blades can be subjected to twice the thrust of wind turbines [26]. Due to the importance of flexural behaviour of composite structures, four-point bending tests were carried out, and the results are illustrated in Figure 10.

Interestingly, the flexural strength of the 64.4% FVF sample set (1.21 GPa) was 3.5% lower than its 59.8% FVF counterpart (1.26 GPa) while having 10.5% greater flexural modulus (115.04 GPa compared to 127.1 GPa). The higher interfacial porosity at very high FVF (see Figure 7) hastens the appearance of compressive buckling behaviour. Indeed, interfacial porosity created localised interfacial stress, resulting in early failure comparatively to less porous samples.

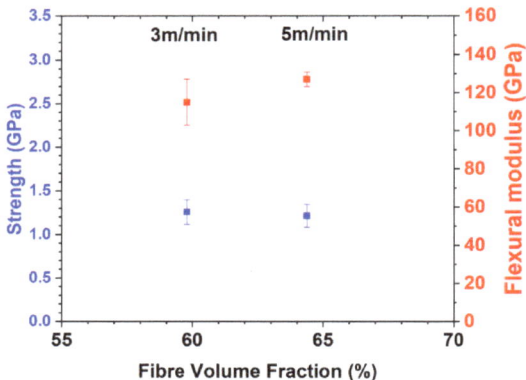

Figure 10. Flexural performance of the UD-CFRP plates.

3.6. Interlaminar Shear Strength (ILSS)

The ILSS test is a cohesive failure test usually used as a quality check in the composites industry due to the low amount of material required and the fast test speed [27]. With this method, material homogeneity, defects, and bulk porosity can be evaluated, especially if it occurs at the centreline of the specimen, where the shear stress is highest.

Figure 11 illustrates the change of ILSS of composite plates at different speeds and FVFs. Samples with 64.4% FVF display poorer interlaminar shear strength compared to the 59.8% FVF samples: 64.24 MPa compared to 58.49 MPa (an 8.95% decrease). The void content of 64.4% FVF samples is also higher than the 59.8% samples (32.2% higher void content). The authors believe that bulk porosity is the main contributing factor for the decrease in ILSS at higher production speed. Moreover, the much higher standard deviation (±6.24 MPa) of the 64.4% FVF sample set points to a less consistent consolidation in different samples due to the lack of resin to reach a high consolidation and therefore reduce bulk porosity.

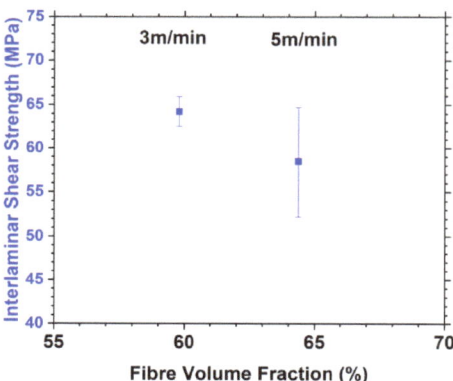

Figure 11. Interlaminar shear strength test results of the UD-CFRP plates.

3.7. Scanning Electron Microscope (SEM) Analysis

SEM micrographs of fractured samples were compared for both production speeds to visualise the fracture surfaces of the samples. The micrograph of the fractured 59.8% FVF sample (3 m/min) (Figure 12a) describes a homogeneous and brittle fracture: the interfaces show river lines, which are typical of cohesive failure and highlight a good fibre–matrix interfacial adhesion. A lateral fracture can be seen, highlighting a sudden stress release behaviour compatible with explosive failure.

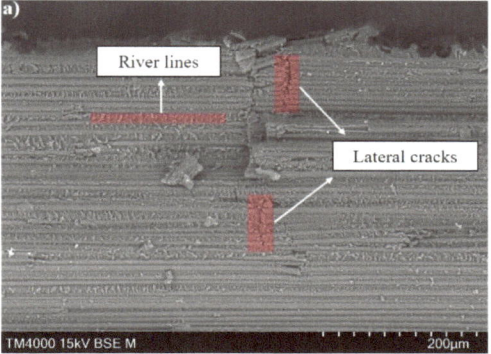 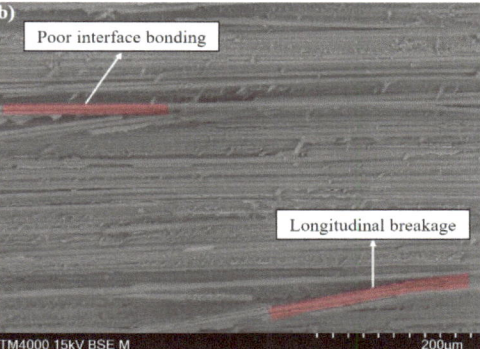

Figure 12. SEM images for UD-CFRP samples fractured in tension with (**a**) 3 m/min (59.8% FVF) (**b**) 5 m/min (64.4% FVF) production speed.

The higher FVF sample set (64.4%) highlights an entirely different behaviour. Clean interfaces can be seen, and the sample highlights longitudinal breakage (Figure 12b). These SEM images justify the conclusion drawn from the mechanical tests, as the lack of molten epoxy to fully coat individual fibre intra-tow regions caused a poorer interfacial stress transfer performance in general and allowed for interfacial localised stresses due to higher interfacial porosity.

4. Conclusions

The suitability of powder–epoxy in the high-speed, low-cost manufacturing of towpregs (prepreg tapes) has been demonstrated recently. These towpregs possess excellent properties for automated composite manufacturing systems such as AFP or filament winding. Controlling the fibre volume fraction (FVF), one of the most important parameters for the composite's performance, is crucial when high production speeds are targeted. With the reported tapeline system, it is possible to produce towpregs with high FVF and excellent mechanical properties; therefore, it offers tremendous potential as an alternative OOA approach. Furthermore, powder–epoxy offers low viscosity, virtually indefinite shelf life, the ability to co-cure multiple parts, and little to no VOC release. This study investigates the relationship between the production speed, FVF, and mechanical performance of the powder–epoxy-based carbon fibre towpregs. It was demonstrated that production at a slower speed (3 m/min) results in better consolidation, increased stress transfer, and better interlaminar adhesion due to the enhanced interfaces between the matrix and the carbon fibre reinforcement. It was observed that increasing the production speed may lead to inferior mechanical characteristics (up to 8.95% decrease in ILSS), although higher FVF values can be achieved. In addition, the interfacial void content of the samples was quantified by analysing optical micrographs with a custom algorithm. Results suggested that higher production speeds in the tapeline system results in higher interfacial voids (3.21% for 3 m/min and 6.07% for 5 m/min). A FVF of ≈60% for UD-CFRP laminates was found to be an optimal value, which can be reached with the proposed tapeline system by carefully controlling the production speed. Investigating the towpreg performance across a larger range and at higher production speeds (>10 m/min) could be of interest in establishing the optimal speed for the tapeline. Adjusting the powder particle flow to the tapeline speed in order to keep the FVF constant seems paramount in order to produce high-quality and homogeneous powder epoxy towpreg, and it is currently being investigated. Finally, a hand layup was shown to be problematic in regard to fibre alignment and resulted in lesser mechanical performances. Therefore, automation of the layup using an automated fibre placement (AFP) system is also of high interest for future investigations.

Author Contributions: Conceptualization, M.Ç., C.M.Ó.B., C.R.; Data curation, M.Ç., C.M.Ó.B., C.R., T.N., F.J., R.J.; Formal analysis, M.Ç., C.R., T.N.; Funding acquisition, C.M.Ó.B. and C.R.; Investigation, M.Ç., C.M.Ó.B., C.R., T.N., F.J., R.J.; Methodology M.Ç., C.R., T.N., F.J., R.J.; Project administration, C.R.; Resources, C.M.Ó.B. and C.R. Software, M.Ç. and T.N.; Supervision, M.Ç., C.M.Ó.B. and C.R.; Writing—original draft, M.Ç., T.N., F.J., R.J. and C.R. Writing—review & editing, M.Ç., C.M.Ó.B. and C.R. All authors have read and agreed to the published version of the manuscript.

Funding: This study was funded by CIMCOMP EPSRC Future Composites Manufacturing Research Hub (EP/P006701/1). The APC was kindly offered by Journal of Composite Science.

Institutional Review Board Statement: Not applicable.

Informed Consent Statement: Not applicable.

Data Availability Statement: Not applicable.

Acknowledgments: The authors would like to thank the CIMComp hub for financing this study, as well as Freilacke, Swiss CMT, and Toray C.A. for providing the needed materials for the investigations.

Conflicts of Interest: The authors declare no conflict of interest.

References

1. Centea, T.; Hubert, P. Out-of-autoclave prepreg consolidation under deficient pressure conditions. *J. Compos. Mater.* **2014**, *48*, 2033–2045. [CrossRef]
2. Maguire, J.M.; Simacek, P.; Advani, S.G.; Ó Brádaigh, C.M. Novel epoxy powder for manufacturing thick-section composite parts under vacuum-bag-only conditions. Part I: Through-thickness process modelling. *Compos. Part A Appl. Sci. Manuf.* **2020**, *136*, 105969. [CrossRef]
3. Maguire, J.M.; Nayak, K.; Ó Brádaigh, C.M. Novel epoxy powder for manufacturing thick-section composite parts under vacuum-bag-only conditions. Part II: Experimental validation and process investigations. *Compos. Part A Appl. Sci. Manuf.* **2020**, *136*, 105970. [CrossRef]
4. Maguire, J.M.; Nayak, K.; Ó Brádaigh, C.M. Characterisation of epoxy powders for processing thick-section composite structures. *Mater. Des.* **2018**, *139*, 112–121. [CrossRef]
5. Mamalis, D.; Murray, J.J.; McClements, J.; Tsikritsis, D.; Koutsos, V.; McCarthy, E.D.; Ó Brádaigh, C.M. Novel carbon-fibre powder-epoxy composites: Interface phenomena and interlaminar fracture behaviour. *Compos. Part B Eng.* **2019**, *174*, 107012. [CrossRef]
6. Floreani, C.; Robert, C.; Alam, P.; Davies, P.; Ó Brádaigh, C.M. Mixed-mode interlaminar fracture toughness of glass and carbon fibre powder epoxy composites—For design of wind and tidal turbine blades. *Materials* **2021**, *14*, 2103. [CrossRef]
7. Robert, C.; Pecur, T.; Maguire, J.M.; Lafferty, A.D.; McCarthy, E.D.; Ó Brádaigh, C.M. A novel powder-epoxy towpregging line for wind and tidal turbine blades. *Compos. Part B Eng.* **2020**, *203*, 108443. [CrossRef]
8. He, H.W.; Gao, F. Effect of Fiber Volume Fraction on the Flexural Properties of Unidirectional Carbon Fiber/Epoxy Composites. *Int. J. Polym. Anal. Charact.* **2015**, *20*, 180–189. [CrossRef]
9. Robert, C.; Mamalis, D.; Alam, P.; Lafferty, A.D.; Cadhain, C.Ó.; Breathnach, G.; McCarthy, E.D.; Ó Brádaigh, C.M. Powder Epoxy Based UD-CFRP Manufacturing Routes for Turbine Blade Application. In Proceedings of the SAMPE Europe Conference 2018, Southampton, UK, 11–13 September 2018; Volume 18, pp. 1–12.
10. Courter, J.; Dustin, J.; Ritchey, A.; Pipes, B.R.; Sargent, L.; Purcell, W. Properties of an out-of-autoclave prepreg material: Oven versus autoclave. In Proceedings of the 41st International SAMPE Technical Conference, Witchita, KS, USA, 19–22 October 2009.
11. Centea, T.; Grunenfelder, L.K.; Nutt, S.R. A review of out-of-autoclave prepregs—Material properties, process phenomena, and manufacturing considerations. *Compos. Part A Appl. Sci. Manuf.* **2015**, *70*, 132–154. [CrossRef]
12. Sutter, J.K.; Kenner, W.S.; Pelham, L.; Miller, S.G.; Polis, D.L.; Nailadi, C.; Hou, T.H.; Guade, D.J.; Lerch, B.D.; Lort, R.D.; et al. Comparison of autoclave and out-of-autoclave composites. In Proceedings of the SAMPE Fall Technical Conference 2010, Salt Lake City, UT, USA, 11–14 October 2010.
13. Noble, T.; Davidson, J.R.; Floreani, C.; Bajpai, A.; Moses, W.; Dooher, T.; McIlhagger, A.; Archer, E.; Ó Brádaigh, C.M.; Robert, C. Powder Epoxy for One-Shot Cure, Out-of-Autoclave Applications: Lap Shear Strength and Z-Pinning Study. *J. Compos. Sci.* **2021**, *5*, 225. [CrossRef]
14. Mamalis, D.; Flanagan, T.; Ó Brádaigh, C.M. Effect of fibre straightness and sizing in carbon fibre reinforced powder epoxy composites. *Compos. Part A Appl. Sci. Manuf.* **2018**, *110*, 93–105. [CrossRef]
15. Athanasopoulos, N.; Kostopoulos, V. Resistive heating of multidirectional and unidirectional dry carbon fibre preforms. *Compos. Sci. Technol.* **2012**, *72*, 1273–1282. [CrossRef]
16. Alam, P. *Composites Engineering: An A–Z Guide*; IOP Publishing: Bristol, UK, 2021.
17. Little, J.E.; Yuan, X.; Jones, M.I. Characterisation of voids in fibre reinforced composite materials. *NDT E Int.* **2012**, *46*, 122–127. [CrossRef]

18. Chang, T.; Zhan, L.; Tan, W.; Li, S. Void content and interfacial properties of composite laminates under different autoclave cure pressure. *Compos. Interf.* **2016**, *24*, 529–540. [CrossRef]
19. Mehdikhani, M.; Gorbatikh, L.; Verpoest, I.; Lomov, S.V. Voids in fiber-reinforced polymer composites: A review on their formation, characteristics, and effects on mechanical performance. *J. Compos. Mater.* **2018**, *53*, 1579–1669. [CrossRef]
20. Dona, K.N.U.G.; Du, E.; Carlsson, L.A.; Fletcher, D.M.; Boardman, R.P. Modeling of water wicking along fiber/matrix interface voids in unidirectional carbon/vinyl ester composites. *Microfluid. Nanofluidics* **2020**, *24*, 31.
21. Judd, N.C.; Wright, W.W. Voids and their effects on mechanical properties of composites—Appraisal. *SAMPE J.* **1978**, *14*, 10–14.
22. Zhang, R.L.; Huang, Y.D.; Li, N.; Liu, L.; Su, D. Effect of the concentration of the sizing agent on the carbon fibers surface and interface properties of its composites. *J. Appl. Polym. Sci.* **2012**, *125*, 425–432. [CrossRef]
23. Sharma, M.; Gao, S.; Mäder, E.; Sharma, H.; Wei, L.Y.; Bijwe, J. Carbon fiber surfaces and composite interphases. *Compos. Sci. Technol.* **2014**, *102*, 35–50. [CrossRef]
24. Ma, Q.; Gu, Y.; Li, M.; Wang, S.; Zhang, Z. Effects of surface treating methods of high-strength carbon fibers on interfacial properties of epoxy resin matrix composite. *Appl. Surf. Sci.* **2016**, *379*, 199–205. [CrossRef]
25. Grogan, D.; Leen, S.; Kennedy, C.; Brádaigh, C. Design of composite tidal turbine blades. *Renew. Energy* **2013**, *57*, 151–162. [CrossRef]
26. Papaelias, M.; Márquez, F.P.G.; Karyotakis, A. *Non-Destructive Testing and Condition Monitoring Techniques for Renewable Energy Industrial Assets*; Butterworth-Heinemann: Oxford, UK, 2020.
27. Chawla, K. *Composite Materials: Science and Engineering*; Springer: New York, NY, USA, 2012.

Article

Performance of Two-Way Concrete Slabs Reinforced with Basalt and Carbon FRP Rebars

Sukanta Kumer Shill, Estela O. Garcez, Riyadh Al-Ameri and Mahbube Subhani *

School of Engineering, Deakin University, Waurn Ponds, Geelong, VIC 3216, Australia; sukanta.shill@deakin.edu.au (S.K.S.); e.garcez@bcrc.com.au (E.O.G.); r.alameri@deakin.edu.au (R.A.-A.)
* Correspondence: mahbube.subhani@deakin.edu.au

Abstract: Fibre-reinforced polymer (FRP) rebars are being increasingly used to reinforce concrete structures that require long-term resistance to a corrosive environment. This study presents structural performance of large scale two-way concrete slabs reinforced with FRP rebars, and their performances were compared against conventional steel reinforced concrete. Both carbon FRP (CFRP) and basalt FRP (BFRP) were considered as steel replacement. Experimental results showed that the CFRP- and BFRP-RC slabs had approximately 7% and 4% higher cracking moment capacities than the steel-RC slab, respectively. The BFRP-RC slabs experienced a gradual decrease in the load capacity beyond the peak load, whereas the CFRP-RC slabs underwent a sharp decrease in load capacity, similar to the steel-RC slab. The BFRP-RC slabs demonstrated 1.72 times higher ductility than CFRP-RC slabs. The steel-RC slab was found to be safe against punching shear but failed due to flexural bending moment. The FRP-RC slabs were adequately safe against bending moment but failed due to punching shear. At failure load, the steel rebars were found to be yielded; however, the FRP rebars were not ruptured. FRP-RC slabs experienced a higher number of cracks and higher deflection compared to the steel-RC slab. However, FRP-RC slabs exhibited elastic recovery while unloading. Elastic recovery was not observed in the steel-RC slab. Additionally, the analytical load carrying capacity was validated against experimental values to investigate the efficacy of the current available standards (ACI 318-14 and ACI 440.1R-15) to predict the capacity of a two-way slab reinforced with CFRP or BFRP. The experimental load capacity of the CFRP-RC slabs was found to be approximately 1.20 times higher than the theoretical ultimate load capacity. However, the experimental load capacity of the BFRP-RC slabs was 6% lower than their theoretical ultimate load capacity.

Keywords: fibre-reinforced polymer (FRP); basalt FRP rebar; carbon FRP rebar; two-way slab; load–deflection behaviour; punching shear; ductility of RC slab

Citation: Shill, S.K.; Garcez, E.O.; Al-Ameri, R.; Subhani, M. Performance of Two-Way Concrete Slabs Reinforced with Basalt and Carbon FRP Rebars. *J. Compos. Sci.* **2022**, *6*, 74. https://doi.org/10.3390/jcs6030074

Academic Editor: Francesco Tornabene

Received: 3 February 2022
Accepted: 28 February 2022
Published: 1 March 2022

Publisher's Note: MDPI stays neutral with regard to jurisdictional claims in published maps and institutional affiliations.

Copyright: © 2022 by the authors. Licensee MDPI, Basel, Switzerland. This article is an open access article distributed under the terms and conditions of the Creative Commons Attribution (CC BY) license (https://creativecommons.org/licenses/by/4.0/).

1. Introduction

Concrete slabs require reinforcement whether they are used as suspended structural members (e.g., floors of a building, bridge deck, or culvert structure) or ground bearing slabs. For suspended floor slabs, the reinforcement design depends on the superimposed load intensity, materials properties, and span length of the slab. Concrete slabs directly placed on the ground also require at least a minimum amount of reinforcement to protect them from shrinkage and temperatures effects. Worldwide, this reinforcing of concrete slabs is conventionally done using mild steel rebars.

A number of outdoor concrete infrastructures, such as marine structures, protective structures in coastal area, airfield rigid pavements, parking areas, bridge decks, railway sleepers, and sewer infrastructures are often subjected to various aggressive environmental exposures, such as de-icing salts, high humidity, elevated temperatures, chloride ions, hydrogen sulphide gas, and other chemicals [1–5]. Exposure to those harsh conditions significantly reduces the alkalinity of the protective layer of concrete to reinforcing steel that results in substantial damage to the steel rebars. As conventional steel rebars are highly prone to oxidation when exposed to moisture and air, they easily become oxidised and produce iron oxides (rust).

Rust usually occupies a higher volume than steel and exerts multidirectional stresses on the surrounding concrete. Consequently, failure in bonds between concrete and steel rebars occurs, and numerous cracks develop [1]. Corrosion of steel rebars eventually leads to the degradation of concrete, reduces the life expectancy of the structures considerably, and demands expensive strengthening/retrofitting works [6,7]. Therefore, the long-term durability of concrete structures subjected to severe conditions is a crucial concern in the construction industry worldwide.

To resolve the issue, various protective measures, e.g., increasing concrete cover, coating steel rebars with epoxy, and improving the permeability of concrete, have been taken into consideration. Nevertheless, not a single method has been fully successful to eliminate the corrosion risk of conventional steel reinforcement [7]. Innovation of fibre-reinforced polymer (FRP) rebars led the construction industry a step ahead to find the solution to the problem at a lower cost. Recently, numerous studies reported that FRP rebars were found to be one of the promising alternatives to conventional steel rebars to reinforce concrete structures because of their excellent corrosion resistance, tensile strength, and lightweight [1,7–17].

FRP bars are usually manufactured from various high tensile strength fibres, such as glass, carbon, basalt, and aramid fibres, which are impregnated using different polymeric resins, fillers, and curing agents. FRP rebars are non-corrosive, almost non-conductive, and possess higher tensile strength [2,8–11]. However, they exhibit a linearly elastic stress–strain relationship (no yield point) with a lower modulus of elasticity compared to conventional steel rebar [16,18–20].

Apart from the benefits of corrosion resistance, FRP-RC structures are lighter in weight than steel-reinforced structures. As a result, the fabrication and installation processes of precast concrete elements are easier. In recent years, FRP rebars have become targeted due to the enormous potential of replacing conventional steel reinforcement in multi-storey buildings, industrial structures, water treatment plants, and other structures. For example, FRP rebars were used in real-life concrete structures around the world where durability and magnetic permeability were the controlling parameters [13,20–22].

Past studies reported that CFRP rebars are effective and appropriate as reinforcements for structural concrete [7,23]. According to Bilotta et al. [24], CFRP-RC slabs perform better than steel-reinforced slabs even when subjected to fire. CFRP was found to be lighter (usually 20% of the mass of conventional steel) and possesses a higher strength-to-weight ratio and tensile strength [25]. In contrast, compared to steel reinforced slabs, CFRP-RC slabs usually require additional shear rebars to improve the punching resistance [26].

A few recent studies investigated the performance of BFRP-RC structures under different loading and environmental conditions [27,28]. Basalt fibre is a relatively new building material, which is composed of minerals such as pyroxene, plagioclase, and olivine [25]. It is environmentally friendly and can be a suitable alternative to glass fibre in the construction industry, as it has better physical and mechanical properties [25,29,30]. Moreover, BFRP rebar has been recognized for its higher elongation at fracture and better chemical resistance, especially in alkaline environments [31–33]. Additionally, BFRP rebars possess a wide range of thermal and UV light resistance, have superior electro-magnetic properties, and are less costly compared to CFRP [30].

The flexural design of FRP-RC members is comparable to that of the concrete structures reinforced with steel rebars [1,34,35]. However, due to the lower modulus of elasticity of FRP rebars, concrete structures reinforced with FRP rebars usually possess lower shear strength and flexural stiffness. As stated, FRP-RC members experience wider and deeper cracks under the same loads when compared with typical RC structures [1,35]. Concrete structures reinforced with FRP exhibit relatively higher deflections and may experience brittle/sudden failure [19,20].

As reported, the structural performance of FRP-RC structures under service conditions are promising [7]. To limit the cracks and deflection of FRP-RC structures, the design is generally governed by the serviceable state limit [1]. In order to avoid catastrophic failures, most of the design codes recommend an over-reinforced flexural design for FRP-RC members [1,36]. Although the design guidelines of FRP-RC members are currently available [1], the limitations of use and design recommendations are still evolving as research progresses.

The present study is motivated by the promising properties of BFRP rebars, which have the potential to substitute steel rebars in corrosive environments. To date, the literature related to the performance of BFRP reinforced two-way concrete slabs is still scarce. Thus, this paper deals with the load–deflection behaviour of large scale two-way concrete slabs reinforced with BFRP and CFRP rebars. It elucidates the ultimate load capacity, modes of failure, flexural stiffness, cracking moment, ultimate bending moment, punching shear capacity, serviceable moment, and strain distribution along the FRP rebars. The experimental load–deflection capacities of the FRP-RC slabs are compared with the theoretical capacities proposed by ACI 440.1R-15 [1]. The findings of this extensive experimental investigation will help practitioners and engineers to design and construct CFRP- and BFRP-RC slabs.

2. Experimental Program

A total of seven simply supported two-way concrete slabs reinforced with CFRP, BFRP, and steel rebar was fabricated and tested to failure. Among the seven concrete slabs, three were reinforced with CFRP, three were reinforced with BFRP, and one was reinforced with typical steel rebar as a control specimen.

2.1. Material Properties

Based on the manufacturer's product data sheet, the physical and mechanical properties of the CFRP and BFRP bars used in this study are listed in Table 1. The surface of the FRP bars and steel reinforcing bars both were deformed (Figure 1a). Locally available ready-mix concrete (Geelong, Australia) was used to cast the concrete slabs. All slabs were cast with the concrete with the same mix design but from two batches of concrete. The ultimate compressive strength of the concrete used in the study is given in Table 2.

Table 1. Properties of the reinforcing bars (taken from manufacturing data sheet).

Parameters	Rebar Type		
	Steel	CFRP	BFRP
Bar diameter (mm)	7.8	6	6
Nominal cross-sectional area (mm^2)	48	28	28
Tensile strength (MPa)	500	2150	1300
Elastic modulus (GPa)	200	140	55
Elongation (%)	2.27 [a]	1.3 [b]	1.8 [b]

[a] Elongation at yielding; [b] elongation at bar rupture.

Table 2. Compressive strength of concrete and details of reinforcement.

Slab Specimen	Concrete Compressive Strength, f'_c (MPa)	Reinforcement Area (mm^2/m)	Effective Depth (mm)	Reinforcement Ratio ρ_f (%)
Steel	29.62	318.56	51.1	0.62
CFRP-1	29.62			
CFRP-2	34.59			
CFRP-3	34.59	188.50	52	0.36
BFRP-1	29.62			
BFRP-2	34.59			
BFRP-3	34.59			

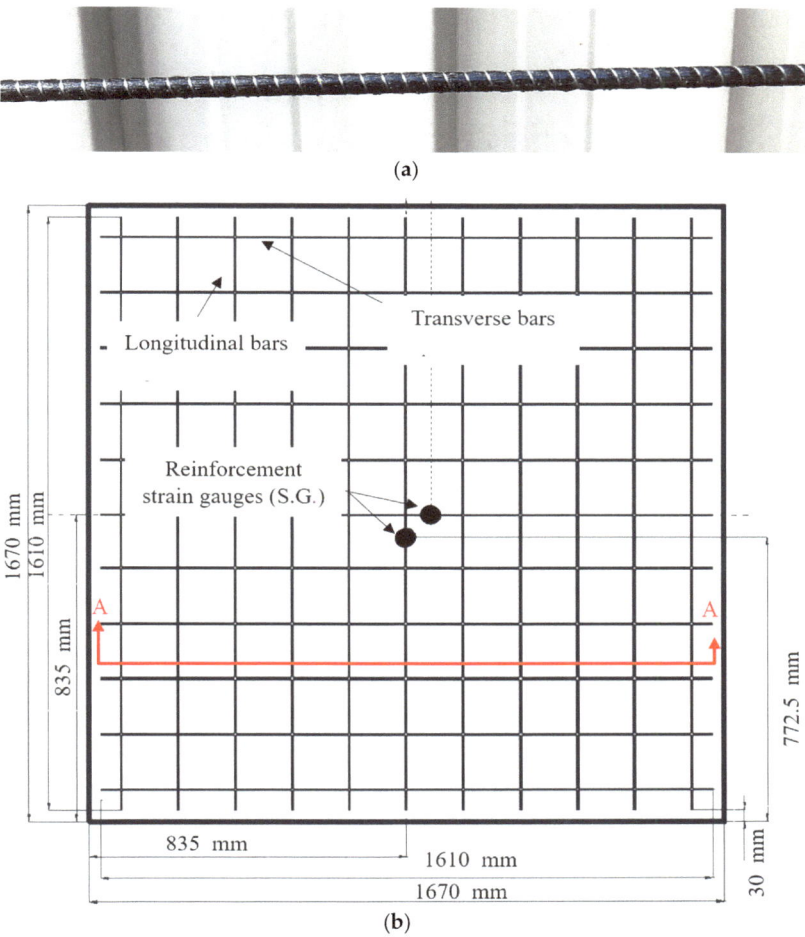

Figure 1. Reinforcement detailing, and the dimensions of the concrete slab. (**a**) deformed BFRP rebar; (**b**) reinforcement details.

2.2. Specimens

Figures 1 and 2 shows the reinforcement details and geometric properties of the fabricated concrete slabs. The length, width, and thickness of the concrete slabs were 1670, 1670, and 75 mm, respectively. Table 2 shows the reinforcement type, cross-sectional area, reinforcement ratio, and concrete compressive strength of each slab specimen. Single-layer reinforcement was provided at the bottom (tension zone) for all concrete slabs, as the slabs were designed to be simply supported from all sides. No reinforcement was placed in the compression zone. The rebar in all seven slabs was spaced at 150 mm centre to centre (c/c) in both directions. Slabs were cast outdoors and covered with plastic sheets for 3 days to prevent moisture loss and ensure adequate curing, followed by air curing for 28 days. Figure 3a shows the casting of concrete slabs, and Figure 3b illustrates the FRP mesh used in the concrete slab.

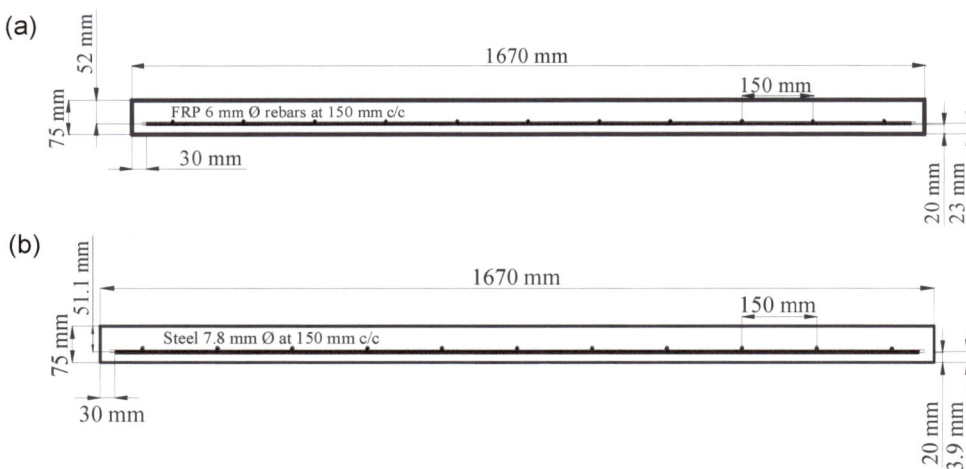

Figure 2. Cross-section (A–A) of (**a**) FRP reinforced slab and (**b**) steel reinforced slab.

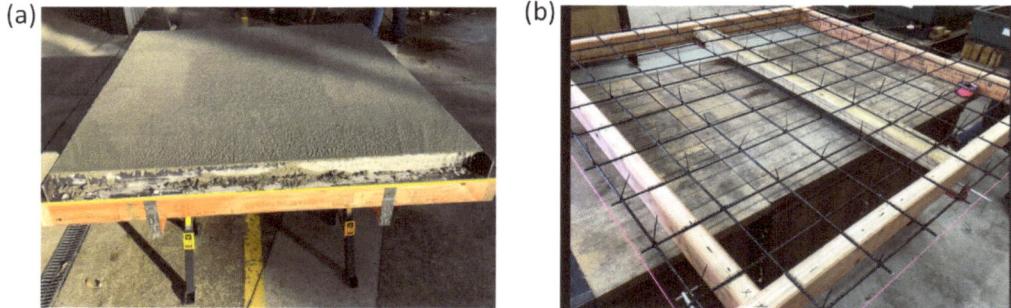

Figure 3. Photos of (**a**) a concrete slab after casting and (**b**) the single-layer reinforcement of a slab.

2.3. Test Setup and Procedure

A monotonic uniformly distributed load (UDL) was imposed on all concrete slabs using a load frame with a capacity of 500 kN. The UDL fixture attached to the load frame is depicted in Figure 4a. The fixture was composed of 16 loading pistons; the diameter of each piston was 175 mm, all acting as loading points to equally distribute the load. A schematic diagram of the loading arrangement is shown in Figure 4b. The slabs were simply supported at the end, resting on a steel frame with a span of 1600 mm between supports. Preloading up to 10 kN was applied at a loading rate of 0.04 kN/s for a few times to allow for the supports' settlements and relieve any residual stresses. After pre-loading, the slabs were initially loaded at a loading rate of 0.08 kN/s up to 35 kN and subsequently loaded up to failure at a rate of 0.6 mm/min. The concrete slabs were carefully investigated after each loading step. To record the structural responses under the applied loads, the slabs were instrumented with one LVDT (placed at the mid-span of the slab), two electrical strain gauges at the top, and two additional electrical strain gauges at the bottom of the slabs.

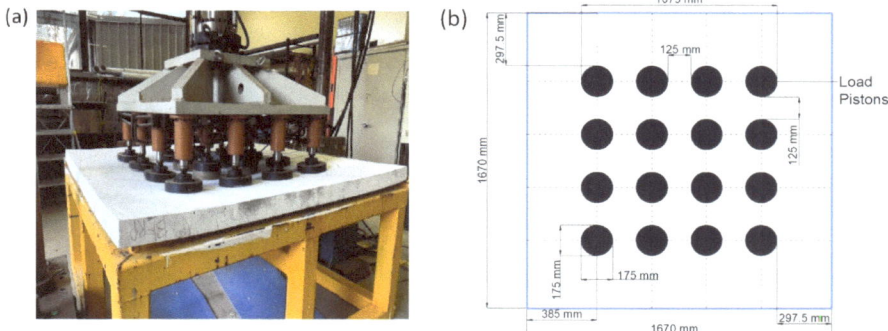

Figure 4. (**a**) Photo of the UDL application on the concrete slab; (**b**) schematic diagram of the loading points on a slab.

3. Test Results and Discussion

This section presents the observed failure modes, load–deflection relationship, flexural stiffness, flexural moment capacity, ultimate bending moment, strain distribution in FRP rebars, and the punching shear capacity of the concrete slabs.

3.1. The Failure Modes

The FRP-RC slabs experienced higher cracks compared to the steel-RC slab during loading. The cracking patterns formed in the concrete slabs are shown in Figure 5. Since the span lengths of all slabs in both directions were equal, they were subjected to an equal amount of bending moment in both directions. Consequently, all slabs developed cracks in both directions almost equally, as shown in Figure 5a–c. The steel-RC slab failed due to the bending moment, as no catastrophic type of failure was observed. Since the steel-RC slab was designed as an under-reinforced member, steel rebars reached yield strain before concrete reached failure strain. FRP reinforced concrete structures are designed to be over-reinforced, as suggested in Canadian (CAN/CSA) and American (ACI) standards. Hence, for FRP-RC slabs, concrete reached its ultimate strength, resulted in punching shear failure. The formation of intermediate cracks in FRP-RC slabs during loading resulted in a progressive drop in shear capacity, which eventually led to shear failure without the rupture of the FRP rebars.

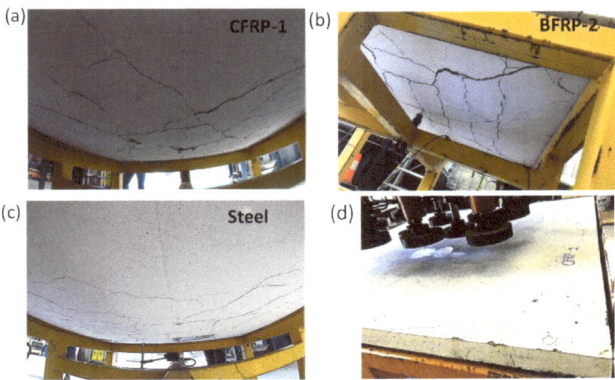

Figure 5. Failures of concrete slabs reinforced with (**a**) CFRP, (**b**) BFRP, (**c**) steel rebars, and (**d**) punching shear failure.

3.2. The Load–Deflection Behaviour

Figure 6 shows the load versus mid-span deflection curves of all the concrete slabs. The uncracked state of the curves, where load–deflection responses are linear, reflect that all concrete slabs performed similarly before cracking. Once applied loads exceeded the rupture strength of the concrete, cracks were developed in the concrete slabs. As loading progressed, slabs developed more cracks and failed when the load reached the peak value. During the loading, whenever a major crack was developed in the concrete slab, RC slabs showed a small drop in load in the load–deflection curve. Compared to the steel-RC slab, both the CFRP- and BFRP-RC slabs exhibited higher deflection but lower ultimate load capacity at failure. It was noted that compared to the CFRP reinforced slabs, the BFRP-RC slabs experienced higher deflection but withstood the lowest ultimate load.

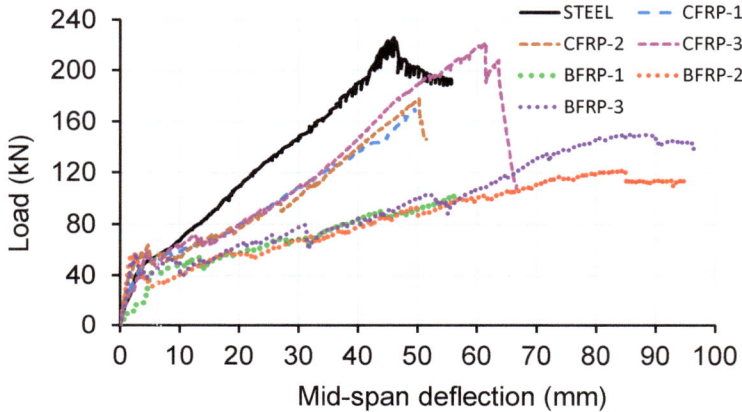

Figure 6. Load vs. mid-span deflection curves for all concrete slabs.

3.3. Stiffness of the Slabs

According to ACI 440.1R-15 [1], the performance of FRP-RC structures is controlled by serviceability criteria. Additionally, ACI 440.1R-15 [1] demonstrated that FRP-RC structures show relatively lower stiffness after cracking when compared against steel-RC structures having the same reinforcement ratio. The present study illustrated the stiffness of the uncracked slabs (within the cracking load) and the stiffness of the cracked sections beyond the serviceable limit. The stiffnesses of all slabs were determined from the slopes of the load–deflection curves [4]. To determine the stiffness of the uncracked and cracked sections of the slabs, three tangents (I, II, III) were drawn on load–deflection curves, as shown in Figure 7. Tangent-I and tangent-II present the linear elastic and plastic behaviours of the slabs under superimposed loads, respectively. The slope of tangent-I and tangent-II presents the stiffness of the uncracked and cracked sections of the slabs, respectively. Tangent-III presents the curves after the fracture point.

Before cracking, most FRP-RC slabs (CFRP-2, CFRP-3, BFRP-2, and BFRP-3) showed similar stiffnesses to the steel-RC slab. After cracking, however, the flexural stiffness of the FRP-RC slabs was reduced significantly, which triggered higher deflections of the slabs under the subsequent loadings. Table 3 list the stiffness values of cracked and uncracked sections of the slabs. It can be noted that the reinforcement ratio of the FRP-RC slabs was 40% less than that of the steel-RC slab, and the axial rigidities of CFRP and BFRP rebars were 41% and 16% of that of the steel rebar. However, CFRP- and BFRP-RC showed 64% and 48% of the cracked stiffness of the steel-RC slab. Within the elastic limit, the stiffness of BFRP and CFRP-RC members was found to be comparable to the steel-RC members. Beyond the cracking load, the stiffness of FRP-RC members dropped substantially due to the formation of numerous cracks.

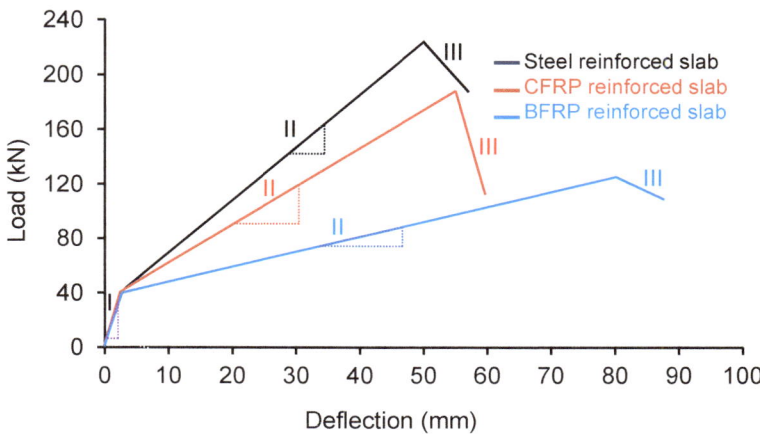

Figure 7. Stiffness of all slabs investigated.

Table 3. Stiffness of all test slabs.

Slab ID	Elastic Modulus of Rebars (GPa)	Area of Reinforcement (mm²/m)	Ratio of the Axial Rigidity ($E_{frp}A_{frp}$)/(E_sA_s)	Stiffness of Uncracked Section (kN/mm)	Uncracked Ratio (FRP/Steel)	Stiffness of Cracked Section (kN/mm)	Cracked Ratio (FRP/Steel)
Steel	200	318.56	1	15.76	1.00	5.45	1.00
CFRP-1				13.53	0.86	3.45	0.63
CFRP-2	140	188.50	0.414	29.57	1.87	3.44	0.63
CFRP-3				16.16	1.02	3.67	0.67
BFRP-1				7.28	0.46	2.47	0.45
BFRP-2	55	188.50	0.162	27.12	1.72	2.65	0.49
BFRP-3				22.31	1.41	2.61	0.48

3.4. Experimental Loads and Moments of the Concrete Slabs

Table 4 summarises the experimental cracking load and moment, serviceable moment, and ultimate positive bending moments of the RC slabs. The moment calculation methods are described in the subsequent sections. Additionally, this study chose two principles to find out the serviceable bending moment (M_s) of the FRP-RC slabs. The first principle was taken from ISIS-07 [37] that states that M_s is the bending moment corresponding to the FRP bar strain of 2000 µε under applied loads. The second principle to determine M_s was taken from Bischoff's study [38], where M_s is estimated as 30% of the ultimate bending moment (M_u).

3.4.1. Cracking Moment

The experimental cracking moment ($M_{cr\text{-}exp}$) was determined from the cracking loads, as described in Equation (1).

$$M_{cr-exp} = 0.036(w_{dl} + w_{cr})l^2 \qquad (1)$$

where M_{cr-exp} = experimental cracking moment in kN-m per meter, w_{cr} = experimental UDL on the slab in kN/m² that created cracks, and l is the effective span length of the slab in m.

The cracking moment capacity of concrete members depends on the modulus of rupture value of concrete and the cross-sectional properties of the members. According to ACI 318-14 [39], the rupture strength of concrete was determined using Equation (2). On average, the moduli of rupture of concrete of the steel, CFRP, and BFRP-RC slabs were 3.37, 3.55, and 3.55 MPa, respectively.

$$f_r = 0.62\sqrt{f'_c} \qquad (2)$$

Table 4. Cracking moment, serviceable moment, and the ultimate bending moment of the concrete slabs studied.

Slab ID	Cracking Load, P_{cr} (kN)	Cracking Moment, $M_{cr\text{-}exp}$ (kN-m)/m	Serviceable Moment, M_s (kN-m)/m		Ultimate Load, P_u (kN)	Ultimate Bending Moment, M_u (kN-m)/m
			2000 με	0.3 M_u		
Steel	50.00	1.96	-	2.45	222.50	8.17
CFRP-1	46.70	1.84	2.00	1.88	169.10	6.25
CFRP-2	56.03	2.18	2.21	1.97	177.50	6.55
CFRP-3	57.95	2.25	2.57	2.29	207.70	7.64
BFRP-1	42.04	1.67	1.60	1.16	103.00	3.87
BFRP-2	56.50	2.19	1.72	1.34	120.10	4.48
BFRP-3	57.60	2.23	1.74	1.61	144.30	5.35

Based on the results, the CFRP- and BFRP-RC slabs showed approximately 6.5 and 3.5% higher cracking moment capacities than the steel-RC slab, respectively. As the cross-sectional properties of all slabs were the same, the variation in the rupture strength of concrete contributed to the higher cracking moment capacities of the CFRP- and BFRP-RC slabs, since the concrete strength was slightly lower in the steel-RC slab (Table 2).

3.4.2. Serviceable State

The flexural bending moment of the FRP-RC member within the service state is the key indicator to assess the performance of the member subjected to out of plane loading [36]. Since FRP bars do not have a risk of corrosion, a larger width of cracks (usually 1.66 times wider than that in typical RC members) are tolerated in FRP-RC member design [40]. Thus, to limit the crack widths in FRP- and steel-RC members subjected to flexure, an upper limit on FRP and steel rebars' strain equal to 2000 με and 1200 με are allowed within the serviceable state, respectively [37,40].

According to ISIS-07 [37], the service load capacities of the CFRP- and BFRP-RC slabs are approximately 58.00 and 43.00 kN/m², respectively. Additionally, based on Bischoff's study [38], the CFRP- and BFRP-RC slabs offer service load capacities of 52.00 and 34.00 kN/m², respectively. The service live load capacities of the CFRP- and BFRP-RC slabs computed from the ISIS-07 [37] method are 1.13 and 1.28 times higher than that of the CFRP- and BFRP-RC slabs estimated following Bischoff's study [38]. These serviceable live load capacities demonstrate the suitability of using CFRP and BFRR rebar in concrete slab.

3.4.3. Ultimate Moment

To determine the ultimate positive bending moment capacity of a simply supported two-way concrete slab, the coefficient method of ACI 318-14 [39] was followed. The ultimate positive bending moment (M_u) in kN-m per unit meter was determined following Equation (3).

$$M_u = 0.036(w_{dl} + w_{ll})l^2 \tag{3}$$

where w_{dl} is the self-weight of the slab, and w_{ll} is the experimental ultimate UDL applied on the slab in kN/m².

Results showed that the CFRP- and BFRP-RC slabs possessed approximately 17 and 45% lower ultimate moment capacities compared to the steel-RC slab, respectively. Furthermore, the ultimate moment capacity of the steel-RC slab was 4.17 times higher than its cracking moment capacity, whereas on average, the ultimate moment capacities of the CFRP- and BFRP-RC slabs was 3.25 and 2.25 times higher than their cracking moment capacities. The reason for that is due to the development of numerous wide cracks in the FRP-RC slabs beyond the cracking load. As the FRP bar has a lower modulus of elasticity compared to steel, the FRP-RC slabs underwent significantly higher deflections under the same load. Consequently, numerous cracks were developed beyond the serviceable limit, and cracking continually increased until the failure. Progressive cracking significantly reduced the flexural capacity of the FRP-RC slabs, as the propagation of cracks into the compressive zone led to the reduction in the effective depth of the slabs. Thus, the FRP-RC slabs showed lower ultimate moment capacity compared to the steel-reinforced slab. Furthermore, the BFRP-RC slabs had around 34% lower ultimate moment capacity than the CFRP-RC slabs, as BFRP bars used in the study had approximately 60% lower modulus of elasticity than that of the CFRP bars.

3.5. Displacement Ductility of the Slabs

The displacement ductility of the FRP and steel-RC slabs was estimated as the ratio of the maximum deflection at ultimate load (Δ_{max}) and the deflection (Δ_x) at the intersection of the cracked and uncracked section observed in the load–deflection curve [41], as shown in Figure 8. The displacement ductility values of the slabs are given in Table 5. On average, the CFRP- and BFRP-RC slabs exhibited 1.26- and 2.18-times higher ductility compared to that of the steel-RC slab. Therefore, FRP-RC members even with a lower reinforcement ratio can have higher ductility compared to typical steel reinforcement slabs with a higher reinforcement ratio (reinforcement ratios in the steel and FRP-RC slabs were 0.0062 and 0.0036, respectively). Moreover, the enhanced ductility properties of the BFRP-RC slabs should not be overlooked, as, on average, the BFRP reinforced slabs demonstrated 1.72 times higher ductility than CFRP reinforced slabs. The reason for this is that BFRP-RC slabs experienced a gradual decrease in the load capacity with incremental displacement, whereas CFRP reinforced slabs underwent a sharp decrease in load capacity. This demonstrates that BFRP bars could effectively be used to reinforce structural concrete where higher ductility of the member is desired.

Table 5. Displacement ductility of the slabs investigated.

Slab ID	Deflection (Δ_x) at the Intersection of the Tangents	Deflection at Ultimate Load	Ductility of the Slabs
	Δ_x (mm)	Δ_{max} (mm)	$\frac{\Delta_{max}}{\Delta_x}$
Steel-RC slab	3.20	46.45	14.51
CFRP-1	2.85	49.58	17.39
CFRP-2	2.20	50.26	22.84
CFRP-3	4.25	63.63	14.97
BFRP-1	4.80	56.79	11.83
BFRP-2	2.15	85.00	39.53
BFRP-3	2.00	87.00	43.504

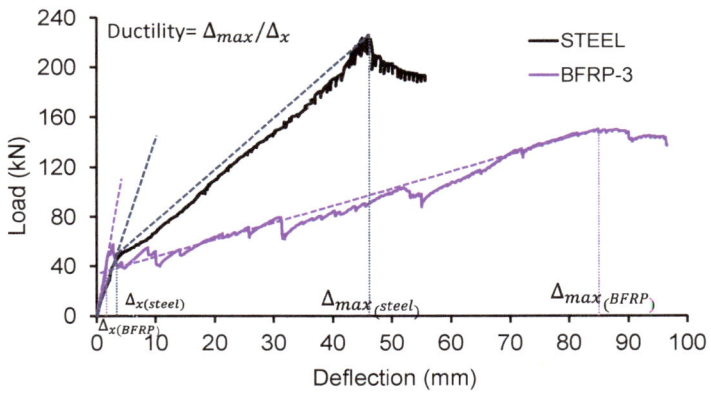

Figure 8. Method of determining displacement ductility of the slabs tested.

3.6. Residual Deflections and Elastic Recovery of the Slabs

Table 6 shows the ultimate deflection at failure load and the residual deflection after load release. This behaviour is consistent with the findings reported by other researchers [22,42,43]. The residual deflection (Δ_R) and the ultimate deflection (Δ_U) ratio (Δ_R/Δ_U) represent the permanent deflection when the applied load was released. Both the CFRP- and BFRP-RC slabs showed a significant elastic recovery when the slabs were unloaded. The elastic recovery of each slab was calculated as the ratio of the difference between ultimate deflection and residual deflection with respect to ultimate deflection. This implies that the FRP bars did not reach their rupture strain under the superimposed loads. However, the elastic recovery in the concrete slab reinforced with steel was not observed. Since the steel-RC slabs are usually designed as under-reinforced members, permanent deformation occurred in the steel of the concrete slab reinforced with steel rebars.

Table 6. The ultimate and residual deflections of the slabs.

Slab ID	Deflection at Serviceability Limit Δ_s (mm)		Ultimate Deflection Δ_U	Ultimate Deflection Ratio (FRP/steel)	Residual Deflection Δ_R	Elastic Recovery (%)
	2000 $\mu\varepsilon$	0.3 M_u				
Steel-RC slab	-	9.97	46.45	1	46.45	0
CFRP-1	20.1	6.65	49.58	1.07	18.00	64
CFRP-2	4.5	4.23	50.26	1.08	22.87	54
CFRP-3	3.35	8.6	63.63	1.37	38.42	40
BFRP-1	5.2	4.8	56.79	1.22	24.69	57
BFRP-2	1.35	1.2	85.00	1.83	38.12	55
BFRP-3	19.6	1.7	87.00	1.87	36.8	57

3.7. Experimental vs. Theoretical Deflections

The amount of deflection at the centre of a two-way RC slab is considered identical irrespective of the direction of a slab. It depends on the modulus of elasticity of concrete, applied load intensity, and geometrical properties, such as moment of inertia and span length of the slab. The theoretical centre point deflection of the RC slabs was calculated using Equation (4).

$$\Delta = \frac{5wl^4}{384EI_e} \quad (4)$$

where $w = (w_{dl} + w_{ll})$, l = effective span length, E = modulus of elasticity of concrete, and I_e = effective moment of inertia of the section, which is related to tangent (II) of Figure 7.

Figure 9a–c shows comparative studies on the experimental and theoretical load–deflection curves obtained based on the guidelines provided in ACI 318-14 [39], ACI 440.1R-15 [1], and Bischoof and Scanlon (2007) [44]. The experimental ultimate load capacities of the steel-, CFRP-, and BFRP-RC slabs were 90.26, 74.95, and 49.68 kN/m^2, respectively. According to ACI 318-14 [39], the theoretical ultimate load capacities of the steel, CFRP, and BFRP-RC slabs are 90.00, 62.50, and 52.70 kN/m^2, respectively, corresponding to the same deflection values obtained under the ultimate experimental loads. The steel-RC slab showed the same load capacity experimentally and theoretically. The experimental load capacity of the CFRP-RC slabs was found to be approximately 1.20 times higher than the theoretical ultimate load capacity. However, the BFRP reinforced concrete slabs showed 6% less experimental load capacity than its theoretical counterpart.

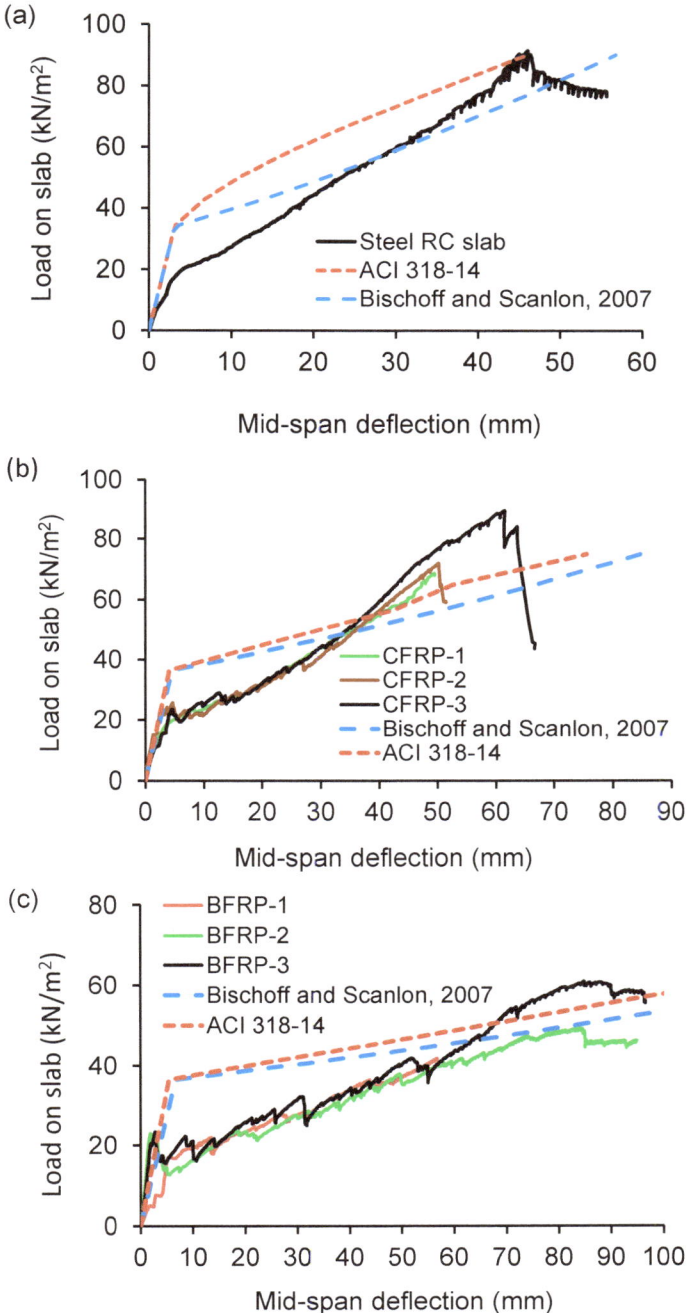

Figure 9. Theoretical and experimental load–deflection curves of (**a**) steel, (**b**) CFRP, and (**c**) BFRP-RC slabs.

3.8. Load–Strain Profile of Rebars Used in the Slabs

Figure 10 shows the strain distributions in the reinforcements of the concrete slabs measured using electrical strain gauges. The dashed lines present strain in the rebar in one direction, while the continuous lines correspond to the strain in the rebar in other (perpendicular) direction (see Figure 1 for directions and the location of strain gauges). The strain value acting in the rebar from one slab of each group is shown in Figure 10. The maximum strain at failure loads in the steel, CFRP, and BFRP rebars were approximately 14,200, 11,000, and 19,000 με, respectively. The maximum measured strain in CFRP rebars was about 85% of its ultimate strain, while the measured strain in the BFRP rebars exceeded the nominal elongation by 2.9%. Although the BFRP rebars showed higher elongation at ultimate load than other rebars types used, the BFRP rebars were not ruptured, as they showed elastic recovery during unloading. Table 7 lists the measured strain values and developed stresses in FRP rebars at failure loads. Additionally, the ultimate strain of concrete for each slab computed using ACI 440.1R-15 [1] is presented. The developed stress in all FRP rebars at ultimate load remained well below the rupture strength. This implies that FRP rebars did not rupture when the slabs failed due to loads. As punching shear of concrete triggers the failure of FRP-RC slabs, this study recommends high strength concrete to be used in concrete structures reinforced with FRP rebars to achieve high resistance against punching shear force.

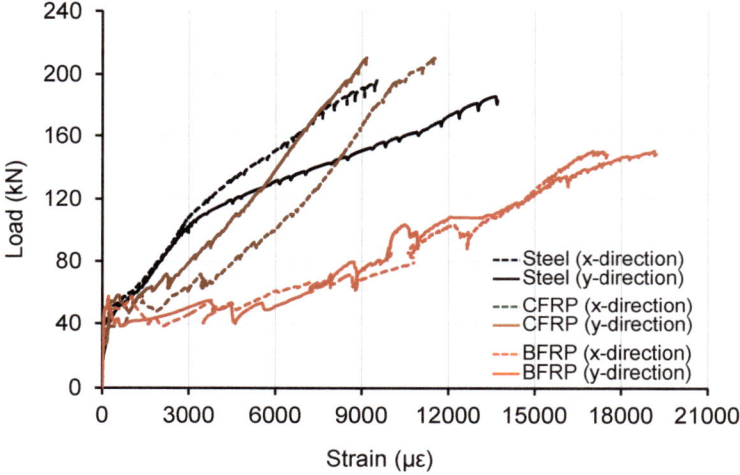

Figure 10. Strain distribution in the reinforcements of the slabs tested.

Table 7. Stress and strain values of the FRP rebars in the slabs at failure load.

Slab ID	Ultimate Strain of Concrete	Strain in Rebars at Failure	Developed Stress in FRP Rebars (MPa)	Ultimate Strength of FRP Rebars (MPa)	Ratio of Stress/Strength of FRP Rebars	Remarks on FRP Rebars
Steel	0.00199	0.01420	-		-	-
CFRP-1	0.00199	0.01042	1458		0.67	
CFRP-2	0.00202	0.01125	1575	2150	0.73	Not ruptured
CFRP-3	0.00202	0.01133	1586		0.73	
BFRP-1	0.00199	0.01752	963		0.74	
BFRP-2	0.00202	0.01887	1037	1300	0.79	Not ruptured
BFRP-3	0.00202	0.01920	1056		0.81	

3.9. Flexural Moment Capacities of the Concrete Slabs

Table 8 shows the resisting moment capacity (M_n) of the concrete slabs calculated following ACI 318-14 [39] and ACI 440.1R-15 [1], and the ultimate positive bending moment at failure load. Additionally, the reinforcement ratio used in the study and the balanced reinforcement ratio for the slabs are listed. Since the resisting moment capacities of FRP-RC slabs are higher than the ultimate positive bending moment at failure, all FRP-RC slabs were safe against flexural failure. However, the steel-RC slab failed due to bending, as the resisting moment capacity of the steel-RC slab was less than the ultimate bending moment. The failure of the steel-RC slab occurred due to the yielding of steel rebars because it was designed as an under-reinforced member.

Table 8. Ultimate resisting moment and bending moment of the slabs.

Slab ID	Reinforcement Ratio	Balanced Reinforcement Ratio	Resisting Moment, M_n (kN-m/m)	Resisting Load, P_n (kN)	Ultimate Bending Moment, M_u (kNm/M)	$\frac{M_n}{M_u}$ Ratio	Remarks
Steel	0.00623	0.02269	7.61	206.95	8.17	0.93	Failed by steel yielding
CFRP-1		0.00241	10.04	274.45	6.25	1.61	Safe against bending moment
CFRP-2	0.00362	0.00248	11.06	302.78	6.55	1.69	
CFRP-3		0.00248	11.06	302.78	7.64	1.45	
BFRP-1		0.00240	6.77	183.61	3.87	1.75	Safe against bending moment
BFRP-2	0.00362	0.00239	7.41	201.39	4.48	1.65	
BFRP-3		0.00239	7.41	201.39	5.35	1.39	

3.10. Punching Shear Capacity of the Slabs

Two critical sections were considered to investigate the punching shear capacity of the RC slabs. Section 1 was around the individual loading piston, and Section 2 was around the total loading considering a point load, as presented in Figure 11a,b, respectively. To determine the theoretical punching shear capacity of the FRP-RC slabs, the punching shear prediction models proposed by ACI 440.1R-15 [1], El-Gamal et al. [35], and Metwally [6] were chosen. All slabs were found to be adequately safe against punching shear along the critical Section 1. The punching shear capacity along the critical Section 2 obtained from the analytical prediction models was comparable with the experimental loads, as listed in Table 9. The ratio $\left(V_{pred}/V_{exp}\right)$ for each theoretical model was calculated for each type of reinforcement. Based on the ACI 318-14 [39] model, the punching shear capacity $\left(V_{pred}\right)$ of the steel-RC slab was 240.10 kN, while the ultimate experimental failure load $\left(V_{exp}\right)$ was 222.50 kN. This infers that the steel-RC slab was safe against punching shear force as $\left(V_{pred}/V_{exp}\right) = 1.08$ for the slab.

Table 9. Experimental and theoretical punching shear capacity of the CFRP- and BFRP-RC slabs based on different analytical models.

Slabs	V_{exp} (kN)	ACI 440.1R-15 [1]		El-Gamal et al. [35]		Metwally [6]	
		V_{pred} (kN)	V_{pred}/V_{exp}	V_{pred} (kN)	V_{pred}/V_{exp}	V_{pred} (kN)	V_{pred}/V_{exp}
CFRP-1	169.10	177.1	1.05	181.5	1.07	202.4	1.20
CFRP-2	177.49	184.9	1.04	196.5	1.11	219.1	1.23
CFRP-3	207.74	184.9	0.89	196.5	0.95	219.1	1.05
BFRP-1	102.96	115.0	1.12	132.9	1.29	148.2	1.44
BFRP-2	120.05	119.9	1.00	143.9	1.20	160.5	1.34
BFRP-3	144.34	119.9	0.83	143.9	1.00	160.5	1.11

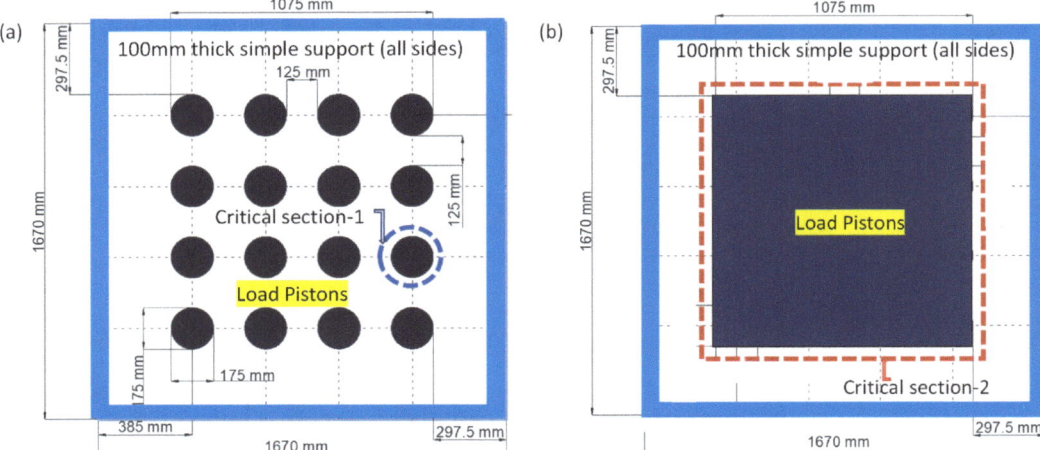

Figure 11. Critical sections to calculate the punching shear force capacity of the slabs. (**a**) Critical Section 1 around individual loading piston; (**b**) Critical Section 2 around total loading piston.

According to ACI 440.1R-15 [1], the CFRP-3, BFRP-2, and BFRP-3 slabs were not safe against punching shear as V_{pred}/V_{exp} was less than 1.00. Although the CFRP-1, CFRP-2, and BFRP-1 slabs showed V_{pred}/V_{exp} ratios slightly higher than 1.00, they were vulnerable to punching shear or were likely to fail due to punching shear.

In contrast, based on the punching shear prediction model proposed by El-Gamal et al., [35], all FRP-RC slabs except CFRP-3 and BFRP-3 were found to be safe against the punching shear. Furthermore, all FRP-RC slabs were found resilient against the punching shear when the prediction model proposed by Metwally [6] was considered. Compared to El-Gamal et al. [35] and Metwally [6], ACI 440.1R-15 [1] underestimates the two-way shear capacity of the FRP-RC slab.

In summary, the punching shear capacity of an RC slab highly depends on the effective depth and compressive strength of concrete. If the effective depth of a slab is compromised due to the formation of cracks, the punching shear capacity is significantly reduced. As the FRP-RC slabs experienced increasingly higher cracking than the steel-RC slab while loading, the effective depths of the slabs were significantly reduced. Consequently, the FRP-RC slabs failed due to punching shear, while the steel-reinforced slab failed due to flexural bending moment.

3.11. Experimental Load and Internal Capacities of the Slabs

According to ACI 440.1R-15 [1] and ACI 318-14 [39], based on the internal resisting moments, punching shear capacities and one-way shear capacities, the load capacities of the RC slabs per square meter are summarised in Table 10. It is evident that the steel-RC slab was safe against shear forces but failed due to flexural bending moment. Nevertheless, all FRP-RC slabs were safe against bending moment and one-way shear but were critical against punching shear. Thus, this study recommends high strength concrete and/or additional shear reinforcement for FRP-RC slabs to mitigate the punching susceptibility of the slabs.

Table 10. Experimental failure load and internal capacities of the RC slabs.

Slab ID	Experimental Failure Load (P_u) of the Slabs in kN/m²	Internal Load Resistance (P_n) of the Slabs in kN/m² Based on			$\frac{P_{n-m}}{P_u}$	$\frac{P_{n-ps}}{P_u}$	$\frac{P_{n-s}}{P_u}$
		Flexural Moment (P_{n-m})	Punching Shear Capacity (P_{n-ps})	One Way Shear Capacity (P_{n-s})			
Steel	90.27	83.96	97.41	122.30	0.93	1.08	1.35
CFRP-1	68.60	111.34	71.85	124.45	1.62	1.05	1.81
CFRP-2	72.01	122.84	75.01	134.50	1.71	1.04	1.87
CFRP-3	84.26	122.84	75.01	134.50	1.46	0.89	1.60
BFRP-1	41.79	74.49	46.66	124.45	1.78	1.12	2.98
BFRP-2	48.72	81.70	48.64	134.50	1.68	1.00	2.76
BFRP-3	58.54	81.70	48.64	134.50	1.40	0.83	2.30

4. Summary and Conclusions

This study demonstrates structural performance of two-way concrete slabs reinforced with CFRP, BFRP, and conventional steel rebars. The load versus mid-span deflection behaviour, cracked and uncracked stiffness, serviceability, flexural moment capacity, and punching and one-way shear capacity of the RC slabs were investigated and considered as indicators of structural performance. Based on the experimental and analytical investigation, the following essential conclusion can be drawn:

- Compared to the typical steel-RC slab, both the CFRP- and BFRP-RC slabs experienced significantly higher deflection and cracking while loading. However, the performance of the FRP-RC slabs was comparable to that of the steel-RC slab within the serviceability limit.
- Although the axial rigidity of CFRP and BFRP rebars are 41% and 16% of that of steel rebar, the CFRP- and BFRP-RC slabs exhibited 64% and 48% the stiffness of the steel-RC slab after cracking.
- Both the CFRP- and BFRP-RC slabs showed significant elastic recovery during unloading, which was not the case in the steel-RC slab. This indicates that the FRP rebars did not reach their rupture strain, as CFRP and BFRP rebars reached 71% and 78% of their rupture strength at failure, respectively.
- Beyond the peak load, the BFRP-RC slabs experienced a gradual decrease in the load capacity with incremental displacement, whereas the CFRP-RC slabs underwent a sharp decrease in load capacity, similar to the steel-RC slab. Consequently, the CFRP- and BFRP-RC slabs exhibited 1.26- and 2.18-times higher displacement-ductility than that of the steel-RC slab. The BFRP-RC slabs demonstrated 1.72-times higher ductility than CFRP-RC slabs because the percentage of elongation of BFRP rebars was higher than that of CFRP rebars.
- The steel-RC slab failed due to flexural tension, and the FRP-RC slabs failed due to punching shear. As the FRP-RC slabs experienced significantly higher cracks, the shear capacity of the slabs dropped gradually with the increase of loading. These cracks resulted in reducing the effective depth of the section. Thus, the FRP-RC slabs failed due to punching shear without any rupture of the FRP rebars.
- Since the design of the FRP-RC flexural member is governed by serviceability criteria, and the performance of CFRP- and BFRP-RC slabs is comparable with that of the steel-RC slab, CFRP- and BFRP both are suitable to reinforce concrete slabs.

Although FRP rebars are more expensive than steel rebars, they can be an alternative to typical steel rebars to reinforce concrete slabs where corrosion resistance is a concern. Therefore, the benefit of using FRP rebars in concrete is to improve durability. This study demonstrated that CFRP and BFRP rebars both can be an alternative to steel rebars as their performance within the serviceability limit is satisfactory and comparable to steel rebars. Additionally, BFRP is cheaper than CFRP rebar and has a competitive price. Hence, BFRP bars can be a choice to reinforce structural concrete where durability and/or higher strength-to-weight ratio is desired. However, additional shear reinforcement and/or higher strength concrete is recommended to improve the punching shear capacity of the BFRP-RC slabs.

Author Contributions: Conceptualization, M.S. and E.O.G.; methodology, M.S. and E.O.G.; software, M.S. and S.K.S.; validation, S.K.S., M.S. and E.O.G.; formal analysis, S.K.S.; investigation, S.K.S. and M.S.; resources, M.S., E.O.G. and R.A.-A.; data curation, E.O.G.; writing—original draft preparation, S.K.S.; writing—review and editing, M.S. and R.A.-A.; visualization, S.K.S.; supervision, M.S.; project administration, E.O.G.; funding acquisition, M.S., E.O.G. and R.A.-A. All authors have read and agreed to the published version of the manuscript.

Funding: This project was funded by internal research grant of School of Engineering at Deakin University.

Institutional Review Board Statement: Not applicable.

Informed Consent Statement: Not applicable.

Conflicts of Interest: The authors declare no conflict of interest.

References

1. ACI PRC-440.1-15; Guide for the Design and Construction of Structural Concrete Reinforced with Fiber-Reinforced Polymer Bars. American Concrete Institute: Farmington Hills, MI, USA, 2015.
2. Wang, X.; Zhang, X.; Ding, L.; Tang, J.; Wu, Z. Punching shear behavior of two-way coral-reef-sand concrete slab reinforced with BFRP composites. *Constr. Build. Mater.* **2020**, *231*, 117113. [CrossRef]
3. Shill, S.K.; Al-Deen, S.; Ashraf, M.; Hossain, M.M. Residual properties of conventional concrete repetitively exposed to high thermal shocks and hydrocarbon fluids. *Constr. Build. Mater.* **2020**, *252*, 119072. [CrossRef]
4. Mirza, O.; Shill, S.K.; Johnston, J. Performance of Precast Prestressed Steel-Concrete Composite Panels Under Static Loadings to Replace the Timber Transoms for Railway Bridge. *Structures* **2019**, *19*, 30–40. [CrossRef]
5. Shill, S.K.; Al-Deen, S.; Ashraf, M.; Elahi, M.A.; Subhani, M.; Hutchison, W. A comparative study on the performance of cementitious composites resilient to airfield conditions. *Constr. Build. Mater.* **2021**, *282*, 122709. [CrossRef]
6. Metwally, I.M. Prediction of punching shear capacities of two-way concrete slabs reinforced with FRP bars. *HBRC J.* **2013**, *9*, 125–133. [CrossRef]
7. Aljazaeri, Z.; Alghazali, H.H.; Myers, J.J. Effectiveness of Using Carbon Fiber Grid Systems in Reinforced Two-Way Concrete Slab System. *ACI Struct. J.* **2020**, *117*, 81–89. [CrossRef]
8. El-Gamal, S.; El-Salakawy, E.; Benmokrane, B. Behavior of Concrete Bridge Deck Slabs Reinforced with Fiber-Reinforced Polymer Bars Under Concentrated Loads. *ACI Struct. J.* **2005**, *102*, 727. [CrossRef]
9. Yost, J.R.; Goodspeed, C.H.; Schmeckpeper, E.R. Flexural Performance of Concrete Beams Reinforced with FRP Grids. *J. Compos. Constr.* **2001**, *5*, 18–25. [CrossRef]
10. Mahroug, M.; Ashour, A.; Lam, D. Tests of continuous concrete slabs reinforced with carbon fibre reinforced polymer bars. *Compos. Part B Eng.* **2014**, *66*, 348–357. [CrossRef]
11. Banthia, N.; Al-Asaly, M.; Ma, S. Behavior of Concrete Slabs Reinforced with Fiber-Reinforced Plastic Grid. *J. Mater. Civ. Eng.* **1995**, *7*, 252–257. [CrossRef]
12. Cai, J.; Pan, J.; Zhou, X. Flexural behavior of basalt FRP reinforced ECC and concrete beams. *Constr. Build. Mater.* **2017**, *142*, 423–430. [CrossRef]
13. Fang, H.; Xu, X.; Liu, W.; Qi, Y.; Bai, Y.; Zhang, B.; Hui, D. Flexural behavior of composite concrete slabs reinforced by FRP grid facesheets. *Compos. Part B Eng.* **2016**, *92*, 46–62. [CrossRef]
14. Zhang, B.; Masmoudi, R.; Benmokrane, B. Behaviour of one-way concrete slabs reinforced with CFRP grid reinforcements. *Constr. Build. Mater.* **2004**, *18*, 625–635. [CrossRef]
15. Rahman, A.H.; Kingsley, C.Y.; Kobayashi, K. Service and Ultimate Load Behavior of Bridge Deck Reinforced with Carbon FRP Grid. *J. Compos. Constr.* **2000**, *4*, 16–23. [CrossRef]
16. Michaluk, C.R.; Rizkalla, S.H.; Tadros, G.; Benmokrane, B. Flexural behavior of one-way concrete slabs reinforced by fiber reinforced plastic reinforcements. *ACI Struct. J.* **1998**, *95*, 353–365.
17. Erfan, A.M.; Elnaby, R.M.A.; Badr, A.A.; El-Sayed, T.A. Flexural behavior of HSC one way slabs reinforced with basalt FRP bars. *Case Stud. Constr. Mater.* **2021**, *14*, e00513. [CrossRef]
18. Mahroug, M.E.M. Behaviour of Continuous Concrete Slabs Reinforced with FRP Bars. Experimental and Computational Investigations on the Use of Basalt and Carbon Fibre Reinforced Polymer Bars in Continuous Concrete Slabs. Ph.D. Thesis, University of Bradford, Bradford, UK, 2014.
19. El-Salakawy, E.; Benmokrane, B. Serviceability of concrete bridge deck slabs reinforced with FRP composite bars. *ACI Struct. J.* **2004**, *101*, 727–736.
20. Taerwe, L. *Non-Metallic (FRP) Reinforcement for Concrete Structures: Proceedings of the Second International RILEM Symposium*; CRC Press: Boca Raton, FL, USA, 2014.
21. Zhang, Q. Behaviour of Two-Way Slabs Reinforced with Cfrp Bars. Ph.D. Thesis, Memorial University of Newfoundland, St. John's, NL, Canada, 2006.
22. Rashid, M.I. *Concrete Slabs Reinforced with GFRP Bars*; Memorial University of Newfoundland: St. John's, NL, Canada, 2004.

23. Karayannis, C.G.; Kosmidou, P.-M.K.; Chalioris, C.E. Reinforced Concrete Beams with Carbon-Fiber-Reinforced Polymer Bars—Experimental Study. *Fibers* **2018**, *6*, 99. [CrossRef]
24. Bilotta, A.; Compagnone, A.; Esposito, L.; Nigro, E. Structural behaviour of FRP reinforced concrete slabs in fire. *Eng. Struct.* **2020**, *221*, 111058. [CrossRef]
25. Amran, Y.H.M.; Alyousef, R.; Rashid, R.S.M.; Alabduljabbar, H.; Hung, C.-C. Properties and applications of FRP in strengthening RC structures: A review. *Structures* **2018**, *16*, 208–238. [CrossRef]
26. Huang, Z.; Zhao, Y.; Zhang, J.; Wu, Y. Punching shear behaviour of concrete slabs reinforced with CFRP grids. *Structures* **2020**, *26*, 617–625. [CrossRef]
27. Yu, X.; Zhou, B.; Hu, F.; Zhang, Y.; Xu, X.; Fan, C.; Zhang, W.; Jiang, H.; Liu, P. Experimental investigation of basalt fiber-reinforced polymer (BFRP) bar reinforced concrete slabs under contact explosions. *Int. J. Impact Eng.* **2020**, *144*, 103632. [CrossRef]
28. Hassan, M.; Benmokrane, B.; ElSafty, A.; Fam, A. Bond durability of basalt-fiber-reinforced-polymer (BFRP) bars embedded in concrete in aggressive environments. *Compos. Part B Eng.* **2016**, *106*, 262–272. [CrossRef]
29. Adhikari, S. Mechanical Properties and Flexural Applications of Basalt Fiber Reinforced Polymer (BFRP) Bars. Ph.D. Thesis, University of Akron, Akron, OH, USA, 2009.
30. Sun, X.; Gao, C.; Wang, H. Bond performance between BFRP bars and 3D printed concrete. *Constr. Build. Mater.* **2021**, *269*, 121325. [CrossRef]
31. Balea, L.; Dusserre, G.; Bernhart, G. Mechanical behaviour of plain-knit reinforced injected composites: Effect of inlay yarns and fibre type. *Compos. Part B Eng.* **2014**, *56*, 20–29. [CrossRef]
32. Deák, T.; Czigány, T. Chemical composition and mechanical properties of basalt and glass fibers: A comparison. *Text. Res. J.* **2009**, *79*, 645–651. [CrossRef]
33. Attia, K.; Alnahhal, W.; Elrefai, A.; Rihan, Y. Flexural behavior of basalt fiber-reinforced concrete slab strips reinforced with BFRP and GFRP bars. *Compos. Struct.* **2019**, *211*, 1–12. [CrossRef]
34. Nanni, A. Flexural Behavior and Design of RC Members Using FRP Reinforcement. *J. Struct. Eng.* **1993**, *119*, 3344–3359. [CrossRef]
35. El-Gamal, S.; El-Salakawy, E.; Benmokrane, B. A new punching shear equation for two-way concrete slabs reinforced with FRP bars. *ACI Spec. Publ.* **2005**, *230*, 877–894.
36. Goonewardena, J.; Ghabraie, K.; Subhani, M. Flexural Performance of FRP-Reinforced Geopolymer Concrete Beam. *J. Compos. Sci.* **2020**, *4*, 187. [CrossRef]
37. Provis, J.L.; Rose, V.; Bernal, S.A.; van Deventer, J.S.J. High-Resolution Nanoprobe X-ray Fluorescence Characterization of Heterogeneous Calcium and Heavy Metal Distributions in Alkali-Activated Fly Ash. *Langmuir* **2009**, *25*, 11897–11904. [CrossRef] [PubMed]
38. Bischoff, P.; Gross, S.; Ospina, C. The story behind proposed changes to ACI 440 deflection requirements for FRP-reinforced concrete. *Spec. Publ.* **2009**, *264*, 53–76.
39. *ACI CODE-318-14: Building Code Requirements for Structural Concrete and Commentary*; American Concrete Institute: Farmington Hills, MI, USA, 2014.
40. Bischoff, P.H. Reevaluation of Deflection Prediction for Concrete Beams Reinforced with Steel and Fiber Reinforced Polymer Bars. *J. Struct. Eng.* **2005**, *131*, 752–767. [CrossRef]
41. Rakhshanimehr, M.; Esfahani, M.R.; Kianoush, M.R.; Mohammadzadeh, B.A.; Mousavi, S.R. Flexural ductility of reinforced concrete beams with lap-spliced bars. *Can. J. Civ. Eng.* **2014**, *41*, 594–604. [CrossRef]
42. Benmokrane, B.; Chaallal, O.; Masmoudi, R. Glass fibre reinforced plastic (GFRP) rebars for concrete structures. *Constr. Build. Mater.* **1995**, *9*, 353–364. [CrossRef]
43. Ospina, C.E.; Alexander, S.D.B.; Cheng, J.J.R. Punching of Two-Way Concrete Slabs with Fiber-Reinforced Polymer Reinforcing Bars or Grids. *Struct. J.* **2003**, *100*, 589–598.
44. Bischoff, P.H.; Scanlon, A. Effective Moment of Inertia for Calculating Deflections of Concrete Members Containing Steel Reinforcement and Fiber-Reinforced Polymer Reinforcement. *ACI Struct. J.* **2007**, *104*, 68. [CrossRef]

Article

Modeling Flexural Failure in Carbon-Fiber-Reinforced Polymer Composites

Thiago de Sousa Burgani [1], Seyedhamidreza Alaie [2] and Mehran Tehrani [1,*]

[1] Walker Department of Mechanical Engineering, The University of Texas at Austin, Austin, TX 78712, USA; thiagoburgani@gmail.com
[2] Department of Mechanical and Aerospace Engineering, New Mexico State University, Las Cruces, NM 88003, USA; alaie@nmsu.edu
* Correspondence: tehrani@utexas.edu

Abstract: Flexural testing provides a rapid and straightforward assessment of fiber-reinforced composites' performance. In many high-strength composites, flexural strength is higher than compressive strength. A finite-element model was developed to better understand this improvement in load-bearing capability and to predict the flexural strength of three different carbon-fiber-reinforced polymer composite systems. The model is validated against publicly available experimental data and verified using theory. Different failure criteria are evaluated with respect to their ability to predict the strength of composites under flexural loading. The Tsai–Wu criterion best explains the experimental data. An expansion in compressive stress limit for all three systems was observed and is explained by the compression from the loading roller and Poisson's effects.

Keywords: carbon fiber; fiber-reinforced polymer composite; flexural strength; modeling; failure mode; finite-element analysis

Citation: Burgani, T.d.S.; Alaie, S.; Tehrani, M. Modeling Flexural Failure in Carbon-Fiber-Reinforced Polymer Composites. *J. Compos. Sci.* 2022, 6, 33. https://doi.org/10.3390/jcs6020033

Academic Editor: Jiadeng Zhu

Received: 23 December 2021
Accepted: 11 January 2022
Published: 19 January 2022

Publisher's Note: MDPI stays neutral with regard to jurisdictional claims in published maps and institutional affiliations.

Copyright: © 2022 by the authors. Licensee MDPI, Basel, Switzerland. This article is an open access article distributed under the terms and conditions of the Creative Commons Attribution (CC BY) license (https://creativecommons.org/licenses/by/4.0/).

1. Introduction

With the desire for high strength-to-weight ratio materials in aerospace, automotive, medical, and energy sectors, carbon-fiber-reinforced polymer composites (CFRP) have been attracting growing attention as a lightweight and strong material. Extensive composite material characterization ensures safety compliance and provides critical design data in the listed industries. Modeling the elastic behavior of orthotropic materials with transverse isotropy usually requires the characterization of five independent elastic constants, namely, $E_1, E_2, G_{12}, \nu_{12}$ and G_{23} (or ν_{23}), which represent Young's modulus in the fiber direction, Young's modulus perpendicular to the fiber direction, the in-plane shear modulus, the in-plane Poisson's ratio, the out-of-plane shear modulus, and the out-of-plane Poisson's ratio, respectively [1]. However, there are few significant mathematical relations for material strength resulting in numerous resource-intensive destructive tests, often providing single failure mode results. Accordingly, it is advantageous to study composite behavior under combined loading, and their dominating failure modes.

A relatively inexpensive and straightforward test for composites is the ASTM D7264, utilized for obtaining flexural properties of polymer-reinforced composites. Though seldom used for design, flexural tests are often used for quality control in composites. Flexural testing requires a simple rectangular sample and uses a three-point bending setup to measure the flexural response of specimens. The dominating flexural failure modes are compression at the top ply (either fiber microbuckling or ply-level buckling) and tension at the bottom-most ply [2]. The flexural strength of different CFRP systems often surpasses their compressive strength (F_{1c}). Flexural strength lies between the fiber direction compressive and tensile strengths in the CFRP systems investigated here [3–5].

Through finite-element analysis (FEA), this study investigates stress interactions occurring in unidirectional, 20 ply CFRP samples loaded through a three-point flexural test, and predicts the flexural strength of each system with reasonable accuracy. Three CFRP systems of interest are investigated where their flexural strengths vary from their compressive strength, an average of the tensile and comprehensive strengths, to their tensile strength. Different failure criteria were evaluated with respect to their ability to predict the flexural strength of the three CFRP systems.

2. Materials and Methods

2.1. ASTM D7264

The evaluated test in this study is Procedure A (3-point bending) of ASTM D7264—Standard Test Method for Flexural Properties of Polymer Matrix Composite Materials [2]. A diagram of the test is shown below in Figure 1. ASTM-recommended dimensions were utilized as shown in Table 1 below.

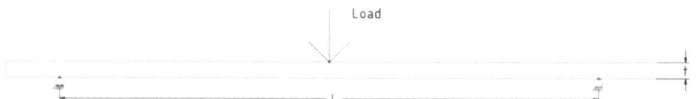

Figure 1. Three-point bending test diagram.

Table 1. ASTM D7264 parameters used in finite-element model.

Parameter	Span Length L (mm)	Thickness t (mm)	Width—Into the Page (mm)	Additional Overhang (%)	Loading Nose Radius (mm)
Value	128	4	13	20	5

The width and the loading/support rollers are not pictured. The resulting span-to-thickness ratio was 32:1, limiting out-of-plane shear deformations such as those experienced by short-beam test ASTM D2344. In the 3-point configuration, the expected failure was directly under the loading nose, as buckling (compression) or through the laminate cracking, originating due to tensile cracking at the bottom-most ply [2].

The chosen stacking sequence was $[0]_{10s}$ to maintain the neutral (bending) axis at the center of the laminate and allow for the application of homogeneous beam theory to estimate flexural strength. Twenty plies were chosen to maintain an averaged ply thickness of 0.2 mm across systems, maintaining 4 mm overall thickness.

2.2. CFRP Systems

Three CFRPs were chosen on the basis of their flexural strength characteristics. Though thermoset composites still dominate the aerospace industry, there has been a recent push to incorporate thermoplastics in the industry [6]. Thermoplastic composites offer superior operational temperatures, weldability, and recyclability in comparison with thermoset composites [6,7]. Toray T700/TC1225 LM-PAEK exhibits flexural strength close to its compressive strength in the fiber direction [3]. Solvay APC-2 PEEK exhibits flexural strength near its tensile strength in the fiber direction [4]. Hexcel's AS4—8552 epoxy exhibits flexural strength approximately halfway between its tensile and compressive strengths [5]. PEEK and LM-PAEK are thermoplastic-based systems, and 8552 is a thermosetting matrix. Toray and Hexcel material properties were obtained from the Wichita State NIAR database [8]. Table 2 summarizes their strength and strain values. Comprehensive test data are publicly available for the above systems (room temperature and dry conditions) [3–5].

Table 2. Strengths and strains of three selected CFRP systems.

Composite System (Fiber/Resin)	Tensile Strength (MPa)	Compressive Strength (MPa)	Flexural Strength (MPa)	Ultimate Tensile Strain (%)	Ultimate Compressive Strain (%)
T700/Toray LM PAEK	2322	1226	1455	1.86	0.98
APC2/PEEK (Solvay)	2070	1360	2000	1.45	1.10
AS4/Hexcel 8552	2205	1530	1889	1.56	1.09

All three systems are present in the aerospace industry to varying degrees, contributing to primary and secondary aircraft structures such as fuselage panels and aircraft pressure bulkheads [4,9,10].

2.3. Failure Criteria

Different failure criteria were chosen to interface with the FEA to study the failure mechanisms observed in the tests. These criteria can be divided into three categories: noninteractive (limit theories), partially interactive, and fully interactive theories. These theories, their respective equations or criterion, and their categories are listed in Table 3.

Table 3. Failure criteria chosen for analysis in this work [1,11].

Theory	Category	Failure Criterion
Maximal stress	Noninteractive	$\sigma_{ij} < F_{ij}$
Maximal strain	Noninteractive	$\varepsilon_{ij} < \varepsilon_{ij}^u$
Hashin–Rotem	Partially interactive	$\frac{\|\sigma_1\|}{F_1} = 1$ $\left(\frac{\sigma_2}{F_2}\right)^2 + \left(\frac{\tau_4}{F_4}\right)^2 + \left(\frac{\tau_6}{F_6}\right)^2 = 1$ $\left(\frac{\sigma_3}{F_3}\right)^2 + \left(\frac{\tau_4}{F_4}\right)^2 + \left(\frac{\tau_5}{F_5}\right)^2 = 1$
Tsai–Wu	Fully interactive	$f_1\sigma_1 + f_2(\sigma_2 + \sigma_3) + f_{11}\sigma_1^2 + f_{22}(\sigma_2^2 + \sigma_3^2)$ $+ f_{44}\tau_4^2$ $+ f_{66}(\tau_5^2 + \tau_6^2) + 2f_{12}(\sigma_1\sigma_2 + \sigma_1\sigma_3) + 2f_{23}\sigma_2\sigma_3$ $= 1$
Hoffman	Fully interactive	Same as Tsai—Wu, coefficient definitions differ.
Tsai–Hill	Fully interactive	$\frac{\sigma_1^2 - \sigma_1\sigma_2 - \sigma_1\sigma_3}{F_1^2} + \frac{\sigma_2^2 + \sigma_3^2 - \sigma_2\sigma_3}{F_2^2} + \frac{\tau_4^2}{F_4^2}$ $+ \frac{\tau_5^2 + \tau_6^2}{F_6^2} = 1$

Limit theories have no explicit interaction between stresses or strains, i.e., failure is predicted when a principal stress or strain reaches its limit. These criteria are expected to have the largest error in predicting flexural strength. Partially interactive criteria consider stress interactions, but have different criteria of failure. For example, Hashin–Rotem have failure coefficients for fiber failure, matrix failure, and delamination. Whichever coefficient is the largest of the three is considered to be the failure mode. Lastly, fully interactive criteria allow for stress interactions in 3D loading. A single failure coefficient is calculated from coupling terms and stresses, and no specific failure mode is predicted [1,12].

These criteria are selected because of their widely available documentation and ease of implementation in ANSYS ACP. The criteria mentioned above are also included in ACP for failure analysis. The inverse reserve factor is often used in place of the Tsai–Wu/Tsai–Hill criterion coefficient, as its calculation is identical when using the respective failure theories [12].

2.4. Prediction of Flexural Strength

To compute nominal flexural strength on the basis of maximal normal stress on either the top or bottom of a specimen, the following equation from the ASTM D7264 standard is employed:

$$\sigma_{flex} = \frac{3PL}{2bh^2} \quad (1)$$

where P is the load, L is the span length, b is the beam width, and h is the beam height. This relationship is valid when failure occurs in a brittle manner, the laminate acts as a homogeneous beam, and with a maximal fiber strain of 2% [2]. Here, P is directly calculated using finite-element analysis (ANSYS), while the other parameters are as explained in Section 2.1. Analysis employed a nonlinear elastic model with large deformations resulting in a maximal strain of under 1.5%, as no progressive damage was considered. Equation (1) was utilized to predict the flexural strength for the FEA.

2.5. Finite-Element Model

2.5.1. Geometry

A quarter symmetry model of the specimen is considered. In this model, thickness remained the same, while width and length were divided in half. Similarly, rollers were cut in half about the width. The rollers were also modeled to consider the effects of contact stresses under the loading nose. The symmetry allows for reducing the computational cost in the quarter-model geometry (Figure 2).

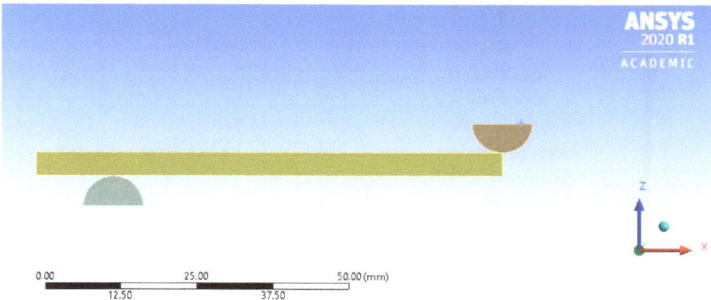

Figure 2. Quarter symmetry model geometry.

2.5.2. Domain Discretization (Meshing)

The pre-ACP mesh consists of two separate meshes, the sample mesh and the roller mesh. The sample must be imported as a surface mesh because of ACP requirements. Two-dimensional (2D) Quad4 and Tri3 elements are utilized for the surface mesh, with finer mesh near the loading (0.5 mm element size) and coarser away from the loading (1 mm element size), resulting in a mesh of 748 elements and 846 nodes. With ACP preprocessing, the sizes are maintained, but elements become Hex8 and Wed6 elements. More elements and nodes are used to capture the stress interactions with reasonable accuracy near the loading nose where failure is expected. To improve contacts between the rollers and the sample, equal-size elements are applied to both (i.e., support roller has a 1 mm element size, and loading has a 0.5 mm size). Figure 3 shows the resultant mesh with the rollers having Hex8 solid elements.

Additional preprocessing was performed in ANSYS ACP. A solid mesh was generated utilizing the surface mesh for element sizing. There were 20 elements through the model's thickness, one per ply, resulting in 14,960 elements and 17,766 nodes in the quarter symmetry model. The complete mesh is shown in Figure 3.

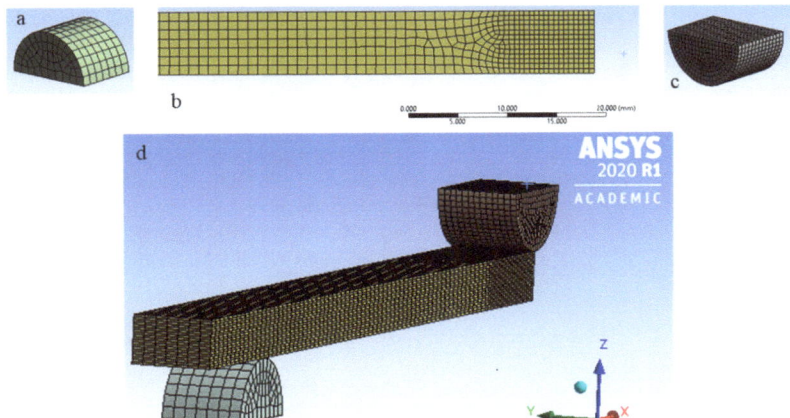

Figure 3. FEA mesh of (**a**) support roller, (**b**) sample, and (**c**) loading roller; (**d**) complete mesh with sample and rollers.

2.5.3. Boundary Conditions and Contact Settings

Contacts were formulated between the sample and roller supports. Additionally, the sample surface was split to limit the contact region accessible to the loading roller, improving computational efficiency, as shown in Figure 4. A frictionless contact was assigned at the support roller, while a no separation contact was assigned to the loading roller. Both utilized a friction coefficient of 0, but the frictionless support was considered to be nonlinear, as it allows for complete separation between the solids to occur. Some slipping can occur while the sample is loaded. However, there must be no separation between the roller and sample on the loading nose for a valid test. The assigned boundary conditions are as shown in Figure 4.

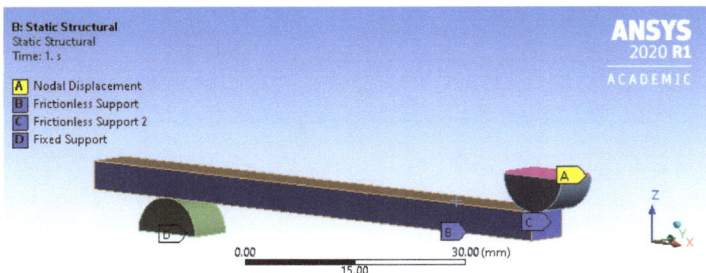

Figure 4. Contact boundary conditions that were assigned in the model.

The symmetry condition imposes the need for frictionless supports at the "cut" surfaces. Frictionless support constraints motion normal to the plane while allowing for motion in the two other principal directions (e.g., at the surface located in the XZ plane, movement in the Y, and rotations about X and Z are constrained). A fixed support was applied at the bottom of the support roller constraining all degrees of freedom at the roller. Lastly, direct nodal displacement was applied at the top of the loading roller. Similar to the fixed support, all degrees of freedom for the roller were constrained with the addition of an imposed nodal displacement in the negative Z direction. The value for this displacement is dynamically changed, such that the sample is loaded just until failure.

With the above boundary conditions and contacts, the model was constrained in certain degrees of freedom, allowing for the proper capture of the physics involved in the 3-point flexural (bending) test.

2.5.4. Solver and Postprocessing

The mechanical APDL solver was used in this study. An iterative solver was chosen given the nonlinearity caused by the contact settings. Large deformations were not considered in the calculation of strains (e.g., infinite small strain tensor), and linear-elastic material property (Hooke's Law) was employed. This was assumed because high-strength CFRP systems show brittle failure [13]. The model, due to the static structural formulation, only considered first-ply failure (FPF); progressive damage was not considered. It was expected that predicted strengths would be underestimated but still be within reasonable accuracy of reported experimental results. The failure was considered to be brittle due to low ultimate strains, corroborated by reported data [3–5].

Postprocessing was performed by both ANSYS Mechanical and ACP Post to estimate stresses across the thickness at the failure point and a point further away from loading. The latter point was used to obtain appropriate transverse shear stresses, as such stresses are irregular near loading, likely due to crushing effects [14].

2.6. Model Verification with Shear Stresses in Beam Bending

The model was verified by observing trends (e.g., compressive stress at the top ply, tensile stress at the bottom ply, as discussed in ASTM D7264) and employing analytical formulas in beam bending in the small strain regime. Beam theory predicts a shear profile as shown in Equation (2), for a point loaded beam with simple supports [15]:

$$\tau = \frac{3V}{2bh}\left(1 - \left(\frac{2y}{h}\right)^2\right) \tag{2}$$

where V, b, h, and y, are the shear force, beam width, beam height, and the distance from the neutral plane, respectively.

Using Bernoulli–Euler beam theory, the theoretical and numerical results were directly compared. Figure 5 shows good agreement between the FEA and the Bernoulli–Euler estimation. This agreement verifies the simulation. The slight difference can be attributed to the effect of shear deformations neglected in Bernoulli–Euler theory [16].

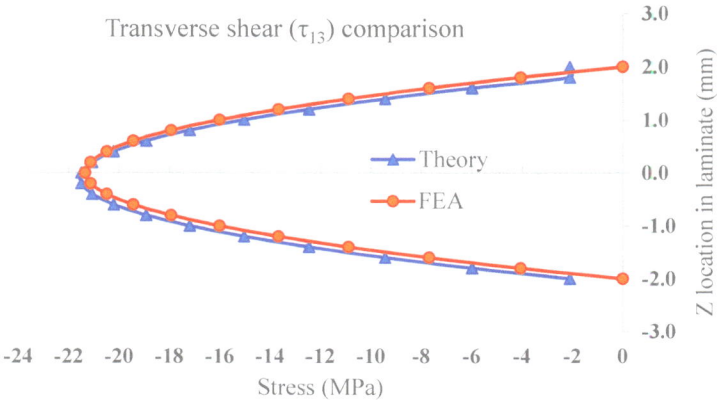

Figure 5. Comparison of FEA-obtained shear to Bernoulli–Euler beam theory.

3. Results

Experimental and FEA strengths for three systems are summarized in Table 4. The FEA results were compared with experimental values obtained from datasheets for tensile, compressive, and flexural strengths. The Tsai–Wu failure criterion was utilized to determine stress at the onset of flexural failure. As mentioned earlier, all numerically obtained strengths were conservative due to nonprogressive damage analysis. The flexural specimen likely did not completely fail even if the Tsai–Wu index reached a value of 1 for a single finite element.

Table 4. Summarized results for three chosen CFRP systems [3–5]. Values in red indicate nearest proximity from flexural strength to either tensile or compressive strength of the material in the fiber direction.

Composite System (Fiber/Resin)	Tensile Strength (MPa)	Compressive Strength (MPa)	Flexural Strength (MPa)	FEA-Flexural Strength (MPa)	% Error between FEA and Experiment
T700/Toray LM PAEK	2322	1226	1455	1367	−6.1
AS4/Hexcel 8552 Epoxy	2205	1530	1889	1717	−9.1
APC2/Solvay PEEK	2070	1360	2000	1725	−13.8

Tsai–Wu failure indices through the thickness the three systems are plotted in Figure 6. Failure occurs for a Tsai–Wu index above 1. An interesting finding was the failure of the Hexcel system in tension at the bottom ply, as opposed to the compressive failure experienced by the other two CFRPs.

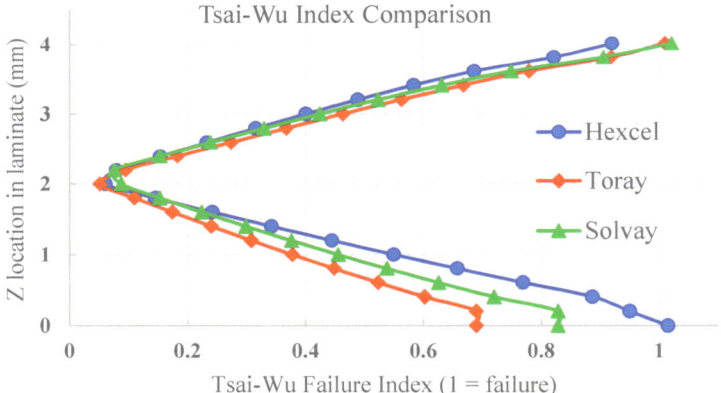

Figure 6. Plotted Tsai–Wu indices across thickness for three systems.

Stresses across the thickness near the loading location are plotted and shown in Figure 7. Only stresses for the Toray system were plotted, as all analyzed systems showed similar trends. Shear stresses τ_{23} and τ_{12} were not plotted, as their values were near zero throughout the thickness, having negligible contribution to laminate failure.

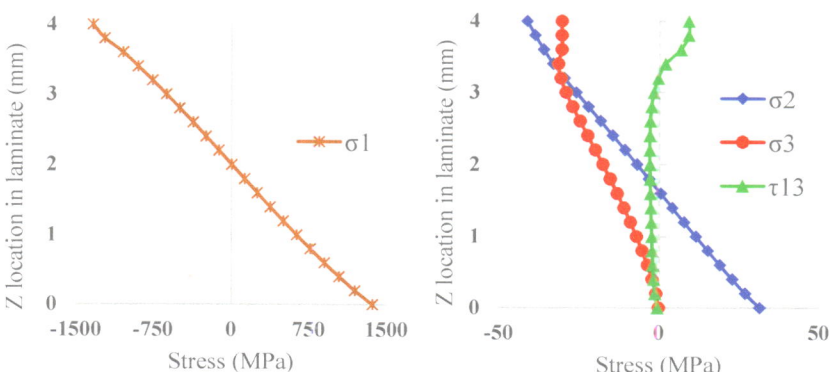

Figure 7. Normal and shear stresses across flexural specimen thickness.

4. Discussion

A key trend in all three systems was that their flexural strength was between their tensile and compressive strengths. This implies that flexural failure occurs because of combined modes instead of a single failure mode. Table 5 summarizes the maximal FEA compressive stresses at the top ply under the loading nose, compared with the reported limits, i.e., compressive strengths.

Table 5. FEA compressive stresses and compressive strengths for studied composite systems [3–5].

	Compressive Strength (MPa)	FEA Maximum Compressive Stress (MPa)	Relative Difference (%)
Toray T700/LM PAEK	1226	1338	9.2
Solvay APC-2/PEEK	1360	1700	25
Hexcel AS4/8552 epoxy	1530	1690	10.5

The Tsai–Wu criterion showed the closest results to experimental data. With compressive interactions occurring at the top ply between σ_1 and σ_2 due to Poisson's effect, the effective stress limit was expanded [1]. Terms from the Tsai–Wu failure criterion are further explored in Table 6 for each analyzed system. Terms in bold font are those contributing to the enhanced load-bearing capacity of the samples.

Table 6. Tsai–Wu index components for three analyzed systems. Bold terms contribute to the enhanced load-bearing capacity of the systems.

	Toray T700/PAEK	Hexcel AS4/8552	Solvay APC2/PEEK
σ_1 (MPa)	-1338	1706	-1700
σ_2 (MPa)	-41	35	-50
τ_{12} (MPa)	0.6	-0.9	0.9
$F_1\sigma_1$	0.603	-0.325	0.440
$F_2\sigma_2$	**-0.225**	0.452	**-0.285**
$F_{11}\sigma_1^2$	0.687	0.947	1.061
$F_{22}\sigma_2^2$	0.135	0.156	0.145
$F_{66}\tau_{12}^2$	0.000	0.000	0.000
$-\sqrt{F_{11}F_{22}}\sigma_1\sigma_2$	-0.197	-0.225	**-0.347**
Total	1.003	1.005	1.014

The Hexcel system, which fails in a tensile combined mode, had the highest reported compressive strength across the considered composite systems. Daniel and Ishai showed that, for similar stress interactions in combined loading, compressive load-bearing capabilities are much better than the tensile capabilities for a similar AS4 system, which could explain the tensile failure obtained by the FEA with the Hexcel AS4 system [1].

Lastly, the flexural finite-element model underpredicted flexural strength across all systems. This could be due to static structural analysis and no progressive damage being considered. The model effectively predicts a first ply failure (FPF), while an ultimate ply failure (UPF) prediction would likely return more accurate results. Nonetheless, the FPF returned values with reasonable accuracy at a lower computational cost to a UPF model.

Error in the prediction of failure consistently increases as flexural strength approaches its tensile strength. This could stem from an increased need for a UPF model, as brittleness could be less pronounced from a tensile failure due to crack propagation. As the system progresses through cracking, large rotations may occur, which in turn can induce significant membrane resultant forces. Equation (1) may not be a valued estimation in such a case, and a more complex theory of beams would be needed (e.g., geometrically exact beam theory) [17]. Lastly, both Hexcel and Solvay systems did not have centralized test data containing all strengths necessary for utilizing the Tsai–Wu criterion and reported flexural strength. Though the resin contents were nearly identical, a possible source of error stems from using mixed test data.

Though the Tsai–Wu failure criterion was chosen due to its fully interactive nature, other failure criteria were also explored. Numerical trends were similar for all systems, and thus only the data for the Toray system are summarized in Table 7. The Tsai–Wu index is provided, and safety factors for other criteria and their respective predicted flexural strength are included.

Table 7. Summary of results from different failure criteria—Toray system.

Failure Criteria	FEA-Predicted Strength (MPa)	Error (%)
Tsai–Wu	1366.6	−6.06
Max strain	1257.3	−13.58
Max stress	1243.6	−14.52
Tsai–Hill	1243.6	−14.52
Hoffman	1352.9	−7.00
Hashin	1243.6	−14.52

The Tsai–Wu failure criterion predicted the strength most accurately from all failure criteria. Hoffman also predicted good results since it is fully interactive, with minor differences from the Tsai–Wu criterion. Partially interactive criterion Hashin had among the lowest predicted accuracy values, likely because of its lack of interaction between the fiber and transverse direction stresses. Noninteractive criteria (max stress, max strain) also underpredicted the flexural strength. Max strain exhibited marginally better performance, as its formulation allows for some Poisson effects in combined loading.

Tsai–Wu and Hoffman predicted failure at the bottom for the Hexcel system, while the other systems predicted a compressive failure occurring in the top ply. This likely stemmed from the lack of interaction and combined loading strength expansion.

5. Conclusions

Different failure criteria were evaluated for their ability to accurately predict failure stemming from combined stresses in a flexural test finite-element model. The model reasonably predicted flexural strength values for three different CFRP systems with different characteristics: one with flexural strength close to its tensile strength, another one with flexural strength close to its compressive strength, and the last one with flexural strength between its tensile and compressive strengths. There was an apparent discrepancy in the maximal compressive stress experienced by the three systems and their respective compressive strengths. This difference is explained by Poisson's effect, causing stress in the in-plane transverse direction. The Tsai–Wu criterion requires many coupling terms. Therefore, comprehensive test data are necessary to interface with the model, including out-of-plane strengths. Such tests can be challenging and costly to perform. Some strength estimations are available, though they introduce additional errors [1] (pp. 116,120–122). The FEM consistently underpredicted strength due to modeling the first ply failure instead of the ultimate ply failure. As progressive damage is not considered, strengths are underestimated for systems that have extensive crack propagation before failure.

Author Contributions: Conceptualization, M.T.; methodology, M.T., S.A.; formal analysis, T.d.S.B.; writing—original draft preparation, T.d.S.B.; writing—review and editing, S.A., M.T.; funding acquisition, M.T. All authors have read and agreed to the published version of the manuscript.

Funding: This research was funded by the Air Force Office of Scientific Research (AFOSR) under award no. FA9550-21-1-0066.

Data Availability Statement: Data will be available upon request.

Acknowledgments: The authors are grateful for the support by the Air Force Office of Scientific Research (AFOSR), under award no. FA9550-21-1-0066.

Conflicts of Interest: The authors declare no conflict of interest.

References

1. Daniel, I.M.; Ishai, O. *Engineering Mechanics of Composite Materials, 2nd ed*; Oxford University Press: New York, NY, USA, 2006.
2. *ASTM D7264/D7264M-07*; Standard Test Method for Flexural Properties of Polymer Matrix Composite Materials. ASTM: West Conshohocken, PA, USA, 2015.
3. Lian, E.; Lovingfoss, R.; Tanoto, V. Medium Toughness PAEK thermoplastics Toray TC1225 (LM PAEK) T700GC 12K T1E Unidirectional Tape Qualification Material Property Data Report. *Natl. Inst. Aviat. Res.* **2021**. Available online: https://www.wichita.edu/industry_and_defense/NIAR/Documents/TorayTC1225UnitapeCAM-RP-2019-036RevA5.10.2021MPDRFinal.pdf (accessed on 15 September 2021).
4. N.A. Solvay. *APC-2 PEEK Datasheet.* 2017. Available online: https://www.solvay.com/en/product/apc-2peeks2 (accessed on 10 February 2021).
5. Marlett, K.; Ng, Y.; Tomblin, J.; By, A.; Hooper, E. Hexcel 8552 AS4 Unidirectional Material Property Data Report. *Natl. Inst. Aviat. Res.* **2011**. Available online: https://www.wichita.edu/industry_and_defense/NIAR/Research/hexcel-8552/AS4-Unitape-2.pdf (accessed on 10 February 2021).
6. Favaloro, M. Thermoplastic Composites in Aerospace—The Future Looks Bright | CompositesWorld. Available online: https://www.compositesworld.com/articles/thermoplastic-composites-in-aerospace-past-present-and-future (accessed on 29 April 2021).
7. Offringa, A.R. Thermoplastic composites—Rapid processing applications. *Compos. Part A Appl. Sci. Manuf.* **1996**, *27 Pt A*, 329–336. [CrossRef]
8. National Institute for Aviation Research. Available online: https://www.wichita.edu/research/NIAR/ (accessed on 1 May 2021).
9. Gardiner, G. Integrating Antennas into Composite Aerostructures. 2021. Available online: https://www.compositesworld.com/articles/integrating-antennas-into-composite-aerostructures (accessed on 1 May 2021).
10. Francis, S. Thermoplastic Composites: Poised to Step Forward. 2019. Available online: https://www.compositesworld.com/articles/thermoplastic-composites-poised-to-step-forward (accessed on 1 May 2021).
11. Schellekens, J.C.J.; de Borst, R. The use of the Hoffman yield criterion in finite element analysis of anisotropic composites. *Comput. Struct.* **1990**, *37*, 1087–1096. [CrossRef]
12. N.A. ANSYS Inc. Failure Criteria for Reinforced Materials—ACP Documentation. 2021. Available online: https://ansyshelp.ansys.com/account/secured?returnurl=/Views/Secured/corp/v221/en/acp_ug/acp_failure_analysis.html (accessed on 15 September 2021).
13. Sung, M.; Jang, J.; Tran, V.L.; Hong, S.T.; Yu, W.R. Increased breaking strain of carbon fiber-reinforced plastic and steel hybrid laminate composites. *Compos. Struct.* **2020**, *235*, 111768. [CrossRef]
14. Fam, A.; Sharaf, T. Flexural performance of sandwich panels comprising polyurethane core and GFRP skins and ribs of various configurations. *Compos. Struct.* **2010**, *92*, 2927–2935. [CrossRef]
15. Budynas, R.G.; Nisbett, J.K. *Shigley's Mechanical Engineering Design*, 11th ed.; McGraw-Hill Education: New York, NY, USA, 2020; p. 119.
16. Sancaktar, E. Mechanics of solids and structures. *ASME Int. Mech. Eng. Congr. Expo. Proc.* **2019**, *10 Pt A*, 53–90.
17. Pai, P.F. Geometrically exact beam theory without Euler angles. *Int. J. Solids Struct.* **2011**, *48*, 3075–3090. [CrossRef]

Article

Damping Properties of Hybrid Composites Made from Carbon, Vectran, Aramid and Cellulose Fibers

Hauke Kröger, Stephan Mock, Christoph Greb * and Thomas Gries

Institut für Textiltechnik, RWTH Aachen University, 52074 Aachen, Germany; hauke.kroeger@ita.rwth-aachen.de (H.K.); stephan.mock@rwth-aachen.de (S.M.); Thomas.Gries@ita.rwth-aachen.de (T.G.)
* Correspondence: christoph.greb@ita.rwth-aachen.de

Abstract: Hybridization of carbon fiber composites can increase the material damping of composite parts. However, there is little research on a direct comparison of different fiber materials—particularly for carbon fiber intraply-hybrid composites. Hence, the mechanical- and damping properties of different carbon fiber intraply hybrids are analyzed in this paper. Quasi unidirectional fabrics made of carbon, aramid, Vectran and cellulose fibers are produced, and their mechanical properties are analyzed. The material tests show an increased material damping due to the use of Vectran and aramid fibers, with a simultaneous reduction in strength and stiffness.

Keywords: composite; hybrid composite; damping; intraply hybrid; carbon fiber

1. Introduction

Vibrations have a harmful effect on the human body. In the field of occupational safety, attention has been drawn to the topic for a long time. However, harmful vibrations also occur in recreational activities when using sports equipment. This also includes bicycles, the use of which can lead to potentially harmful vibrations for the user [1]. Carbon fiber reinforced composites are a popular material for the construction of bicycles due to their excellent lightweight properties. Vibrations are transmitted to the cyclists via composite components but can be reduced by increasing the material damping of the composites. Material damping causes a conversion of vibrational energy into (mostly) thermal energy by internal friction [2].

The material damping of composites is determined by the damping of the matrix and the fibers, as well as their arrangement, connection and interphase [3]. The matrix has a greater influence on the damping of the composite than the fibers [4]. Nonetheless, many applications do not allow for a change in matrix material or change of the interphase between fibers and matrix. An increase in damping can therefore only be achieved by the fibers. This requires a change in fiber type [3], fiber orientation, respectively, laminate construction [5–7] or fiber volume content [8].

In particular, the use of fiber types with high inherent self-damping offers great potential for increasing the material damping of carbon fiber composites. Yet, the fibers in question have too little stiffness or strength to produce a lightweight composite consisting entirely out of high damping fibers. This problem can be addressed by hybridization: ordinary reinforcing fibers (e.g., carbon fibers) are combined with fibers with high self-damping within a composite.

A hybrid composite consists of two or more different types of fibers. A distinction is made between hybridization on the laminate level (*interply*), on the roving level (*intraply*) or on the filament level (*intrayarn*), the schematic structure is shown in Figure 1 [9]. In addition, mixed forms of the above-mentioned hybridization types are possible. The aim of hybridization is to improve certain properties (e.g., increased damping) disproportionately to the loss of other properties (e.g., decreased stiffness) [10], which is referred to as positive synergy effects.

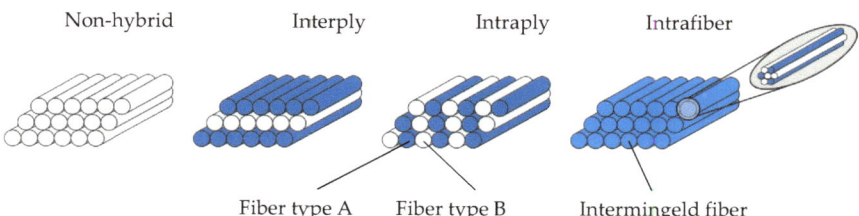

Figure 1. Hybridization types of composites in comparison.

A main characteristic of hybridization is the dispersion of the different fiber types; the higher the dispersion, the higher the synergy effects achieved [11]. The dispersion depends largely on the type of hybridization used, but can also be influenced by the fiber diameter or the ply thickness [12,13]. For a given lay-up and fiber diameter, the best dispersion is achieved by an intrafiber hybrid. Interply hybridization offers the lowest dispersion, while intraply is a good trade-off between Intrayarn and interply hybrids [10,14].

The main focus of research in the field of hybrid composites has been on the following areas: Increasing the elongation at break or creating a pseudo ductile fracture behavior [12,15,16], improving static strength [17] and fatigue strength [18–20], cost reduction through the integration of low-cost fiber types [21] and improvements in impact properties [14,22–24]. There is comparatively less research activity in the area of damping enhancement through hybridization [25]. In this context, the most research results were reported on investigations of the damping properties of natural fiber hybrid composites:

Ashtworth et al. demonstrated an increase in damping by using jute fibers in an interply hybrid with carbon fibers [26]. Assarar et al. and Guen et al. achieved an increase in damping using a flax-carbon interply hybrid [27,28]. Other authors showed an increase in damping by hybridizing glass fiber laminates with flax fibers [29,30] or kenaf fibers [31]. Although natural fibers allow an increase in material damping, they also have central deficits that make their use in high-performance composites—such as bicycles—unattractive. In particular, the inconsistent fiber properties and the absorption of the resin are major deficits [32]. Synthetic fibers with good damping properties could overcome the disadvantages of natural fibers. However, there is very little research activity on increasing the damping of hybrid (carbon fiber) composites using synthetic fibers. Especially the increase of material damping by intraply hybridization has been investigated very little. The increase of material damping by using certain synthetic fibers have been shown by different authors, although not within the context of hybrid composites: Studies showed an increase in damping by using aramid fibers [33] and cellulose fibers [34]. Liquid crystalline polymer (LCP) materials are also known for their good damping properties [35]. Therefor the scope of this work is to investigate the manufacturability and mechanical properties of intraply hybrid composites made of carbon and synthetic fibers. The aim is to use the novel hybrid composites as a material for the manufacture of bicycle components in order to reduce the vibration load on the human body.

2. Materials and Methods

2.1. Materials

Three different unidirectional hybrids and a non-hybrid reference made from carbon fibers were investigated. The investigations commenced with the production of quasi unidirectional fabrics, which were further processed into composites using the vacuum infusion process. An intraply level hybridization was chosen, as this allows a good compromise between dispersion and efficient and economical production. The fabrics were produced on a weaving machine. As shown in Figure 2 all reinforcing fibers were oriented in warp direction and connected in weft direction by a 10 tex polyester fiber.

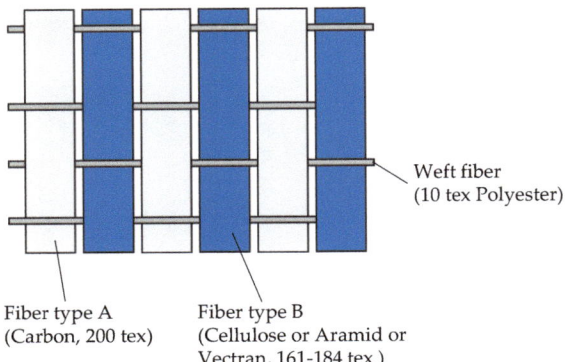

Figure 2. Construction of unidirectional hybrid textiles.

The weft thread only serves as a fixation of the warp threads and leads to very low ondulations, which are not comparable to the ondulations of a classical woven fabric. Therefore, the final product is a fabric with a unidirectional (UD) arrangement of the reinforcing fibers. The hybrid fabrics produced each consist of two types of fibers: on the one hand, carbon fibers to generate strength and stiffness and, on the other hand, functional fibers to generate the damping effect. The functional fibers investigated are aramid, cellulose and fibers made of liquid crystalline polymer (LCP) (Vectran®). The properties of the fibers are shown in Table 1.

Table 1. Properties of the fibers used.

Material	Manufacturer and Product Name	Density [g/cm^3]	Tensile Modulus [GPa]	Tensile Strength [MPa]
Carbon	Teijin HTS 45	1.76	245	4500
Aramid	Teijin Twaron® 2200	1.45	100	2240
LCP/Vectran	Kuraray Vectran® HTME	1.4	75	3200
Cellulose	Cordenka 700	1.5	14.4	778

The design of the UD-fabric was based on HTS 45 type carbon fibers with a fineness of 200 tex. The fineness of the functional fibers was selected in such a way that (ideally) a ratio of 50 volume% carbon fiber to 50 volume% functional fibers is achieved in the hybrid UD fabric. This ensures the best possible dispersion under the given boundary conditions (fineness of the carbon fibers). The properties of the UD fabrics are shown in Table 2.

The different fiber types were arranged in warp direction according to the following pattern: A, B, A, B, A, ... A plain weave was used, with one weft thread per centimeter of warp length. The low weft density resulted in an ondulation of the fabric in the weft direction. This does not influence the properties in fiber direction, respectively, warp direction—in warp direction the fibers are parallel and unidirectional. Figure 3 shows the UD-fabrics produced.

Table 2. Properties of the dry UD-fabrics.

Textile Name	Fiber A	Fiber B	Volume Content Fiber A [%]	Volume Content Fiber B [%]	Arial Weight [g/m^2]
CaCa	Carbon	Carbon	50	50	200
CaAr	Carbon	Aramid	50.4	49.6	180
CaVe	Carbon	Vectran	48.8	51.2	185
CaCe	Carbon	Cellulose	48.1	51.9	192

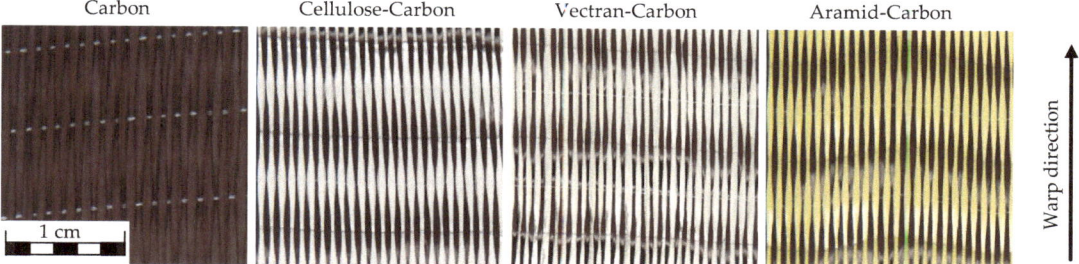

Figure 3. Hybrid and non-hybrid UD-fabrics produced.

The UD fabrics were then used to make composite panels using a vacuum infusion process. An epoxy thermoset resin of the type of Hexion RIM R426/RIM H35 was used. For the tensile and damping tests, plates were made from five layers of each hybrid fabric type, while the bending test specimens were made from nine layers of each UD fabric type. All layers were arranged in the 0° direction. After the infusion, the plates were tempered at 80 °C for two hours, followed by the final cutting of the test specimens. Table 3 shows the properties of the composite specimens for tensile and damping testing. The density was calculated from the specimen size and weight. The fiber volume content was calculated from the layup, the size of the composite plates and the densities of the individual components.

Table 3. Properties of specimens for tensile and damping testing.

Composite Name	Layup	Density [g/cm^3]	Thickness [mm]	Fiber Volume Content [Vol%]
Carbon	$5 \times [CaCa_{0°}]$	1.41	1.28	47
Aramid-Carbon	$5 \times [CaAr_{0°}]$	1.33	1.36	49
Vectran-Carbon	$5 \times [CaVe_{0°}]$	1.24	1.38	46
Cellulose-Carbon	$5 \times [CaCe_{0°}]$	1.35	1.35	46

For quality control of the manufacturing process, interlaminar shear strength (ILSS) tests were carried out, which showed a low interlaminar strength of samples made of Vectran-Carbon UD-fabrics. The Vectran-Carbon fabrics were therefor plasma treated, to improve the surface properties of the Vectran fibers regarding the adhesion of the epoxy matrix.

2.2. Methods

The material properties of the composites were determined by tensile-, flexural- and damping tests. The tensile tests were carried out in accordance with DIN EN ISO 527-5, with tabs were attached to the specimens. The elongation was measured using a video extensometer. Only plasma-treated Vectran-Carbon hybrids were used in the tensile tests. Test specimens for tensile tests were 250 mm long, 20 mm wide and approximately one millimeter thick.

The flexural tests were carried out in accordance with DIN EN ISO 14125. The dimensions were length = 100 mm, width = 15 mm, and thickness = 2.1 mm to 2.6 mm (depending on the fabric type). The distance between the supports was 91 mm. ILSS tests were carried out according to DIN EN ISO 14130, but were only used for quality control, hence the results are not discussed in further detail.

The material damping was measured via the logarithmic decrement (Λ), which is determined from the step response of a cantilever beam in free-fixed configuration. The logarithmic decrement is defined as the ratio of the amplitudes (y) of two adjacent amplitudes of a damped oscillation (see Equation (1)).

$$\Lambda = \ln\frac{y_i}{y_{i+1}} \qquad (1)$$

The determination was carried out on an apparatus as shown in Figure 4 that is based on the development of Romano [36]. The specimens are firmly clamped at one end, while the opposite side is freely movable. The specimens are excited to vibrate by a single deflection and then oscillate at the natural frequency of the vibration system. The deflection is caused by a cam, which is operated by a hand lever and always deflects the specimen by the same distance. The (decaying) vibration amplitude and period duration are then recorded using a laser distance sensor and evaluated in MATLAB®.

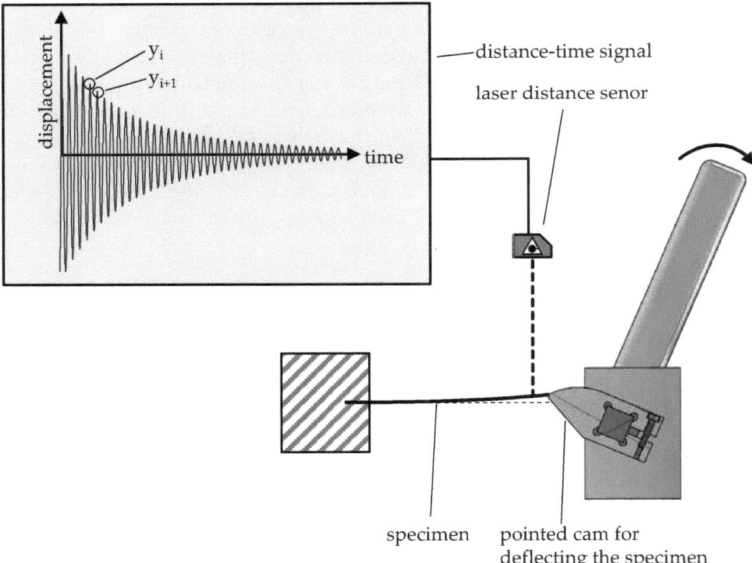

Figure 4. Apparatus for measuring material damping.

The damping was determined in the direction of the fibers. Since the vibration response (and thus also the determined damping) depends to a large extent on the stiffness of the beams, all samples were tested with the same spring stiffness. By changing the free length, the spring stiffness of the specimen is changed. The spring stiffness is based on the material stiffness, the free length and the test specimen cross-section. It can be easily determined by

Equation (2), with the spring stiffness (c), the acting force (F) and the resulting maximum deflection (s).

$$c = F/s \qquad (2)$$

To adjust the spring stiffness, the free ends of the test specimens were weighted with a (constant) weight and the free length was changed until a deflection of 10.4 mm was achieved.

3. Results and Discussion

All results presented below are found to be significantly different (alpha = 5%). All error bars describe the extent of one standard deviation of the respective test. Seven test specimens were tested per test series.

3.1. Strength

The following Figure 5 shows the tensile strength and flexural strength of the composite specimens in fiber direction. In the following assessment, the specimens are referred to by the types of fibers used, e.g., *Carbon* refers to composite specimens made from a plain carbon fabric, *Vectran-Carbon* refers to specimens made from a hybrid vectran-carbon fabric.

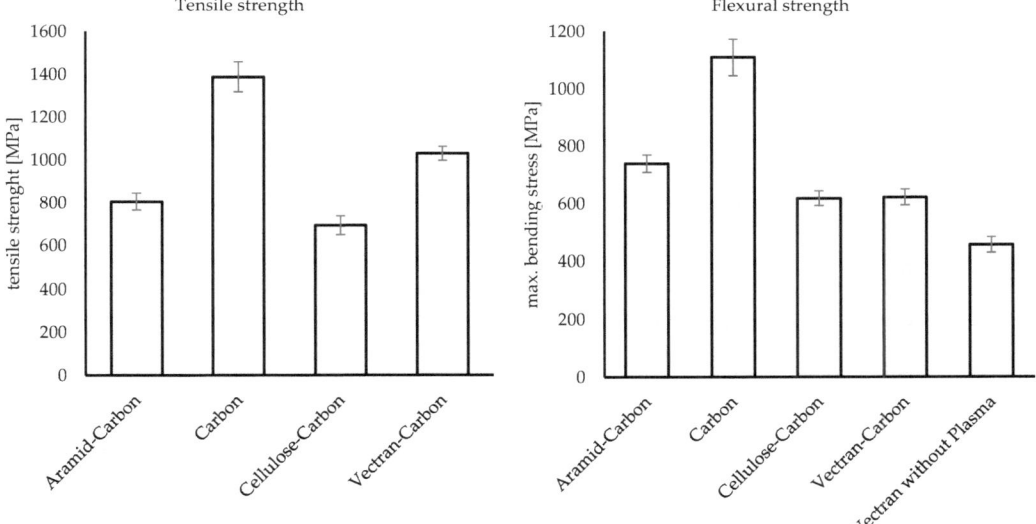

Figure 5. Strength in tensile and flexural direction.

Carbon shows the highest tensile- and flexural strength, Cellulose-Carbon the lowest. The tensile strengths of the composite specimens correlates with the tensile strengths of the individual fibers (see Table 1). The results for the tensile strength compared to the flexural strength show a different material behavior in relation to aramid- and (plasma treated) Vectran fibers.

Vectran-Carbon has the second highest tensile strength but only the third highest flexural strength, while Aramid-Carbon has the third highest tensile strength and the second highest flexural strength. These differences are particularly evident in Vectran-Carbon, which has a 25% lower tensile strength than Carbon but a 44% lower flexural strength. This material behavior could be an indication of a reduced compressive strength of the Vectran fibers or—despite plasma treatment—a low fiber-matrix adhesion of the Vectran fibers.

The fracture behavior in the comparison between tensile- and flexural-tests is consistent. Carbon and Cellulose-Carbon samples break brittle. Whereas Aramid-Carbon and Vectran-Carbon have a certain residual strength, as the majority of the aramid- or Vectran fibers are still intact after the failure of the specimen. Figure 6 shows the stress-displacement curve of representative specimens of each material during flexural tests.

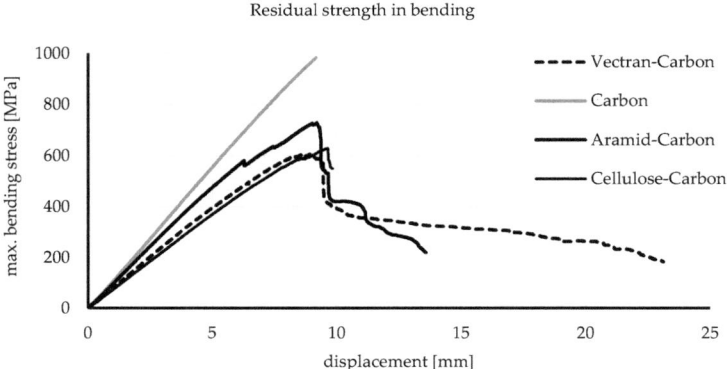

Figure 6. Fracture behavior under flexural loading.

The brittle failure of Carbon and Cellulose-Carbon is evident. After reaching the maximum stress, the whole specimen breaks suddenly and brittle. In the case of Vectran-Carbon and Aramid-Carbon, the carbon fibers break first when the maximum force is reached. The Vectran or aramid fibers, however, do not break yet due to their higher elongation at break and allow the composite to retain a certain residual strength.

3.2. Surface Treament of Vectran Fibers

Plasma treatment of Vectran fibers has produced a significant improvement in fiber-matrix adhesion. This is shown by the result of the bending strength of Vectran-Carbon vs. Vectran(-Carbon) without plasma; the use of plasma treatment improved the bending strength by 35%. A comparison of computer tomography (CT) scans supports this thesis. Figure 7 shows CT scans in the thickness direction of the bending test specimens made of (plasma) treated and untreated Vectran-Carbon in comparison to specimens from Carbon textiles.

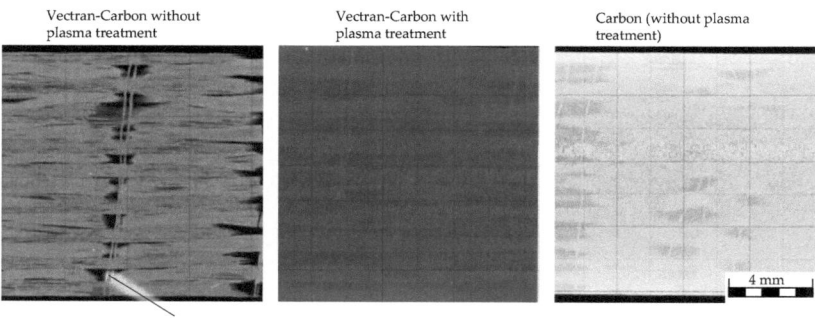

Figure 7. CT scans of Vectran-Carbon composite and Carbon composites.

Untreated Vectran-Carbon shows non-impregnated areas. Whereas plasma-treated Vectran-Carbon shows an impregnation comparable to that of Carbon.

3.3. Stiffness

The following Figure 8 shows the tensile moduli in fiber direction of the different hybrids and the Carbon reference.

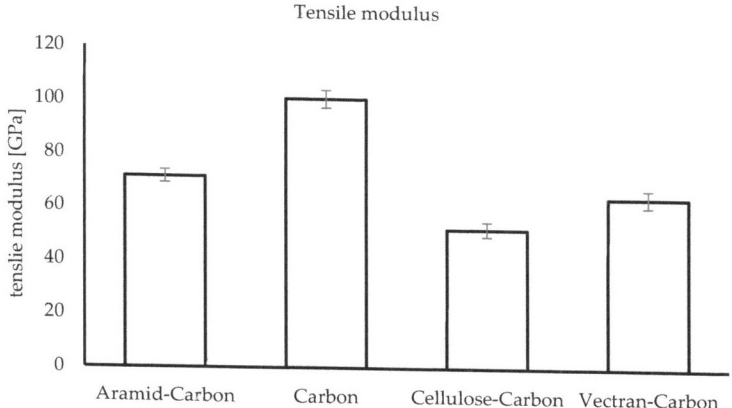

Figure 8. Tensile modulus in fiber direction.

A Carbon reinforcement leads to the highest stiffness, Cellulose-Carbon to the lowest. The tensile moduli of the composites also show a correlation with the moduli of the individual fibers.

3.4. Damping

The logarithmic decrement of the different samples—with the same spring stiffness—is shown below in Figure 9.

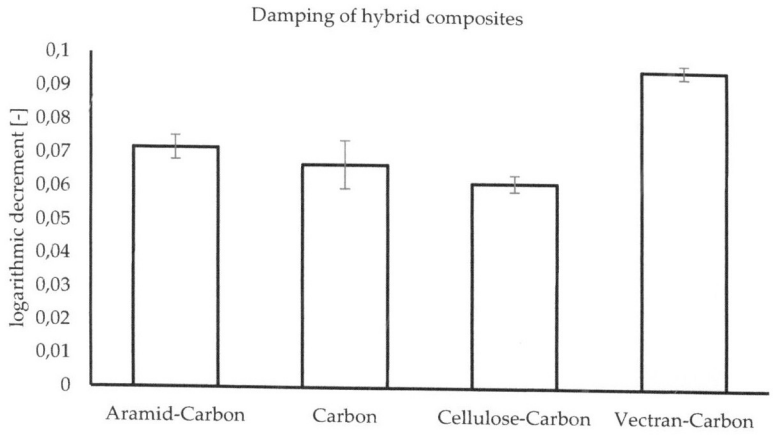

Figure 9. Logarithmic decrement with the same spring stiffness.

Specimens from Vectran-Carbon archive significantly higher damping than the Carbon reference. Aramid-Carbon and Cellulose-Carbon achieve similar values to Carbon. Whereby the damping of Aramid-Carbon is slightly higher than that of Carbon, while the damping of Cellulose-Carbon is slightly smaller than that of Carbon. To achieve the same spring stiffness (162 N/m), the tests were carried out at the free lengths as displayed in Table 4.

Table 4. Free length and damped natural frequency.

Material	Free Length [mm]	Frequency [Hz]
Aramid-Carbon	161	52
Carbon	164	53
Cellulose-Carbon	145	56
Vectran-Carbon	149	59

3.5. Density Specific Properties

The hybridization results in a reduction in the strength and stiffness of the composite, but at the same time also in a lower density of the overall composite compared to Carbon. Since the composites investigated are used for lightweight applications, the density must also be considered; the lower the density, the more of a material can be used. Through a density-specific consideration of the strength and stiffness, the decrease in strength and stiffness is less pronounced. Figure 10 shows the comparison between absolute and (density-) specific properties of the investigated materials. The area of the circles represents the amount of material damping. All data has been normalized to the respective properties of Carbon and is therefore given as a percentage.

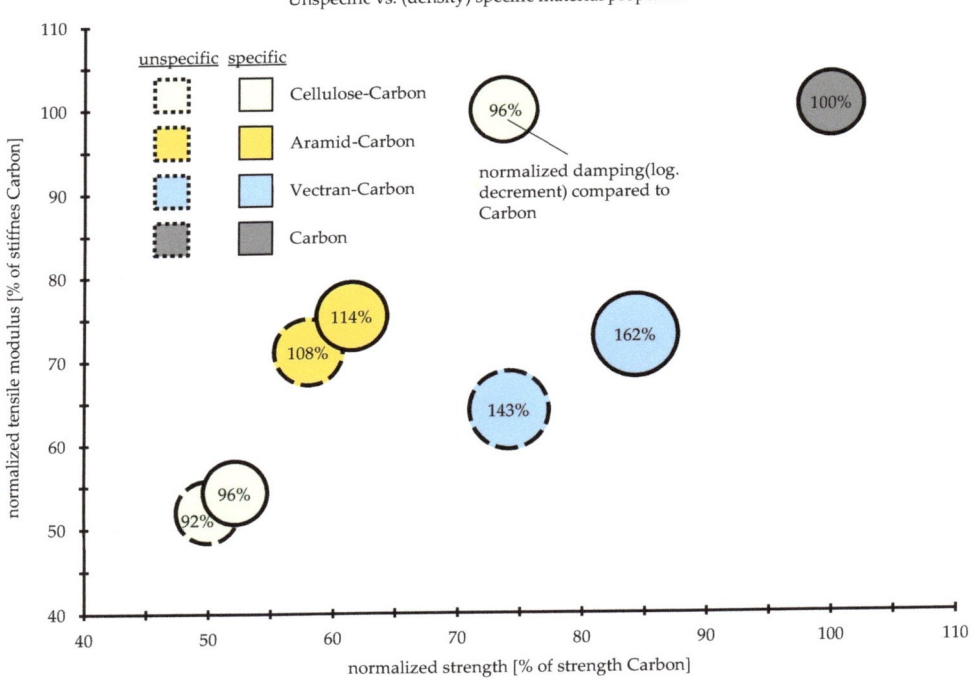

Figure 10. Effect of density on stiffness, strength and damping in fiber direction.

The specific consideration of strength, stiffness and damping reduces the difference between Carbon and the hybrid composites. The lower the density, the more pronounced the effect. By using hybrid composites, the damping can be increased over-proportionally to the *loss* of the mechanical properties.

3.6. Discussion of Mechanical Properties

The reason for the low strength of hybrid composites compared to carbon composites lies in the different stiffnesses and elongations at break of the individual fiber materials. The strength of the composites is dominated by carbon fibers, as these carry the higher load due to their high stiffness. Vectran-Carbon and Aramid-Carbon exhibit the same behavior: When subjected to a load, the carbon fibers and the aramid or Vectran fibers are elongated by the same amount. With increasing elongation, the tensile strength of the carbon fibers is exceeded first before the tensile strength of the Vectran or aramid fibers is reached. The reason for this is the lower stiffness and higher elongation at break of the Vectran or aramid fibers. The carbon fibers therefore fail first, with the result that the applied loads are now carried by the Vectran or aramid fibers. Due to the lower stiffness of the aramid and Vectran fibers, a drop in stress occurs in the specimens (see Figure 6). The specimens can be elongated even further until total failure. Overall, the Vectran-Carbon and Aramid-Carbon specimens achieve a higher elongation at break than the carbon specimens. However, the absolute tensile strength of the hybrid composite is lower than that of the pure carbon specimens since the hybrid specimens contain fewer—load bearing—carbon fibers overall.

The same behavior should theoretically apply to the Cellulose-Carbon specimens as the cellulose fibers have a very low stiffness as well as a very elongation at break. At the strain that leads to failure of the carbon fibers, stress in the cellulose fibers is still very much below their tensile strength. After the failure of the carbon fibers, there should be a stress drop in the specimen and the specimen should subsequently still have residual strength until failure. However, this is not the case, as the Cellulose-Carbon specimens fail brittly. The reason for this could be the interaction between the control of the testing machine and the strength difference between carbon and cellulose fibers. When the carbon fibers break, the testing machine must adjust the (displacement controlled) drive within a very short time in order to keep the strain rate of the specimen constant. Presumably, the machine's control system is not fast enough, so that the cellulose fibers experience a very high load and the specimen fails. Macroscopically, this results in a brittle fracture behavior of the cellulose-carbon samples.

The stiffness of the hybrid composites can be modeled as a good approximation as springs with different stiffnesses connected in parallel. The composite stiffness of the hybrid specimens follows very closely the mixing ratio (see Table 2) between carbon and functional fibers. The respective fiber type contributes proportionally to its amount and its fiber Young's modulus to the total stiffness of the composite. The calculation according to the rule of mixture showed a total fiber volume content of 42%. The fiber volume content calculated in this way is lower than the fiber volume content calculated using the aerial weight of the textile layers (see Table 3), since in the latter case imperfections and voids are not considered.

The changes in the material damping of the hybrid composites can be attributed to the use of the additional fiber types, since both the matrix and the manufacturing and testing conditions were kept constant. As the damping was measured at small deformations, damage and viscoplastic material behavior can be excluded as the cause of the damping. The damping is based on the viscoelastic properties of the functional fibers [4]. From the results it can be concluded that Vectran and aramid fibers have a more pronounced viscoelastic material behavior than cellulose and carbon fibers. As a result, the damping is increased through the integration of aramid and Vectran fibers.

4. Conclusions

Intraply hybridization of carbon fibers with Vectran fibers increases the material damping by up to 60% compared to samples made of pure carbon fibers. Hybridization with aramid fibers results in a small increase in damping, hybridization with cellulose fibers results in a decrease in damping compared to specimens from pure carbon fibers.

However, hybridization also leads to reduced stiffness and strength compared to a reference from pure carbon fiber: Aramid-Carbon shows the highest flexural strength of all hybrids, Aramid-Vectran the highest tensile strength. The tensile modulus of the hybrid composites is also lower than that of the non-hybrid carbon fiber reference. On the other hand, a density-specific consideration of the strength and stiffness reduces the difference between the hybrid composites and the non-hybrid carbon fiber reference. Vectran-Carbon in particular shows promising results. If the hybrid laminates are only used at selected locations in a laminate, e.g., areas with high shear [37], bicycle components can be realized with a minimal increase in weight but significantly increased material damping.

Author Contributions: Conceptualization, H.K. and C.G.; Investigation, H.K. and S.M.; Supervision, C.G. and T.G.; Writing—original draft, H.K. All authors have read and agreed to the published version of the manuscript.

Funding: This research received no external funding.

Conflicts of Interest: The authors declare no conflict of interest.

References

1. Chiementin, X.; Rigaut, M.; Crequy, S.; Bolaers, F.; Bertucci, W. Hand–arm vibration in cycling. *J. Vib. Control* **2013**, *19*, 2551–2560. [CrossRef]
2. Rivin, E.I. *Handbook on Stiffness & Damping in Mechanical Design*; American Society of Mechanical Engineers: New York, NY, USA, 2010.
3. Treviso, A.; van Genechten, B.; Mundo, D.; Tournour, M. Damping in composite materials: Properties and models. *Compos. Part B Engineering* **2015**, *78*, 144–152. [CrossRef]
4. Chandra, R.; Singh, S.; Gupta, K. Damping studies in fiber-reinforced composites—A review. *Compos. Struct.* **1999**, *46*, 41–51. [CrossRef]
5. Adams, R.D.; Bacon, D. Effect of Fibre Orientation and Laminate Geometry on the Dynamic Properties of CFRP. *J. Compos. Mater.* **1973**, *7*, 402–428. [CrossRef]
6. Adams, R.D.; Maheri, M.R. Dynamic flexural properties of anisotropic fibrous composite beams. *Compos. Sci. Technol.* **1994**, *50*, 497–514. [CrossRef]
7. Hanselka, H.; Hoffmann, U. Damping Characteristics of Fibre Reinforced Polymers. *Tech. Mech.* **1998**, *10*, 91–101.
8. Wright, G.C. The dynamic properties of glass and carbon fibre reinforced plastic beams. *J. Sound Vib.* **1972**, *21*, 205–212. [CrossRef]
9. Kretsis, G. A review of the tensile, compressive, flexural and shear properties of hybrid fibre-reinforced plastics. *Composites* **1987**, *18*, 13–23. [CrossRef]
10. Swolfs, Y.; Gorbatikh, L.; Verpoest, I. Fibre hybridisation in polymer composites: A review. *Compos. Part A Appl. Sci. Manuf.* **2014**, *67*, 181–200. [CrossRef]
11. Swolfs, Y.; McMeeking, R.M.; Verpoest, I.; Gorbatikh, L. The effect of fibre dispersion on initial failure strain and cluster development in unidirectional carbon/glass hybrid composites. *Compos. Part A Appl. Sci. Manuf.* **2015**, *69*, 279–287. [CrossRef]
12. Czél, G.; Wisnom, M.R. Demonstration of pseudo-ductility in high performance glass/epoxy composites by hybridisation with thin-ply carbon prepreg. *Compos. Part A Appl. Sci. Manuf.* **2013**, *52*, 23–30. [CrossRef]
13. Suwarta, P.; Fotouhi, M.; Czél, G.; Longana, M.; Wisnom, M.R. Fatigue behaviour of pseudo-ductile unidirectional thin-ply carbon/epoxy-glass/epoxy hybrid composites. *Compos. Struct.* **2019**, *224*, 110996. [CrossRef]
14. Pegoretti, A.; Fabbri, E.; Migliaresi, C.; Pilati, F. Intraply and interply hybrid composites based on E-glass and poly(vinyl alcohol) woven fabrics: Tensile and impact properties. *Polym. Int.* **2004**, *53*, 1290–1297. [CrossRef]
15. Fuller, J.D.; Wisnom, M.R. Pseudo-ductility and damage suppression in thin ply CFRP angle-ply laminates. *Compos. Part A Appl. Sci. Manuf.* **2015**, *69*, 64–71. [CrossRef]
16. Manders, P.W.; Bader, M.G. The strength of hybrid glass/carbon fibre composites. *J. Mater. Sci.* **1981**, *16*, 2233–2245. [CrossRef]
17. Fukunaga, H.; Chou, T.-W.; Fukuda, H. Strength of Intermingled Hybrid Composites. *J. Reinf. Plast. Compos.* **1984**, *3*, 145–160. [CrossRef]
18. Dickson, R.; Fernando, G.; Adam, T.; Reiter, H.; Harris, B. Fatigue behaviour of hybrid composites: Part 2 Carbon-glass hybrids. *J. Mater. Sci.* **1989**, *24*, 227–233. [CrossRef]
19. Fernando, G.; Dickson, R.; Adam, T.; Reiter, H.; Harris, B. Fatigue behaviour of hybrid composites: Part 1 Carbon/Kevlar hybrids. *J. Mater. Sci.* **1988**, *23*, 3732–3743. [CrossRef]
20. Peijs, A.; de Kok, J. Hybrid composites based on polyethylene and carbon fibres. Part 6: Tensile and fatigue behaviour. *Composites* **1993**, *24*, 19–32. [CrossRef]
21. Giancaspro, J.; Papakonstantino, C.G.; Balaguru, P.N. Flexural Response of Inorganic Hybrid Composites with E-Glass and Carbon Fibers. *J. Eng. Mater. Technol.* **2010**, *132*, 021005. [CrossRef]
22. Hosur, M.V.; Adbullah, M.; Jeelani, S. Studies on the low-velocity impact response of woven hybrid composites. *Compos. Struct.* **2005**, *67*, 253–262. [CrossRef]

23. Muñoz, R.; Martínez-Hergueta, F.; Gálvez, F.; González, C.; Llorca, J. Ballistic performance of hybrid 3D woven composites: Experiments and simulations. *Compos. Struct.* **2015**, *127*, 141–151. [CrossRef]
24. Dorey, G.; Sidey, G.R.; Hutchings, J. Impact properties of carbon fibre/Kevlar 49 fibre hybrid composites. *Composites* **1978**, *9*, 25–32. [CrossRef]
25. Swolfs, Y.; Verpoest, I.; Gorbatikh, L. Recent advances in fibre-hybrid composites: Materials selection, opportunities and applications. *Int. Mater. Rev.* **2019**, *64*, 181–215. [CrossRef]
26. Ashworth, S.; Rongong, J.; Wilson, P.; Meredith, J. Mechanical and damping properties of resin transfer moulded jute-carbon hybrid composites. *Compos. Part B Eng.* **2016**, *105*, 60–66. [CrossRef]
27. Berthelot, J.-M.; Assarar, M.; Sefrani, Y.; Mahi, A.E. Damping analysis of composite materials and structures. *Compos. Struct.* **2008**, *85*, 189–204. [CrossRef]
28. Le Guen, M.J.; Newman, R.H.; Fernyhough, A.; Emms, G.W.; Staiger, M.P. The damping–modulus relationship in flax–carbon fibre hybrid composites. *Compos. Part B Eng.* **2016**, *89*, 27–33. [CrossRef]
29. Cheour, K.; Assarar, M.; Scida, D.; Ayad, R.; Gong, X.-L. Effect of Stacking Sequences on the Mechanical and Damping Properties of Flax Glass Fiber Hybrid. *J. Renew. Mater.* **2019**, *7*, 877–889. [CrossRef]
30. Cihan, M.; Sobey, A.J.; Blake, J. Mechanical and dynamic performance of woven flax/E-glass hybrid composites. *Compos. Sci. Technol.* **2019**, *172*, 36–42. [CrossRef]
31. Davoodi, M.M.; Sapuan, S.M.; Ahmad, D.; Ali, A.; Khalina, A.; Jonoobi, M. Mechanical properties of hybrid kenaf/glass reinforced epoxy composite for passenger car bumper beam. *Mater. Des.* **2010**, *31*, 4927–4932. [CrossRef]
32. Pickering, K.L.; Efendy, M.A.; Le, T.M. A review of recent developments in natural fibre composites and their mechanical performance. *Compos. Part A Appl. Sci. Manuf.* **2016**, *83*, 98–112. [CrossRef]
33. Berthelot, J.-M.; Sefrani, Y. Damping analysis of unidirectional glass and Kevlar fibre composites. *Compos. Sci. Technol.* **2004**, *64*, 1261–1278. [CrossRef]
34. Adusumalli, R.B.; Venkateshan, K.C.; Gindl-Altmutter, W. Micromechanics of Cellulose Fibres and Their Composites. In *Wood Is Good*; Pandey, K.K., Ramakantha, V., Chauhan, S.S., Arun Kumar, A., Eds.; Springer: Singapore, 2017; pp. 299–321.
35. Buravalla, V.R.; Remillat, C.; Rongong, J.A.; Tomlinson, G.R. Advances in damping materials and technology. *Smart Mater. Bull.* **2001**, *2001*, 10–13. [CrossRef]
36. Romano, M. Charakterisierung von Gewebeverstärkten Einzellagen aus Kohlenstofffaserverstärktem Kunststoff (CFK) mit Hilfe Einer Mesomechanischen Kinematik Sowie Strukturdynamischen Versuchen. PhD Thesis, Universität der Bundeswehr München, Neubiberg, Germany, 2016.
37. Adams, R.D.; Maheri, M.R. Damping in advanced polymer–matrix composites. *J. Alloys Compd.* **2003**, *355*, 126–130. [CrossRef]

MDPI AG
Grosspeteranlage 5
4052 Basel
Switzerland
Tel.: +41 61 683 77 34

Journal of Composites Science Editorial Office
E-mail: jcs@mdpi.com
www.mdpi.com/journal/jcs

Disclaimer/Publisher's Note: The statements, opinions and data contained in all publications are solely those of the individual author(s) and contributor(s) and not of MDPI and/or the editor(s). MDPI and/or the editor(s) disclaim responsibility for any injury to people or property resulting from any ideas, methods, instructions or products referred to in the content.

www.ingramcontent.com/pod-product-compliance
Lightning Source LLC
LaVergne TN
LVHW072315090526
838202LV00019B/2291